Agencies, Organizations and Associations Dealing with Safety and Health

American Public Health Association
1015 15th St. NW
Washington, DC 20005

Bureau of Labor Statistics
U.S. Department of Labor
Washington, DC 20212

Bureau of National Affairs, Inc.
Occupational Safety and Health Reporter
1231 25th St. NW
Washington, DC 20037

Commerce Clearing House
Employee Safety and Health Guide
4205 W. Peterson Ave.
Chicago, IL 60646

Environmental Protection Agency
401 M St. SW
Washington, DC 20001

Mine Safety and Health Administration
4015 Wilson Blvd.
Rm. 601
Arlington, VA 22203

National Institute for Occupational Safety and Health (NIOSH)
4676 Columbia Pkwy.
Cincinnati, OH 45226

Occupational Safety and Health Administration
U.S. Department of Labor
200 Constitution Ave.
Washington, DC 20210

Occupational Safety and Health Review Committee (OSHRC)
Washington, DC 20210

U.S. Consumer Product Safety Commission
Washington, DC 20207

THE BASICS OF OCCUPATIONAL SAFETY

David L. Goetsch
President and CEO, The Institute for Workforce Transformation

Prentice Hall
Upper Saddle River, New Jersey
Columbus, Ohio

Library of Congress Cataloging-in-Publication Data
Goetsch, David L.
The basics of occupational safety / David L. Goetsch.
p. ; cm.
Includes bibliographical references and index.
ISBN-13: 978-0-13-502613-7
ISBN-10: 0-13-502613-X
1. Industrial safety. 2. Industrial hygiene. I. Title.
[DNLM: 1. Occupational Health—United States. 2. Accidents, Occupational—prevention & control—United States. 3. Occupational Diseases—prevention & control—United States. 4. Safety Management—methods—United States. WA 400 G599b 2008]
T55.G5848 2010
363.11—dc22 2008043267

Vice President and Executive Publisher: Vernon R. Anthony
Acquisitions Editor: Eric Krassow
Editorial Assistant: Sonya Kottcamp
Production Manager: Wanda Rockwell
Creative Director: Jayne Conte
Cover Design and Art: Aaron Dixon
Director of Marketing: David Gesell
Senior Marketing Coordinator: Alicia Dysert

This book was set in TimesTen Roman by Aptara®, Inc., and was printed and bound by Hamilton Printer. The cover was printed by DPC.

Pearson Education Ltd.
Pearson Education Singapore Pte. Ltd.
Pearson Education Canada, Ltd.
Pearson Education—Japan
Pearson Education Australia Pty. Limited
Pearson Education North Asia Ltd.
Pearson Educación de Mexico, S.A. de C.V.
Pearson Education Malaysia Pte. Ltd.

Prentice Hall
is an imprint of

www.pearsonhighered.com

9 8 7 6 5 4 3 2 1
ISBN-13: 978-0-13-502613-7
ISBN-10: 0-13-502613-X

Preface

Background

The field of occupational safety and health has undergone significant change over the past two decades. There are many reasons for this. Some of the more prominent include the following: technological changes that have introduced new hazards in the workplace; proliferation of health and safety legislation and corresponding regulations; increased pressure from regulatory agencies; realization by executives that workers in a safe and healthy workplace are typically more productive; health care and workers' compensation cost increases; increased pressure from environmental groups and the public; a growing interest in ethics and corporate social responsibility; professionalization of health and safety occupations; increased pressure from labor organizations and employees in general; rapidly mounting costs associated with product safety and other types of litigation; and increasing incidents of workplace violence.

All of these factors, when taken together, have made the job of the modern safety and health professional more challenging and more important than it has ever been. These factors have also created a need for an up-to-date book on the basics of workplace safety and health that contains the latest information needed by people who will have responsible positions in today's fast-paced, competitive workplace.

Why Was This Book Written and for Whom?

This book was written to fulfill the need for an up-to-date, practical teaching resource that focuses on the basic safety-related needs of people in the workplace. It is intended for use in universities, colleges, community colleges, and corporate training settings that offer programs, courses, workshops, and seminars in occupational safety and health. Educators and students in such disciplines as industrial technology, manufacturing technology, industrial engineering, safety engineering, engineering technology, occupational safety, management, and supervision will find this book both valuable and easy to use. The direct, straightforward presentation of material focuses on making the theories and principles of occupational safety and health practical and useful in a real-world setting. Up-to-date research has been integrated throughout in a down-to-earth manner.

Organization of the Book

The text contains 20 chapters, each focusing on a major area of concern in workplace safety and health. The chapters are presented in an order that is compatible with the typical organization of a college-level safety and health course. A standard chapter format is used throughout the book. Each chapter begins with a list of major topics and ends with a comprehensive summary. Following the summary, most chapters include review questions, key terms and concepts, and endnotes. Within each chapter are case studies to promote classroom discussion, as well as at least one safety fact or myth. These materials are provided to encourage review, stimulate additional thought, and provide opportunities for applying what has been learned.

Instructor Resources

To access supplementary materials online, instructors need to request an instructor access code. Go to **www.pearsonhighered.com/irc**, where you can register for an instructor access code. Within 48 hours after registering, you will receive a confirming e-mail, including an instructor access code. Once you have received your code, go to the site and log on for full instructions on downloading the materials you wish to use.

How This Book Differs from Others

This book was written because in the age of global competition, safety and health in the workplace have changed drastically. Many issues, concerns, and factors relating specifically to modern workplace environments have been given more attention, greater depth of coverage, and more illumination here than other textbooks. Some of the areas receiving more attention and specific occupational examples include the following:

- The OSH Act and OSHA
- Standards and codes
- Laws and liability
- Stress-related problems
- Life safety and fire hazards
- The evolving roles of health and safety professionals
- Health and safety training
- Human factors in safety
- Computers, robots, and automation
- Bloodborne pathogens in the workplace
- Ergonomics and safety
- Workers' compensation
- Repetitive strain injuries (RSIs)

About the Author

David L. Goetsch is Vice President of Okaloosa-Walton College and Provost of the joint campus of the University of West Florida and Okaloosa-Walton College in Fort Walton Beach, Florida. He is also a professor of safety, quality, and environmental management. Dr. Goetsch administers the state of Florida's Center for Manufacturing Competitiveness that is located on this campus and the Corporate Training Center of Okaloosa-Walton College. In addition, Dr. Goetsch is president and CEO of the Institute for Workforce Transformation (IWT), a private consulting firm dedicated to the continual improvement of organizational competitiveness, safety, and quality. Dr. Goetsch is cofounder of The Quality Institute, a partnership of the University of West Florida, Okaloosa-Walton College, and the Okaloosa Economic Development Council. He currently serves on the executive board of the institute.

Acknowledgments

The author acknowledges the invaluable assistance of the following people in developing this book: Dr. Lissa Galbraith, Florida A&M/Florida State University, for the material she contributed on electrical and fire hazards in the first edition; and the following reviewers for their invaluable input.

Introduction

Safety versus Health

The title of this book intentionally includes the words *safety* and *health.* Throughout the text, the titles "safety and health professional" and "safety and health manager" are used. This, too, is done by design. This approach underscores the point that the field of occupational safety has been broadened to encompass both safety and health. Consequently, managers, technical personnel, and engineers in this field must be knowledgeable about safety and health and be prepared to oversee a program that encompasses both areas of responsibility.

Safety and health, although closely related, are not the same. One view is that safety is concerned with injury-causing situations, whereas health is concerned with disease-causing conditions. Another view is that safety is concerned with hazards to humans that result from sudden severe conditions; health deals with adverse reactions to prolonged exposure to dangerous, but less intense, hazards. Both of these views are generally accurate in portraying the difference between safety and health. However, the line between these two concepts is not always clearly marked.

For example, stress is a hazard that can cause both psychological and physiological problems over a prolonged period. In this case, it is a health concern. On the other hand, an overly stressed worker may be more prone to unintentionally overlook safety precautions and thus may cause an accident. In this case, stress is a safety concern.

Because personnel in this evolving profession are likely to be responsible for safety *and* health, it is important that they have a broad academic background covering both. This book attempts to provide that background.

Brief Contents

Contents

CHAPTER 1

Safety and Health Movement: An Overview

Major Topics

- Developments before the Industrial Revolution
- Milestones in the Safety Movement
- Tragedies That Have Changed the Safety Movement
- Role of Organized Labor
- Role of Specific Health Problems
- Development of Accident Prevention Programs
- Development of Safety Organizations
- Safety and Health Movement Today
- Integrated Approach to Safety and Health
- New Materials, New Processes, and New Problems
- Rapid Growth in the Profession

The **safety movement** in the United States has developed steadily since the early 1900s. In that time period, industrial accidents were common place in this country; for example, in 1907 over 3,200 people were killed in mining accidents. Legislation, precedent, and public opinion all favored management. There were few protections for workers' safety.

Working conditions for industrial employees today have improved significantly. The chance of a worker being killed in an industrial accident is less than half of what it was 60 years ago.[1] According to the National Safety Council (NSC), the current death rate from work-related injuries is approximately 4 per 100,000, or less than a third of the rate 50 years ago.[2]

Improvements in safety until now have been the result of pressure for legislation to promote safety and health, the steadily increasing costs associated with accidents and injuries, and the professionalization of safety as an occupation. Improvements in the future are likely to come as a result of greater awareness of the cost effectiveness and resultant competitiveness gained from a safe and healthy workforce.

This chapter examines the history of the safety movement in the United States and how it has developed over the years. Such a perspective will help practicing and prospective safety professionals form a better understanding of both their roots and their future.

Developments Before the Industrial Revolution

It is important for students of occupational health and safety to first study the past. Understanding the past can help safety and health professionals examine the present and future with a sense of perspective and continuity. Modern developments in health and safety are neither isolated nor independent. Rather, they are part of the long continuum of developments in the safety and health movement.

The continuum begins with the days of the ancient Babylonians. During that time, circa 2000 BC, their ruler Hammurabi developed his **Code of Hammurabi**. The code encompassed all the laws of the land at that time, showed Hammurabi to be a just ruler, and set a precedent followed by other Mesopotamian kings. The significance of the code from the perspective of safety and health is that it contained clauses dealing with injuries, allowable fees for physicians, and monetary damages assessed against those who injured others.[3] This clause from the code illustrates Hammurabi's concern for the proper handling of injuries: "If a man has caused the loss of a gentleman's eye, his own eye shall be caused to be lost."[4]

This movement continued and emerged in later Egyptian civilization. As evidenced from the temples and pyramids that still remain, the Egyptians were an industrious people. Much of the labor was provided by slaves, and there is ample evidence that slaves were not treated well; that is, unless it suited the needs of the Egyptian taskmasters.

One such case occurred during the reign of Rameses II (circa 1500 BC), who undertook a major construction project, the Ramesseum. To ensure the maintenance of a workforce sufficient to build this huge temple bearing his name, Rameses created an industrial medical service to care for the workers. They were required to bathe daily in the Nile and were given regular medical examinations. Sick workers were isolated.[5]

The Romans were vitally concerned with safety and health, as can be seen from the remains of their construction projects. The Romans built aqueducts, sewerage systems, public baths, latrines, and well-ventilated houses.[6]

As civilization progressed, so did safety and health developments. In 1567, Philippus Aureolus produced a treatise on the pulmonary diseases of miners. Titled *On the Miners' Sickness and Other Miners' Diseases,* the treatise covered diseases of smelter workers and metallurgists and diseases associated with the handling of and exposure to mercury. Around the same time, Georgius Agricola published his treatise *De Re Metallica,* emphasizing the need for ventilation in mines and illustrating various devices that could be used to introduce fresh air into mines.[7]

The eighteenth century saw the contributions of Bernardino Ramazzini, who wrote *Discourse on the Diseases of Workers*. Ramazzini drew conclusive parallels between diseases suffered by workers and their occupations. He related occupational diseases to the handling of harmful materials and to irregular or unnatural movements of the body. Much of what Ramazzini wrote is still relevant today.[8]

The Industrial Revolution changed forever the methods of producing goods. According to LaDou, the changes in production brought about by the Industrial Revolution can be summarized as follows:

- Introduction of **inanimate power** (i.e., steam power) to replace people and animal power.
- Substitution of machines for people.
- Introduction of new methods for converting raw materials.
- Organization and specialization of work, resulting in a division of labor.[9]

These changes necessitated a greater focusing of attention on the safety and health of workers. Steam power increased markedly the potential for life-threatening injuries, as did machines. The new methods used for converting raw materials also introduced new risks of injuries and diseases. Specialization, by increasing the likelihood of boredom and inattentiveness, also made the workplace a more dangerous environment.

Milestones in the Safety Movement

Just as the United States traces its roots to Great Britain, the safety movement in this country traces its roots to England. During the Industrial Revolution, child labor in factories was common. The hours were long, the work hard, and the conditions often unhealthy and unsafe. Following an outbreak of fever among the children working in their cotton mills, the people of Manchester, England, began demanding better working conditions in the factories. Public pressure eventually forced a government response, and in 1802 the Health and Morals of Apprentices Act was passed. This was a milestone piece of legislation: it marked the beginning of governmental involvement in workplace safety.

When the industrial sector began to grow in the United States, hazardous working conditions were commonplace. Following the Civil War, the seeds of the safety movement were sown in this country. Factory inspection was introduced in Massachusetts in 1867. In 1868, the first barrier safeguard was patented. In 1869, the Pennsylvania legislature passed a mine safety law requiring two exits from all mines. The Bureau of Labor Statistics (BLS) was established in 1869 to study industrial accidents and report pertinent information about those accidents.

The following decade saw little new progress in the safety movement until 1877, when the Massachusetts legislature passed a law requiring safeguards for hazardous machinery. This year also saw passage of the Employer's Liability Law, establishing the potential for **employer liability** in workplace accidents. In 1892, the first recorded safety program was established in a Joliet, Illinois, steel plant in response to a scare caused when a flywheel exploded. Following the explosion, a committee of managers was formed to investigate and make recommendations. The committee's recommendations were used as the basis for the development of a safety program that is considered to be the first safety program in American industry.

Around 1900, Frederick Taylor began studying efficiency in manufacturing. His purpose was to identify the impact of various factors on efficiency, productivity, and profitability. Although safety was not a major focus of his work, Taylor did draw a connection between lost personnel time and management policies and procedures. This connection between safety and management represented a major step toward broad-based safety consciousness.

In 1907, the U.S. Department of the Interior created the Bureau of Mines to investigate accidents, examine health hazards, and make recommendations for improvements. Mining workers definitely welcomed this development since over 3,200 of their fellow workers were killed in mining accidents in 1907 alone.[10]

One of the most important developments in the history of the safety movement occurred in 1908 when an early form of **workers' compensation** was introduced in the United States. Workers' compensation actually had its beginnings in Germany. The practice soon spread throughout the rest of Europe. Workers' compensation as a concept made great strides in the United States when Wisconsin passed the first effective workers' compensation law in 1911. In the same year, New Jersey passed a workers' compensation law that withstood a court challenge.

The common thread among the various early approaches to workers' compensation was that they all provided some amount of compensation for on-the-job injuries regardless of who was at fault. When the workers' compensation concept was first introduced in the United States, it covered a very limited portion of the workforce and provided only minimal benefits. Today, all 50 states have some form of workers' compensation that requires the payment of a wide range of benefits to a broad base of workers. Workers' compensation is examined in more depth in Chapter 6.

The Association of Iron and Steel Electrical Engineers (AISEE), formed in the early 1900s, pressed for a national conference on safety. As a result of the AISEE's efforts, the first meeting of the **Cooperative Safety Congress (CSC)** took place in Milwaukee in 1912. What is particularly significant about this meeting is that it planted the seeds for the eventual establishment of the National Safety Council. A year after the initial meeting of the CSC, the **National Council of Industrial Safety (NCIS)** was established in Chicago. In 1915, this organization changed its name to the National Safety Council. It is now the premier safety organization in the United States.

From the end of World War I (1918) through the 1950s, safety awareness grew steadily. During this period, the federal government encouraged contractors to implement and maintain a safe work environment. Also during this period, industry in the United States arrived at two critical conclusions: (1) there is a definite connection between quality and safety, and (2) off-the-job accidents have a negative impact on productivity. The second conclusion became painfully clear to manufacturers during World War II when the call-up and deployment of troops had employers struggling to meet their labor needs. For these employers, the loss of a skilled worker due to an injury or for any other reason created an excessive hardship.[11]

The 1960s saw the passage of a flurry of legislation promoting workplace safety. The Service Contract Act of 1965, the Federal Metal and Non-metallic Mine Safety Act, the Federal Coal Mine and Safety Act, and the Contract Workers and Safety Standards Act all were passed during the 1960s. As their names indicate, these laws applied to a limited audience of workers. According to the Society of Manufacturing Engineers (SME), more significant legislation than that enacted in the 1960s was needed:

> Generally, the state legislated safety requirements only in specific industries, had inadequate safety and health standards, and had inadequate budgets for enforcement. . . . The injury and death toll due to industrial mishaps was still . . .

> too high. In the late 1960s, more than 14,000 employees were killed annually in connection with their jobs. . . . Work injury rates were taking an upward swing.[12]

These were the primary reasons behind passage of the **Occupational Safety and Health Act (OSH Act)** of 1970 and the Federal Mine Safety Act of 1977. These federal laws, particularly the OSH Act, represent the most significant legislation to date in the history of the safety movement. Figure 1-1 summarizes some significant milestones in the development of the safety movement in the United States.

FIGURE 1-1 Milestones in the safety movement.

1867	Massachusetts introduces factory inspection.
1868	Patent is awarded for first barrier safeguard.
1869	Pennsylvania passes law requiring two exits from all mines, and the Bureau of Labor Statistics is formed.
1877	Massachusetts passes law requiring safeguards on hazardous machines, and the Employer's Liability Law is passed.
1892	First recorded safety program is established.
1900	Frederick Taylor conducts first systematic studies of efficiency in manufacturing.
1907	Bureau of Mines is created by U.S. Department of the Interior.
1908	Concept of workers' compensation is introduced in the United States.
1911	Wisconsin passes the first effective workers' compensation law in the United States, and New Jersey becomes the first state to uphold a workers' compensation law.
1912	First Cooperative Safety Congress meets in Milwaukee.
1913	National Council of Industrial Safety is formed.
1915	National Council of Industrial Safety changes its name to National Safety Council.
1916	Concept of negligent manufacture is established (product liability).
1936	National Silicosis Conference convened by the U.S. Secretary of Labor.
1970	Occupational Safety and Health Act passes.
1977	Federal Mine Safety Act passes.
1986	Superfund Amendments and Reauthorization Act pass.
1990	Amended Clean Air Act of 1970 passes.
1996	Total safety management (TSM) concept is introduced.
2000	U.S. firms begin to pursue ISO 14000 registration for environmental safety management.
2003	Workplace terrorism is an ongoing concern of safety and health professionals.
2007	Safety of older people reentering the workplace becomes an issue.

Tragedies That Have Changed the Safety Movement

Safety and health tragedies in the workplace have greatly accelerated the pace of the safety movement in the United States. Three of the most significant events in the history of the safety and health movement were the Hawk's Nest, asbestos, and Bhopal tragedies. This section explains these three milestone events and their lasting effects on the safety and health movement in the United States.

Hawk's Nest Tragedy

In the 1930s, the public began to take notice of the health problems suffered by employees who worked in dusty environments. The Great Depression was indirectly responsible for the attention given to an occupational disease that came to be known as *silicosis*. As the economic crash spread, business after business shut down and laid off its workers. Unemployed miners and foundry workers began to experience problems finding new jobs when physical examinations revealed they had lung damage from breathing silica. Cautious insurance companies recommended preemployment physicals as a way to prevent future claims based on preexisting conditions. Applicants with silica-damaged lungs were refused employment. Many of them sued. This marked the beginning of industrywide interest in what would eventually be called the "king" of occupational diseases.

Lawsuits and insurance claims generated public interest in silicosis, but it was the **Hawk's Nest tragedy** that solidified public opinion in favor of protecting workers from this debilitating disease.[13] A company was given a contract to drill a passageway through a mountain located in the Hawk's Nest region of West Virginia (near the city of Gauley Bridge). Workers spent as many as 10 hours per day breathing the dust created by drilling and blasting. It turned out that this particular mountain had an unusually high silica content. Silicosis is a disease that normally takes 10 to 30 years to show up in exposed workers. At Hawk's Nest, workers began dying in as little time as a year. By the time the project was completed, hundreds had died. To make matters even worse, the company often buried an employee who died from exposure to silica in a nearby field without notifying the family. Those who inquired were told that their loved one left without saying where he was going.

A fictitious account of the Gauley Bridge disaster titled *Hawk's Nest,* by Hubert Skidmore, whipped the public outcry into a frenzy, forcing Congress to respond.

This tragedy and the public outcry that resulted from it led a group of companies to form the Air Hygiene Foundation to conduct research and develop standards for working in dusty environments. Soon thereafter, the U.S. Department of Labor provided the leadership necessary to make silicosis a compensable disease under workers' compensation in most states. Today, dust-producing industries use a wide variety of administrative controls, engineering controls, and personal protective equipment to protect workers in dusty environments. However, silicosis is still a problem. Approximately one million workers in the United States are still exposed to silica every year, and 250 people die annually from silicosis.

Asbestos Menace

Asbestos was once considered a "miracle" fiber, but in 1964 Dr. Irving J. Selikoff told 400 scientists at a conference on the biological effects of asbestos that this widely used material was killing workers. This conference changed how Americans viewed not just asbestos, but workplace hazards in general. Selikoff was the first to link asbestos to lung cancer and respiratory diseases.[14]

At the time of Selikoff's findings, asbestos was one of the most widely used materials in the United States. It was found in homes, schools, offices, factories, ships, and even in the filters of cigarettes. Selikoff continued to study the effects of asbestos exposure from 1967 to 1986. During this time, he studied the mortality rate of 17,800 workers who had been exposed to asbestos. He found asbestos-related cancer in the lungs, gastrointestinal tract, larynx, pharynx, kidneys, pancreas, gall bladder, and bile ducts of workers.

Finally, in the 1970s and 1980s, asbestos became a controlled material. Regulations governing the use of asbestos were developed, and standards for exposure were established. Asbestos-related lawsuits eventually changed how industry dealt with this tragic material. In the 1960s, industry covered up or denied the truth about asbestos. Now, there is an industrywide effort to protect workers who must remove asbestos from old buildings and ships during remodeling, renovation, or demolition projects.

Bhopal Tragedy

On the morning of December 3, 1984, over 40 tons of methyl isocyanate (MIC) and other lethal gases, including hydrogen cyanide, leaked into the northern end of Bhopal, killing more than 3,000 people in its aftermath.[15] After the accident, it was discovered that the protective equipment that could have halted the impending disaster was not in full working order. The refrigeration system that should have cooled the storage tank was shut down, the scrubbing system that should have absorbed the vapor was not immediately available, and the flare system that would have burned any vapor that got past the scrubbing system was out of order.[16]

The International Medical Commission visited Bhopal to assess the situation and found that as many as 50,000 other people were exposed to the poisonous gas and may still suffer disability as a result. This disaster shocked the world. Union Carbide, the owner of the chemical plant in Bhopal, India, where the incident occurred, was accused of many things, including the following:

- Criminal negligence.
- Corporate prejudice. Choosing poverty-stricken Bhopal, India, as the location for a hazardous chemical plant on the assumption that few would care if something went wrong.
- Avoidance. Putting its chemical plant in Bhopal, India, to avoid the stricter safety and health standards of the United States and the Occupational Safety and Health Administration (OSHA) in particular.

In February 1989, India's Supreme Court ordered Union Carbide India, Ltd., to pay $470 million in compensatory damages. The funds were paid to the Indian government to be used to compensate the victims. This disaster provided the impetus for the passage of stricter safety legislation worldwide. In the United States, it led to passage of the Emergency Planning and Community Right-to-Know Act (EPCRA) of 1986.

Role of Organized Labor

Organized labor has played a crucial role in the development of the safety movement in the United States. From the outset of the Industrial Revolution in this country, organized labor has fought for safer working conditions and appropriate compensation for workers injured on the job. Many of the earliest developments in the safety movement were the result of long and hard-fought battles by organized labor.

Although the role of unions in promoting safety is generally acknowledged, one school of thought takes the opposite view. Proponents of this dissenting view hold that union involvement actually slowed the development of the safety movement. Their theory is that unions allowed their demands for safer working conditions to become entangled with their demands for better wages and, as a result, they met with resistance from management. Regardless of the point of view, there is no question that working conditions in the earliest years of the safety movement were often reflective of an insensitivity to safety concerns on the part of management.

Among the most important contributions of organized labor to the safety movement was their work to overturn antilabor laws relating to safety in the workplace. These laws were the fellow servant rule, the statutes defining contributory negligence, and the concept of assumption of risk.[17] The **fellow servant rule** held that employers were not liable for workplace injuries that resulted from the negligence of other employees. For example, if Worker X slipped and fell, breaking his back in the process, because Worker Y spilled oil on the floor and left it there, the employer's liability was removed. In addition, if the actions of employees contributed to their own injuries, the employer was absolved of any liability. This was the doctrine of **contributory negligence**. The concept of **assumption of risk** was based on the theory that people who accept a job assume the risks that go with it. It says employees who work voluntarily should accept the consequences of their actions on the job rather than blame the employer.

Because the overwhelming majority of industrial accidents involve negligence on the part of one or more workers, employers had little to worry about. Therefore, they had little incentive to promote a safe work environment. Organized labor played a crucial role in bringing deplorable working conditions to the attention of the general public. Public awareness and, in some cases, outrage eventually led to these **employer-biased laws** being overturned in all states except one. In New Hampshire, the fellow servant rule still applies.

Role of Specific Health Problems

Specific health problems that have been tied to workplace hazards have played significant roles in the development of the modern safety and health movement. These health problems contributed to public awareness of dangerous and unhealthy working conditions that, in turn, led to legislation, regulations, better work procedures, and better working conditions.

Lung disease in coal miners was a major problem in the 1800s, particularly in Great Britain, where much of the Western world's coal was mined at the time. Frequent contact with coal dust led to a widespread outbreak of anthrocosis among Great Britain's coal miners. Also known as the *black spit,* this disease persisted from the early 1800s, when it was first identified, until around 1875, when it was finally eliminated by such safety and health measures as ventilation and decreased work hours

In the 1930s, Great Britain saw a resurgence of lung problems among coal miners. By the early 1940s, British scientists were using the term *coal-miner's pneumoconiosis,* or CWP, to describe a disease from which many miners suffered. Great Britain designated CWP a separate and compensable disease in 1943. However, the United States did not immediately follow suit, even though numerous outbreaks of the disease had occurred among miners in this country.

The issue was debated in the United States until Congress finally passed the Coal Mine Health and Safety Act in 1969. The events that led up to passage of this act were tragic. An explosion in a coal mine in West Virginia in 1968 killed 78 miners. This tragedy focused attention on mining health and safety, and Congress responded by passing the Coal Mine Health and Safety Act. The act was amended in 1977 and again in 1978 to broaden the scope of its coverage.

Over the years, the diseases suffered by miners were typically lung diseases caused by the inhalation of coal dust particulates. However, health problems were not limited to coal miners. Other types of miners developed a variety of diseases, the most common of which was silicosis. Once again, it took a tragic event—the Gauley Bridge disaster, discussed earlier—to focus attention on a serious workplace problem.

Congress held a series of hearings on the matter in 1936. That same year, representatives from business, industry, and government attended the National Silicosis Conference, convened by the U.S. Secretary of Labor. Among other outcomes of this conference was a finding that silica dust particulates did, in fact, cause silicosis.

Mercury poisoning is another health problem that has contributed to the evolution of the safety and health movement by focusing public attention on unsafe conditions in the workplace. The disease was first noticed among the citizens of a Japanese fishing village in the early 1930s. A disease with severe symptoms was common in Minamata, but extremely rare throughout the rest of Japan. After much investigation into the situation, it was determined that a nearby chemical plant periodically dumped methyl mercury into the bay that was the village's primary source of food. Consequently, the citizens of this small village ingested hazardous dosages of mercury every time they ate fish from the bay.

Mercury poisoning became an issue in the United States after a study was conducted in the early 1940s that focused on New York City's hat-making industry. During that time, many workers in this industry displayed the same types of symptoms as the citizens of Minamata, Japan. Because mercury nitrate was used in the production of hats, enough suspicion was aroused to warrant a study. The study linked the symptoms of workers with the use of mercury nitrate. As a result, the use of this hazardous chemical in the hat-making industry was stopped, and a suitable substitute—hydrogen peroxide—was found.

As discussed earlier, asbestos was another important substance in the evolution of the modern safety and health movement. By the time it was determined that asbestos is a hazardous material, the fibers of which can cause asbestosis or lung cancer (mesothelioma), thousands of buildings contained the substance. As these buildings began to age, the asbestos, particularly that used to insulate pipes, began to break down. As asbestos breaks down, it releases dangerous microscopic fibers into the air. These fibers are so hazardous that removing asbestos from old buildings has become a highly specialized task requiring special equipment and training.

Development of Accident Prevention Programs

In the modern workplace, there are many different types of **accident prevention** programs ranging from the simple to the complex. Widely used accident prevention techniques include failure minimization, fail-safe designs, isolation, lockouts, screening, personal protective equipment, redundancy, timed replacements, and many others. These techniques are individual components of broader safety programs. Such programs have evolved since the late 1800s.

In the early 1800s, employers had little concern for the safety of workers and little incentive to be concerned. Consequently, organized safety programs were nonexistent, a situation that continued for many years. However, between World War I and World War II, industry discovered the connection between quality and safety. Then, during World War II, troop call-ups and deployments created severe labor shortages. Faced with these shortages, employers could not afford to lose workers to accidents or for any other reason. This realization created a greater openness toward giving safety the serious consideration that it deserved. For example, according to the Society of Manufacturing Engineers, around this time industry began to realize the following:

- Improved engineering could prevent accidents.
- Employees were willing to learn and accept safety rules.
- Safety rules could be established and enforced.
- Financial savings from safety improvement could be reaped by savings in compensation and medical bills.[18]

With these realizations came the long-needed incentive for employers to begin playing an active role in creating and maintaining a safe workplace. This, in turn, led to the development of organized safety programs sponsored by management. Early safety programs were based on the **three E's of safety**: engineering, education, and enforcement (see Figure 1-2). The engineering aspects of a safety program involve making design improvements to both product and process. By altering the design of a product, the processes used to manufacture it can be simplified and, as a result, made less dangerous. In addition, the manufacturing processes for products can be engineered in ways that decrease potential hazards associated with the processes.

FIGURE 1-2 Three E's of safety.

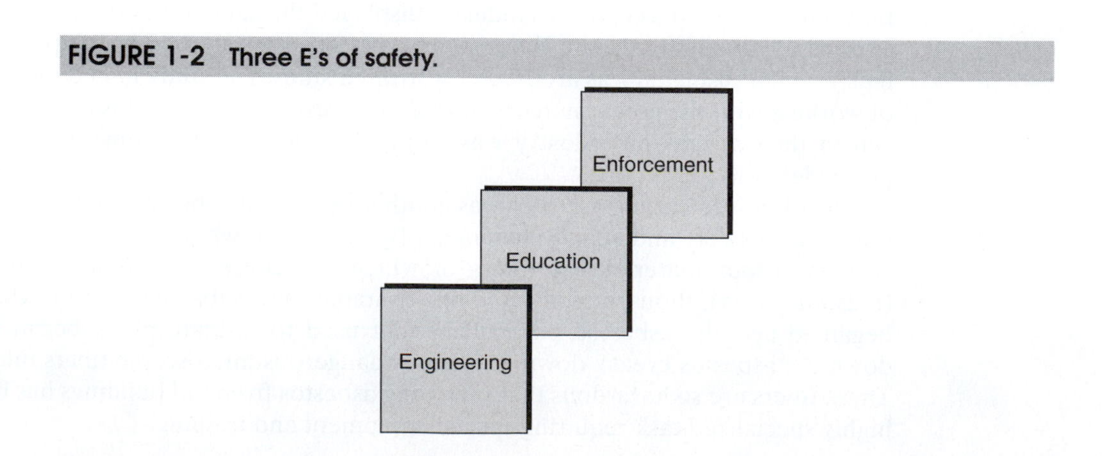

The education aspect of a safety program ensures that employees know how to work safely, why it is important to do so, and that safety is expected by management. Safety education typically covers the what, when, where, why, and how of safety.

The enforcement aspect of a safety program involves making sure that employees abide by safety policies, rules, regulations, practices, and procedures. Supervisors and fellow employees play a key role in the enforcement aspects of modern safety programs.

Development of Safety Organizations

Today, numerous organizations are devoted in full, or at least in part, to the promotion of safety and health in the workplace. Figure 1-3 lists organizations with workplace safety as part of their missions. Figure 1-4 lists several governmental agencies and two related organizations concerned with safety and health. These lists are extensive now, but this has not always been the case. Safety organizations in this country had humble beginnings

The grandfather of them all is the National Safety Council. The Society of Manufacturing Engineers traces the genesis of this organization as follows:

> The Association of Iron and Steel Electrical Engineers was organized in the first decade of the 20th century and devoted much attention to safety problems in its industry. In 1911, a request came from this association to call a national industrial safety conference. The first Cooperative Safety Congress met in

FIGURE 1-3 Organizations concerned with workplace safety.

Alliance for American Insurers
American Board of Industrial Hygiene
American Conference of Government Industrial Hygienists
American Industrial Hygiene Association
American Insurance Association
American National Standards Institute
American Occupational Medical Association
American Society for Testing and Materials
American Society of Mechanical Engineers
American Society of Safety Engineers
Chemical Transportation Emergency Center
Human Factors Society
National Fire Protection Association
National Safety Council
National Safety Management Society
Society of Automotive Engineers
System Safety Society
Underwriters Laboratories, Inc.

FIGURE 1-4 Government agencies and other organizations concerned with workplace safety.

American Public Health Association*
Bureau of Labor Statistics
Bureau of National Affairs
Commerce Clearing House*
Environmental Protection Agency
National Institute for Standards and Technology (formerly National Bureau of Standards)
National Institute for Occupational Safety and Health
Occupational Safety and Health Administration
Superintendent of Documents, U.S. Government Printing Office
U.S. Consumer Product Safety Commission

*Not a government agency.

Milwaukee in 1912. A year later, at a meeting in New York City, the National Council of Industrial Safety was formed. It began operation in a small office in Chicago. At its meeting in 1915, the organization's name was changed to the National Safety Council (NSC).[19]

Today, the NSC is the largest organization in the United States devoted solely to safety and health practices and procedures. Its purpose is to prevent the losses, both direct and indirect, arising out of accidents or from exposure to unhealthy environments. Although it is chartered by an act of Congress, the NSC is a nongovernmental, not-for-profit, public service organization.

The **Occupational Safety and Health Administration (OSHA)** is the government's administrative arm for the Occupational Safety and Health Act (OSH Act). Formed in 1970, OSHA sets and revokes safety and health standards, conducts inspections, investigates problems, issues citations, assesses penalties, petitions the courts to take appropriate action against unsafe employers, provides safety training, provides injury prevention consultation, and maintains a database of health and safety statistics.

Another governmental organization is the **National Institute for Occupational Safety and Health (NIOSH)**. This organization is part of the Centers for Disease Control and Prevention (CDC) of the Department of Health and Human Services. NIOSH is required to publish annually a comprehensive list of all known toxic substances. NIOSH will also provide on-site tests of potentially toxic substances so that companies know what they are handling and what precautions to take.

Safety and Health Movement Today

The safety and health movement has come a long way since the Industrial Revolution. Today, there is widespread understanding of the importance of providing a safe and healthy workplace. The tone was set during and after World War II when all the various practitioners of occupational health and safety began to see the need for cooperative efforts. These practitioners included safety engineers, safety managers, industrial hygienists, occupational health nurses, and physicians.

One of the earliest and most vocal proponents of the cooperative or integrated approach was H. G. Dyktor. He proposed the following objectives of integration:

- Learn more through sharing knowledge about health problems in the workplace, particularly those caused by toxic substances.
- Provide a greater level of expertise in evaluating health and safety problems.
- Provide a broad database that can be used to compare health and safety problems experienced by different companies in the same industry.
- Encourage accident prevention.
- Make employee health and safety a high priority.[20]

Integrated Approach to Safety and Health

This integrated approach has become the norm that typifies the safety and health movement of today. By working together and drawing on their own respective areas of expertise, safety and health professionals are better able to identify, predict, control, and correct safety and health problems.

OSHA reinforces the integrated approach by requiring companies to have a plan for doing at least the following: (1) providing appropriate medical treatment for injured or ill workers, (2) regularly examining workers who are exposed to toxic substances, and (3) having a qualified first-aid person available during all working hours.

Smaller companies may contract out the fulfillment of these requirements. Larger companies often maintain a staff of safety and health professionals. According to Hamilton and Hardy, the health and safety staff in a modern industrial company may include the following positions:

- **Industrial hygiene chemist and/or engineer.** Companies that use toxic substances may employ **industrial hygiene chemists** periodically to test the work environment and the people who work in it. In this way, unsafe conditions or hazardous levels of exposure can be identified early and corrective or preventive measures can be taken. Dust levels, ventilation, and noise levels are also monitored by individuals serving in this capacity.
- **Radiation control specialist.** Companies that use or produce radioactive materials employ **radiation control specialists** who are typically electrical engineers or physicists. These specialists monitor the radiation levels to which workers may be exposed, test workers for levels of exposure, respond to radiation accidents, develop companywide plans for handling radiation accidents, and implement decontamination procedures when necessary.
- **Industrial safety engineer or manager.** Individuals serving as **industrial safety engineers or managers** are safety and health generalists with specialized education and training. In larger companies, they may be devoted to safety and health matters. In smaller companies, they may have other duties in addition to safety and health. In either case, they are responsible for developing and carrying out the company's overall safety and health program including accident prevention, accident investigation, and education and training.[21]

Other professionals who may be part of a company's safety and health team include occupational nurses, physicians, psychologists, counselors, educators, and dietitians.

New Materials, New Processes, and New Problems

The job of the safety and health professional is more complex than it has ever been. The materials out of which products are made have become increasingly complex and exotic. Engineering metals now include carbon steels, alloy steels, high-strength low-alloy steels, stainless steels, managing steels, cast steels, cast irons, tungsten, molybdenum, titanium, aluminum, copper, magnesium, lead, tin, zinc, and powdered metals. Each of these metals requires its own specialized processes.

Nonmetals are more numerous and have also become more complex. Plastics, plastic alloys and blends, advanced composites, fibrous materials, elastomers, and ceramics also bring their own potential hazards to the workplace.

In addition to the more complex materials being used in modern industry and the new safety and health concerns associated with them, modern industrial processes are also becoming more complex. As these processes become automated, the potential hazards associated with them often increase. Computers; lasers; industrial robots; nontraditional processes such as explosive welding, photochemical machining, laser beam machining, ultrasonic machining, and chemical milling; automated material handling; water-jet cutting expert systems; flexible manufacturing cells; and computer-integrated manufacturing have all introduced new safety and health problems in the workplace and new challenges for the safety and health professional.

Chapter 19 is devoted to coverage of the special safety and health problems associated with computers, robots, and automation. In addition, coverage of specific aspects of these problems is provided in different chapters throughout this book.

Rapid Growth in the Profession

The complexities of the modern workplace have made safety and health a growing profession. Associate and baccalaureate degree programs in industrial technology typically include industrial safety courses. Some engineering degree programs have safety and health tracks. Several colleges and universities offer full degrees in occupational safety and health.

The inevitable result of the increased attention given to safety and health is that more large companies are employing safety and health professionals and more small companies are assigning these duties to existing employees. This is a trend that is likely to continue as employers see their responsibilities for safety and health spread beyond the workplace to the environment, the community, the users of their products, and the recipients of their by-products and waste.

Key Terms and Concepts

- Accident prevention
- Asbestos menace
- Assumption of risk
- Bhopal tragedy
- Code of Hammurabi
- Contributory negligence
- Cooperative Safety Congress (CSC)
- Employer-biased laws
- Employer liability
- Fellow servant rule
- Hawk's Nest tragedy

- Inanimate power
- Industrial hygiene chemist
- Industrial safety engineer
- Industrial safety manager
- National Council of Industrial Safety (NCIS)
- National Institute for Occupational Safety and Health (NIOSH)
- Occupational Safety and Health Act (OSH Act)
- Occupational Safety and Health Administration (OSHA)
- Organized labor
- Radiation control specialist
- Safety movement
- Three E's of safety
- Workers' compensation

REVIEW QUESTIONS

1. To what cause(s) can the improvements in workplace safety made to date be attributed?
2. Explain the significance of the Code of Hammurabi in terms of the safety movement.
3. Describe the circumstances that led to the development of the first organized safety program.
4. What is Frederick Taylor's connection to the safety movement?
5. Explain the development of the National Safety Council.
6. What impact did labor shortages in World War II have on the safety movement?
7. Explain how workplace tragedies have affected the safety movement. Give examples.
8. Explain the primary reasons behind the passage of the OSH Act.
9. Summarize briefly the role that organized labor has played in the advancement of the safety movement.
10. Define the following terms: fellow servant rule, contributory negligence, and assumption of risk.
11. Explain the three E's of safety.
12. Explain the term *integration* as it relates to modern safety and health.

ENDNOTES

1. S. Minter, "The Birth of OSHA," *Occupational Hazards,* July 1998, 59.
2. National Safety Council, *Accident Facts* (Chicago: National Safety Council, 1998).
3. J. LaDou, ed., *Introduction to Occupational Health and Safety* (Chicago: National Safety Council, 1997), 28.
4. Ibid., 28.
5. A. Soubiran, "Medical Services under the Pharaohs," *Abbottempo* 1: 19–23.
6. LaDou, *Occupational Health and Safety,* 31.
7. Ibid., 34.
8. Ibid., 35.
9. Ibid., 37.
10. S. Minter and V. Sutcliff, "Fighting Two Wars," *Occupational Hazards,* July 1998, 41–42.
11. Ibid., 41.
12. Ibid., 42.
13. Ibid., 41–42.
14. L. Finnegan, "Asbestos Becomes a Menace," *Occupational Hazards,* July 1998, 53–54.
15. L. Finnegan, "Reform and Reinvention," *Occupational Hazards,* July 1998, 67–68.
16. Ibid., 67.
17. Minter and Sutcliff, "Fighting Two Wars," 41.
18. Ibid., 41.
19. Ibid., 42.
20. H. G. Dyktor, "Integration of Industrial Hygiene with Industrial Medicine," *Industrial Medicine* 9, no. 4 (1940): 193.
21. A. Hamilton and H. Hardy, *Industrial Toxicology* (Boston: John Wright, 1983).

CHAPTER 2

Accidents and Their Effects

Major Topics

- Costs of Accidents
- Accidental Deaths in the United States
- Accidents versus Other Causes of Death
- Work Accident Costs and Rates
- Time Lost Because of Work Injuries
- Deaths in Work Accidents
- Work Injuries by Type of Accident
- Death Rates by Industry
- Parts of the Body Injured on the Job
- Chemical Burn Injuries
- Heat Burn Injuries
- Repetitive Strain/Soft Tissue Injuries
- Estimating the Cost of Accidents
- Global Impact of Accidents and Injuries

There is a long history of debate in this country concerning the effect of **accidents** on industry (the workers and the companies) and the cost of preventing accidents. Historically, the prevailing view was that **accident prevention** programs were too costly. The more contemporary view is that accidents are too costly and that accident prevention makes sense economically. As a result, accident prevention, which had been advocated on a moral basis, is now justified in economic terms.

Accidents are the fourth leading cause of death in this country after heart disease, cancer, and strokes. This ranking is based on all types of accidents including motor vehicle accidents, drownings, fires, falls, natural disasters, and work-related accidents.

Although deaths from **natural disasters** tend to be more newsworthy than workplace deaths, their actual impact is substantially less. For example, natural disasters in the United States cause fewer than 100 deaths per year on average. **Workplace**

accidents, on the other hand, cause more than 10,000 deaths every year in the United States.[1] The following quote from the National Safety Council (NSC) puts workplace accidents and deaths in the proper perspective, notwithstanding their apparent lack of newsworthiness.

> While you make a 10-minute speech—2 persons will be killed and about 170 will suffer a disabling injury. Costs will amount to $2,800,000. On the average, there are 11 accidental deaths and about 1,030 disabling injuries every hour during the year.[2]

This chapter provides prospective and practicing **safety and health professionals** with the information they need to have a full understanding of workplace accidents and their effect on industry in the United States. Such an understanding will help professionals play a more effective role in keeping both management and labor focused appropriately on safety and health in the workplace.

Costs of Accidents

To gain a proper perspective on the economics of workplace accidents, we must view them in the overall context of all accidents. The overall cost of accidents in the United States is approximately $150 billion. These costs include such factors as **lost wages**, **medical expenses**, **insurance administration**, **fire-related losses**, motor vehicle **property damage**, and **indirect costs**.

Figure 2-1 breaks down this overall amount by categories of accidents. Figure 2-2 breaks them down by cost categories. Notice in Figure 2-1 that workplace accidents rank second behind motor vehicle accidents in cost. Figure 2-2 shows that the highest cost category is wages lost by workers who are either injured or killed. The category

FIGURE 2-1 Accident costs by accident type (in billion, in a typical year).

Accident type	Cost
Motor vehicle accidents	$72
Workplace accidents	48
Home accidents	18
Public accidents	12

FIGURE 2-2 Accident costs by categories (in billions, in a typical year).

Category	Cost
Wages lost	$37
Medical expenses	24
Insurance administration	29
Property damage (motor vehicle)	27
Fire losses	10
Indirect losses for work accidents	23

of indirect losses from work accidents consists of costs associated with responding to accidents (i.e., giving first aid, filling out accident reports, handling production slowdowns).

Clearly, accidents on and off the job cost U.S. industry dearly. Every dollar that is spent responding to accidents is a dollar that could have been reinvested in modernization, research and development, facility upgrades, and other competitiveness-enhancing activities.

Accidental Deaths in the United States

Accidental deaths in the United States result from a variety of causes, including motor vehicle accidents, falls, poisoning, drowning, fire-related injuries, **suffocation** (ingested object), firearms, **medical complications**, air transport accidents, machinery, mechanical suffocation, and the impact of falling objects. The NSC periodically computes death totals and death rates in each of these categories. The statistics for a typical year are as follows:

- **Motor vehicle accidents.** Motor vehicle accidents are the leading cause of accidental deaths in the United States each year. They include deaths resulting from accidents involving mechanically or electrically powered vehicles (excluding rail vehicles) that occur on or off the road. In a typical year, there are approximately 47,000 such deaths in the United States.
- **Falls.** This category includes all deaths from **falls** except those associated with transport vehicles. For example, a person who is killed as the result of falling while boarding a bus or train would not be included in this category. In a typical year, there are approximately 13,000 deaths in the United States from falls.
- **Poisoning.** The **poisoning** category is divided into two subcategories: (1) poisoning by solids and liquids, and (2) poisoning by gases and vapors. The first category includes deaths that result from the ingestion of drugs, medicine, widely recognized solid and liquid poisons, mushrooms, and shellfish. It does not include poisoning from spoiled food or salmonella. The second category includes deaths caused by incomplete combustion (for example, gas vapors from an oven or unlit pilot light) or from carbon monoxide (for example, exhaust fumes from an automobile). In a typical year, there are approximately 6,000 deaths in the first category and 1,000 in the second.
- **Drowning.** This category includes work-related and non-work-related **drownings** but excludes those associated with floods or other natural disasters. In a typical year, there are approximately 5,000 deaths from drowning in the United States.
- **Fire-related injuries.** This category includes deaths from burns, asphyxiation, falls, and those that result from falling objects in a fire. In a typical year, there are over 4,000 fire-related deaths in the United States.
- **Suffocation (ingested object).** This category includes deaths from the ingestion of an object that blocks the air passages. In many such deaths, the ingested object is food. In a typical year, there are approximately 4,000 suffocation deaths in the United States.
- **Firearms.** This category includes deaths that result when recreational activities involving firearms or household accidents involving firearms result in death. For

FIGURE 2-3 Causes of accidents (ages 25 to 44 years in a typical year).

Cause	Deaths
Accidents	27,500
Cancer	20,300
Motor vehicle	16,500
Heart disease	16,000
Poison (solid, liquid)	2,700
Drowning	1,500
Falls	1,100
Fire related	900

example, a person killed in the home while cleaning a firearm would be included in this category. However, a person killed in combat would not be. In a typical year, there are approximately 2,000 deaths in this category.

- **Others.** This category includes deaths resulting from medical complications arising out of mistakes made by health care professionals, air transport injuries, interaction with machinery, **mechanical suffocation**, and the impact of falling objects. In a typical year, there are over 14,000 deaths in these subcategories.[3]

Accidents versus Other Causes of Death

Although there are more deaths every year from **heart disease**, **cancer**, and **strokes** than from accidents, these causes tend to be concentrated among people at or near retirement age. Among people 37 years of age or younger—prime working years—accidents are the number one cause of death. Figure 2-3 summarizes the causes of death for persons from 25 to 44 years of age. Notice that the leading cause is accidents.

Figure 2-3 shows that accidents represent a serious detriment to productivity, quality, and competitiveness in today's workplace. Yet accidents are the one cause of death and injury that companies can most easily control. Although it is true that companies may have some success in decreasing the incidence of heart disease and stroke among their employees through such activities as corporate wellness programs, their impact in this regard is limited. However, employers can have a significant impact on preventing accidents.

Work Accident Costs and Rates

Workplace accidents cost employers millions every year. Consider the following examples from the recent past. Arco Chemical Company was ordered to pay $3.48 million in fines as a result of failing to protect workers from an explosion at its petrochemical plant in Channelview, Texas. The steel-making division of USX paid a $3.25 million fine to settle numerous health and safety violation citations. BASF Corporation agreed to pay a fine of $1.06 million to settle OSHA citations associated with an explosion at a Cincinnati chemical plant that caused 2 deaths and 17 injuries.

These examples show the costs of fines only. In addition to fines, these employers incurred costs for safety corrections, medical treatment, survivor benefits, death and burial costs, and a variety of indirect costs. Clearly, work accidents are expensive. However, the news is not all bad. The trend in the rate of accidents is downward.

Work **accident rates** in this century are evidence of the success of the safety movement in the United States. As the amount of attention given to workplace safety and health has increased, the accident rate has decreased. According to the NSC,

> Between 1912 and 1998, accidental work deaths per 100,000 population were reduced 81 percent, from 21 to 4. In 1912, an estimated 18,000 to 21,000 workers' lives were lost. In 1998, in a workforce more than triple in size and producing 11 times the goods and services, there were approximately 10,000 work deaths.[4]

As Figure 2-1 shows, the cost of these 10,000 work deaths and work injuries was $48.5 billion. This translates into a cost of $420 per worker in the United States, computed as the value-added required per worker to offset the cost of work injuries. It translates further into $610,000 per death and $18,000 per disabling injury.[5]

Although statistics are not available to document the supposition, many safety and health professionals believe that the major cost of accidents and injuries on the job results from damage to morale. Employee morale is a less tangible factor than documentable factors such as lost time and medical costs. However, it is widely accepted among management professionals that few factors affect productivity more than employee morale. Employees with low morale do not produce up to their maximum potential. This is why so much time and money are spent every year to help supervisors and managers learn different ways to help improve employee morale.

Because few things are so detrimental to employee morale as seeing a fellow worker injured, accidents can have a devastating effect on morale. Whenever an employee is injured, his or her colleagues silently think, "That could have been me," in addition to worrying about the employee. Morale is damaged even more if the injured employee is well liked and other employees know his or her family.

Time Lost Because of Work Injuries

An important consideration when assessing the effect of accidents on industry is the amount of **lost time** due to **work injuries**.[6] According to the NSC, approximately 35,000,000 hours are lost in a typical year as a result of accidents. This is actual time lost from disabling injuries and does not include additional time lost for medical checkups after the injured employee returns to work. Accidents that occurred in previous years often continue to cause lost time in the current year.

Deaths in Work Accidents

Deaths on the job have decreased markedly over the years. However, they still occur. For example, in a typical year, there are 10,400 work deaths in the United States. The causes of death in the workplace vary. They include those related to motor vehicles, falls, electric current, drowning, fires, air transport, poisoning, water transport, machinery, falling objects, rail transport, and mechanical suffocation.[7] Figure 2-4 gives a complete breakdown of the percentages for the various categories of causes.

FIGURE 2-4 Work deaths by cause for a typical year.

Cause	Percent
Motor vehicle related	37.2%
Falls	12.5
Electric current	3.7
Drowning	3.2
Fire related	3.1
Air transport related	3.0
Poison (solid, liquid)	2.7
Water transport related	1.6
Poison (gas, vapor)	1.4
Other	31.6

Work Injuries by Type of Accident

Work injuries can be classified by the type of accident from which they resulted. The most common causes of work injuries are:

- Overexertion
- Impact accidents
- Falls
- Bodily reaction (to chemicals)
- Compression
- Motor vehicle accidents
- Exposure to radiation or caustics
- Rubbing or abrasions
- Exposure to extreme temperatures

Overexertion, the result of employees working beyond their physical limits, is the leading cause of work injuries. According to the NSC, almost 31 percent of all work injuries are caused by overexertion. **Impact accidents** involve a worker being struck by or against an object. The next most prominent cause of work injuries is falls.[8] The remaining accidents are distributed fairly equally among the other causes just listed.

Death Rates by Industry

A variety of agencies and organizations including the Bureau of Labor Statistics, the National Center for Health Statistics, and the National Safety Council collect data on **death rates** within industrial categories.[9] Such information can be used in a variety of ways, not the least of which is in assigning workers' compensation rates. The most widely used industrial categories are agriculture, including farming, forestry, and fishing; mining/quarrying, including oil and gas drilling and extraction; construction; manufacturing; transportation/public utilities; trade, both wholesale and retail; services, including finance, insurance, and real estate; and federal, state, and local government.

When death rates are computed on the basis of the number of deaths per 100,000 workers in a given year, the industry categories rank as follows (from highest death rate to lowest):

1. Mining/quarrying
2. Agriculture
3. Construction
4. Transportation/public utilities
5. Government
6. Manufacturing
7. Services
8. Trade

The rankings sometimes change slightly from year to year. For example, agriculture and mining/quarrying may exchange the first and second ranking in any given year. This is also true at the low end of the rankings with services and trade. However, generally, the typical ranking is as listed.

Parts of the Body Injured on the Job

In order to develop and maintain an effective safety and health program, it is necessary to know not only the most common causes of death and injury but also the parts of the body most frequently injured. The NSC stated the following:

> Disabling work injuries in the entire nation totaled approximately 1.75 million in 1998. Of these, about 10,400 were fatal and 60,000 resulted in some permanent impairment. Injuries to the back occurred most frequently, followed by thumb and finger injuries and leg injuries.[10]

Typically, the most frequent injuries to specific parts of the body are as follows (from most frequent to least):

1. Back
2. Legs and fingers
3. Arms and multiple parts of the body
4. Trunk
5. Hands
6. Eyes, head, and feet
7. Neck, toes, and body systems

The back is the most frequently injured part of the body. Legs and fingers are injured with approximately the same frequency, as are arms and multiple parts of the body; the hands are next in frequency, followed by the eyes, the head, and feet; and neck, toes, and body systems. This ranking shows that one of the most fundamental components of a safety and health program should be instruction on how to lift without hurting the back (see Chapter 11).

Chemical Burn Injuries

Chemical burn injuries are a special category with which prospective and practicing safety professionals should be familiar. The greatest incidence of chemical burns (approximately one-third) occurs in manufacturing.[11] Other high-incidence industries are services, trade, and construction.

The chemicals that most frequently cause chemical burn injuries include acids and alkalies; soaps, detergents, and cleaning compounds; solvents and degreasers; calcium hydroxide (a chemical used in cement and plaster); potassium hydroxide (an ingredient in drain cleaners and other cleaning solutions); and sulfuric acid (battery acid). Almost 46 percent of all chemical burn injuries occur while workers are cleaning equipment, tools, and vehicles.[12]

What is particularly disturbing about chemical burn injuries is that a high percentage of them occur in spite of the use of personal protective equipment, the provision of safety instruction, and the availability of treatment facilities. In some cases, the personal protective equipment is faulty or inadequate. In others, it is not properly used in spite of instructions.

Preventing chemical burn injuries presents a special challenge to safety and health professionals. The following strategies are recommended:

- Familiarize yourself, the workers, and their supervisors with the chemicals that will be used and their inherent dangers.
- Secure the proper personal protection equipment for each type of chemical that will be used.
- Provide instruction on the proper use of personal protection equipment and then make sure that supervisors confirm that the equipment is used properly every time.
- Monitor that workers are wearing personal protection equipment and replace it when it begins to show wear.

Heat Burn Injuries

Heat burn injuries present a special challenge to safety and health professionals in the modern workplace. Almost 40 percent of all such injuries occur in manufacturing every year. The most frequent causes are flame (this includes smoke inhalation injuries), molten metal, petroleum asphalts, steam, and water. The most common activities associated with heat burn injuries are welding, cutting with a torch, and handling tar or asphalt.[13]

Following are several factors that contribute to heat burn injuries in the workplace. Safety and health professionals who understand these factors will be in a better position to prevent heat burn injuries.

- Employer has no health and safety policy regarding heat hazards.
- Employer fails to enforce safety procedures and practices.
- Employees are not familiar with the employer's safety policy and procedures concerning heat hazards.
- Employees fail to use or improperly use personal protection equipment.
- Employees have inadequate or worn personal protection equipment.
- Employees work in a limited space.
- Employees attempt to work too fast.
- Employees are careless.
- Employees have poorly maintained tools and equipment.[14]

These factors should be considered carefully by safety and health professionals when developing accident prevention programs. Employees should be familiar with the hazards, should know the appropriate safety precautions, and should have and use the proper personal protection equipment. Safety professionals should monitor to ensure that safety rules are being followed, that personal protection equipment is being used correctly, and that it is in good condition.

Repetitive Strain/Soft-Tissue Injuries

Repetitive strain injury (RSI) is a broad and generic term that encompasses a variety of injuries resulting from cumulative trauma to the soft tissues of the body, including tendons, tendon sheaths, muscles, ligaments, joints, and nerves. Such injuries are typically associated with the soft tissues of the hands, arms, neck, and shoulders.

Carpal tunnel syndrome (CTS) is the most widely known repetitive strain injury. There are also several other repetitive strain injuries to the body's soft tissues. The carpal tunnel is the area inside the wrist through which the median nerve passes. It is formed by the wrist bones and a ligament. CTS is typically caused by repeated and cumulative stress on the median nerve. Symptoms of CTS include numbness, a tingling sensation, and pain in the fingers, hand, and/or wrist.

Stress placed on the median nerve typically results from repeated motion while the hands and fingers are bent in an unnatural position. However, sometimes the stress results from a single traumatic event such as a sharp blow to the wrist.

Evidence suggests a higher incidence of CTS among women than men. Scientists have found that the incidence rate for CTS among women was 1.96 per 1,000 full-time equivalent (FTE) workers and 1.58 per 1,000 FTE for men. The overall incidence rate for CTS is increasing at a rate of more than 15 percent per year.[15]

A common misconception about RSI is that it is synonymous with carpal tunnel syndrome. It isn't. In fact, carpal tunnel syndrome is relatively rare among RSI patients. Following is a list of the broad classifications of RSI and the types commonly associated with each classification:

MUSCLE AND TENDON DISORDERS

- Tendinitis
- Muscle damage
- Tenosynovitis
- Stenosing tenosynovitis

DEQUERVAIN'S DISEASE

- Flexor tenosynovitis (trigger finger)
- Shoulder tendinitis
- Forearm tendinitis
- Cervical radioculopathy
- Epicondylitis
- Ganglion cysts

TUNNEL SYNDROMES

- Carpal tunnel syndrome
- Radial tunnel syndrome

ULNAR NERVE DISORDERS

- Sulcus ulnaris syndrome
- Cubital tunnel syndrome
- Guyon's canal syndrome

NERVE AND CIRCULATION DISORDERS

- Thoracic outlet syndrome
- Raynaud's disease

OTHER ASSOCIATED DISORDERS

- Reflex sympathetic dysfunction
- Focal dystonia (writer's cramp)
- Degenerative joint disorder (osteoarthritis)
- Fibromyalgia
- Dupuytren's contracture[16]

RSI and its manifestations are explained in greater detail in Chapter 10.

Estimating the Cost of Accidents

Even decision makers who support accident prevention must consider the relative costs of such efforts. Clearly, accidents are expensive. However, to be successful, safety and health professionals must be able to show that accidents are more expensive than prevention. To do this, they must be able to estimate the cost of accidents. The procedure for estimating costs set forth in this section was developed by Professor Rollin H. Simonds of Michigan State College working in conjunction with the Statistics Division of the National Safety Council.

Cost Estimation Method

Professor Simonds states that in order to have value, a cost estimate must relate directly to the specific company in question. Applying broad industry cost factors will not suffice. To arrive at company-specific figures, Simonds recommends that costs associated with an accident be divided into *insured* and *uninsured* costs.[17]

Determining the insured costs of accidents is a simple matter of examining accounting records. The next step involves calculating the uninsured costs. Simonds recommends that accidents be divided into the following four classes:

- **Class 1 accidents.** Lost workdays, permanent partial disabilities, and temporary total disabilities.
- **Class 2 accidents.** Treatment by a physician outside the company's facility.
- **Class 3 accidents.** Locally provided first aid, property damage of less than $100, or the loss of fewer than eight hours of work time.
- **Class 4 accidents.** Injuries that are so minor they do not require the attention of a physician, result in property damage of $100 or more, or cause eight or more work hours to be lost.[18]

Average uninsured costs for each class of accident can be determined by pulling the records of all accidents that occurred during a specified period and sorting the records according to class. For each accident in each class, record every cost that was not covered by insurance. Compute the total of these costs by class of accident and divide by the total number of accidents in that class to determine an average uninsured cost for each class, specific to the particular company.

FIGURE 2-5 Uninsured costs worksheet.

Class of Accident	Accident Number							
Class 1	**1**	**2**	**3**	**4**	**5**	**6**	**7**	**8**
Cost A	$ 16.00	$ 6.95	$ 15.17	$ 3.26				
Cost B	72.00	103.15	97.06	51.52				
Cost C	26.73	12.62	—	36.94				
Cost D	—	51.36	—	38.76				
Cost E	—	11.17	—	24.95				
Cost F	—	—	—	–13.41				
Cost G	—	—	—	—				
Total	114.73	185.25	112.23	142.02				

Grand Total: $554.23

Average Cost per Accident: $138.56 (Grand Total ÷ Number of Accidents)

Signature: ______________________________ Date: __________

Figure 2-5 is an example of how the average cost of a selected sample of Class 1 accidents can be determined. In this example, there were four Class 1 accidents in the pilot test. These four accidents cost the company a total of $554.23 in uninsured costs, or an average of $138.56 per accident. Using this information, accurate cost estimates of an accident can be figured, as can accurate predictions.

Other Cost Estimation Methods

The costs associated with workplace accidents, injuries, and incidents fall into broad categories such as the following:

- Lost work hours
- Medical costs
- Insurance premiums and administration
- Property damage
- Fire losses
- Indirect costs

Calculating the direct costs associated with lost work hours involves compiling the total number of lost hours for the period in question and multiplying the hours times the applicable loaded labor rate. The loaded labor rate is the employee's hourly rate plus benefits. Benefits vary from company to company but typically inflate the hourly wage by 20 to 35 percent. A sample cost-of-lost-hours computation follows:

Employee Hours Lost (4th quarter) × Average Loaded Labor Rate = Cost 386 × 13.48 = $5,203.28

In this example, the company lost 386 hours due to accidents on the job in the fourth quarter of its fiscal year. The employees who actually missed time at work formed a pool of people with an average loaded labor rate of $13.48 per hour ($10.78 average hourly wage plus 20 percent for benefits). The average loaded labor rate multiplied times the 386 lost hours reveals an unproductive cost of $5,203.28 to this company.

By studying records that are readily available in the company, a safety professional can also determine medical costs, insurance premiums, property damage, and fire losses for the time period in question. All these costs taken together result in a subtotal cost. This figure is then increased by a standard percentage to cover indirect costs to determine the total cost of accidents for a specific time period. The percentage used to calculate indirect costs can vary from company to company, but 20 percent is a widely used figure.

Estimating Hidden Costs

Safety professionals often use the *iceberg analogy* when talking about the real costs of accidents. Accident costs are like an iceberg in that their greatest portion is hidden from view.[19] In the case of icebergs, the larger part is hidden beneath the surface of the water. In the case of an accident, the larger part of the actual cost is also hidden beneath the surface.

According to Daniel Corcoran,

> When a serious accident occurs, there is usually a great deal of activity associated with the accident. There may be a slowdown in production near the site of the accident, for instance. There also will be a need to replace the injured worker, at least temporarily, and there will be costs associated with the learning curve of the replacement worker.

The supervisor and the accident investigation team probably will need to spend time conducting an investigation, and there will be a lot of time spent on the administration of paper work related to the accident.[20]

There are many different models that can be used for estimating both the direct and indirect costs of accidents. Some of these models are so complex that their usefulness is questionable. The checklist in Figure 2-6 is a simple and straightforward tool that can be used to estimate the hidden costs of accidents.

Global Impact of Accidents and Injuries

According to the International Labour Organization (ILO) of the United Nations, approximately 2.2 million people die every year of work-related injuries and occupational diseases.[21] Actually, due to poor record keeping and reporting in underdeveloped countries, it is estimated that this figure is low. Rapid development and the pressure of global competition are resulting in increased workplace fatalities in China and the Pacific Rim countries.

What is missing in many of the developing countries that are becoming industrialized is a safety and health infrastructure. Such an infrastructure would include government-enforced safety and health regulations, company-sponsored safety and health training, record-keeping and reporting systems, and management practices that make occupational safety and health a fully integrated component of work processes and the

FIGURE 2-6 Some accident costs that might be overlooked.

	Checklist for Estimating the Hidden Costs of Accidents
√	Paid time to the injured employee on the day of the accident.
√	Paid time of any emergency-responder personnel involved (including ambulance driver).
√	Paid time of all employees who were interviewed as part of the accident investigation.
√	Paid time of the safety personnel who conducted the accident investigation.
√	Paid time of the human resources personnel who handled the workers' compensation and medical aspects of the accident.
√	Paid time of the supervisor involved in the accident investigation and accident response.
√	Paid time to employees near the accident working (or slowed down) temporarily as a result of the accident.
√	Paid time to employees who spent time talking about the accident as news of it spread through the company's grapevine.

competitive philosophy of organizations. Occupational safety and health must come to be seen as a strategy for sustaining economic growth and social development in emerging countries.

There is much to be done if developing countries are going to provide safe and healthy working conditions for their citizens. The ILO reports the following:

- Record-keeping and reporting systems in developing countries are deteriorating instead of improving. Consequently, only a fraction of the real toll of workplace accidents and injuries is being reported.
- Men in developing countries tend to die as the result of accidents, lung diseases, and work-related cancers such as those caused by asbestos. Women in developing countries suffer more from musculoskeletal disorders, communicable diseases, and psychosocial problems.
- Occupational injuries in developing countries are more prevalent in such high-risk industries as mining, construction, and agriculture.
- Younger workers in developing countries are more likely to suffer nonfatal injuries, while older workers are more likely to suffer fatal injuries.
- In developing countries, more than half of retirements are taken early to collect pensions based on work-related disabilities rather than normal retirement.

Key Terms and Concepts

- Accident prevention
- Accident rates
- Accidents
- Cancer
- Carpal tunnel syndrome (CTS)
- Chemical burn injuries
- Death rates
- Drownings
- Falls
- Fire-related losses
- Heart disease
- Heat burn injuries
- Impact accidents
- Indirect costs
- Insurance administration
- Lost time
- Lost wages
- Mechanical suffocation
- Medical complications
- Medical expenses
- Natural disasters
- Overexertion
- Poisoning
- Property damage
- Repetitive strain injury (RSI)
- Safety and health professionals
- Strokes
- Suffocation
- Work injuries
- Workplace accidents

Review Questions

1. What are the leading causes of death in the United States?
2. When the overall cost of an accident is calculated, what elements make up the cost?
3. What are the five leading causes of accidental deaths in the United States?
4. What are the leading causes of death in the United States of people between the ages of 25 and 44?
5. Explain how today's rate of accidental work deaths compares with the rate in the early 1900s.
6. What are the five leading causes of work deaths?
7. What are the five leading causes of work injuries by type of accident?
8. When death rates are classified by industry type, what are the three leading industry types?
9. Rank the following body parts according to frequency of injury from highest to lowest: neck, fingers, trunk, back, and eyes.
10. Name three chemicals that frequently cause chemical burns in the workplace.
11. Identify three factors that contribute to heat burn injuries in the workplace.
12. Explain the difference between RSI and carpal tunnel syndrome.
13. Explain the reasons for high accident rates in developing countries.

Endnotes

1. National Safety Council, Accident Facts (Chicago: National Safety Council, 2005), 37.
2. Ibid., 25.
3. Ibid., 4–5.
4. Ibid., 34.
5. Ibid., 35.
6. Ibid., 35.
7. Ibid., 36.
8. Ibid., 36.
9. Ibid., 37.
10. Ibid., 38.
11. Ibid., 39.
12. Ibid., 40.
13. Ibid., 41.
14. Ibid., 41.
15. ErgoOutfitters.com, http://www.ergooutfitters.com, March 31, 2006.
16. Ibid.
17. National Safety Council, Accident Prevention Manual for Industrial Operations: Administration and Programs (Chicago: National Safety Council, 1997), 158.
18. Ibid., 158.
19. Daniel Corcoran, "The Hidden Value of Safety." Occupational Health & Safety 71, no. 6: 20–22.
20. Corcoran, "The Hidden Value of Safety," 22.
21. Occupational Health & Safety Online, "Study: 2.2 Million People Die Worldwide of Work-Related Accidents, Occupational Diseases." www.ohsonline.com, September 20, 2005, 1–3.

CHAPTER 3

Theories of Accident Causation

Major Topics

- Domino Theory of Accident Causation
- Human Factors Theory of Accident Causation
- Accident/Incident Theory of Accident Causation
- Epidemiological Theory of Accident Causation
- Systems Theory of Accident Causation
- Combination Theory of Accident Causation
- Behavioral Theory of Accident Causation
- Drugs and Accident Causation
- Depression and Accident Causation
- Management Failures and Accident Causation
- Obesity and Accident Causation

Each year, work-related accidents cost the United States almost $50 billion.[1] This figure includes costs associated with lost wages, medical expenses, insurance costs, and indirect costs. The number of persons injured in **industrial place accidents** in a typical year is 7,128,000, or 3 per 100 persons per year.[2] In the workplace, there is one accidental death approximately every 51 minutes and one injury every 19 seconds.[3]

Why do accidents happen? This question has concerned safety and health decision makers for decades, because in order to prevent accidents we must know why they happen. Over the years, several theories of accident causation have evolved that attempt to explain why accidents occur. Models based on these theories are used to predict and prevent accidents.

The most widely known theories of accident causation are the domino theory, the human factors theory, the accident/incident theory, the epidemiological theory, the systems theory, the combination theory, and the behavioral theory. This chapter provides practicing and prospective safety professionals with the information they need to understand fully and apply these theories.

Domino Theory of Accident Causation

An early pioneer of accident prevention and industrial safety was Herbert W. Heinrich, an official with the Travelers Insurance Company. In the late 1920s, after studying the reports of 75,000 industrial accidents, Heinrich concluded that

- 88 percent of industrial accidents are caused by unsafe acts committed by fellow workers.
- 10 percent of industrial accidents are caused by unsafe conditions.
- 2 percent of industrial accidents are unavoidable.[4]

Heinrich's study laid the foundation for his *Axioms of Industrial Safety* and his theory of accident causation, which came to be known as the **domino theory**. So much of Heinrich's theory has been discounted by more contemporary research that it is now considered outdated. However, because some of today's more widely accepted theories can be traced back to Heinrich's theory, students of industrial safety should be familiar with his work.

Heinrich's Axioms of Industrial Safety

Heinrich summarized what he thought health and safety decision makers should know about industrial accidents in 10 statements he called **Axioms of Industrial Safety**. These axioms can be paraphrased as follows:

1. Injuries result from a completed series of factors, one of which is the accident itself.
2. An accident can occur only as the result of an unsafe act by a person and/or a physical or mechanical hazard.
3. Most accidents are the result of unsafe behavior by people.
4. An unsafe act by a person or an unsafe condition does not always immediately result in an accident/injury.
5. The reasons why people commit unsafe acts can serve as helpful guides in selecting corrective actions.
6. The severity of an accident is largely fortuitous, and the accident that caused it is largely preventable.
7. The best accident prevention techniques are analogous with the best quality and productivity techniques.
8. Management should assume responsibility for safety because it is in the best position to get results.
9. The supervisor is the key person in the prevention of industrial accidents.
10. In addition to the direct costs of an accident (e.g., compensation, liability claims, medical costs, and hospital expenses), there are also hidden or indirect costs.[5]

According to Heinrich, these axioms encompass the fundamental body of knowledge that must be understood by decision makers interested in preventing accidents. Any accident prevention program that takes all 10 axioms into account is more likely to be effective than a program that leaves out one or more axioms.

Heinrich's Domino Theory

Perhaps you have stood up a row of dominoes, tipped the first one over, and watched as each successive domino topples the one next to it. This is how Heinrich's theory of accident causation works. According to Heinrich, there are five factors in the sequence of events leading up to an accident. These factors can be summarized as follows:

1. **Ancestry and social environment.** Negative character traits that may lead people to behave in an unsafe manner can be inherited **(ancestry)** or acquired as a result of the **social environment**.
2. **Fault of person.** Negative character traits, whether inherited or acquired, are why people behave in an unsafe manner and why hazardous conditions exist.
3. **Unsafe act/mechanical or physical hazard.** **Unsafe acts** committed by people and **mechanical** or **physical hazards** are the direct causes of accidents.
4. **Accident.** Typically, accidents that result in injury are caused by falling or being hit by moving objects.
5. **Injury.** Typical injuries resulting from accidents include lacerations and fractures.[6]

Heinrich's theory has two central points: (1) Injuries are caused by the action of **preceding factors**; and (2) removal of the **central factor** (unsafe act/**hazardous condition**) negates the action of the preceding factors and, in so doing, prevents accidents and injuries.

Domino Theory in Practice

Construction Products Company (CPC) is a distributor of lumber, pipe, and concrete products. Its customers are typically small building contractors. CPC's facility consists of an office in which orders are placed and several large warehouses. Contractors place their orders in the office. They then drive their trucks through the appropriate warehouses to be loaded by CPC personnel.

Because the contractors are small operations, most of their orders are also relatively small and can be loaded by hand. Warehouse personnel go to the appropriate bins, pull out the material needed to fill their orders, and load the materials on customers' trucks. Even though most orders are small enough to be loaded by hand, many of the materials purchased are bulky and cumbersome to handle. Because of this, CPC's loaders are required to wear such personal protection gear as hard hats, padded gloves, steel-toed boots, and lower-back-support belts.

For years CPC's management team had noticed an increase in minor injuries to warehouse personnel during the summer months. Typically, these injuries consisted of nothing worse than minor cuts, scrapes, and bruises. However, this past summer had been different. Two warehouse workers had sustained serious back injuries. These injuries have been costly to CPC both financially and in terms of employee morale.

An investigation of these accidents quickly identified a series of events and a central causal behavior that set up a *domino effect* that, in turn, resulted in the injuries. The investigation revealed that CPC's warehouses became so hot during the summer months that personal protection gear was uncomfortable. As a result, warehouse personnel simply discarded it. Failure to use appropriate personal protection gear in the summer months had always led to an increase in injuries. However, because the injuries were minor in nature, management had never paid much attention to the situation. It was probably inevitable that more serious injuries would occur eventually.

To prevent a recurrence of the summer-injury epidemic, CPC's management team decided to remove the causal factor—failure of warehouse personnel to use their personal protection gear during the summer months. To facilitate the removal of this factor, CPC's management team formed a committee consisting of one executive manager, one warehouse supervisor, and three warehouse employees.

The committee made the following recommendations: (1) provide all warehouse personnel with training on the importance and proper use of personal protection gear; (2) require warehouse supervisors to monitor the use of personal protection gear more closely; (3) establish a company policy that contains specific and progressive disciplinary measures for failure to use required personal protection gear; and (4) implement several heat reduction measures to make warehouses cooler during the summer months.

CPC's management team adopted all the committee's recommendations. In doing so, it removed the central causal factor that had historically led to an increase in injuries during the summer months.

Human Factors Theory of Accident Causation

The **human factors theory** of accident causation attributes accidents to a chain of events ultimately caused by **human error**. It consists of the following three broad factors that lead to human error: overload, inappropriate response, and inappropriate activities (see Figure 3-1). These factors are explained in the following paragraphs.

Overload

Overload amounts to an imbalance between a person's capacity at any given time and the load that person is carrying in a given state. A person's capacity is the product of such factors as his or her natural ability, training, state of mind, fatigue, stress, and physical condition. The load that a person is carrying consists of tasks for which he or she is responsible and added burdens resulting from **environmental factors** (noise, distractions, and so on), **internal factors** (personal problems, emotional stress, and worry), and **situational factors** (level of risk, unclear instructions, and so on). The state in which a person is acting is the product of his or her motivational and arousal levels.

FIGURE 3-1 Factors that cause human errors.

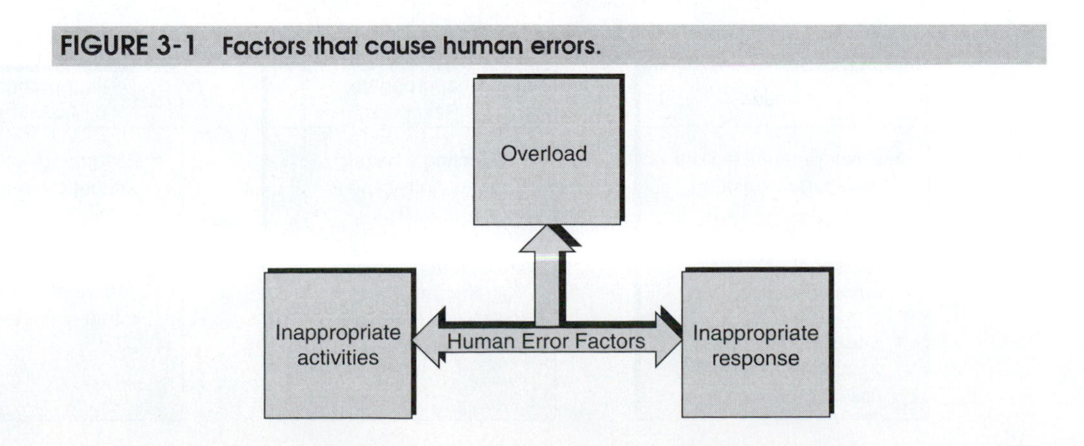

Inappropriate Response and Incompatibility

How a person responds in a given situation can cause or prevent an accident. If a person detects a hazardous condition but does nothing to correct it, he or she has responded inappropriately. If a person removes a safeguard from a machine in an effort to increase output, he or she has responded inappropriately. If a person disregards an established safety procedure, he or she has responded inappropriately. Such responses can lead to accidents. In addition to **inappropriate responses**, this component includes workstation incompatibility. The incompatibility of a person's workstation with regard to size, force, reach, feel, and similar factors can lead to accidents and injuries.

Inappropriate Activities

Human error can be the result of **inappropriate activities**. An example of an inappropriate activity is a person who undertakes a task that he or she doesn't know how to do. Another example is a person who misjudges the degree of risk involved in a given task and proceeds based on that misjudgment. Such inappropriate activities can lead to accidents and injuries. Figure 3-2 summarizes the various components of the human factors theory.[7]

Human Factors Theory in Practice

Kitchenware Manufacturing Incorporated (KMI) produces aluminum kitchenware for commercial settings. After ten years of steady, respectable growth in the U.S. market, KMI suddenly saw its sales triple in less than six months. This rapid growth was the result of KMI's successful entry into European and Asian markets.

The growth in sales, although welcomed by both management and employees, quickly overloaded and, before long, overwhelmed the company's production facility. KMI responded by adding a second shift of production personnel and approving unlimited overtime for highly skilled personnel. Shortly after the upturn in production, KMI began to experience a disturbing increase in accidents and injuries. During his

FIGURE 3-2 Human factors theory.

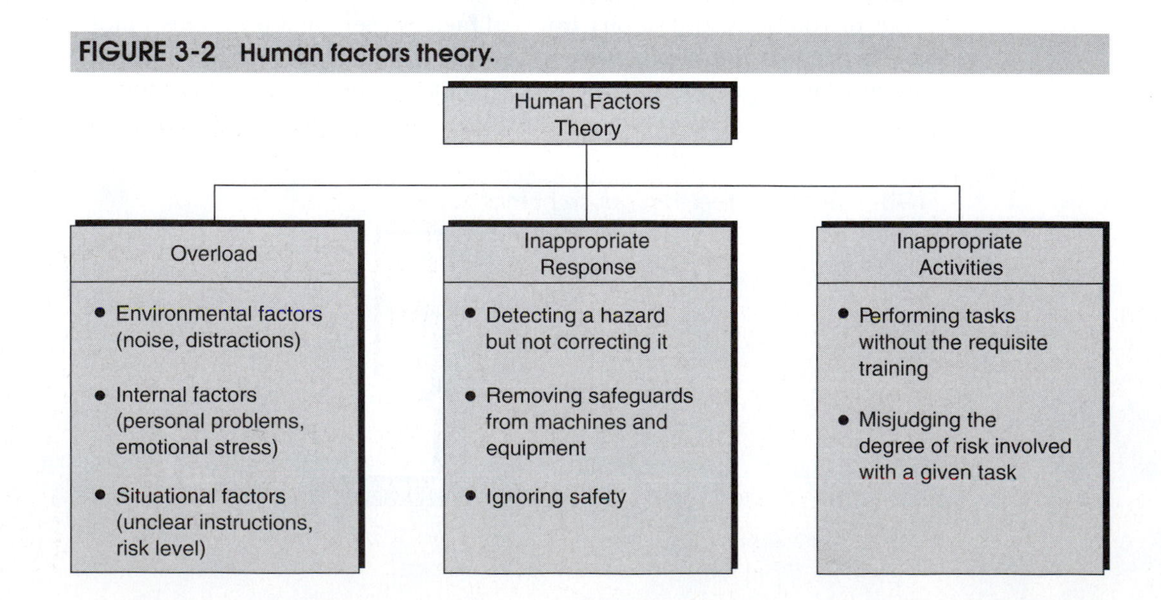

accident investigations, KMI's safety manager noticed that human error figured prominently in the accidents. He grouped all the human errors identified into three categories: (1) overload, (2) inappropriate response, and (3) inappropriate activities.

In the category of *overload,* he found that the rush to fill orders was pushing production personnel beyond their personal limits in some cases, and beyond their capabilities in others. Stress, insufficient training of new employees, and fatigue all contributed to the overload. In the category of *inappropriate response,* the safety manager determined that many of KMI's production personnel had removed safeguards from their machines in an attempt to speed up production. All the machines involved in accidents had had safeguards removed.

In the category of *inappropriate activities*, the safety manager found that new employees were being assigned duties for which they weren't yet fully trained. As a result, they often misjudged the amount of risk associated with their work tasks.

With enough accident investigations completed to identify a pattern of human error, the safety manager prepared a presentation containing a set of recommendations for corrective measures for KMI's executive management team. His recommendations were designed to prevent human-error-oriented accidents without slowing production.

Accident/Incident Theory of Accident Causation

The **accident/incident theory** is an extension of the human factors theory. It was developed by Dan Petersen and is sometimes referred to as the Petersen accident/incident theory.[8] Petersen introduced such new elements as **ergonomic traps**, the decision to err, and systems failures, while retaining much of the human factors theory. A model based on his theory is shown in Figure 3-3.

In this model, overload, ergonomic traps, or a decision to err lead to human error. The decision to err may be conscious and based on logic, or it may be unconscious. A variety of pressures such as deadlines, peer pressure, and budget factors can lead to **unsafe behavior**. Another factor that can influence such a decision is the "It won't happen to me" syndrome.

The systems failure component is an important contribution of Petersen's theory. First, it shows the potential for a **causal relationship** between management decisions or management behavior and safety. Second, it establishes management's role in accident prevention as well as the broader concepts of safety and health in the workplace.

Following are just some of the different ways that systems can fail, according to Petersen's theory:

- Management does not establish a comprehensive safety policy.
- Responsibility and authority with regard to safety are not clearly defined.
- Safety procedures such as measurement, inspection, correction, and investigation are ignored or given insufficient attention.
- Employees do not receive proper orientation.
- Employees are not given sufficient safety training.

Accident/Incident Theory in Practice

Poultry Processing Corporation (PPC) processes chickens and turkeys for grocery chains. Poultry processing is a labor-intensive enterprise involving a great deal of handwork.

FIGURE 3-3 Accident/incident theory.

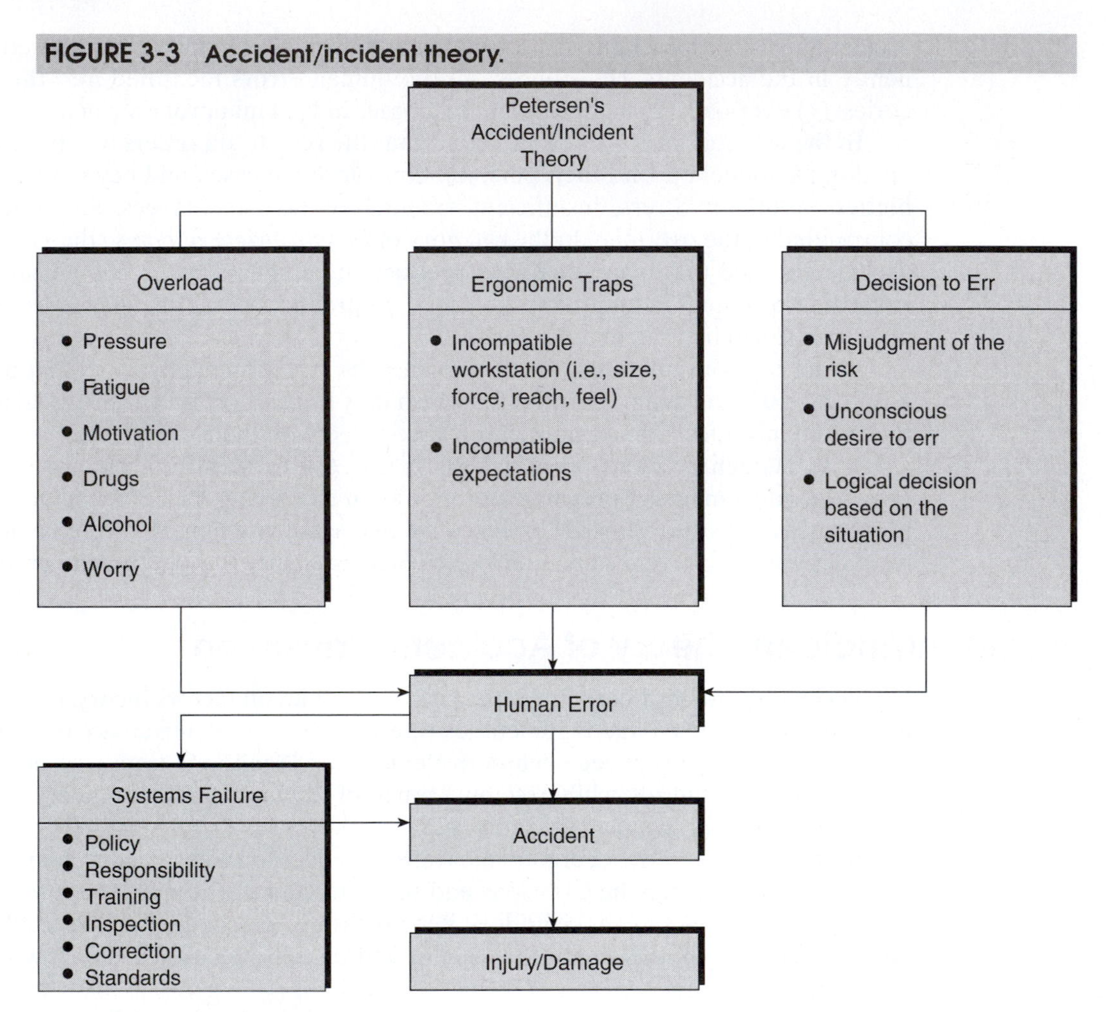

A variety of different knives, shears, and cleavers are used. Much of the work is monotonous and repetitive. Selected parts of the overall process must be done in cold conditions.

PPC has gone to great lengths to ensure that workstations are ergonomically sound, that personal protection gear is used as appropriate, and that adequate precautions are taken to prevent illness and injuries. As a result, PPC is an award-winning company in the area of workplace safety and health.

Consequently, the poultry-processing industry was shocked when a class action lawsuit was filed against PPC on behalf of over 50 employees, all of whom claimed to be suffering from carpal tunnel syndrome. Because of PPC's excellent safety and health record, most observers felt sure that the company would be vindicated in the end.

The company's policies and procedures relating to safety and health were investigated thoroughly by consultants brought in by both PPC and the attorney for the plaintiffs. Over 100 witnesses gave depositions, and several preliminary hearings were held. By the time the trial finally rolled around, both sides had accumulated mountains of paper and filing cabinets full of evidence. Then, suddenly and without advance notice, PPC offered a substantial financial settlement, which the plaintiffs accepted.

It was one of PPC's outside consultants who discovered what had caused the increased incidence of carpal tunnel syndrome. The company had always used a centralized approach to managing safety and health. Responsibility for such tasks as measurement, inspection, correction, and investigation was assigned to the safety manager, Joe Don Huttle. Huttle had an excellent record during his 20 years in the poultry-processing industry, with the last 5 spent at PPC. In fact, he was so well respected in the industry that his peers had elected him president of a statewide safety organization. This, as it turned out, is where PPC's troubles began.

When Huttle took it over, the safety organization had experienced a three-year decline in membership and was struggling to stay afloat financially. He had been elected as "the man who could save the organization." Intending to do just that, Huttle went right to work. For months at a time he worked seven days a week, often spending as much as two weeks at a time on the road. When he was in his office at PPC, Huttle was either on the telephone or doing paperwork for the safety organization.

Within six months, he had reversed the organization's downhill slide, but not without paying a price at home. During the same six-month period, his duties at PPC were badly neglected. Measurement of individual and group safety performance had come to a standstill. The same was true of inspection, correction, investigation, and reporting.

It was during this time of neglect that the increased incidence of carpal tunnel syndrome occurred. Safety precautions that Huttle had instituted to guard against this particular problem were no longer observed properly once the workers realized that he had stopped observing and correcting them. Measurement and inspection may also have prevented the injuries had Huttle maintained his normal schedule of these activities.

PPC's consultant, in a confidential report to executive managers, cited the *accident/incident theory* in explaining his view of why the injuries occurred. In this report, the consultant said that Huttle was guilty of applying "it won't happen here" logic when he made a conscious decision to neglect his duties at PPC in favor of his duties with the professional organization. Of course, the employees themselves were guilty of not following clearly established procedures. However, because Huttle's neglect was also a major contributing factor, PPC decided to settle out of court.

Epidemiological Theory of Accident Causation

Traditionally, safety theories and programs have focused on accidents and the resulting injuries. However, the current trend is toward a broader perspective that also encompasses the issue of industrial hygiene. **Industrial hygiene** concerns environmental factors that can lead to sickness, disease, or other forms of impaired health.

This trend has, in turn, led to the development of an epidemiological theory of accident causation. Epidemiology is the study of causal relationships between environmental factors and disease. The **epidemiological theory** holds that the models used for studying and determining these relationships can also be used to study causal relationships between environmental factors and accidents or diseases.[9]

Figure 3-4 illustrates the epidemiological theory of accident causation. The key components are **predispositional characteristics** and **situational characteristics**. These characteristics, taken together, can either result in or prevent conditions that may result in an accident. For example, if an employee who is particularly susceptible to peer pressure

FIGURE 3-4 Epidemiological theory.

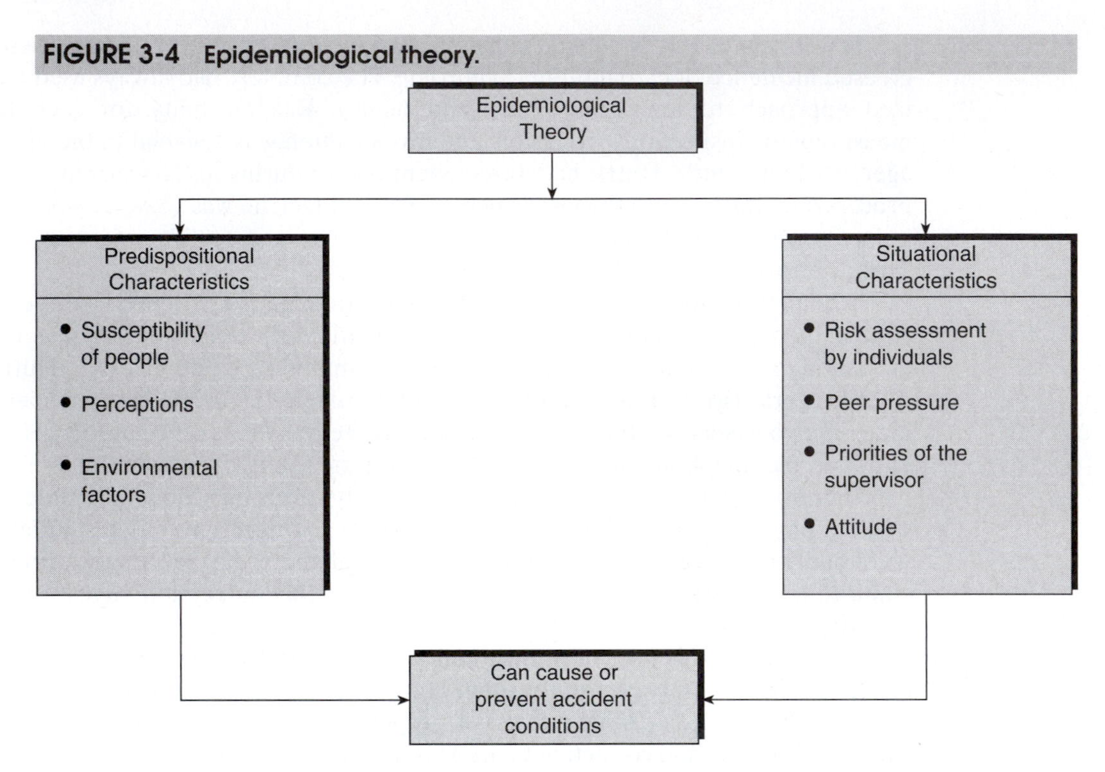

(predispositional characteristic) is pressured by his coworkers (situational characteristic) to speed up his operation, the result will be an increased probability of an accident.

Epidemiological Theory in Practice

Jane Andrews was the newest member of the loading unit for Parcel Delivery Service (PDS). She and the other members of her unit were responsible for loading 50 trucks every morning. It was physically demanding work, and she was the first woman ever selected by PDS to work in the loading unit. She had gotten the job as part of the company's upward mobility program. She was excited about her new position because within PDS, the loading unit was considered a springboard to advancement. Consequently, she was eager to do well. The responsibility she felt toward other female employees at PDS only intensified her anxiety. Andrews felt that if she failed, other women might not get a chance to try in the future.

Before beginning work in the loading unit, employees must complete two days of training on proper lifting techniques. The use of back-support belts is mandatory for all loading dock personnel. Consequently, Andrews became concerned when the supervisor called her aside on her first day in the unit and told her to forget what she had learned in training. He said, "Jane, nobody wants a back injury, so be careful. But the key to success in this unit is speed. The lifting techniques they teach in that workshop will just slow you down. You've got the job, and I'm glad you're here. But you won't last long if you can't keep up."

Andrews was torn between following safety procedures and making a good impression on her new supervisor. At first, she made an effort to use proper lifting techniques.

However, when several of her coworkers complained that she wasn't keeping up, the supervisor told Andrews to "keep up or get out of the way." Feeling the pressure, she started taking the same shortcuts she had seen her coworkers use. Positive results were immediate, and Andrews received several nods of approval from fellow workers and a "good job" from the supervisor. Before long, Andrews had won the approval and respect of her colleagues.

However, after two months of working in the loading unit, she began to experience persistent lower-back pain. Andrews felt sure that her hurried lifting techniques were to blame, but she valued the approval of her supervisor and fellow workers too much to do anything that might slow her down. Finally, one day while loading a truck, Andrews fell to the pavement in pain and could not get up. Her back throbbed with intense pain, and her legs were numb. She had to be rushed to the emergency room of the local hospital. By the time Andrews checked out of the hospital a week later, she had undergone major surgery to repair two ruptured discs.

Jane Andrews's situation can be explained by the *epidemiological theory* of accident causation. The predispositional factor was her susceptibility to peer pressure from her coworkers and supervisor. The applicable situational factors were peer pressure and the priorities of the supervisor. These factors, taken together, caused the accident.

Systems Theory of Accident Causation

A *system* is a group of regularly interacting and interrelated components that together form a unified whole. This definition is the basis for the **systems theory** of accident causation. This theory views a situation in which an accident may occur as a system comprised of the following components: person (host), machine (agency), and **environment**.[10] The likelihood of an accident occurring is determined by how these components interact. Changes in the patterns of interaction can increase or reduce the probability of an accident.

For example, an experienced employee who operates a numerically controlled five-axis machining center in a shop environment may take a two-week vacation. Her temporary replacement may be less experienced. This change in one component of the system (person/host) increases the probability of an accident. Such a simple example is easily understood. However, not all changes in patterns of interaction are this simple. Some are so subtle that their analysis may require a team of people, each with a different type of expertise.

The primary components of the systems model are the person/machine/environment, information, decisions, risks, and the task to be performed.[11] Each of the components has a bearing on the probability that an accident will occur. The systems model is illustrated in Figure 3-5.

As this model shows, even as a person interacts with a machine within an environment, three activities take place between the system and the task to be performed. Every time a task must be performed, there is the risk that an accident may occur. Sometimes the risks are great; at other times, they are small. This is where information collection and decision making come in.

Based on the information that has been collected by observing and mentally noting the current circumstances, the person weighs the risks and decides whether to perform the task under existing circumstances. For example, say a machine operator is working on a rush order that is behind schedule. An important safety device has

FIGURE 3-5 Systems theory model.

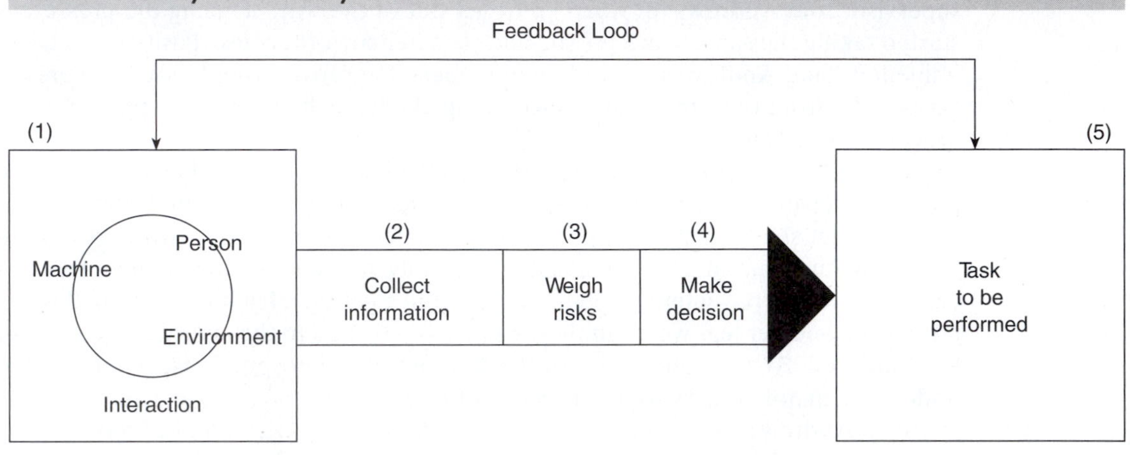

malfunctioned on his machine. Simply taking it off will interrupt work for only five minutes, but it will also increase the probability of an accident. However, replacing it could take up to an hour. Should the operator remove the safety guard and proceed with the task or take the time to replace it? The operator and his supervisor may assess the situation (collect information), weigh the risks, and make a decision to proceed. If their information was right and their assessment of the risks accurate, the task will probably be accomplished without an accident.

However, the environment in which the machine operator is working is unusually hectic, and the pressure to complete an order that is already behind schedule is intense. These factors are **stressors** that can cloud the judgment of those collecting information, weighing risks, and making the decision. When stressors are introduced between points 1 and 3 in Figure 3-5, the likelihood of an accident increases.

For this reason, Firenzie recommends that five factors be considered before beginning the process of collecting information, weighing risks, and making a decision:

- Job requirements
- The workers' abilities and limitations
- The gain if the task is successfully accomplished
- The loss if the task is attempted but fails
- The loss if the task is not attempted[12]

These factors can help a person achieve the proper perspective before collecting information, weighing risks, and making a decision. It is particularly important to consider these factors when stressors such as noise, time constraints, or pressure from a supervisor may tend to cloud one's judgment.

Systems Theory in Practice

Precision Tooling Company (PTC) specializes in difficult orders that are produced in small lots, and in making corrections to parts that otherwise would wind up as expensive rejects in the scrap bin. In short, PTC specializes in doing the types of work that other companies cannot, or will not, do. Most of PTC's work comes in the form of subcontracts from larger manufacturing companies. Consequently, living up to its reputation as a high-performance, on-time company is important to PTC.

Because much of its work consists of small batches of parts to be reworked, PTC still uses several manually operated machines. The least experienced machinists operate these machines. This causes two problems. The first problem is that it is difficult for even a master machinist to hold to modern tolerance levels on these old machines. Consequently, apprentice machinists find holding to precise tolerances quite a challenge. The second problem is that the machines are so old that they frequently break down.

Complaints from apprentice machinists about the old machines are frequent. However, their supervisors consider time on the old "ulcer makers" to be one of the rites of passage that upstart machinists must endure. Their attitude is, "We had to do it, so why shouldn't you?" This was where things stood at PTC when the company won the Johnson contract.

PTC had been trying for years to become a preferred supplier for H. R. Johnson Company. PTC's big chance finally came when Johnson's manufacturing division incorrectly produced 10,000 copies of a critical part before noticing the problem. Simply scrapping the part and starting over was an expensive solution. Johnson's vice president for manufacturing decided to give PTC a chance.

PTC's management team was ecstatic! Finally, they had an opportunity to partner with H. R. Johnson Company. If PTC could perform well on this one, even more lucrative contracts were sure to follow. The top managers called a companywide meeting of all employees. Attendance was mandatory. The CEO explained the situation as follows:

> Ladies and gentlemen, we are faced with a great opportunity. I've just signed a contract with H. R. Johnson Company to rework 10,000 parts that their manufacturing folks produced improperly. The rework tasks are not that complicated, but every part has got to go through several manual operations at the front end of the rework process. This means our manual machining unit is going to have to supply the heroes on this job. I've promised the manufacturing VP at Johnson that we would have his parts ready in 90 days. I know that's a lot to do in so short a period of time, but Johnson is in a real bind here. If we can produce on this one, they won't forget us in the future.

This put PTC's apprentice machinists on the spot. If PTC didn't perform on this contract, it would be their fault. They cursed their old machines and got to work. The CEO had said the rework tasks would not be "that complicated," but, as it turned out, the processes weren't that simple either. The problem was tolerances. Holding to the tolerances specified in the Johnson contract took extra time and a special effort on every single part. Before long, the manual machining unit was behind schedule, and management was getting nervous. The situation was made even worse by the continual breakdowns and equipment failures experienced. The harder the unit supervisor pushed, the more stressed the employees and machines became.

Predictably, it wasn't long before safety procedures were forgotten, and unreasonable risks were being taken. The pressure from management, the inexperience of the apprentice machinists, and the constant equipment failures finally took their toll. In a hurry to get back on schedule, and fearing that his machine would break down again, one machinist got careless and ran his hand into the cutter on his milling machine. By the time the machine had been shut down, his hand was badly mutilated. In the aftershock of this accident, PTC was unable to meet the agreed-upon completion schedule. Unfortunately, PTC did not make the kind of impression on H. R. Johnson's management team that it had hoped.

This accident can be explained by the systems theory. The *person-machine-environment* chain has a direct application in this case. The person involved was relatively inexperienced. The machine involved was old and prone to breakdowns. The environment was especially stressful and pressure packed. These three factors, taken together, resulted in this serious and tragic accident.

Combination Theory of Accident Causation

There is often a degree of difference between any theory of accident causation and reality. The various models presented with their corresponding theories in this chapter attempt to explain why accidents occur. For some accidents, a given model may be very accurate. For others, it may be less so. Often the cause of an accident cannot be adequately explained by just one model or theory. Thus, according to the **combination theory**, the actual cause may combine parts of several different models. Safety personnel should use these theories as appropriate both for accident prevention and accident investigation. However, they should avoid the tendency to try to apply one model to all accidents.

Combination Theory in Practice

Crestview Grain Corporation (CGC) maintains ten large silos for storing corn, rice, wheat, barley, and various other grains. Because stored grain generates fine dust and gases, ventilation of the silos is important. Consequently, all of CGC's silos have several large vents. Each of these vents uses a filter similar to the type used in home air conditioners that must be changed periodically.

There is an element of risk involved in changing the vent filters because of two potential hazards. The first hazard comes from unvented dust and gases that can make breathing difficult, or even dangerous. The second hazard is the grain itself. Each silo has a catwalk that runs around its inside circumference near the top. These catwalks give employees access to the vents that are also near the top of each silo. The catwalks are almost 100 feet above ground level, they are narrow, and the guardrails on them are only knee high. A fall from a catwalk into the grain below would probably be fatal.

Consequently, CGC has well-defined rules that employees are to follow when changing filters. Because these rules are strictly enforced, there had never been an accident in one of CGC's silos—that is, not until the Juan Perez tragedy occurred. Perez was not new to the company. At the time of his accident, he had worked at CGC for over five years. However, he was new to the job of silo maintenance. His inexperience, as it turned out, would prove fatal.

It was time to change the vent filters in silo number 4. Perez had never changed vent filters himself. He hadn't been in the job long enough. However, he had served as the required "second man" when his supervisor, Bao Chu Lai, had changed the filters in silos 1, 2, and 3. Because Chu Lai was at home recuperating from heart surgery and would be out for another four weeks, Perez decided to change the filters himself. Changing the filters was a simple enough task, and Perez had always thought the second-man concept was overdoing it a little. He believed in taking reasonable precautions as much as the next person, but in his opinion, CGC was paranoid about safety.

Perez collected his safety harness, respirator, and four new vent filters. Then he climbed the external ladder to the entrance–exit platform near the top of silo number 4. Before going in, Perez donned his respirator and strapped on his safety harness. Opening the hatch cover, he stepped inside the silo onto the catwalk. Following procedure, Perez attached a lifeline to his safety harness, picked up the new vent filters, and headed for the first vent. He changed the first two filters without incident. It was while he was changing the third filter that tragedy struck.

The filter in the third vent was wedged in tightly. After several attempts to pull it out, Perez became frustrated and gave the filter a good jerk. When the filter suddenly broke loose, the momentum propelled Perez backward and he toppled off the catwalk. At first it appeared that his lifeline would hold, but without a second person to pull him up or call for help, Perez was suspended by only the lifeline for over 20 minutes. He finally panicked, and in his struggle to pull himself up, knocked open the buckle of his safety harness. The buckle gave way, and Perez fell over 50 feet into the grain below. The impact knocked off his respirator, the grain quickly enveloped him, and Perez was asphyxiated.

The accident investigation that followed revealed that several factors combined to cause the fatal accident—the combination theory. The most critical of these factors were as follows:

- Absence of the supervisor
- Inexperience of Perez
- A conscious decision by Perez to disregard CGC's safety procedures
- A faulty buckling mechanism on the safety harness
- An unsafe design (only a knee-high guardrail on the catwalk)

Behavioral Theory of Accident Causation

The behavioral theory of accident causation and prevention is often referred to as **behavior-based safety (BBS)**. BBS has both proponents and critics. One of the most prominent proponents of BBS is E. Scott Geller, a senior partner of Safety Performance Solutions, Inc. and a professor of psychology. It is appropriate that Geller is a professional psychologist because BBS is the application of behavioral theories from the field of psychology to the field of occupational safety.

According to Geller, there are seven basic principles of BBS: (1) intervention that is focused on employee behavior; (2) identification of external factors that will help understand and improve employee behavior (from the perspective of safety in the workplace); (3) direct behavior with activators or events antecedent to the desired behavior, and motivation of the employee to behave as desired with incentives and rewards that will follow the desired behavior; (4) focus on the positive consequences that will result from the desired behavior as a way to motivate employees; (5) application of the scientific method to improve attempts at behavioral interventions; (6) use of theory to integrate information rather than to limit possibilities; and (7) planned interventions with the feelings and attitudes of the individual employee in mind.[13]

Those who have studied psychology will recognize BBS as an innovative and practical application of standard behavioral theory to the field of occupational safety. These theories are relevant in any situation in which certain types of human behaviors are desired while others are to be avoided. Positive reinforcement in the form of incentives and rewards is used to promote the desired (safe) behaviors and to discourage undesirable (unsafe) behaviors.

Proponents of BBS use the "ABC" model to summarize the concept of understanding human behavior and developing appropriate interventions when the behavior is undesirable (unsafe). Geller explains the model as follows:

> Behavior-based safety trainers and consultants teach the ABC model (or three-term contingency) as a framework to understand and analyze behavior or to develop interventions for improving behavior. As given in BBS principle 3 . . . the "A" stands for activators or antecedent events that precede behavior ("B") and "C" refers to the consequences following behavior or produced by it. Activators direct behavior, whereas consequences motivate behavior.[14]

Two other advocates of BBS, Bruce Fern and Lori Alzamora, propose the expansion of the ABC model to ABCO.[15] The "O" stands for outcomes. They explain the addition as follows:

> "*Outcome*" refers to the longer-term results of engaging in safe or unsafe behavior. For example, an antecedent of a sign requiring employees to wear safety goggles could produce the behavior of putting on the goggles, the consequence of avoiding an eye injury, and the outcome of being able to continue working and enjoying time with the family. On the other hand, the consequence of not wearing goggles could be an eye injury with a potential outcome of blindness, time off the job, and a reduced quality of life. Failure to address the issue of outcomes represents a lost opportunity to give employees a good reason for engaging in safe behaviors.[16]

Behavioral Theory in Action

Mark Potter is the safety manager for Excello Corporation. Several months ago, he became concerned because employees seemed to have developed a lax attitude toward wearing hard hats. What really troubled Potter was that there is more than the usual potential for head injuries because of the type of work done in Excello's plant, and he had personally witnessed two near misses in less than a week. An advocate of behavior-based safety (BBS), he decided to apply the ABC model in turning around this unsafe behavior pattern.

His first step was to remove all the old "Hard Hat Area" signs from the plant and replace them with newer, more noticeable signs. Then he scheduled a brief seminar on head injuries and cycled all employees through it over a two-week period. The seminar took an unusual approach. It told a story of two employees. One was in a hospital bed surrounded by family members he did not even recognize. The other was shown enjoying a family outing with happy family members. The clear message of the video was "the difference between these two employees is a hard hat." These two activities were the antecedents to the behavior he hoped to produce (all employees wearing hard hats when in a hard hat area).

The video contained a powerful message and it had the desired effect. Within days, employees were once again disciplining themselves to wear their hard hats (the desired behavior). The consequence was that near misses stopped and no head injuries have occurred at Excello in months. The outcome of this is that Excello's employees have been able to continue enjoying the fruits of their labor and the company of loved ones.

Drugs and Accident Causation

One of the most pernicious causes of accidents on the job is chemicals—but not the kind industrial hygienists generally concern themselves with.[17] The chemicals alluded to here are the illicit drugs and alcohol used by employees. Drugs and alcohol are the root cause or contributing cause of many accidents on the job every year. Consequently, safety professionals need to be on guard for employees who are drug and alcohol abusers.

According to Stephen Minter,

> The workplace cannot be separated from the society around it, and substance abuse continues to be a serious and costly health and safety issue for employers. According to surveys by the Department of Health and Human Services, some 77 percent of drug users are employed—more than 9 million workers. An estimated 6.5 percent of full-time and 8.6 of part-time workers use illicit drugs. More than a third of all workers between the ages of 18 and 25 are binge drinkers. . . . Alcoholism alone causes 500 million lost days annually (125 million days are lost each year due to work-related injuries). . . . Some 20 percent of workers report that they have been put in danger or injured, or had to work harder, redo work or cover for a co-worker, as a result of a co-worker's drinking.[18]

These discouraging statistics are why so many companies implement drug-free workplace programs. In fact, since 1989 federal contractors have been required to do so. Such programs typically include the following components: drug-free workplace policy, supervisory training, employee education, employee assistance programs, and alcohol and drug testing.

Establishing drug-free workplace programs is typically the responsibility of the human resources department. However, safety and health professionals should be aware of the workplace problems that can be caused by alcohol and drug abuse. Further, if a cross-functional team of representatives from various departments is convened by the human resources department for the purpose of developing a drug-free workplace program, the chief safety and health professional for the organization should be a member of that team.

Depression and Accident Causation

An invisible problem in today's workplace is **clinical depression**. People who suffer from clinical depression are seriously impaired and, as a result, they pose a clear and present safety risk to themselves, fellow workers, and their employer.[19] Mental health professionals estimate that up to 10 percent of the adult population in the United States suffers from clinical depression. This translates to 1 in every 20 people on the job.

The causes of clinical depression are many and varied, but the most common causes are biological (too few or too many of the brain chemicals known as

neurotransmitters), cognitive (negative thought processes), genetic (family history of depression), and concurring illnesses (strokes, cancer, heart disease, Alzheimer's, and other diseases can increase the incidence of depression).

According to Todd Nighswonger,

> Depression results in more than 200 million lost workdays and costs the U.S. economy $43.7 billion annually. Much of that cost is hidden, including $23.8 billion lost to U.S. businesses in absenteeism and lost productivity. Beyond productivity issues, studies suggest that depressed workers may be more prone to accidents. Stephen Heidel, M.D., MBA, an occupational psychiatrist in San Diego, notes a lack of concentration, fatigue, failing memory and slow reaction time as reasons that workers who are depressed may not work safely.[20]

Warning Signs

Safety and health professionals are not mental health professionals and should not attempt to play that role. However, they should be alert to the warning signs of clinical depression in employees. These signs are as follows:

- Persistent dreary moods (sadness, anxiety, nervousness)
- Signs of too little sleep
- Sleeping on the job or persistent drowsiness
- Sudden weight loss or gain
- General loss of interest, especially in areas of previous interest
- Restlessness, inability to concentrate, or irritability
- Chronic physical problems (headaches, digestive disorders, etc.)
- Forgetfulness or an inability to make simple decisions
- Persistent feelings of guilt
- Feelings of low self-worth
- Focus on death or talk of suicide

Safety and health professionals who recognize any or all of these symptoms in an employee should avoid the natural human tendency to *help the employee deal with the problems*. Rather, the appropriate action is to get the employee into the hands of competent mental health professionals right away. The best way to do this is to approach the employee's supervisor and recommend that he or she refer the employee to the organization's employee assistance program (EAP) or to the human resources department. If the supervisor is uncomfortable approaching the employee in question or does not know how to go about it, recommend that he or she use the following statement suggested by the Society for Human Resource Management:

> I'm concerned that recently you've been late to work often and are not meeting your performance objectives. I'd like to see you get back on track. I don't know whether this is the case for you, but if personal issues are affecting your work, you can speak confidentially to one of our employee assistance counselors. The service was set up to help employees. Our conversation today and appointments with the counselor will be kept confidential. Whether or not you contact this service, you will still be expected to meet your performance goals.[21]

Sources of Help

Because clinical depression in employees has become such an all-pervasive problem that increases the risk of accidents and injuries on the job, safety and health professionals need to learn all they can about this problem and keep up-to-date with the latest information concerning it. The following sources may help:

Employee Assistance Professionals Association, 703–522–6272, http://www.eapa.org

National Institute of Mental Health, 800–421–4211, http://www.nimh.nih.gov

National Mental Health Association, 800–969–NMHA, http://www.nmha.org

Management Failures and Accident Causation

One of the leading causes of accidents in the workplace is the failure of management to do its part to ensure a safe and healthy work environment.[22] Different levels of management have different levels of responsibility. The level of management with the most direct, hands-on, day-to-day responsibility for workplace safety and health is the supervisory level. Supervisors play a critical role in making sure that employees work in a safe and healthy environment.

Role of the Supervisor in Workplace Safety and Health

Safety and health professionals cannot do their jobs effectively without the full cooperation and day-to-day assistance of first-line supervisors. Supervisors and safety professionals must be partners when it comes to providing a safe and healthy workplace for employees. Supervisors should be assigned responsibility for the work environment and for the safety of employees in their units. Safety and health professionals should be readily available to help supervisors fulfill this responsibility.

Key responsibilities of supervisors relating to workplace safety and health include the following:

- Orienting new employees to the safe way to do their jobs.
- Ensuring that new and experienced employees receive the safety and health training they need on a continual basis.
- Monitoring employee performance and enforcing safety rules and regulations.
- Assisting safety and health professionals in conducting accident investigations.
- Assisting safety and health professionals in developing accident reports.
- Keeping up-to-date on safety issues.
- Setting a positive example for employees that says "the safe way is the right way."

Typical Management Failures That Cause Accidents

Management failures represent a major cause of accidents on the job. If management is serious about providing a safe and healthy work environment for employees it must (1) show employees that safe and healthy work practices are expected by including such practices in job descriptions, monitoring employee work practices, and setting an example of safe and healthy work practices; (2) provide training in how to work safely, including orientation training for new employees as well as ongoing updated training for experienced employees; (3) include safe and healthy work practices as criteria in the periodic performance appraisals of employees; and (4) reinforce safe and healthy work practices by rewarding and recognizing employees who use them. Common examples of management failures include the following:

POOR HOUSEKEEPING OR IMPROPER USE OF TOOLS, EQUIPMENT, OR FACILITIES

Manage-ment either has not developed the necessary requirements, or has but does not enforce them. The management failure in this case could be lack of safety procedures (failure to let employees know the expectations), lack of training (failure to give employees the knowledge and skills they need to work safely), or failure to properly supervise (failure to monitor employee actions).

PRESSURE TO MEET DEADLINES

Sometimes management has developed a good safety and health policy, established good safety and health procedures, built safety and health expectations into job descriptions and performance appraisals, and provided the necessary training only to put all this aside when a rush order comes in. This may be the most problematic of the many different types of management failures that can occur because it can undermine all of the organization's safety and health efforts. When management allows safety and health procedures to be ignored or, worse yet, encourages them to be ignored to speed up production in the short run, employees soon get the message that safety and health are important only when there is no rush. This is an example of management failing to set the proper example.

Obesity and Accident Causation

Researchers at Ohio State University found that extremely obese people are more likely than normal-weight people to injure themselves.[23] This is bad news for what an Australian study conducted at Queensland University calls sedentary workplaces—those that involve a lot of sitting at desks. It is bad news because the Australian study concluded that the more people sit at desks during the workday, the more likely they are to be overweight.

Obesity has long been associated with such chronic diseases as high blood pressure, coronary heart disease, diabetes, and certain types of cancer, but these studies now tie it to workplace injuries too. The ramifications for safety and health professionals are profound. The World Health Organization estimates that there are more than 300 million obese people worldwide. In the industrialized nations of the world—nations such as the United States—the number of people considered obese is growing rapidly. These studies used a standard body mass index (BMI) score of 30 or above to define obesity.

In the study conducted by Ohio State University, researchers collected data on more than 2,500 adults. The data show that 26 percent of extremely obese male subjects reported personal injuries. The percentage for extremely obese women was only slightly lower at approximately 22 percent. Researchers compared these percentages with those for normal-weight people and found a noticeable difference in the percentages of males and females who reported injuries (17 percent for males and 12 percent for women). Researchers used a BMI of 18.5 to 24.9 to define "normal weight."

The most common causes of injuries to obese people were the result of overexertion (35.2 percent) and falls (29.9) percent. Underweight people—BMI of 18.5 or less—reported the fewest number of injuries. According to the study's author, Huiyun Xiang, "There is undeniably a link between obesity and injury risk in adults. Efforts to promote optimal body weight may reduce not only the risk of chronic diseases, but also the risk of unintentional injuries."[24]

Key Terms and Concepts

- Accident/incident theory
- Ancestry
- Axioms of Industrial Safety
- Behavior-based safety (BBS)
- Causal relationship
- Central factor
- Clinical depression
- Combination theory
- Domino theory
- Environment
- Environmental factors
- Epidemiological theory
- Ergonomic traps
- Fault of person
- Hazardous condition
- Human error
- Human factors theory
- Inappropriate activities
- Inappropriate responses
- Industrial hygiene
- Industrial place accidents
- Internal factors
- Mechanical hazards
- Neurotransmitters
- Obesity
- Overload
- Physical hazards
- Preceding factors
- Predispositional characteristics
- Situational characteristics
- Situational factors
- Social environment
- Stressors
- Systems theory
- Unsafe acts
- Unsafe behavior

Review Questions

1. Explain the domino theory of accident causation, including its origin and its impact on more modern theories.
2. What were the findings of Herbert W. Heinrich's 1920s study of the causes of industrial accidents?
3. List five of Heinrich's Axioms of Industrial Safety.
4. Explain the following concepts in the domino theory: preceding factor; central factor.
5. What are the three broad factors that lead to human error in the human factors theory? Briefly explain each.
6. Explain the systems failure component of the accident/incident theory.
7. What are the key components of the epidemiological theory? How does their interaction affect accident causation?
8. Explain the systems theory of accident causation.
9. What impact do stressors have in the systems theory?
10. List five factors to consider before making workplace decisions that involve risk.
11. Explain the principles of behavior-based safety.
12. What is the role of the safety and health professional with regard to handling employees who might be drug or alcohol abusers?
13. List the warning signs of clinical depression.
14. What must management do if it is serious about providing a safe and healthy work environment for employees?
15. Explain the connection between obesity and injuries.

Endnotes

1. National Safety Council, *Accident Facts* (Chicago: National Safety Council, 2005), 23.
2. Ibid., 24.
3. Ibid., 26.
4. Industrial Foundation for Accident Prevention, www.ifap.asn.au, March 31, 2006.
5. Ibid.
6. Ibid.
7. Ibid.
8. Ibid.
9. M. A. Topf, "Chicken/Egg/Chegg!", *Occupational Health Safety* 68, no. 6: 60–66.

10. D. L. Goetsch, *Implementing Total Safety Management* (Upper Saddle River, NJ: Prentice Hall, 1998), 227.
11. Ibid.
12. Ibid.
13. E. S. Geller, "Behavior-Based Safety: Confusion, Controversy, and Clarification," *Occupational Health & Safety* 68, no. 1: 40–49.
14. Ibid., 44.
15. B. Fern and L. P. Alzamora, "How and Why Behavioral Safety Needs to Change," *Occupational Health & Safety* 68, no. 9: 69.
16. Ibid.
17. Stepher G. Minter, "The Safety Threat from Within," *Occupational Hazards* 64, no. 4: 8.
18. Ibid.
19. Todd Nighswonger, "Depression: The Unseen Safety Risk," *Occupational Hazards* 64, no. 4: 38–42.
20. Ibid., 40.
21. Ibid., 42.
22. Retrieved from http://online.misu.kodak.edu/19577/AccCautrac.htm.
23. Occupational Health & Safety Online, "Obesity Studies Focus on Injuries, Sedentary Workplaces." Retrieved from http://www.ohsonline.com/stevens/ohspub.nsf/d3d5b4f938b22b6e8625670c006dbc58/d9, July 25, 2005, 1–3.
24. Ibid., 3.

CHAPTER 4

Roles and Professional Certifications for Safety and Health Professionals

Major Topics

- Modern Safety and Health Teams
- Safety and Health Manager
- Engineers and Safety
- Industrial Hygienist
- Health Physicist
- Occupational Physician
- Occupational Health Nurse
- Risk Manager
- Certification of Safety

This book was designed for use by prospective and practicing safety and health managers. People with such titles are typically responsible to higher management for the safety and health of a company's workforce. Modern safety and health managers seldom work alone. Rather, they usually head a team of specialists that may include engineers, physicists, industrial hygienists, occupational physicians, and occupational health nurses.

It is important for safety and health managers today to understand not only their roles but also the roles of all members of the safety and health team. This chapter provides the information that prospective and practicing safety and health managers need to know about the roles of safety personnel in the age of high technology and the certifications that they need.

FIGURE 4-1 A modern safety and health team.

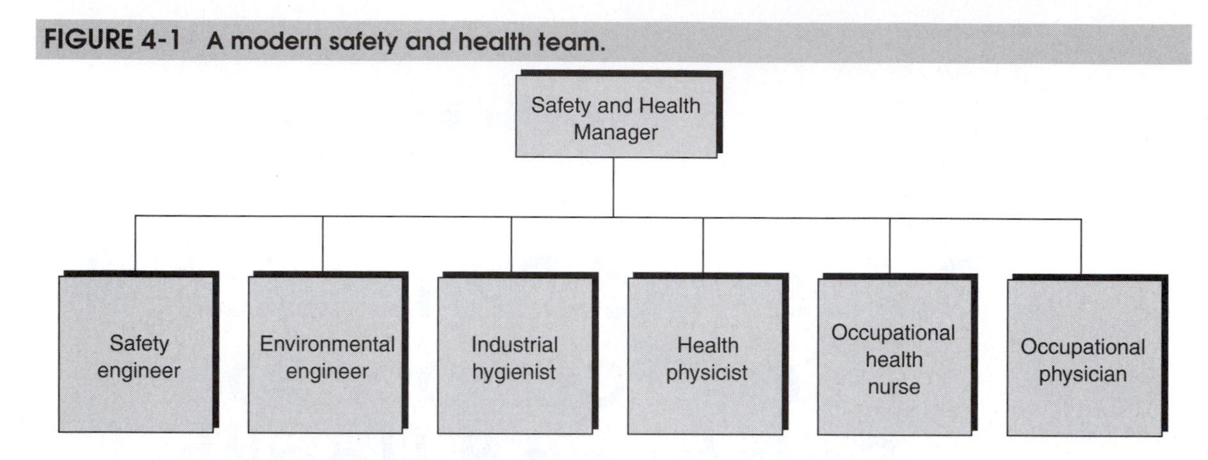

Modern Safety and Health Teams

The issues that concern modern safety and health managers are multifaceted and complex. They include such diverse issues as stress; explosives; laws, standards, and codes; radiation; AIDS; product safety and liability; ergonomics; ethics; automation; workers' compensation; and an ever-changing multitude of others.

It would be unreasonable to expect one person to be an expert in all the many complex and diverse issues faced in the modern workplace. For this reason, the practice of safety and health management in the age of high technology has become a team sport. Figure 4-1 illustrates the types of positions that may comprise a safety and health team. In the remaining sections of this chapter, the roles, duties, responsibilities, and relationships of members of the safety and health team are described. In small companies, one person may have to wear several of these hats and contract for nursing services.

Safety and Health Manager

The most important member of the safety and health team is its manager. Companies that are committed to providing a safe and healthy workplace employ a **safety and health manager** at an appropriate level in the corporate hierarchy. The manager's position in the hierarchy is an indication of the company's commitment and priorities. This, more than anything else, sets the tone for a company's safety and health program.

In times past, companies with a highly placed safety and health manager were rare. However, passage of the OSH Act in 1970 (see Chapter 6) began to change this. The OSH Act, more than any other single factor, put teeth in the job descriptions of safety and health professionals. OSHA standards, on-site inspections, and penalties have encouraged a greater commitment to safety and health. Environmental, liability, and workers' compensation issues have also had an impact, as has the growing awareness that providing a safe and healthy workplace is the right thing to do from both an ethical and a business perspective.

Job of the Safety and Health Manager

The job of the safety and health manager is complex and diverse. Figure 4-2 is an example of a job description for such a position. The description attests to the diverse nature of the job. Duties range from hazard analysis, accident reporting, standards and compliance, record keeping, and training to emergency planning and so on.

The minimum educational requirement set by Poultry Processing, Inc., (PPI) is an occupational certificate from a community college with the full associate of science or applied science degree preferred. Preference is given to applicants with a bachelor's degree in specifically identified fields.

Role in the Company Hierarchy

The safety and health manager described in Figure 4-2 reports to PPI's local plant manager and has line authority over all other members of the safety and health team.

FIGURE 4-2 **Safety and health manager job description.**

POULTRY PROCESSING, INC.
Highway 90 West
Crestview, FL 32536
Vacancy Announcement

Position Title: Safety and Health Manager

Position Description: The Safety and Health Manager for PPI is responsible for establishing, implementing, and managing the company's overall safety and health program. The position reports to the local plant manager. Specific duties include the following:

- Establish and maintain a comprehensive companywide safety and health program.
- Assess and analyze all departments, processes, and materials for potential hazards.
- Work with appropriate personnel to develop, implement, monitor, and evaluate accident prevention and hazard control strategies.
- Ensure companywide compliance with all applicable laws, standards, and codes.
- Coordinate the activities of all members of the company's safety and health program.
- Plan, implement, and broker, as appropriate, safety and health–related training.
- Maintain all required safety and health–related records and reports.
- Conduct accident investigations as necessary.
- Develop and maintain a companywide emergency action plan (EAP).
- Establish and maintain an ongoing safety promotion effort.
- Analyze the company's products from the perspectives of safety, health, and liability.

Qualifications Required: The following qualifications have been established by the PPI management team with input from all levels and all departments.

- *Minimum Education.* Applicants must have at least a one-year community college certificate or an associate of science or applied science degree in industrial safety or a closely related degree (AS holders will begin work at a salary 15 percent higher than a certificate graduate).
- *Preferred Education.* Applicants with a bachelor's degree in any of the following major fields of study will be given first priority: industrial safety and health, industrial technology, industrial management, manufacturing technology, engineering technology, and related. Degree programs in these fields must include at least one three-semester or five-quarter-hour course in industrial or occupational safety and health.

This and the duties set forth in the job description are evidence of the company's commitment to safety and health and that PPI is large enough to have a dedicated safety and health manager.

In some companies, the safety and health manager may also have other duties such as production manager or personnel manager. In these cases, the other members of the safety and health team, like those shown in Figure 4-1, are not normally company employees. Rather, they are available to the company on a part-time or consultative basis as needed. The safety and health manager's role in a company depends in part on whether his or her safety and health duties are full time or are in addition to other duties.

Another role determinant is the issue of authority. Does the safety and health manager have line or staff authority? **Line authority** means that the safety and health manager has authority over and supervises certain employees (i.e., other safety and health personnel). **Staff authority** means that the safety and health manager is the staff person responsible for a certain function, but he or she has no line authority over others involved with that function.

Those occupying staff positions operate like internal consultants—that is, they may recommend, suggest, and promote, but they do not have the authority to order or mandate. This is typically the case with safety and health managers. Even managers with line authority over other safety and health personnel typically have a staff relationship with other functional managers (e. g., personnel, production, or purchasing). For example, consider the following safety and health–related situations:

- A machine operator continually creates unsafe conditions by refusing to practice good housekeeping.
- A certain process is associated with an inordinately high number of accidents.
- A new machine is being purchased that has been proven to be unsafe at other companies.

In the first example, the safety and health manager could recommend that the employee be disciplined but could not normally undertake or administer disciplinary measures. In the second example, the safety and health manager could recommend that the process be shut down until a thorough analysis could be conducted, hazards identified, and corrective measures taken. However, the manager would rarely have the authority to order the process to be shut down. In the final example, the safety and health manager could recommend that an alternative machine be purchased, but he or she would not normally have the authority to stop the purchase.

Maintaining a safe and healthy workplace while playing the role of internal consultant is often the greatest challenge of safety and health managers. It requires managers to be resourceful, clever, astute with regard to corporate politics, good at building relationships, persuasive, adept at trading for favors, credible, and talented in the development and use of influence.

Problems Safety and Health Managers Face

As if the diversity and complexity of the job were not enough, there are a number of predictable problems that safety and health managers are likely to face. These problems are discussed in the following paragraphs.

Lack of Commitment

Top management may go along with having a companywide safety and health program because they see it as a necessary evil. The less enthusiastic may even see safety and health as a collection of government regulations that interfere with **profits**. Although this is less often true now than it has been in the past, safety and health professionals should be prepared to confront a less than wholehearted commitment in some companies.

Production versus Safety

Industrial firms are in business to make a profit. They do this by producing or processing products. Therefore, anything that interferes with production or processing is likely to be looked on unfavorably. At times, a health or safety measure will be viewed by some as interfering with productivity. A common example is removal of safety devices from machines as a way to speed production. Another is running machines until the last possible moment before a shift change rather than shutting down with enough time left to perform routine maintenance and housekeeping tasks.

The modern marketplace has expanded globally and, therefore, become intensely competitive. To survive and succeed, today's industrial firm must continually improve its **productivity**, quality, cost, image, response time, and service. This sometimes puts professionals who are responsible for safety and health at odds with others who are responsible for productivity, quality, cost, and response time.

Sometimes, this cannot be avoided. At other times, it is the fault of a management team that is less than fully committed to safety and health. However, sometimes the fault rests squarely on the shoulders of the safety and health manager. This is because one of the most important responsibilities of this person is to convince higher management, middle management, supervisors, and employees that, in the long run, the safe and healthy way of doing business is also the competitive, profitable way of doing business. The next section explains several strategies for making this point.

Companywide Commitment to Safety and Health

In many cases, safety and health managers have been their own worst enemy when it comes to gaining a companywide commitment. The most successful are those who understand the goals of improved productivity, quality, cost, image, service, and response time and are able to convey the message that a safe and healthy workplace is the best way to accomplish these goals. The least successful are those who earn a reputation for being grumpy in-house bureaucrats who quote government regulations chapter and verse but know little and care even less about profits. Unfortunately, in the past, there have been too many safety and health managers who fall into the latter category. This is not the way to gain a companywide commitment to safety and health, but it is a sure way to engender resentment.

Lack of Resources

Safety and health managers are like other managers in an organization in that they must compete for the resources needed to do the job. Often they find that their departments rank lower in priority than the production and operations departments (at least until a disaster occurs). Safety and health managers need to become proficient in showing the financial benefits of a safe workplace.

FIGURE 4-3 Factors that produce competitiveness.

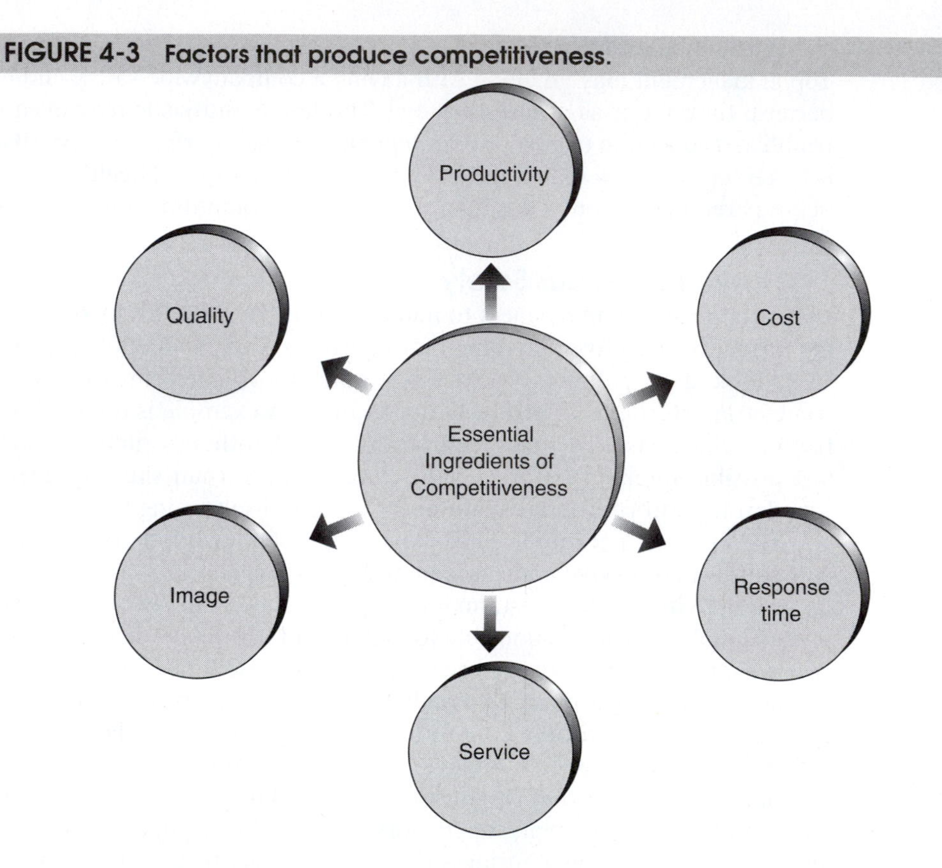

Today's safety and health manager must understand the bottom-line concerns of management, supervisors, and employees and be able to use these concerns to gain a commitment to safety and health. Figure 4-3 illustrates the essential message that **competitiveness** comes from continually improving a company's productivity, quality, cost, image, service, and response time. These continual improvements can be achieved and maintained best in a safe and healthy work environment.

Safety and health managers should use this message to gain a commitment from management and employees. Following are some strategies that can be used to get the point across.

Productivity, Quality, Cost, and Response Time

These four factors, taken together, are the key to productivity in the age of high technology and global competitiveness. The most productive company is the one that generates the most output with the least input. Output is the company's product. Input is any resource—time, talent, money, technology, and so on—needed to produce the product. **Quality** is a measure of reliability and customer satisfaction. **Cost** is the amount of money required to purchase the item. If all other factors are equal, customers will select the product that costs less. **Response time** is the amount of time that elapses between an order being placed and the product being delivered.

To compete in the global marketplace, industrial companies must continually improve these four factors. At the most fundamental level, successfully competing in the global marketplace means having the best people and the best technology and getting the most out of both by applying the best management strategies.

Safety and health managers who understand this can use their knowledge to gain a commitment to their programs. In attempting to do so, the following five points are helpful:

- If it is important to attract and keep the best people, a safe and healthy workplace will help.
- If it is important to get the most out of talented people, it must be important to keep them safe and healthy so that they are functioning at peak performance levels (for example, the best technician in the world can't help when he or she is out of action as the result of an injury or illness).
- Employees cannot concentrate fully on quality when they are concerned for their safety and health.
- Keeping industrial technologies up-to-date requires the continual investment of funds. Profits that are siphoned off to pay the costs associated with accidents, emergencies, and health problems cannot be reinvested in the latest technologies needed to stay competitive.
- With the skyrocketing costs of medical care, workers' compensation, and litigation, it costs less to prevent accidents than to pay for them.

Image and Service

Image and **service** are also important factors in the competitiveness equation. Of the two, image relates more directly to safety and health.

In today's intensely competitive marketplace, a company's image, internal and external, can be a deciding factor in its ability to succeed. Companies that establish a solid internal image in terms of safety and health will find it easier to attract and keep the best employees. Companies that establish a solid external image with regard to environmental and product safety issues will find it easier to attract and retain customers.

Gaining a full and real commitment to safety and health in the workplace is one of the most important roles of the safety and health manager. Traditionally, safety and health managers have argued their cases from the perspectives of ethics or government mandates. The ethical argument is as valid now as it has always been and should continue to be used.

However, in today's competition-driven workplace, managers responsible for the bottom line may resent arguments that are based on government mandates and regulations. On the other hand, these same managers may respond positively if they can be shown that resources invested in safety and health can actually improve a company's competitiveness. Using the points made earlier about productivity, quality, cost, image, and response time can go a long way in helping to gain management commitment to safety and health.

Education and Training for Safety and Health Managers

Advances in technology, new federal legislation, the potential for costly litigation, and a proliferation of standards have combined to make the job of safety and health professionals more complex than ever before. These factors have correspondingly increased the importance of education and training for safety and health managers. The ideal formula for safety and health professionals is **formal education** prior to entering the profession supplemented by **in-service training** on a lifelong basis afterward.

Universities, colleges, and community colleges across the country have responded to the need for formal education for safety and health managers as well as other safety and health personnel. Many community colleges offer occupational certificates and **associate degrees** in applied science or science degrees with such program titles as industrial safety, occupational safety, environmental technology, safety and health management, and industrial hygiene.

Universities have responded to the need for formal education by making safety and health–related courses either optional or required parts of such **bachelor's (or baccalaureate) degree** programs as industrial technology, manufacturing technology, engineering technology, industrial engineering technology, industrial management, and **industrial engineering**. Some universities offer bachelor's degrees in industrial safety and health, occupational safety management, and industrial hygiene.

Formal education provides the foundation of knowledge needed to enter the profession. After one has begun a career as a safety and health manager, the next challenge is keeping up as the laws, regulations, standards, and overall body of knowledge relating to safety and health grow, change, and evolve.

In-service training, ongoing interaction with professional colleagues, and continued reading of professional literature are effective ways to stay current. New safety and health managers should move immediately to get themselves "plugged into" the profession. This means joining the appropriate professional organizations, becoming familiar with related government agencies, and establishing links with relevant standards organizations. The next section covers agencies and organizations that can be particularly helpful to safety and health managers.

Helpful Agencies and Organizations

Numerous agencies and organizations are available to help the safety and health manager keep up-to-date. These agencies and organizations provide databases, training, and professional literature. There are professional societies, trade associations, scientific organizations, certification boards, service organizations, and emergency service organizations. Those listed here represent only a portion of those available.

- **Certification boards.** Professional certification is an excellent way to establish one's status in the field of safety and health. To qualify to take a certification examination, safety and health managers must have the required education and experience and submit letters of recommendation as specified by the **certification board**. Figure 4-4 contains the names and addresses of certification boards of interest to safety and health managers. Certification is covered in greater detail later in this chapter.
- **Professional societies.** **Professional societies** are typically formed to promote professionalism, adding to the body of knowledge, and forming networks among colleagues in a given field. Numerous professional societies focus on various safety and health issues. Figure 4-5 summarizes some of these.
- **Scientific standards and testing organizations.** Scientific **standards and testing organizations** conduct research, run tests, and establish standards that identify the acceptable levels for materials, substances, conditions, and mechanisms to which

FIGURE 4-4 Professional certification boards.

- **Board of Certified Safety Professionals of America**
  208 Burwash Ave.
  Savoy, IL 61874
  217-359-9265
  http://www.bcsp.org/
- **American Board of Industrial Hygiene**
  6015 W. St. Joseph, Suite 102
  Lansing, MI 48917-3980
  517-321-2638
  http://www.abih.org/
- **Board of Certification in Professional Engineering**
  PO Box 2811
  Bellingham, WA 98227-2811
  360-671-7601
  http://www.bcpe.org/
- **The Institute of Industrial Engineers**
  25 Technology Park
  Norcross, GA 30092
  770-449-0460
  http://www.iienet.org
- **American Board for Occupational Health Nurses**
  201 East Ogden Ave., Suite 114
  Hinsdale, IL 60521
  630-789-5799
  http://www.abohn.org

FIGURE 4-5 Professional societies.

American Academy of Industrial Hygiene
302 S. Waverly Rd.
Lansing, MI 48917

American Industrial Hygiene Association
475 Wolf Ledges Pkwy.
Akron, OH 44311

American Occupational Medical Association
2340 S. Arlington Heights Rd.
Arlington Heights, IL 60005

American Society of Safety Engineers
1900 E. Oakton St.
Des Plaines, IL 60016

National Safety Council
1121 Spring Lake Dr.
Itasca, IL 60143

Society of Toxicology
1133 I St. NW, Suite 800
Washington, DC 20005

FIGURE 4-6 Scientific standards and testing organizations.

American National Standards Institute (ANSI)
1430 Broadway
New York, NY 10018

International Atomic Energy Agency
Wagramstrasse 5
A01400 Vienna, Austria

National Fire Protection Association (NFPA)
Batterymarch Park
Quincy, MA 02269

Underwriters Laboratories, Inc. (UL)
333 Pfingsten Rd.
New York, NY 10017

American Society of Mechanical Engineers (ASME)
345 E. 47th St.
New York, NY 10017

American Society for Testing and Materials (ASTM)
1916 Race St.
Philadelphia, PA 19103

people might be exposed in the modern workplace. Figure 4-6 summarizes those of critical importance to safety and health managers.

- **Government agencies.** Many **government agencies** are concerned with various aspects of workplace-related safety and health. Some of the most helpful agencies for the safety and health manager are listed in Figure 4-7.
- **Trade associations.** The purpose of a **trade association** is to promote the trade that it represents. Consequently, material produced by trade associations can be somewhat self-serving. Even so, trade associations can be valuable sources of information and training for safety and health managers. Figure 4-8 lists some of the trade associations that can be particularly helpful.

Engineers and Safety

Engineers can make a significant contribution to safety. Correspondingly, they can cause, inadvertently or through incompetence, accidents that result in serious injury and property damage. The engineer has more potential to affect safety in the workplace than any other person does. The following example illustrates this:

> A car-pooler transports himself and three fellow employees to work each day. He is not a particularly safe driver and does not insist that his passengers use seatbelts. After running a red light, he crashes into the side of a building while swerving to avoid another vehicle. The two passengers wearing seatbelts are not hurt, but the driver and one other passenger, neither of whom were wearing seatbelts, are critically injured.

FIGURE 4-7 Government agencies.

Bureau of Mines
Department of the Interior
2401 E St. NW
Washington, DC 20241

National Institute for Occupational Safety and Health
Department of Health and Human Services
Parklawn Bldg.
5600 Fishers La. Rd. 1401
Rockville, MD 20857

Center for Devices and Radiological Health
Food and Drug Administration
8757 Georgia Ave.
Silver Springs, MD 20910

Consumer Product Safety Commission
1111 18th St. NW
Washington, DC 20207

Department of Labor
200 Constitution Ave. NW
Washington, DC 20210

Environmental Protection Agency
401 M St. SW
Washington, DC 20460

Mine Safety and Health Administration
Department of Labor
4015 Wilson Blvd.
Arlington, VA 22203

National Bureau of Standards Headquarters
Route I-270 and Quince Orchard Rd.
Gaithersburg, MD 20899

National Center for Toxicological Research
Food and Drug Administration
5600 Fishers La.
Rockville, MD 20857

National Technical Information Service
Department of Commerce
5285 Port Royal Rd.
Springfield, VA 22161

Nuclear Regulatory Commission
1717 H St. NW
Washington, DC 20555

Occupational Safety and Health Commission
Department of Labor
200 Constitution Ave. NW
Washington, DC 20210

Occupational Safety and Health Review Committee
1825 K St. NW
Washington, DC 20006

Office of Energy Research
Department of Energy
1000 Independence Ave. SW
Washington, DC 20585

Office of Hazardous Materials Transportation
400 7th St. SW
Washington, DC 20590

U.S. Fire Administration
Federal Emergency Management Agency
16825 S. Seton Ave.
Emmitsburg, MD 21727

This brief story illustrates an accident that has two things in common with many workplace accidents. The first is a careless worker—in this case, the driver. The second is other workers who do not follow prescribed safety rules—in this case, failing to use seatbelts. Employees such as these can and do cause many workplace accidents, but even the most careless employee cannot cause a fraction of the problems caused by a careless engineer. The following example illustrates this point:

> An engineer is charged with the responsibility for designing a new seatbelt that is comfortable, functional, inexpensive, and easy for factory workers to install. He designs a belt that meets all of these requirements, and it is installed in 10,000 new cars. As the cars are bought and accidents begin to occur, it becomes apparent that the new seatbelt fails in crashes involving speeds over 36 miles per hour. The engineer who designed the belt took all factors into consideration except one: safety.

FIGURE 4-8 Trade associations.

- Alliance of American Insurers
- American Foundrymen's Association
- American Insurance Association
- American Iron and Steel Institute
- American Metal Stamping Association
- American Petroleum Institute
- American Welding Society
- Associated General Contractors of America
- Compressed Gas Association
- Industrial Safety Equipment Association
- Institute of Makers of Explosives
- Lead Industrial Association
- National Electrical Manufacturers Association
- National LP-Gas Association
- National Machine Tool Builders Association
- Scaffolding, Shoring, and Forming Institute
- Soap and Detergent Association
- Technical Association of the Pulp and Paper Industry

Note: Complete addresses are available in the reference section of most college and university libraries. Most have websites that can be found on the Internet.

This brief story illustrates how far-reaching an engineer's impact can be. With a poorly designed seatbelt installed in 10,000 automobiles, the engineer has inadvertently endangered the lives of as many as 40,000 people (estimating a maximum of four passengers per automobile).

The engineer's opportunity for both good and bad comes during the design process. The process is basically the same regardless of whether the product being designed is a small toy, an industrial machine, an automobile, a nuclear power plant, a ship, a jumbo jetliner, or a space vehicle. Safety and health professionals should be familiar with the design process so that they can more fully understand the role of engineers concerning workplace safety.

Not all engineers are **design engineers**. However, engineers involved in design are usually in the aerospace, electrical, mechanical, and nuclear fields. The following paragraphs give an overview of these design-oriented engineering fields as seen in the course descriptions of a college catalog.

AEROSPACE ENGINEERING

The Bachelor of Science in Engineering (Aerospace Engineering) program incorporates a solid foundation of physical and mathematical fundamentals which provides the basis for the development of the engineering principles essential to the understanding of both atmospheric and extra-atmospheric flight. Aerodynamics, lightweight structures, flight propulsion, and related subjects typical of aeronautical engineering are included. Other courses introduce problems associated with space flight and its requirements. Integration of fundamental principles with useful applications is made in design work in the junior and senior years. Thus, the program prepares the student to contribute to future technological growth, which promises exciting and demanding careers in aerospace engineering. . . . Examples of concentration areas are aerodynamics; design; flight propulsion; flight structures; space technology; stability, control, and guidance.[1]

ELECTRICAL ENGINEERING

Electrical Engineering is a science-oriented branch of engineering primarily concerned with all phases and development of the transmission and utilization of electric power and intelligence. The study of electrical engineering can be conveniently divided into the academic areas of circuits, electronics, electromagnetics, electric energy systems, communications, control, and computer engineering.[2]

MECHANICAL ENGINEERING

Mechanical Engineering is the professional field that is concerned with motion and the processes whereby other energy forms are converted into motion. Mechanical engineers are the people who are responsible for conceiving, designing, manufacturing, testing, and marketing devices and systems that alter, transfer, transform, and utilize the energy forms that ultimately cause motion. Thus mechanical engineers . . . are the people who make the engines that power ships, trains, automobiles, and spacecraft; they design the power plants which convert the energy in fuels, atoms, waterfalls, and sunlight into useful mechanical forms; and they construct intelligent machines and robots as well as the gears, cams, bearings, and couplings that facilitate and control all kinds of mechanical motion.[3]

INDUSTRIAL ENGINEERING

Industrial Engineering is concerned with the design, improvement, and installation of integrated systems that include people, material, equipment, and energy. This field of engineering draws upon specialized knowledge and skills in the mathematical, physical, and social sciences in concert with the principles and methods of engineering analysis and design to specify, predict, and evaluate the results to be obtained from such systems.

NUCLEAR ENGINEERING

Nuclear Engineering Sciences comprises those fields of engineering and science directly concerned with the release, control, and safe utilization of nuclear energy. Applications range over such broad topics as the design, development, and operation of nuclear reactor power systems to the applications of radiation in medicine, space, industry, and other related areas. The nuclear engineer, by virtue of his/her engineering and science-based training, is in a unique position to contribute to the many diverse aspects of this major component of the energy radiation field.[4]

Design Process

Professor William S. Chalk describes the **design process**:

> The design process is a plan of action for reaching a goal. The plan, sometimes labeled problem-solving strategy, is used by engineers, designers, drafters, scientists, technologists, and a multitude of professionals.[5]
>
> The design process proceeds in five sequential steps:

1. **Problem identification.** Engineers draft a description of the problem. This involves gathering information, considering constraints, reviewing specifications, and combining all of these into a clear and concise description of the problem.
2. **Synthesis.** Engineers combine or synthesize systematic, scientific procedures with creative techniques to develop initial solutions to the problem identified in Step 1. At this point, several possible solutions may be considered.
3. **Analysis and evaluation.** All potential solutions developed in the previous step are subjected to scientific analysis and careful evaluation. Such questions as the following are asked: Will the proposed solution satisfy the functional requirements? Will it meet all specifications? Can it be produced quickly and economically?
4. **Document and communicate.** Engineering drawings, detailed calculations, and written specifications are prepared. These document the design and communicate its various components to interested parties. It is common to revise the design at this point based on feedback from different reviewers.
5. **Produce and deliver.** Shop or detail drawings are developed, and the design is produced, usually as a prototype. The prototype is analyzed and tested. Design changes are made if necessary. The product is then produced and delivered.

The design process gives engineers unparalleled opportunities to contribute significantly to safety in the workplace and in the marketplace by producing products with safety built into them. However, in too many cases, the design process does not serve this purpose. There are two primary reasons for this:

- In analyzing and evaluating designs, engineers consider such factors as function, cost, life span, and manufacturability. All too often safety is not even considered or is only a secondary consideration.
- Even when engineers do consider safety in analyzing and evaluating designs, many are insufficiently prepared to do so effectively.

Engineers who design products may complete their entire college curriculum without taking even one safety course. Safety courses, when available to engineering students in design-oriented disciplines, often tend to be electives. This limits the contributions that design engineers can make to both product and workplace safety. Figure 4-9 is a typical core-course listing for a mechanical engineering student.

Safety Engineer

The title **safety engineer** is often a misnomer in the modern workplace. It implies that the person filling the position is a degreed engineer with formal education and/or special training in workplace safety. Although this is sometimes the case, typically the title

FIGURE 4-9 Typical required core courses for mechanical and industrial engineering degrees.

Engineering mechanics: Statics
Engineering mechanics: Dynamics
Mechanics of materials
Kinematics and dynamics of machinery
Manufacturing processes
Control of mechanical engineering systems
Mechanical vibrations
Machine analysis and design
Thermodynamics
Heat transfer
Fluid dynamics

is given to the person who has overall responsibility for the company's safety program (the safety manager) or to a member of the company's safety team. This person is responsible for the traditional aspects of the safety program, such as preventing mechanical injuries; falls, impact, and acceleration injuries; heat and temperature injuries; electrical accidents; fire-related accidents; and so on.

In the former case, the person should be given a title that includes the term manager. In the latter case, the title "safety engineer" is appropriate. However, persons with academic credentials in areas other than engineering should be encouraged to seek such positions because they are likely to be at least as well prepared and possibly even better prepared than persons with engineering degrees. These other educational disciplines include industrial technology, industrial engineering technology, manufacturing technology, engineering technology, industrial management, and industrial safety technology (bachelor's or associate degree).

There are signs that engineering schools are becoming more sensitive to safety and health issues. Graduate degrees in such areas as nuclear physics and nuclear engineering now often require safety courses. The federal government sponsors postgraduate studies in safety. However, the following quote summarizes clearly and succinctly the current status of safety engineering:

> Four states now have registration of professional engineers in a Safety Engineer discipline. Registration gives the registrant the right to use the title "Safety Engineer," but the enabling law has no other requirement that the services of such an engineer be used. The contracting office of one military service responsible for development of advanced high-tech systems does require certain hazard analyses and documents to be signed off and approved at specific points in the designs. Such approval will be valid only under the signature of a registered safety engineer or other engineer shown to have had extensive experience in safety programs. The principal problem is that in a new advanced design project, there may be 400 engineers with no training in accident avoidance who may make critical errors and only one or two safety engineers to find them.[6]

Industrial Engineers and Safety

Industrial engineers are the most likely candidates from among the various engineering disciplines to work as safety engineers. Their knowledge of industrial systems, both manual and automated, can make them valuable members of a design team, particularly one that designs industrial systems and technologies. They can also contribute after the fact as a member of a company's safety team by helping design job and plant layouts for both efficiency and safety.

The industrial engineering discipline can be described as follows:

> Industrial growth has created unusual opportunities for the industrial and systems engineer. Automation and the emphasis on increased productivity coupled with higher levels of systems sophistication are providing impetus to the demand for engineering graduates with a broad interdisciplinary background. The industrial engineering option prepares the student for industrial practice in such areas as product design, process design, plant operation, production control, quality control, facilities planning, work system analysis and evaluation, and economic analysis of operational systems.[7]

Although industrial engineers are more likely to work as safety engineers than are engineers from other disciplines, they are not much more likely to have safety courses as a required part of their program of study. However, their focus on industrial systems and the integration of people and technology does give industrial engineers a solid foundation for additional learning through either in-service training or graduate work.

Environmental Engineers and Safety

A relatively new discipline (when compared with more traditional disciplines such as mechanical engineering) is environmental engineering. This discipline may be described as follows:

> Environmental Engineering Sciences is a field in which the application of engineering and scientific principles is used to protect and preserve human health and the well-being of the environment. It embraces the broad field of the general environment including air and water quality, solid and hazardous wastes, water resources and management, radiological health, environmental biology and chemistry, systems ecology, and water and waste-water treatment.[8]

With the addition of health concerns to the more traditional safety concerns, **environmental engineers** will be sought as members of corporate safety and health teams. The course work they take is particularly relevant since all of it relates either directly or indirectly to health. Figure 4-10 shows the types of courses typically required of environmental engineering students.

A person with the type of formal education shown in Figure 4-10 would be a valuable addition to the safety and health team of any modern industrial firm. Environmental engineers typically report to the overall safety and health manager and are responsible for those elements of the program relating to hazardous waste management, atmospheric pollution, indoor air pollution, water pollution, and wastewater management.

FIGURE 4-10 Typical required core courses for environmental engineering students.

- Environmental biology
- Environmental chemistry
- Water chemistry
- Atmospheric pollution
- Solid waste management
- Water and wastewater
- Hazardous waste control
- Environmental resources management
- Air pollution control design
- Hydraulic systems design

Chemical Engineers and Safety

Increasingly, industrial companies are seeking **chemical engineers** to fill the industrial hygiene role on the safety and health team. Their formal education makes people in this discipline well equipped to serve in this capacity. Chemical engineering may be described as follows:

> Although chemical engineering has existed as a field of engineering for only about 80 years, its name is no longer completely descriptive of this dynamic, growing profession. The work of the chemical engineer is neither restricted to the chemical industry nor limited to chemical changes or chemistry. Instead, modern chemical engineers, who are also called process engineers, are concerned with all the physical and chemical changes of matter to produce economically a product or result that is useful to mankind. Such a broad background has made the chemical engineer extremely versatile and capable of working in a wide variety of industries: chemical, petroleum, aerospace, nuclear, materials, microelectronics, sanitation, food processing, and computer technology. The chemical industry alone provides an opportunity for the chemical engineer to participate in the research, development, design, or operation of plants for the production of new synthetic fibers, plastics, chemical fertilizers, vitamins, antibiotics, rocket fuels, nuclear fuels, paper pulp, photographic products, paints, fuel cells, transistors, and the thousands of chemicals that are used as intermediates in the manufacture of the above products.[9]

Industrial Hygienist

Industrial hygiene is defined by the American Industrial Hygiene Association as the "science and art devoted to the recognition, evaluation, and control of those environmental factors or stresses, arising in and from the workplace, which may cause sickness, impaired health and well-being, or significant discomfort and inefficiency among workers or among citizens of the community."[10] The National Safety Council (NSC) describes the job of the **industrial hygienist** as follows:

> An industrial hygienist is a person having a college or university degree or degrees in engineering, chemistry, physics, medicine, or related physical and biological sciences who, by virtue of special studies and training, has acquired competence in industrial hygiene. Such special studies and training must have been sufficient in all of the above cognate sciences to provide the abilities: (a) to recognize environmental factors and to understand their effect on humans and their well-being; (b) to evaluate, on the basis of experience and with the aid of quantitative measurement techniques, the magnitude of these stresses in terms of ability to impair human health and well-being; and (c) to prescribe methods to eliminate, control, or reduce such stresses when necessary to alleviate their effects.[11]

Industrial hygienists are primarily concerned about the following types of hazards: solvents, particulates, noise, dermatoses, radiation, temperature, ergonomics, toxic substances, biological substances, ventilation, gas, and vapors. In a safety and health team, the industrial hygienist typically reports to the safety and health manager.

Health Physicist

Health physicists are concerned primarily with radiation in the workplace. Consequently, they are employed by companies that generate or use nuclear power. Their primary duties include the following: monitoring radiation inside and outside the facility, measuring the radioactivity levels of biological samples, developing the radiation components of the company's emergency action plan, and supervising the decontamination of workers and the workplace when necessary.

Nuclear engineering and nuclear physics are the two most widely pursued fields of study for health physicists. A study conducted by Moeller and Eliassen gave the following breakdown of the academic preparation of practicing health physicists:

Associate degree	5.41 percent
Baccalaureate degree	28.38 percent
Master's degree	42.43 percent
Doctorate degree	19.46 percent[12]

The remaining practitioners are nondegreed personnel who have completed various types of noncollege credit training. This breakdown shows that graduate study is particularly important for health physicists. Professionals in this field may be certified by the American Board of Health Physics (ABHP).

Occupational Physician

Occupational medicine as a specialized field dates back to World War II, when the United States experienced unprecedented industrial expansion. Production of manufactured goods skyrocketed, and workplace-related medical needs followed suit. Occupational medicine was not classified as a medical specialty by the American Board of

Preventive Medicine until 1955, however. The NSC describes the main concerns of the **occupational physician** as follows:

- Appraisal, maintenance, restoration, and improvement of the workers' health through application of the principles of preventive medicine, emergency medical care, rehabilitation, and environmental medicine.
- Promotion of a productive and fulfilling interaction of the worker and the job, via application of principles of human behavior.
- Active appreciation of the social, economic, and administrative needs and responsibilities of both the worker and work community.
- Team approach to safety and health, involving cooperation of the physician with occupational or industrial hygienists, occupational health nurses, safety personnel, and other specialties.[13]

Occupational physicians are fully degreed and licensed medical doctors. In addition they must have completed postgraduate work in the following areas of safety:

> biostatistics and epidemiology, industrial toxicology, work physiology, radiation (ionizing and nonionizing), noise and hearing conservation, effects of certain environmental conditions such as high altitude and high pressures (hyperbaric and hypobaric factors), principles of occupational safety, fundamentals of industrial hygiene, occupational aspects of dermatology, psychiatric and psychological factors, occupational respiratory diseases, biological monitoring, ergonomics, basic personnel management functions, record and data collection, governmental regulations, general environmental health (air, water, ground pollution, and waste management control).[14]

Bernardino Ramazzini is widely thought of as being the first occupational physician. This is primarily as a result of his study of the work-related problems of workers in Modena, Italy, and a subsequent book he authored titled The Diseases of Workers (1700).[15] The first leading occupational physician in the United States was Alice Hamilton, MD. According to the NSC,

> In 1910 Dr. Hamilton became managing director of the Illinois Occupational Disease Commission in the United States. Members of this commission were given one year to study and report on health hazards in Illinois industries. Dr. Hamilton investigated lead poisoning in workers who manufactured white lead carbonate, ceramics, railway carriages, automobiles, batteries, printing and mining, and smelting of the metal itself. She fought with company officials for changes in their plants; she lectured to the public; she worked with the Hull House community; she wrote books and magazine articles on occupational disease; she was appointed to Harvard Medical School in 1921 as assistant professor of medicine (the first woman to hold a teaching position there); and she was the only woman member of the League of Nations Health Committee in the 1920 s.[16]

Pioneers such as Ramazzini and Hamilton paved the way for the approximately 3,000 occupational physicians practicing today. Whereas in the past the primary role of the occupational physician was treatment, today the primary role is prevention. This means more analytical, diagnostic, and intervention-related work than in the past and a much more proactive rather than passive approach.

One major difference in the role of modern occupational physicians compared with those in the past is in their relationship with employers. In the past, the occupational physician tended to be an in-house physician. Over 80 percent of the Fortune 100 companies still have at least one in-house occupational physician. However, the trend is away from this approach to the contracted approach in all but the largest companies. Companies contract with a private physician, clinic, or hospital to provide specific medical services. The health care provider is typically placed on a retainer in much the same manner as an attorney.

This contracted approach presents a new challenge for the safety and health manager. Including the occupational physician in planning, analyzing, assessing, monitoring, and other prevention-related duties can be more difficult when the physician is not in-house and readily available. For this reason, the safety and health manager must work with higher management to develop a contract that builds in time for prevention activities and compensates the health care provider appropriately.

The NSC makes the following recommendations concerning occupational physicians (OPs):

> Whether your company uses an in-house OP, an OP that works as a consultant, or a clinic that provides all health care, remember:
>
> - There should be a written medical program available to all management and employees.
> - The OP should understand the workplace and the chemicals used and produced.
> - Periodic tours of all facilities are necessary for an understanding of possible work-related injuries, and also to aid in job accommodation.
> - The OP should be familiar with OSHA and NIOSH health mandates.
> - The OP should be the leader of other medical personnel.
> - And, the OP should understand what the company expects and what the OP expects from the company.[17]

Occupational Health Nurse

Occupational health nurses have long been important members of corporate safety and health programs. According to the American Association of Occupational Health Nurses (AAOHN),

> Occupational health nursing is the application of nursing principles in conserving the health of workers in all occupations. It involves prevention, recognition, and treatment of illness and injury, and requires special skills and knowledge in the areas of health education and counselling, environmental health, rehabilitation, and human relations.[18]

Like occupational physicians, today's occupational health nurses have seen their profession evolve over the years. The shift in emphasis is away from after-the-fact treatment to prevention-related activities such as analysis, monitoring, counseling, and education.

The AAOHN defines the objectives of occupational nurses as follows:

- To adapt the nursing program to meet the specific needs of the individual company.
- To give competent nursing care for all employees.
- To ensure that adequate resources are available to support the nursing program.

- To seek out competent medical direction if it is not available on-site.
- To establish and maintain an adequate system of records relating to workplace health care.
- To plan, prepare, promote, present, and broker educational activities for employees.
- To establish and maintain positive working relationships with all departments within the company.
- To maintain positive working relationships with all components of the local health care community.
- To monitor and evaluate the nursing program on a continual basis and adjust accordingly.[19]

Like all members of the safety and health team, occupational nurses are concerned about cost containment. Cost containment is the driving force behind the emphasis on activities promoting health and accident prevention.

Occupational nurses typically report to an occupational physician when he or she is part of the on-site safety and health team. In companies that contract for off-site physician services, occupational nurses report to the overall safety and health manager.

Risk Manager

Organizations are at risk every time they open their doors for business. On any given day, an employee may be injured, a customer may have an accident, or a consumer may be injured using the organization's product. *Risk* is defined as a specific contingency or peril. Because the situations that put organizations at risk can be so expensive when they do occur, many organizations employ risk managers.

Risk management consists of the various activities and strategies that an organization can use to protect itself from situations, circumstances, or events that may undermine its security. You are, yourself, a risk manager. You take action every day to protect your personal and economic security. Take, for instance, the act of driving an automobile—a risky undertaking. Every time you drive, you put yourself at risk of incurring injuries, medical bills, lawsuits, and property damage.

You manage the risk associated with driving using two broad strategies: **reduction** and **transference**. The risk associated with driving an automobile can be reduced by wearing a seatbelt, driving defensively, and obeying traffic laws. The remaining risk is managed by transferring it to an insurance company by purchasing a policy that covers both collision and liability.

The same approach—managing risk by using reduction and transference strategies—applies in the workplace. Risk managers work closely with safety and health personnel to reduce the risk of accidents and injuries on the job. They also work closely with insurance companies to achieve the most effective transference possible.

Certification of Safety and Health Professionals

Professional certification is an excellent way to establish credentials in the safety, health, and environmental profession. The most widely pursued accreditations are the following:

- Certified Safety Professional, awarded by the Board of Certified Safety Professionals (BCSP)

- Certified Industrial Hygienist, awarded by the American Board of Industrial Hygiene
- Certified Professional **Ergonomist**, awarded by the Board of Certification in Professional Ergonomics
- Certified Occupational Health Nurse, awarded by the American Board for Occupational Health Nurses

Certified Safety Professional

To quality for the Certified Safety Professional title, applicants must follow these steps:

1. Apply to the Board of Certified Safety Professionals.
2. Meet an academic requirement.
3. Meet a professional safety experience requirement.
4. Pass the Safety Fundamentals Examination.
5. Pass the Comprehensive Practice Examination.[20]

Academic Requirement

The model educational background for a candidate for the **Certified Safety Professional (CSP)** is a bachelor's degree in safety from a program accredited by the Accreditation Board for Engineering and Technology (ABET). Because many people enter the safety profession from other educational backgrounds, candidates for the CSP may substitute other degrees plus professional safety experience for an accredited BA in safety. A CSP candidate must meet one of the following minimum educational qualifications:

- Associate degree in safety and health, or
- Bachelor's degree in any field.

Experience Requirement

In addition to the academic requirement, CSP candidates must have four years of professional safety experience in addition to any experience used to meet the academic requirement. Professional safety experience must meet all the following criteria to be considered acceptable by the Board of Certified Safety Professionals:

- The professional safety function must be the primary function of the position. Collateral duties in safety are not considered the primary function.
- The position's primary responsibility must be the prevention of harm to people, property, and the environment, rather than responsibility for responding to harmful events.
- The professional safety function must be at least 50 percent of the position duties.
- The position must be full time (defined by the BCSP as at least 35 hours per week).
- The position must be at the professional level. This is determined by evaluating the degree of responsible charge and reliance by peers, employers, or clients and on the person's ability to defend analytical approaches used in professional practice and the recommendations made for controlling hazards through engineering or administrative approaches.
- The position must have breadth of duties. This is determined by evaluating the variety of hazards about which a candidate must advise and the range of skills involved in recognizing, evaluating, and controlling hazards. Examples of skills are analysis, synthesis, design, investigation, planning, administration, and communication.

Examination Requirements

The process to achieve the Certified Safety Professional designation typically involves passing two examinations: Safety Fundamentals and Comprehensive Practice.

- Safety Fundamentals Examination. The Safety Fundamentals Examination covers basic knowledge appropriate to professional safety practice. Candidates who meet the academic standard (achieve 48 points through an associate or bachelor's degree plus experience) may sit for the Safety Fundamentals Examination. Upon passing this examination, candidates receive the Associate Safety Professional (ASP) title to denote their progress toward the CSP.
- Comprehensive Practice Examination. All CSP candidates must acquire 96 points and pass the Comprehensive Practice Examination. To take this examination, a candidate must meet both the academic and experience requirements and have passed or waived the Safety Fundamentals Examination. The total credit for academic degrees at all levels plus the months of professional safety experience must equal or exceed 96 points. After passing this examination, a candidate receives the Certified Safety Professional title. The address, telephone numbers, and website for the Board of Certified Safety Professionals are shown in Figure 4-4.

Certified Industrial Hygienist

In the mid-1950 s, a group of industrial hygienists from a national organization recommended that a voluntary certification program be established for industrial hygiene practitioners. In 1960, an independent corporation was established from the two national membership organizations—American Industrial Hygiene Association (AIHA) and American Council of Government Industrial Hygienists (ACGIH)—to establish a national examination process to certify a minimum level of knowledge in industrial hygiene.[21] Because the program was voluntary, it did not restrict the practice of individuals calling themselves "industrial hygienists." Indeed, today there are many competent persons practicing the profession of industrial hygiene who have not even sought certification. However, since its establishment, the program has proven to be a hallmark of achievement that provides an indicator of success in the field. It measures to a defined standard the knowledge of a practicing industrial hygienist in 16 technical areas of practice (called *rubrics*).

The intended purpose of the examination for **Certified Industrial Hygienist (CIH)** is to ensure that professionals working in this field have the skills and knowledge needed in the practice of industrial hygiene. The intent of the board is that the examinations are fair, reasonable, current, and representative of a broad range of industrial hygiene topics. Successful completion of the examination, meeting the educational requirements, and meeting the comprehensive, professional-level industrial hygiene experience requirements are the means by which the board attempts to identify those practitioners who have demonstrated the necessary skills to be an industrial hygienist.

Certification Process

What is the process for becoming certified? First, there is the obvious need for technical knowledge. There are various methods for gathering this information, but no one system has been identified as being the most effective. Review courses are available to prepare aspiring applicants. Conventional wisdom tends to support knowledge gained through experience and the watchful guidance of a competent mentor. Many individuals take the examination to find out their weak areas and study up on those until they manage to pass.

The professional reference questionnaire (PRQ) is important as well. The board's requirement for experience is based on activity on a professional or apprentice level. This causes a sticking point for many who find they are in a position with titles such as "industrial hygienist" or "project manager" when their scope of practice relies on strict adherence to a regulatory interpretation or exercise of very little independent judgment. Others with titles of "IH technician" or "specialist" may be acting totally independently and practicing with a scope of extreme variability and many unknowns. This should be portrayed in the PRQ to be fair to the applicant. Details are important to the board in making their evaluation of an applicant.

The American Board of Industrial Hygiene (ABIH) also has a process established for reviewing each application and preparing each examination. Examinations are prepared in minute detail. Each question is evaluated by a group of practicing Certified Industrial Hygienists to ensure its correctness and relevance to the practice of IH. Each item is rated on difficulty for its target audience (CORE or Comprehensive), and this is used to set the passing score for each test. Each question is also rated by professional testers to ensure its validity as a question for an examination. Questions are selected for use on an examination based on the latest survey of the practice of industrial hygiene, in both rubric areas and domains of practice, to achieve a balance indicative of the current practice and some historical knowledge. After each presentation of the examination, questions are again reviewed for validity. The examination itself is the subject of an effort to ensure that it adheres to the standardized evaluation method.

Certification Maintenance

After a candidate has managed to meet the established standard for certification, maintenance becomes the issue. The board requires that all Certified Industrial Hygienists demonstrate that they are active in the field and have continued to improve their knowledge. Seven categories of practice are noted for the accumulation of "points" toward the 40 required every five years. Some portion of these points are gathered for active practice, technical committee work, publications, education and meetings, teaching, retest, or other work. This requires that some program of continuing education must be pursued. The address, telephone number, and website for the American Board of Industrial Hygiene is shown in Figure 4-4.

Certified Professional Ergonomist

The examination for **Certified Professional Ergonomist (CPE)** is administered by the Board of Certification in Professional Ergonomics. Details concerning the examination are explained in the following paragraphs.[22]

To take the examination, individuals must meet the following requirements:

- Academic requirements. Applicants should have graduated from a regionally accredited college or university with a master's degree, or equivalent, in one of the correlative fields of ergonomics, such as biomechanics, human factors/ergonomics, industrial engineering, industrial hygiene, kinesiology, psychology, or systems engineering. Not everyone trained in these fields necessarily has the capabilities required for certification. The board uses other criteria to determine whether an applicant's ergonomics education has been sufficiently broad.

- Work experience. Applicants must have completed at least four years of ergonomic work experience. Appropriateness of work experience is determined from the applicant's employment history and evidence of participation in projects requiring ergonomic expertise.
- Work product. One work sample must be submitted with the application. This work sample must demonstrate a breadth of ergonomic knowledge and the ability to use ergonomic methods successfully. The work sample must demonstrate, at a minimum, the candidate's capabilities in the application of ergonomic principles to the design of a product, system, or work environment. Allowable work products are technical reports, design papers, analysis reports, evaluation reports, patent applications, or a thorough written description of the project.

Associate-Level Certification

The Associate Ergonomics Professional (AEP) category is a precursor to the CPE designation and is available to a candidate who

- Meets the education requirements for BCPE certification;
- Has passed Part 1 of the examination (on basic knowledge of human factors ergonomics); and
- Is currently working toward fulfilling the BCPE requirement of four years' practical experience as a human factors and ergonomics professional.

Bachelor's Degree Certification

A credential is available that recognizes a candidate for achieving the following levels of knowledge, skills, and experience in ergonomics practice:

- A bachelor's degree from a recognized university.
- At least 200 hours of ergonomics training.
- At least two full years practicing ergonomics.
- A satisfactory score on the four-hour, two-part, multiple-choice examination on ergonomics foundations and ergonomics practice methods.

This certification may be obtained by candidates with a BA in engineering, health care/rehabilitation, industrial hygiene, and psychology. The address, telephone number, and website for the Board of Certification in Professional Ergonomics is shown in Figure 4-4.

Certified Occupational Health Nurse

The American Board for Occupational Health Nurses (ABOHN) offers several certifications relating to safety and health:

- **Certified Occupational Health Nurse (COHN)**
- Certified Occupational Health Nurse–Specialist (COHN-S)
- Certified Occupational Health Nurse/Case Manager (COHN/CM)
- Certified Occupational Health Nurse–Specialist/Case Manager (COHN-S/CM)
- Certified Occupational Health Nurse–Safety Manager (COHN-SM)

All the certifications offered by the ABOHN require that individuals first pass either the COHN or COHN-S certification tests. The other certifications are subspecialties.

Academic and Experience Requirements

In order to sit for the COHN certification tests, individuals must be registered nurses holding an associate degree or higher degree or their international equivalents. The focus of this credential is the nurse's role as a clinician, adviser, coordinator, and case manager. To sit for the COHN-S certification test, individuals must be registered nurses who hold a bachelor's degree or higher. The focus of this credential is on the nurse's role in direct care, management, education, consulting, and case management. The BA required for this credential need not be in nursing.

In addition to the academic requirements, those who wish to achieve COHN or COHN-S certification must have 4,000 or more hours of work experience in occupational health and 50 or more contact hours of continuing education completed during the past five years. Additional testing is required for the various subspecialty credentials.

The newest certification offered by the ABOHN is the COHN-SM or Safety Manager credential. The focus of this credential is on occupational health nurses who plan, organize, implement, and evaluate hazard control activities to help organizations reduce or eliminate risks in the workplace. In order to sit for this subspecialty test, an individual must hold the COHN or COHN-S credential, hold a position that requires at least 25 percent of the time be spent on safety management activities, have completed 50 or more hours of safety-related continuing education during the past five years, and have 1,000 or more hours of safety management experience earned during the past five years.

Key Terms and Concepts

- Aerospace engineering
- Analysis and evaluation
- Associate degree
- Bachelor's degree
- Certification boards
- Certified Industrial Hygienist (CIH)
- Certified Occupational Health Nurse (COHN)
- Certified Professional Ergonomist (CPE)
- Certified Safety Professional (CSP)
- Chemical engineer
- Competitiveness
- Cost
- Design engineers
- Design process
- Document and communicate
- Electrical engineering
- Environmental engineering
- Ergonomist
- Formal education
- Government agencies
- Health physicist
- Image
- In-service training
- Industrial engineering
- Industrial hygienist
- Lack of commitment
- Line authority
- Mechanical engineering
- Nuclear engineering
- Occupational health nurse
- Occupational physician
- Problem identification
- Produce and deliver
- Production versus safety
- Productivity
- Professional societies
- Profits
- Quality
- Reduction
- Response time
- Risk management
- Safety and health manager
- Safety engineer
- Service
- Staff authority
- Standards and testing organizations
- Synthesis
- Trade associations
- Transference

Review Questions

1. What types of positions may be included in a modern safety and health team?
2. Briefly explain the impact that such issues as workers' compensation and the environment have had on the commitment of corporate management to safety and health.
3. What is the difference between a staff and a line position?

4. Explain the types of problems that safety and health managers can expect to confront in attempting to implement their programs.
5. Briefly explain what a company must do to succeed in today's competitive global marketplace.
6. How can safety and health managers use the competitiveness issue to gain a commitment to their programs?
7. List five different college majors that can lead to a career as a safety and health manager.
8. Explain the importance of ongoing in-service training for modern safety and health managers and how to get it.
9. How can safety and health managers become certified in their profession?
10. Name three professional societies that a modern safety and health manager may join.
11. What is meant by the statement "If a physician makes an error, he may harm one person, but an engineer who errs may harm hundreds"?
12. Explain how the design process can affect safety.
13. What types of engineers are most likely to work as design engineers?
14. Why is the title "safety engineer" sometimes a misnomer?
15. What specific strengths may industrial engineers bring to bear as safety engineers?
16. What specific strengths may environmental engineers bring to the safety and health team?
17. What specific strengths may chemical engineers bring to the safety and health team?
18. Describe the job of the industrial hygienist.
19. What is a health physicist?
20. Describe the job of the occupational physician.
21. Describe the job of the occupational health nurse.
22. Explain the concept of risk management.
23. Explain the role of the ergonomist.
24. Explain how to achieve each of the following certifications: Certified Safety Professional, Certified Industrial Hygienist, and Certified Professional Ergonomist.
25. Explain how to achieve the following certification: Certified Occupational Health Nurse–Safety Manager.

ENDNOTES

1. The University Record, Undergraduate Catalog of the University of Florida, 2006, 67.
2. Ibid., 70.
3. Ibid., 74.
4. Ibid., 75.
5. D. L. Goetsch, J. Nelson, and William S. Chalk, Technical Drawing, 6th ed. (Albany: Delmar, 2006), 791.
6. W. Hammer, Occupational Safety Management and Engineering, 5th ed. (Upper Saddle River, NJ: Prentice Hall, 2001), 791.
7. The University Record, 73.
8. Ibid., 72.
9. Ibid., 68.
10. American Industrial Hygiene Association (AIHA), Engineering Field Reference Manual. Retrieved from http://www.aiha.org, April 4, 2006.
11. C. Berry (revised by Barbara A. Plog), "The Industrial Hygienist," in Fundamentals of Industrial Hygiene (Chicago: National Safety Council, 1998), 571.
12. Ibid., 572.
13. C. Zenz, "The Occupational Physician," in Fundamentals of Industrial Hygiene, 607.
14. Ibid.
15. National Safety Council, "Occupational Physicians: What Is Their New Role?" Retrieved from http://www.nsc.org, April 4, 2006.
16. Ibid.
17. Ibid., 38.
18. American Association of Occupational Health Nurses (AAOHN), A Guide for Establishing an Occupational Health Nursing Service. Retrieved from http://www.aaoha.org, April 5, 2006.
19. Ibid., 79.
20. Board of Certified Safety Professionals (BCSP), "The CSP and Specialty Certifications," July 2000, 1–6, http://www.bcsp.org/.
21. American Board of Industrial Hygiene (ABIH), "Certified Industrial Hygienist (CIH)—What It Means," July 2000, 1–5, http://www.abih.org/.
22. Board of Certified Professional Ergonomists (BCPE), "BCPE—Frequently Asked Questions," July 2000, 1–5, http://www.bcpe.org/.

CHAPTER 5

The OSH Act, Standards, and Liability

Major Topics

- Rationale for the OSH Act
- OSHA's Mission and Purpose
- OSH Act Coverage
- OSHA Standards
- OSHA's Record Keeping and Reporting
- Keeping Employees Informed
- Workplace Inspections and Enforcement
- OSHA's Enhanced Enforcement Policy
- Citations and Penalties
- Appeals Process
- State-Level OSHA Programs
- Services Available from OSHA
- Employer Rights and Responsibilities
- Employee Rights and Responsibilities
- Keeping Up-to-Date on OSHA
- Problems with OSHA
- Other Agencies and Organizations
- OSHA's General Industry Standards
- OSHA's Maritime Standards
- OSHA's Construction Standards
- Standards and Codes
- Laws and Liability

Since the early 1970s, the amount of legislation passed—and the number of subsequent regulations—concerning workplace safety and health have increased markedly. Of all the legislation, by far the most significant has been the **Occupational Safety and Health**

Act (OSH Act) of 1970. Prospective and practicing safety and health professionals must be knowledgeable about the OSH Act and the agency established by it—the **Occupational Safety and Health Administration (OSHA)**. This chapter provides students with the information they need about the OSH Act, OSHA, and other pertinent federal legislation and agencies.

Rationale for the OSH Act

Perhaps the most debilitating experience one can have on the job is to be involved in, or exposed to, a work-related accident or illness. Such an occurrence can be physically and psychologically incapacitating for the victim, psychologically stressful for the victim's fellow workers, and extraordinarily expensive for the victim's employer. In spite of this, until 1970, laws governing workplace safety were limited and sporadic. Finally, in 1970, Congress passed the Occupational Safety and Health Act (OSH Act) with the following stated purpose: "to assure so far as possible every working man and woman in the nation safe and healthful working conditions and to preserve our human resources."[1]

According to the U.S. Department of Labor, in developing this comprehensive and far-reaching piece of legislation, Congress considered the following statistics:

- Every year, an average of 14,000 deaths was caused by **workplace accidents**.
- Every year, 2.5 million workers were disabled in workplace accidents.
- Every year, approximately 300,000 new cases of **occupational diseases** were reported.[2]

Clearly, a comprehensive, uniform law was needed to help reduce the incidence of work-related injuries, illnesses, and deaths. The OSH Act of 1970 addressed this need. It is contained in Title 29 of the Code of Federal Regulations, Parts 1900 through 2200. The act also establishes the Occupational Safety and Health Administration (OSHA), which is part of the U.S. Department of Labor and is responsible for administering the OSH Act.

OSHA's Mission and Purpose

According to the Department of Labor, OSHA's mission and purpose can be summarized as follows:

- Encourage employers and employees to reduce workplace hazards.
- Implement new safety and health programs.
- Improve existing safety and health programs.
- Encourage research that will lead to innovative ways of dealing with workplace safety and health problems.
- Establish the rights of employers regarding the improvement of workplace safety and health.
- Establish the rights of employees regarding the improvement of workplace safety and health.
- Monitor job-related illnesses and injuries through a system of reporting and record keeping.

- Establish training programs to increase the number of safety and health professionals and to improve their competence continually.
- Establish mandatory workplace safety and health standards and enforce those standards.
- Provide for the development and approval of state-level workplace safety and health programs.
- Monitor, analyze, and evaluate state-level safety and health programs.[3]

OSH Act Coverage

The OSH Act applies to most employers. If an organization has even one employee, it is considered an employer and must comply with applicable sections of the act. This includes all types of employers from manufacturing and construction to retail and service organizations. There is no exemption for small businesses, although organizations with ten or fewer employees are exempted from OSHA inspections and the requirement to maintain injury and illness records.

Although the OSH Act is the most comprehensive and far-reaching piece of safety and health legislation ever passed in this country, it does not cover all employers. In general, the OSH Act covers employers in all 50 states, the District of Columbia, Puerto Rico, and all other territories that fall under the jurisdiction of the U.S. government. Exempted employers include the following:

- Persons who are self-employed
- Family farms that employ only immediate members of the family
- Federal agencies covered by other federal statutes (in cases where these other federal statutes do not cover working conditions in a specific area or areas, OSHA standards apply)
- State and local governments (except to gain OSHA's approval of a state-level safety and health plan, states must provide a program for state and local government employees that is at least equal to its private sector plan)
- Coal mines (coal mines are regulated by mining-specific laws)

Federal government agencies are required to adhere to safety and health standards that are comparable to and consistent with OSHA standards for private sector employees. OSHA evaluates the safety and health programs of federal agencies. However, OSHA cannot assess fines or monetary damages against other federal agencies as it can against private sector employers.

There are many OSHA requirements to which employers must adhere. Some apply to all employers—except those exempted—whereas others apply only to specific types of employers. These requirements cover areas of concern such as the following:

- Fire protection
- Electricity
- Sanitation
- Air quality
- Machine use, maintenance, and repair
- Posting of notices and warnings
- Reporting of accidents and illnesses

- Maintaining written compliance programs
- Employee training

In addition to these, other more important and widely applicable requirements are explained later in this chapter.

OSHA Standards

The following statement by the U.S. Department of Commerce summarizes OSHA's responsibilities relating to standards:

> In carrying out its duties, OSHA is responsible for promulgating legally enforceable standards. OSHA standards may require conditions, or the adoption or use of one or more practices, means, methods, or processes reasonably necessary and appropriate to protect workers on the job. It is the responsibility of employers to become familiar with standards applicable to their establishments and to ensure that employees have and use personal protective equipment when required for safety.[4]

The general duty clause of the OSH Act requires that employers provide a workplace that is free from hazards that are likely to harm employees. This is important because the general duty clause applies when there is no specific OSHA standard for a given situation. Where OSHA standards do exist, employers are required to comply with them as written.

How Standards Are Developed

OSHA develops standards based on its perception of need and at the request of other federal agencies, state and local governments, other standards-setting agencies, labor organizations, or even individual private citizens. OSHA uses the committee approach for developing standards, both standing committees within OSHA and special ad hoc committees. Ad hoc committees are appointed to deal with issues that are beyond the scope of the standing committees.

OSHA's standing committees are the **National Advisory Committee on Occupational Safety and Health (NACOSH)** and the Advisory Committee on Construction Safety and Health. NACOSH makes recommendations on standards to the secretary of health and human services and to the secretary of labor. The Advisory Committee on Construction Safety and Health advises the secretary of labor on standards and regulations relating specifically to the construction industry.

The **National Institute for Occupational Safety and Health (NIOSH)**, like OSHA, was established by the OSH Act. Whereas OSHA is part of the Department of Labor, NIOSH is part of the Department of Health and Human Services. NIOSH has an education and research orientation. The results of this agency's research are often used to assist OSHA in developing standards.

OSHA Standards versus OSHA Regulations

OSHA issues both standards and regulations. Safety and health professionals need to know the difference between the two. OSHA standards address specific hazards such as working in confined spaces, handling hazardous waste, or working with dangerous chemicals. Regulations are more generic in some cases than standards and more specific in others. However, even when they are specific, regulations do not apply to specific hazards. Regulations do not require the rigorous review process that standards go through. This process is explained in the next section.

How Standards Are Adopted, Amended, or Revoked

OSHA can adopt, amend, or revise standards. Before any of these actions can be undertaken, OSHA must publish its intentions in the *Federal Register* in either a notice of proposed rule making or an advance notice of proposed rule making. The **notice of proposed rule making** must explain the terms of the new rule, delineate proposed changes to existing rules, or list rules that are to be revoked. The advance notice of proposed rule making may be used instead of the regular notice when it is necessary to solicit input before drafting a rule.

After publishing notice, OSHA must conduct a public hearing if one is requested. Any interested party may ask for a public hearing on a proposed rule or rule change. When this happens, OSHA must schedule the hearing and announce the time and place in the *Federal Register*.

The final step, according to the Department of Labor, is as follows:

> After the close of the comment period and public hearing, if one is held, OSHA must publish in the Federal Register the full, final text of any standard amended or adopted and the date it becomes effective, along with an explanation of the standard and the reasons for implementing it. OSHA may also publish a determination that no standard or amendment needs to be issued.[5]

How to Read an OSHA Standard

OSHA standards are typically long and complex and are written in the language of lawyers and bureaucrats, making them difficult to read. However, reading OSHA standards can be simplified somewhat if one understands the system.

OSHA standards are part of the Code of Federal Regulations (CFR) published by the Office of the Federal Register. The regulations of all federal-government agencies are published in the CFR. Title 29 contains all the standards assigned to OSHA. Title 29 is divided into several parts, each carrying a four-number designator (such as Part 1901, Part 1910). These parts are divided into sections, each carrying a numerical designation. For example, 29 CFR 1910.1 means *Title 29, Part 1910, Section 1, Code of Federal Regulations.*

The sections are divided into four different levels of subsections, each with a particular type of designator as follows:

First Level:	Alphabetically using lowercase letters in parentheses: (a) (b) (c) (d)
Second Level:	Numerically using numerals in parentheses: (1) (2) (3) (4)
Third Level:	Numerically using roman numerals in parentheses: (i) (ii) (iii) (iv)
Fourth Level:	Alphabetically using uppercase letters in parentheses: (A) (B) (C) (D)

Occasionally, the standards go beyond the fourth level of subsection. In these cases, the sequence just described is repeated with the designator shown in parentheses underlined. For example: (a), (1), (i), (A).

Understanding the system used for designating sections and subsections of OSHA standards can guide readers more quickly to the specific information needed. This helps reduce the amount of cumbersome reading needed to comply with the standards.

Temporary Emergency Standards

The procedures described in the previous section apply in all cases. However, OSHA is empowered to pass **temporary emergency standards** on an emergency basis without undergoing normal adoption procedures. Such standards remain in effect only until permanent standards can be developed.

To justify passing temporary standards on an emergency basis, OSHA must determine that workers are in imminent danger from exposure to a hazard not covered by existing standards. Once a temporary standard has been developed, it is published in the *Federal Register.* This step serves as the notification step in the permanent adoption process. At this point, the standard is subjected to all the other adoption steps outlined in the preceding section.

How to Appeal a Standard

After a standard has been passed, it becomes effective on the date prescribed. This is not necessarily the final step in the **appeals process**, however. A standard, either permanent or temporary, may be appealed by any person who is opposed to it.

An appeal must be filed with the U.S. Court of Appeals serving the geographic region in which the complainant lives or does business. Appeal paperwork must be initiated within 60 days of a standard's approval. However, the filing of one or more appeals does not delay the enforcement of a standard unless the court of appeals handling the matter mandates a delay. Typically, the new standard is enforced as passed until a ruling on the appeal is handed down.

Requesting a Variance

Occasionally, an employer may be unable to comply with a new standard by the effective date of enforcement. In such cases, the employer may petition OSHA at the state or federal level for a variance. Following are the different types of variances that can be granted.

Temporary Variance

When an employer advises that it is unable to comply with a new standard immediately but may be able to if given additional time, a **temporary variance** may be requested. OSHA may grant such a variance for up to a maximum of one year. To be granted a temporary variance, employers must demonstrate that they are making a concerted effort to comply and taking the steps necessary to protect employees while working toward compliance.

Application procedures are very specific. Prominent among the requirements are the following: (1) identification of the parts of the standard that cannot be complied with; (2) explanation of the reasons why compliance is not possible; (3) detailed explanations of the steps that have been taken so far to comply with the standard; and (4) explanation of the steps that will be taken to comply fully.

According to the U.S. Department of Labor, employers are required to keep their employees informed. They must "certify that workers have been informed of the variance application, that a copy has been given to the employees' authorized representative, and that a summary of the application has been posted wherever notices are normally posted. Employees also must be informed that they have the right to request a hearing on the application."[6]

Variances are not granted simply because an employer cannot afford to comply. For example, if a new standard requires employers to hire a particular type of specialist but there is a shortage of people with the requisite qualifications, a temporary variance might be granted. However, if the employer simply cannot afford to hire such a specialist, the variance will probably be denied. Once a temporary variance is granted, it may be renewed twice. The maximum period of each extension is six months.

Permanent Variance

Employers who feel they already provide a workplace that exceeds the requirements of a new standard may request a **permanent variance**. They present their evidence, which is inspected by OSHA. Employees must be informed of the application for a variance and notified of their right to request a hearing. Having reviewed the evidence and heard testimony (if a hearing has been held), OSHA can award or deny the variance. If a permanent variance is awarded, it comes with a detailed explanation of the employer's ongoing responsibilities regarding the variance. If, at any time, the company does not meet these responsibilities, the variance can be revoked.

Other Variances

In addition to temporary and permanent variances, an experimental variance may be awarded to companies that participate in OSHA-sponsored experiments to test the effectiveness of new health and safety procedures. Variances also may be awarded in cases where the secretary of labor determines that a variance is in the best interest of the country's national defense.

When applying for a variance, employers are required to comply with the standard until a decision has been made. If this is a problem, the employer may petition OSHA for an interim order. If granted, the employer is released from the obligation to comply until a decision is made. In such cases, employees must be informed of the order.

Typical of OSHA standards are the confined space and hazardous waste standards. Brief profiles of these standards provide an instructive look at how OSHA standards are structured and the extent of their coverage.

Confined Space Standard

This standard was developed in response to the approximately 300 work-related deaths that occur in confined spaces each year.[7] The standard applies to a broad cross section of industries that have employees working in spaces with the following characteristics: limited openings for entry or exit, poor natural ventilation, and a design not intended to accommodate continuous human occupancy. Such spaces as manholes, storage tanks, underground vaults, pipelines, silos, vats, exhaust ducts, boilers, and degreasers are typically considered confined spaces.

The key component in the standard is the *permit requirement.* Employers are required to develop an in-house program under which employees must have a permit to enter confined spaces. Through such programs, employers must do the following:

- Identify spaces that can be entered only by permit.
- Restrict access to identified spaces to ensure that only authorized personnel may enter.
- Control hazards in the identified spaces through engineering, revised work practices, and other methods.
- Continually monitor the identified spaces to ensure that any known hazards remain under control.

The standard applies to approximately 60 percent of the workers in the United States. Excluded from coverage are federal, state, and local government employees; agricultural workers; maritime and construction workers; and employees of companies with ten or fewer workers.

Hazardous Waste Standard

This standard specifically addresses the safety of the estimated 1.75 million workers who deal with hazardous waste: hazardous waste workers in all situations including treatment, storage, handling, and disposal; firefighters; police officers; ambulance personnel; and hazardous materials response team personnel.[8]

The requirements of this standard are as follows:

- Each hazardous waste site employer must develop a safety and health program designed to identify, evaluate, and control safety and health hazards, and provide for emergency response.
- There must be preliminary evaluation of the site's characteristics prior to entry by a trained person to identify potential site hazards and to aid in the selection of appropriate employee protection methods.
- The employer must implement a site control program to prevent contamination of employees. At a minimum, the program must identify a site map, site work zones, site communications, safe work practices, and the location of the nearest medical assistance. Also required in particularly hazardous situations is the use of the buddy system so that employees can keep watch on one another and provide quick aid if needed.
- Employees must be trained before they are allowed to engage in hazardous waste operations or emergency response that could expose them to safety and health hazards.
- The employer must provide medical surveillance at least annually and at the end of employment for all employees exposed to any particular hazardous substance at or above established exposure levels or those who wear approved respirators for 30 days or more on-site.
- Engineering controls, work practices, and personal protective equipment, or a combination of these methods, must be implemented to reduce exposure below established exposure levels for the hazardous substances involved.

- There must be periodic air monitoring to identify and quantify levels of hazardous substances and to ensure that proper protective equipment is being used.
- The employer must set up an informational program with the names of key personnel and their alternates responsible for site safety and health, and the requirements of the standard.
- The employer must implement a decontamination procedure before any employee or equipment leaves an area of potential hazardous exposure; establish operating procedures to minimize exposure through contact with exposed equipment, other employees, or used clothing; and provide showers and change rooms where needed.
- There must be an emergency response plan to handle possible on-site emergencies prior to beginning hazardous waste operations. Such plans must address personnel roles; lines of authority, training, and communications; emergency recognition and prevention; safe places of refuge; site security; evacuation routes and procedures; emergency medical treatment; and emergency alerting.
- There must be an off-site emergency response plan to better coordinate emergency action by local services and to implement appropriate control actions.[9]

OSHA's Record Keeping and Reporting

One of the breakthroughs of the OSH Act was the centralization and systematization of **record keeping**. This has simplified the process of collecting health and safety statistics for the purpose of monitoring problems and taking the appropriate steps to solve them.

Over the years, OSHA has made substantial changes to its record-keeping and reporting requirements. Employers had complained for years about the mandated injury and illness record-keeping system. Their complaints can be summarized as follows:

- Original system was cumbersome and complicated.
- OSHA record-keeping rule had not kept up with new and emerging issues.
- There were too many interpretations in many of the record-keeping documents.
- Record-keeping forms were too complex.
- Guidelines for record keeping were too long and difficult to understand.

In response to these complaints, OSHA initiated a dialogue among stakeholders to improve the record-keeping and reporting process. Input was solicited and received from employers, unions, trade associations, record keepers, OSHA staff, state occupational safety and health personnel, and state consultation program personnel. OSHA's goals for the new record-keeping and reporting system were as follows:

- Simplify all aspects of the process.
- Improve the quality of records.
- Meet the needs of a broad base of stakeholders.
- Improve access for employees.
- Minimize the regulatory burden.
- Reduce vagueness—give clear guidance.
- Promote the use of data from the new system in local safety and health programs.

In recording and reporting occupational illnesses and injuries, it is important to have common definitions. The Department of Labor uses the following definitions for record-keeping and reporting purposes:

> An occupational injury is any injury such as a cut, fracture, sprain, or amputation that results from a work-related accident or from exposure involving a single incident in the work environment. An occupational illness is any abnormal condition or disorder other than one resulting from an occupational injury caused by exposure to environmental factors associated with employment. Included are acute and chronic illnesses or diseases which may be caused by inhalation, absorption, ingestion, or direct contact with toxic substances or harmful agents.[10]

Reporting Requirements

All occupational illnesses and injuries must be reported if they result in one or more of the following:

- Death of one or more workers.
- One or more days away from work.
- Restricted motion or restrictions to the work that an employee can do.
- Loss of consciousness of one or more workers.
- Transfer of an employee to another job.
- Medical treatment beyond in-house first aid (if it is not on the first-aid list, it is considered medical treatment).
- Any other condition listed in Appendix B of the rule.[11]

Record-Keeping Requirements

Employers are required to keep injury and illness records for each location where they do business.[12] For example, an automobile manufacturer with plants in several states must keep records at each individual plant for that plant. Records must be maintained on an annual basis using special forms prescribed by OSHA. Computer or electronic copies can replace paper copies. Records are not sent to OSHA. Rather, they must be maintained locally for a minimum of three years. However, they must be available for inspection by OSHA at any time.

All records required by OSHA must be maintained on the following forms or forms based on them:

> OSHA's Form 300. Form 300 is used to record information about every work-related death and every work-related injury or illness that involves loss of consciousness, restricted work activity, job transfer, days away from work, or medical treatment beyond first aid. First aid is defined as follows:
>
> - Using nonprescription medication at nonprescription doses.
> - Administering tetanus immunizations.
> - Cleaning, flushing, or soaking wounds on the surface of the skin.
> - Using wound coverings such as bandages, gauze pads, and so on.
> - Using hot or cold therapy.
> - Using totally nonrigid means of support, such as elastic bandages, wraps, nonrigid backbelts, and so forth.

- Using temporary immobilization devices such as splints, slings, neck collars, or back boards while transporting an accident victim.
- Drilling a fingernail or toenail to relieve pressure, or draining fluids from blisters.
- Using eye patches.
- Using irrigation, tweezers, cotton swab, or other simple means to remove splinters or foreign material from areas other than the eye.
- Using finger guards.
- Using massages.
- Drinking fluids to relieve heat stress.

Injuries that require no more than these first-aid procedures do not have to be recorded. Form 300 is also used to record significant work-related injuries and illnesses that are diagnosed by a physician or licensed health care professional as well as those that meet any of the specific recording criteria set forth in 29 CFR Part 1904.12. Figure 5-1 is an example of a log based on OSHA's Form 300.

OSHA's Form 300A. All organizations covered by 29 CFR Part 1904 must complete **Form 300A**, even if there have been no work-related injuries or illnesses during the year in question. This form is used to summarize all injuries and illnesses that appear on OSHA's Form 300.

OSHA's Form 301. **Form 301** is used for every incidence of a recordable injury or illness. Form 301 must be completed within seven calendar days of learning that a recordable injury or illness has occurred. Figure 5-2 is an example of an incident report based on OSHA's Form 301.

Reporting and Record-Keeping Summary

Reporting and record-keeping requirements appear as part of several different OSHA standards. Not all of them apply in all cases. Following is a summary of the most widely applicable OSHA reporting and record-keeping requirements.

- 29 CFR 1903.2(a) OSHA Poster. **OSHA Poster 2203**, which advises employees of the various provisions of the OSH Act, must be conspicuously posted in all facilities subject to OSHA regulations.
- 29 CFR 1903.16(a) Posting of OSHA Citations. Citations issued by OSHA must be clearly posted for the information of employees in a location as close as possible to the site of the violation.
- 29 CFR 1904.2 Injury/Illness Log. Employers are required to maintain a log and summary of all *recordable* (Form 300) injuries and illnesses of their employees.
- 29 CFR 1904.4 Supplementary Records. Employers are required to maintain supplementary records (OSHA Form 301) that give more complete details relating to all recordable injuries and illnesses of their employees.
- 29 CFR 1904.5 Annual Summary. Employers must complete and post an annual summary (OSHA Form 300A) of all recordable illnesses and injuries. The summary must be posted from February 1 to April 30 every year.
- 29 CFR 1904.6 Lifetime of Records. Employers are required to keep injury and illness records on file for three years.

FIGURE 5-1 Log based on OSHA's Form 300.

You must record information about every work-related death and about every work-related injury or illness that involves loss of consciousness, restricted work activity or job transfer, days away from work, or medical treatment beyond first aid. You must also record significant work-related injuries and illnesses that are diagnosed by a physician or licensed health care professional. You must also record work-related injuries and illnesses that meet any; of the specific recording criteria listed in 29 CFR Part 1904.8 through 1904.12. Feel free to use two lines for a single case if you need to. You must complete an Injury and Illness Incident Report (OSHA Form 301) or equivalent form for each injury or illness recorded on this form. If you're not sure whether a case is recordable, call your local OSHA office for help.

Company name:______________________

City:____________________ State:________

IDENTIFY THE PERSON			DESCRIBE THE CASE			CLASSIFY THE CASE											
(A) Case No.	(B) Employee's name	(C) Job title (e.g., Welder)	(D) Date of injury or onset of illness	(E) Where the event occurred (e.g., Landing dock north end)	(F) Describe injury or illness, parts of body affected, and object/substance that directly injured or made person ill (e.g., Second degree burns on right forearm from acetylene torch)	Using these four categories, check ONLY the most serious result for each case:				Enter the number of days the injured or ill worker was:		Check the "Injury" column or choose one type of illness: (M)					
						Death	Days away from work	Remained at work: Job transfer or restriction	Remained at work: Other recordable cases	On job transfer or restriction	Away from work	Injury	Skin disorder	Hearing Injury	Respiratory condition	Poisoning	All other\
						(G)	(H)	(I)	(J)	(K)	(L)	(1)	(2)	(3)	(4)	(5)	(6)
						☐	☐	☐	☐	days	days	☐	☐	☐	☐	☐	☐
						☐	☐	☐	☐	days	days	☐	☐	☐	☐	☐	☐
						☐	☐	☐	☐	days	days	☐	☐	☐	☐	☐	☐
						☐	☐	☐	☐	days	days	☐	☐	☐	☐	☐	☐
						☐	☐	☐	☐	days	days	☐	☐	☐	☐	☐	☐
						☐	☐	☐	☐	days	days	☐	☐	☐	☐	☐	☐
						☐	☐	☐	☐	days	days	☐	☐	☐	☐	☐	☐
						☐	☐	☐	☐	days	days	☐	☐	☐	☐	☐	☐
						☐	☐	☐	☐	days	days	☐	☐	☐	☐	☐	☐
						☐	☐	☐	☐	days	days	☐	☐	☐	☐	☐	☐
						☐	☐	☐	☐	days	days	☐	☐	☐	☐	☐	☐
						☐	☐	☐	☐	days	days	☐	☐	☐	☐	☐	☐
						☐	☐	☐	☐	days	days	☐	☐	☐	☐	☐	☐
						☐	☐	☐	☐	days	days	☐	☐	☐	☐	☐	☐
						☐	☐	☐	☐	days	days	☐	☐	☐	☐	☐	☐
					Page Totals ➢												

FIGURE 5-2 Accident report based based on OSHA's Form 301.

Within 7 calendar days after you receive information that a recordable work-related injury or illness has occurred, you must fill out this form or an equivalent. Some state workers' compensation, insurance, or other reports may be acceptable substitutes. To be considered an equivalent form, any substitute must contain all the information asked for on this form.

According to Public Law 91-596 and 29 CFR 1904, OSHA's recordkeeping rule, you must keep this form on file for five years following the year to which it pertains.

Completed by: ____________________

Title ____________________

Phone ( ) _____ - _____

Date _____/_____/_____

INFORMATION ABOUT THE EMPLOYEE

1) Full Name ____________________

2) Street ____________________

City ____________ State ______ Zip ________

3) Date of birth _____/_____/_____

4) Date hired _____/_____/_____

5) ❑ Male

❑ Female

INFORMATION ABOUT THE PHYSICIAN OR OHER HEALTH CARE PROFESSIONAL

6) Name of physician or other health care professional

____

7) If treatment was given away from the worksite, where was it given?

Facility ____________________

Street ____________________

City ______________ State _____ Zip______

8) Was employee treated in an emergency room?

❑ Yes

❑ No

9) Was employee hospitalized overnight as an in-patient?

❑ Yes

❑ No

INFORMATION ABOUT THE CASE

10) Case number from *Log* ______________ *(Transfer the number from the Log after you record the case.)*

11) Date of injury or illness _____/_____/_____

12) Time employee began work ______________ AM / PM

13) Time of event ______________ AM / PM

❑ Check if time cannot be determined

14) **What was the employee doing just before the incident occurred?** Describe the activity, as well as the tools, equipment, or material the employee was using. Be specific. *Examples*: "climbing a ladder while carrying roofing materials"; "spraying chlorine from hand sprayer"; "daily computer key-entry."

15) **What happened?** Tell us how the injury occurred. *Examples*: "When ladder slipped on wet floor, worker fell 20 feet"; "Worker was sprayed with chlorine when gasket broke during replacement"; "Worker developed soreness in wrist over time."

16) **What was the injury or illness?** Tell us the part of the body that was affected and how it was affected; be more specific than "hurt," "pain," or "sore." *Examples*: "strained back"; "chemical burn, hand"; "carpal tunnel syndrome."

17) **What object or substance directly harmed the employee?** *Examples*: "concrete floor"; "chlorine"; "radial arm saw." If this question does not apply to the incident, leave it blank.

18) **If the employee died, when did death occur?**

Date _____/_____/_____

- 29 CFR 1904.7 Access to Records. Employers are required to give employees, government representatives, and former employees and their designated representatives access to their own individual injury and illness records.
- 29 CFR 1904.8 Major Incident Report. A major incident is the death of one employee or the hospitalization of five or more employees in one incident. All such incidents must be reported to OSHA.

- 29 CFR 1904.11 Change of Ownership. When a business changes ownership, the new owner is required to maintain the OSHA-related records of the previous owner.
- 29 CFR 1910.1020(d) Medical/Exposure Records. Employers are required to maintain medical and/or exposure records for the duration of employment plus 30 years (unless the requirement is superseded by another OSHA standard).
- 29 CFR 1910.1020(e) Access to Medical/Exposure Records. Employers that keep medical and/or exposure records are required to give employees access to their own individual records.
- 29 CFR 1910.1020(g)(1) Toxic Exposure. Employees who will be exposed to toxic substances or other harmful agents in the course of their work must be notified when first hired and reminded continually on at least an annual basis thereafter of their right to access their own individual medical records.
- 29 CFR 1910.1020(g)(2) Distribution of Materials. Employers are required to make a copy of OSHA's *Records Access Standard* (29 CFR 1910.1020) available to employees. They are also required to distribute to employees any informational materials provided by OSHA.

OSHA makes provisions for awarding record-keeping variances to companies that wish to establish their own record-keeping systems (see variance regulation 1905). Application procedures are similar to those described earlier in this chapter for standard variances. To be awarded a variance, employers must show that their record-keeping system meets or exceeds OSHA's requirements.

Incidence Rates

Two concepts can be important when completing OSHA 300 forms: *incidence rates* and *severity rates.* On occasion, it is necessary to calculate the total injury and illness incident rate of an organization in order to complete an OSHA Form 300. This calculation must include fatalities and all injuries requiring medical treatment beyond mere first aid.

The formula for determining the total injury and illness incident rate follows:

$$IR5N \times 200{,}000 \div T$$

IR = Total injury and illness incidence rate
N = Number of injuries, illnesses, and fatalities
T = Total hours worked by all employees during the period in question

The number 200,000 in the formula represents the number of hours that 100 employees work in a year (40 hours per week times 50 weeks equals 50,000 hours per year per employee). Using the same basic formula with only minor substitutions, safety managers can calculate the following types of incidence rates:

- Injury rate
- Illness rate
- Fatality rate
- Lost workday cases rate
- Number of lost workdays rate
- Specific hazard rate
- Lost workday injuries rate

The *number of lost workdays rate,* which does not include holidays, weekends, or any other days that employees would not have worked anyway, takes the place of the old severity rate calculation.

Record-Keeping and Reporting Exceptions

Among the exceptions to OSHA's record-keeping and reporting requirements, the two most prominent are as follows:

- Employers with ten or fewer employees (full- or part-time in any combination).
- Employers in one of the following categories: real estate, finance, retail trade, or insurance.

There are also partial exceptions to OSHA's record-keeping and reporting requirements. Most businesses that fall into Standard Industrial Classifications (SIC) codes 52–89 are exempt from all record-keeping and reporting requirements except in the case of a fatality or an incident in which five or more employees are hospitalized. The following types of organizations are partially exempt. They are not required to maintain OSHA injury and illness records unless asked by OSHA to do so. OSHA's request must be in writing.

Advertising services
Apparel and accessory stores
Barber shops
Beauty shops
Bowling centers
Candy, nut, and confectionary stores
Child day care services
Computer and data processing services
Credit reporting and collection services
Dairy products stores
Dance studios, schools, and halls
Depository institutions
Drug and proprietary stores
Eating and drinking establishments
Educational institutions and services
Engineering, accounting, and related
Funeral service and crematories
Gasoline service stations
Hardware stores
Health services, not elsewhere classified
Holding and other investment offices
Individual and family services
Insurance agents, brokers, services
Insurance carriers
Legal services
Liquor stores
Meat and fish markets
Mailing, reproduction, and stenographic
Medical and dental laboratories
Membership organizations
Miscellaneous business services
Miscellaneous food stores
Miscellaneous personnel services
Miscellaneous shopping goods stores
Motion picture
Motorcycle dealers
Museums and art galleries
New and used car dealers
Non-depository institutions
Offices and clinics of dentists
Offices of osteopathic physicians
Offices and clinics of medical doctors
Offices of other health practitioners
Photographic studios, portrait
Producers, orchestras, entertainers
Radio, television, and computer stores
Real estate agents and managers

Retail stores, not elsewhere classified
Research, management, and related
Retail bakeries
Re-upholstery and furniture repair
Security and commodity brokers
Services, not elsewhere classified
Shoe repair and shoeshine parlors
Social services, not elsewhere classified
Title abstract offices
Used car dealers

Keeping Employees Informed

One of the most important requirements of the OSH Act is *communication.* Employers are required to keep employees informed about safety and health issues that concern them. Most of OSHA's requirements in this area concern the posting of material. Employers are required to post the following material at locations where employee information is normally displayed:

- OSHA Poster 2203, which explains employee rights and responsibilities as prescribed in the OSH Act. The state version of this poster may be used as a substitute.
- Summaries of variance requests of all types.
- Copies of all OSHA citations received for failure to meet standards. Unlike other informational material, citations must be posted near the site of the violation. They must remain until the violation is corrected or for a minimum of three days, whichever period is longer.
- OSHA Form 300A (Summary of Workplace Injuries and Illnesses). Each year the new summary must be posted by February 1 and must remain posted until April 30.

In addition to the posting requirements, employers must also provide employees who request them with copies of the OSH Act and any OSHA rules that may concern them. Employees must be given access to records of exposure to hazardous materials and medical surveillance that has been conducted.

Workplace Inspections and Enforcement

One of the methods OSHA uses for enforcing its rules is the **workplace inspection**. OSHA personnel may conduct workplace inspections unannounced, and except under special circumstances, giving an employer prior notice is a crime punishable by fine, imprisonment, or both.

When OSHA compliance officers arrive to conduct an inspection, they are required to present their credentials to the person in charge. Having done so, they are authorized to enter, at reasonable times, any site, location, or facility where work is taking place. They may inspect, at reasonable times, any condition, facility, machine, equipment, materials, and so on. Finally, they may question, in private, any employee or other person formally associated with the company.

Under special circumstances, employers may be given up to a maximum of 24 hours' notice of an inspection. These circumstances are

- When imminent danger conditions exist;
- When special preparation on the part of the employer is required;
- When inspection must take place at times other than during regular business hours

- When it is necessary to ensure that the employer, employee representative, and other pertinent personnel will be present; and
- When the local area director for OSHA advises that advance notice will result in a more effective inspection.

Employers may require that OSHA have a judicially authorized warrant before conducting an inspection. The U.S. Supreme Court handed down this ruling in 1978 (*Marshall v. Barlow's Inc.*).[13] However, having obtained a legal warrant, OSHA personnel must be allowed to proceed without interference or impediment.

The OSH Act applies to approximately six million work sites in the United States. Sheer volume dictates that OSHA establish priorities for conducting inspections. These priorities are as follows: imminent danger situations, catastrophic fatal accidents, employee complaints, planned high-hazard inspections, and follow-up inspections.

After being scheduled, the inspection proceeds in the following steps:

1. The OSHA compliance officer presents his or her credentials to a company official.
2. The compliance officer conducts an **opening conference** with pertinent company officials and employee representatives. The following information is explained during the conference: why the plant was selected for inspection, the purpose of the inspection, its scope, and applicable standards.
3. After choosing the route and duration, the compliance officer makes the **inspection tour**. During the tour, the compliance officer may observe, interview pertinent personnel, examine records, take readings, and make photographs.
4. The compliance officer holds a **closing conference**, which involves open discussion between the officer and company and employee representatives. OSHA personnel advise company representatives of problems noted, actions planned as a result, and assistance available from OSHA.

OSHA's Enhanced Enforcement Policy

Organizations that receive OSHA citations for high-gravity violations are subject to enhanced enforcement measures. Components of the enhanced enforcement policy include follow-up inspections, programmed inspections, public awareness, settlements, and Section 11(b) summary enforcement orders.

Follow-Up Inspections

High-gravity violations include high-gravity willful violations, multiple high-gravity serious violations, repeat violations at the originating establishment, failure-to-abate notices, and serious or willful violations related to a workplace fatality. Organizations that commit any of these types of violations will receive on-site follow-up inspections. In addition, OSHA's area directors are empowered to conduct follow-up inspections to verify compliance if there is reason to suspect that a required abatement has not occurred.

Programmed Inspections

OSHA's *site-specific targeting process* uses objective selection criteria to schedule programmed inspections. As part of this process, OSHA records the name of the overall corporate entity during inspections and prioritizes within its site-specific targeting list all branch and affiliated facilities under the entity's broad corporate umbrella that have received high-gravity violations.

Public Awareness

When an organization receives a high-gravity violation of the type listed above under "Follow-up Inspections," OSHA makes the public aware of the violation and all applicable enforcement actions taken by issuing press releases through local and national media. In addition, high-gravity violations at branch and affiliated facilities are made known to the company's corporate headquarters through official notification by OSHA of the citation and corresponding penalties.

Settlements

Provisions for high-gravity violation settlement agreements

1. Require the organization in question to hire consultants to develop a feasible process for changing the health and safety culture in the facility where the violations occurred;
2. Apply the violation settlement not just to the facility in question but also corporatewide;
3. Include information about other job sites of the organization in question;
4. Require the organization to consent in advance to report all serious injuries and illnesses that require outside medical care and to receive OSHA inspections based on the report; and
5. Agree to consent to entry of a court enforcement order under Section 11(b) of the OSH Act.

Section 11(b) Summary Enforcement Orders

In cases of the kinds of high-gravity violations listed earlier in this section, OSHA applies to the appropriate federal court of appeal for orders that summarily enforce the citations in question under Section 11(b) of the OSH Act. This means that once a Section 11(b) order has been entered, organizations that fail to comply may be held in contempt of court—an action OSHA pursues.

Citations and Penalties

Based on the findings of the compliance officer's workplace inspections, OSHA is empowered to issue citations and/or assess penalties. A citation informs the employer of OSHA violations. Penalties are typically fines assessed as the result of citations. The types of citations and their corresponding penalties are as follows:

- **Other-than-serious violation.** A violation that has a direct relationship to job safety and health, but probably would not cause death or serious physical harm. A **proposed penalty** of up to $7,000 for each violation is discretionary. A penalty for an other-than-serious violation may be adjusted downward by as much as 95 percent, depending on the employer's good faith (demonstrated efforts to comply with the act), history of previous violations, and size of business.

- **Serious violation.** A violation in which there is a high probability that death or serious physical injury may result and that the employer knew or should have known of the hazard. OSHA proposes a mandatory penalty for each serious violation. The actual penalty may be adjusted downward depending on the employer's good faith, history of prior violations, and the gravity of the alleged violation.
- **Willful violation.** A violation that the employer intentionally and knowingly commits. The employer either knows that what he or she is doing constitutes a violation or is aware that a hazardous condition exists and has made no reasonable effort to eliminate it. Penalties may be proposed for each willful violation, with a minimum penalty of $5,000 for each violation. A proposed penalty for a willful violation may be adjusted downward, depending on the size of the business and its history of previous violations. Usually, no credit is given for good faith. If an employer is convicted of a willful violation of a standard that has resulted in the death of an employee, the offense is punishable by a court-imposed fine or by imprisonment for up to six months, or both. A fine of up to $250,000 for an individual or $500,000 for a corporation may be imposed for a criminal conviction.
- **Repeat violation.** A violation of any standard, regulation, rule, or order where, upon reinspection, a substantially similar violation is found. **Repeat violations** can result in a fine for each such violation. To be the basis of a repeat citation, the original citation must be final; a citation under contest may not serve as the basis for a subsequent repeat citation.
- **Failure to abate prior violation.** Failure to correct a prior violation may bring a civil penalty for each day that the violation continues beyond the prescribed abatement date.
- **De minimis violation.** Violations of standards that have no direct bearing on safety and health. **De minimis violations** are documented like any other violation, but they are not included in citations.[14]

In addition to the citations and penalties described in the preceding paragraphs, employers may also be penalized by additional fines and/or prison if convicted of any of the following offenses: (1) falsifying records or any other information given to OSHA personnel; (2) failing to comply with posting requirements; and (3) interfering in any way with OSHA compliance officers in the performance of their duties.

Examples of OSHA Citations and Fines

Private sector companies in the United States experience more than five million workplace injuries and illness every year. In just the manufacturing and construction sectors alone, OSHA documents more than 75,000 violations annually and issues penalties amounting to more than $70 million. This section provides examples of companies that received citations for various violations and, as a result, were assessed penalties in the form of fines.

> Failure to properly use machine guards. A wire manufacturing company was fined $47,700 when an employee who was operating two wire spinning machines was caught around the neck by a loop of wire and choked to death. The company was cited for failure to properly guard machine pulleys, use lockout/tagout procedures, and properly guard rotating machine parts (the spools on the wire spinning machines).[15]

Selected through OSHA's site-specific targeting program. A metal manufacturing company came under OSHA's scrutiny because it had a higher-than-average number of lost day injuries. OSHA conducted an on-site inspection and found numerous violations including the following: failure to remove dangerous electrical wiring, provide personal protection gear and ensure employees wear it, reduce hazardous noise levels, guard machines, train employers on hazardous communication, and properly store oxygen tanks. As a result of these violations, OSHA fined the company $288,000.[16]

Failure to train employees in the application of proper safety procedures. A paper manufacturing company was fined $258,000 when it allowed employees to apply improper and unsafe procedures when attempting to light a boiler. As a result of these improper procedures, the boiler exploded. Two of the three employees were killed (fire and chemical burns). The company was also cited for failure to remove debris from the facility, repair numerous structural hazards (loose bricks, concrete, and glass), and provide fall protection.[17]

Failure to abate previous hazards. A woodworking company was fined $27,800 when it failed to correct hazardous situations that had been previously cited. The conditions that led to the fine included failure to protect employees from paint and lacquer finish vapors, properly store flammable materials, fit-test and train employees in the proper use of respirators, and train employees on fire safety.[18]

Appeals Process

Employee Appeals

Employees may not contest the fact that citations were or were not awarded or the amounts of the penalties assessed. However, they may appeal the following aspects of OSHA's decisions regarding their workplace: (1) the amount of time **(abatement period)** given an employer to correct a hazardous condition that has been cited, and (2) an employer's request for an extension of an abatement period. Such appeals must be filed within ten working days of a posting. Although opportunities for formal appeals by employees are unlimited, employees may request an informal conference with OSHA officials to discuss any issue relating to the findings and results of a workplace inspection.

Employer Appeals

Employers may appeal a citation, an abatement period, or the amount of a proposed penalty. Before actually filing an appeal, however, an employer may ask for an informal meeting with OSHA's area director. The area director is empowered to revise citations, abatement periods, and penalties in order to settle disputed claims. If the situation is not resolved through this step, an employer may formalize the appeal. Formal appeals are of two types: (1) a petition for modification of abatement, or (2) a notice of contest. The specifics of both are explained in the following paragraphs.

Petition for Modification of Abatement

The **Petition for Modification of Abatement (PMA)** is available to employers who intend to correct the situation for which a citation was issued, but who need more time.

As a first step, the employer must make a good-faith effort to correct the problem within the prescribed timeframe. Having done so, the employer may file a petition for modification of abatement. The petition must contain the following information:

- Descriptions of steps taken so far to comply.
- Length of additional time is needed for compliance and why.
- Descriptions of the steps being taken to protect employees during the interim.
- Verification that the PMA has been posted for employee information and that the employee representative has been given a copy.

Notice of Contest

An employer who does not wish to comply may contest a citation, an abatement period, and/or a penalty. The first step is to notify OSHA's area director in writing. This is known as filing a **notice of contest**. It must be done within 15 federal working days of receipt of a citation or penalty notice. The notice of contest must clearly describe the basis for the employer's challenge and contain all of the information about what is being challenged (i.e., amount of proposed penalty or abatement period, and so on).

Once OSHA receives a notice of contest, the area director forwards it and all pertinent materials to the **Occupational Safety and Health Review Commission (OSHRC)**. OSHRC is an independent agency that is associated with neither OSHA nor the Department of Labor. The Department of Labor describes how OSHRC handles an employer's claim:

> The commission assigns the case to an administrative law judge. The judge may disallow the contest if it is found to be legally invalid, or a hearing may be scheduled for a public place near the employer's workplace. The employer and the employees have the right to participate in the hearing; the OSHRC does not require that they be represented by attorneys. Once the administrative law judge has ruled, any party to the case may request further review by OSHRC. Any of the three OSHRC commissioners also may, at his or her own motion, bring a case before the Commission for review. Commission rulings may be appealed to the appropriate U.S. Court of Appeals.[19]

Employer appeals are common. In fact, each issue of *Occupational Hazards* carries a special section titled "Contested Cases." A review of examples of contested cases can be instructive in both process and outcomes. Two such cases follow on the next page.

State-Level OSHA Programs

States are allowed to develop their own safety and health programs.[20] In fact, the OSH Act encourages it. As an incentive, OSHA will fund up to 50 percent of the cost of operating a state program for states with approved plans. States may develop comprehensive plans covering public and private employers or limit their plans to coverage of public employers only. In such cases, OSHA covers employers not included in the state plan.

To develop an OSHA-approved safety and health plan, a state must have adequate legislative authority and must demonstrate the ability to develop standards-setting, enforcement, and appeals procedures within three years; public employee protection; a sufficient number of qualified enforcement personnel; and education, training, and technical assistance programs. When a state satisfies all these requirements and accomplishes all developmental steps,

OSHA then certifies that a state has the legal, administrative, and enforcement means necessary to operate effectively. This action renders no judgment on how well or poorly a state is actually operating its program, but merely attests to the structural completeness of its program. After this certification, there is a period of at least one year to determine if a state is effectively providing safety and health protection.[21]

Figure 5-3 lists the states that currently have OSHA-approved safety and health plans. Connecticut and New York cover the public sector only. The Virgin Islands and Puerto Rico account for 2 of the 23 plans.

FIGURE 5-3 States with approved safety and health plans.

Alaska
PO Box 21149
Juneau, AL 99802-1149
907-465-2700

Arizona
800 W. Washington
Phoenix, AZ 85007
602-542-5795

California
395 Oyster Pt.Blvd., 3rd Flr.
Wing C
San Francisco, CA 94080
415-737-2960

Connecticut
200 Folly Brook Blvd.
Wethersfield, CT 06109
203-566-5123

Hawaii
830 Punchbowl St.
Honolulu, HI 96813
808-548-3150

Indiana
State Office Bldg. 1013
100 N. Senate Ave.
Indianapolis, IN 46204-2287
317-232-2665

Iowa
1000 E. Grand Ave.
Des Moines, IA 50319
515-281-3447

Kentucky
U.S. Hwy. 127 South
Frankfort, KY 40601
502-564-3070

Maryland
501 St. Paul Pl., 15th Flr.
Baltimore, MD 21202
301-333-4179

Michigan (Labor)
309 N. Washington Sq.
PO Box 30015
Lansing, MI 48909
517-335-8022

Michigan (Public Health)
3423 N. Logan St., Box 30195
Lansing, MI 48909
517-335-8022

Minnesota
443 Lafayette Rd.
St. Paul, MN 55155
651-296-2342

Nevada
1370 S. Curry St.
Carson City, NV 89710
702-885-5240

New Mexico
1190 St. Francis Dr., N2200
Santa Fe, NM 87503-0968
702-885-5240

New York
One Main St.
Brooklyn, NY 11201
518-457-3518

North Carolina
4 W. Edenton St.
Raleigh, NC 27601
919-733-7166

Oregon
21 Labor & Ind. Bldg.
Salem, OR 97310
503-378-3304

Puerto Rico
Prudencio Rivera Martinez
Bldg.
505 Munoz Rivera Ave.
Hato Rey, PR 00918
809-654-2119-22

South Carolina
PO Box 11329
Columbia, SC 29211-1329
803-734-9594

Vermont
120 State St.
Montpelier, VT 05602
802-828-2765

Virginia
PO Box 12064
Richmond, VA 23241-0064
804-786-2376

Virgin Islands
Box 890, Christiansted
St. Croix, VI 00820
809-773-1994

Washington
Gen. Admin. Bldg.
Rm. 334-AX-31
Olympia, WA 98504-0631
206-753-6307

Services Available from OSHA

In addition to setting standards and inspecting for compliance, OSHA provides services to help employers meet the latest safety and health standards. The services are typically offered at no cost. Three categories of services are available from OSHA: consultation, voluntary protection programs, and training and education services.

Consultation Services

Consultation services provided by OSHA include assistance in (1) identifying hazardous conditions; (2) correcting identified hazards; and (3) developing and implementing programs to prevent injuries and illnesses. To arrange consultation services, employers contact the consultation provider in their state (see Figure 5-4).

The actual services are provided by professional safety and health consultants, who are not OSHA employees. They typically work for state agencies or universities and provide consultation services on a contract basis; OSHA provides the funding. OSHA publication 3047, titled *Consultation Services for the Employer,* may be obtained from the nearest OSHA office.

Voluntary Protection Programs

OSHA's **Voluntary Protection Programs (VPPs)** serve the following three basic purposes:

- To recognize companies that have incorporated safety and health programs into their overall management system.
- To motivate companies to incorporate health and safety programs into their overall management system.
- To promote positive, cooperative relationships among employers, employees, and OSHA.

OSHA currently operates three programs under the VPP umbrella. These programs are discussed in the following paragraphs.

Star Program

The **Star Program** recognizes companies that have incorporated safety and health into their regular management system so successfully that their injury rates are below the national average for their industry. This is OSHA's most strenuous program. To be part of the Star Program, a company must demonstrate

- Management commitment;
- Employee participation;
- An excellent work-site analysis program;
- A hazard prevention and control program; and
- A comprehensive safety and health training program.[22]

Merit Program

The **Merit Program** is less strenuous than the Star Program. It is seen as a steppingstone to recognize companies that have made a good start toward Star Program recognition. OSHA works with such companies to help them take the next step and achieve Star Program recognition.

FIGURE 5-4 State consultation project directory.

State	Telephone	State	Telephone
Alabama	205-348-3033	Nebraska	401-471-4717
Alaska	907-264-2599	Nevada	701-789-0546
Arizona	602-255-5795	New Hampshire	603-271-3170
Arkansas	501-682-4522	New Jersey	609-984-3517
California	415-557-2870	New Mexico	505-827-2885
Colorado	303-491-6151	New York	518-457-5468
Connecticut	203-566-4550	North Carolina	919-733-3949
Delaware	302-571-3908	North Dakota	701-224-2348
District of Columbia	202-576-6339	Ohio	614-644-2631
Florida	850-488-3044	Oklahoma	405-528-1500
Georgia	404-894-8274	Oregon	503-378-3272
Guam	9-011 671-646-9246	Pennsylvania	800-381-1241 (Toll-free) 412-357-2561
Hawaii	808-548-7510	Puerto Rico	809-754-2134-2171
Idaho	208-385-3283	Rhode Island	401-277-2438
Illinois	312-917-2339	South Carolina	803-734-9579
Indiana	317-232-2688	South Dakota	605-688-4101
Iowa	515-281-5352	Tennessee	615-741-7036
Kansas	913-296-4386	Texas	512-458-7254
Kentucky	502-564-6895	Utah	801-530-6868
Louisiana	504-342-9601	Vermont	801-828-2765
Maine	207-289-6460	Virginia	804-367-1986
Maryland	301-333-4219	Virgin Islands	809-772-1315
Massachusetts	616-727-3463	Washington	206-586-0961
Michigan	517-335-8250 (Health) 517-322-1814 (Safety)	West Virginia	304-348-7890
Minnesota	612-297-2393	Wisconsin	608-266-8579 (Health) 414-512-5063 (Safety)
Mississippi	601-987-3961	Wyoming	307-777-7786
Missouri	314-751-3403		
Montana	406-444-6401		

Demonstration Program

The Department of Labor describes the **Demonstration Program** as follows: "for companies that provide Star-quality worker protection in industries where certain Star requirements can be changed to include these companies as Star participants."[23]

Companies participating in any of the VPPs are exempt from regular programmed OSHA inspections. However, employee complaints, accidents that result in serious injury, or major chemical releases will be "handled according to routine enforcement procedures."[24]

Training and Education Services

Training and education services available from OSHA take several forms. OSHA operates a training institute in Des Plaines, Illinois, that offers a wide variety of services to safety and health personnel from the public and private sectors. The institute has a full range of facilities including classrooms and laboratories in which it offers more than 60 courses.

To promote training and education in locations other than the institute, OSHA awards grants to nonprofit organizations. Colleges, universities, and other nonprofit organizations apply for funding to cover the costs of providing workshops, seminars, or short courses on safety and health topics currently high on OSHA's list of priorities. Grant funds must be used to plan, develop, and present instruction. Grants are awarded annually and require a match of at least 20 percent of the total grant amount.

Employer Rights and Responsibilities

OSHA is very specific in delineating the rights and responsibilities of employers regarding safety and health. These rights and responsibilities, as set forth in OSHA publication 2056, are summarized in this section.

Employer Rights

The following is a list of **employer rights** under the OSH Act. Employers have the right to

- Seek advice and consultation as needed by contacting or visiting the nearest OSHA office;
- Request proper identification of the OSHA compliance officer prior to an inspection;
- Be advised by the compliance officer of the reason for an inspection;
- Have an opening and closing conference with the compliance officer in conjunction with an inspection;
- Accompany the compliance officer on the inspection;
- File a notice of contest with the OSHA area director within 15 working days of receipt of a notice of citation and proposed penalty;
- Apply for a temporary variance from a standard if unable to comply because the materials, equipment, or personnel needed to make necessary changes within the required time are not available;
- Apply for a permanent variance from a standard if able to furnish proof that the facilities or methods of operation provide employee protection at least as effective as that required by the standard;
- Take an active role in developing safety and health standards through participation in OSHA Standards Advisory Committees, through nationally recognized standards-setting organizations, and through evidence and views presented in writing or at hearings;
- Be assured of the confidentiality of any trade secrets observed by an OSHA compliance officer during an inspection; and
- Ask NIOSH for information concerning whether any substance in the workplace has potentially toxic effects.[25]

Employer Responsibilities

In addition to the rights set forth in the previous subsection, employers have prescribed responsibilities. The following is a list of **employer responsibilities** under the OSH Act. Employers must

- Meet the general duty responsibility to provide a workplace free from hazards that are causing or are likely to cause death or serious physical harm to employees and to comply with standards, rules, and regulations issued under the OSH Act;
- Be knowledgeable of mandatory standards and make copies available to employees for review upon request;
- Keep employees informed about OSHA;
- Continually examine workplace conditions to ensure that they conform to standards;
- Minimize or reduce hazards;
- Make sure employees have and use safe tools and equipment (including appropriate personal protective equipment) that is properly maintained;
- Use color codes, posters, labels, or signs as appropriate to warn employees of potential hazards;
- Establish or update operating procedures and communicate them so that employees follow safety and health requirements;
- Provide medical examinations when required by OSHA standards;
- Provide the training required by OSHA standards;
- Report to the nearest OSHA office within eight hours any fatal accident or one that results in the hospitalization of three or more employees;
- Keep OSHA-required records of injuries and illnesses and post a copy of OSHA Form 300 from February 1 through April 30 each year (this applies to employers with 11 or more employees);
- At a prominent location within the workplace, post OSHA Poster 2203 informing employees of their rights and responsibilities;
- Provide employees, former employees, and their representatives access to the Log of Work-Related Injuries and Illnesses (OSHA Form 300) at a reasonable time and in a reasonable manner;
- Give employees access to medical and exposure records;
- Give the OSHA compliance officer the names of authorized employee representatives who may be asked to accompany the compliance officer during an inspection;
- Not discriminate against employees who properly exercise their rights under the act;
- Post OSHA citations at or near the work site involved (each citation or copy must remain posted until the violation has been abated or for three working days, whichever is longer); and
- Abate cited violations within the prescribed period.[26]

Employee Rights and Responsibilities

Employee Rights

Section 11(c) of the OSH Act delineates **employee rights**. These rights are actually protection against punishment for employees who exercise their right to pursue any of the following courses of action:

- Complain to an employer, union, OSHA, or any other government agency about job safety and health hazards.

- File safety or health grievances;
- Participate in a workplace safety and health committee or in union activities concerning job safety and health;
- Participate in OSHA inspections, conferences, hearings, or other OSHA-related activities.[27]

Employees who feel they are being treated unfairly because of actions they have taken in the interest of safety and health have 30 days in which to contact the nearest OSHA office. Upon receipt of a complaint, OSHA conducts an investigation and makes recommendations based on its findings. If an employer refuses to comply, OSHA is empowered to pursue legal remedies at no cost to the employee who filed the original complaint.

In addition to those just set forth, employees have a number of other rights. Employees may

- Expect employers to make review copies available of OSHA standards and requirements;
- Ask employers for information about hazards that may be present in the workplace;
- Ask employers for information on emergency procedures;
- Receive safety and health training;
- Be kept informed about safety and health issues;
- Anonymously ask OSHA to conduct an investigation of hazardous conditions at the work site;
- Be informed of actions taken by OSHA as a result of a complaint;
- Observe during an OSHA inspection and respond to the questions asked by a compliance officer;
- See records of hazardous materials in the workplace;
- See their medical record;
- Review the annual Log of Work-Related Injuries and Illnesses (OSHA Form 300);
- Have an exit briefing with the OSHA compliance officer following an OSHA inspection;
- Anonymously ask NIOSH to provide information about toxicity levels of substances used in the workplace;
- Challenge the abatement period given employers to correct hazards discovered in an OSHA inspection;
- Participate in hearings conducted by the Occupational Safety and Health Review Commission;
- Be advised when an employer requests a variance to a citation or any OSHA standard;
- Testify at variance hearings;
- Appeal decisions handed down at OSHA variance hearings; and
- Give OSHA input concerning the development, implementation, modification, and/or revocation of standards.[28]

Employee Responsibilities

Employees have a number of specific responsibilities. The following list of **employee responsibilities** is adapted from OSHA 2056, 1991 (Revised). Employees must

- Read the OSHA poster at the job site and be familiar with its contents;
- Comply with all applicable OSHA standards;

- Follow safety and health rules and regulations prescribed by the employer and promptly use personal protective equipment while engaged in work;
- Report hazardous conditions to the supervisor;
- Report any job-related injury or illness to the employer and seek treatment promptly;
- Cooperate with the OSHA compliance officer conducting an inspection; and
- Exercise their rights under the OSH Act in a responsible manner.[29]

Keeping Up-To-Date on OSHA

OSHA's standards, rules, and regulations are always subject to change. The development, modification, and revocation of standards is an ongoing process. It is important for prospective and practicing safety and health professionals to stay up-to-date with the latest actions and activities of OSHA. Following is an annotated list of strategies that can be used to keep current:

- Establish contact with the nearest regional or area OSHA office and periodically request copies of new publications or contact the OSHA Publications Office at the following address:

  OSHA Publications Office

  200 Constitution Ave. NW

  Room N–3101

  Washington, DC 20210

  http://www.osha.gov/pls/publications/pubindex.list
- Review professional literature in the safety and health field. Numerous periodicals carry OSHA updates that are helpful.
- Establish and maintain relationships with other safety and health professionals for the purpose of sharing information, and do so frequently.
- Join professional organizations, review their literature, and attend their conferences.

Problems with OSHA

Federal agencies are seldom without their detractors. Resentment of the federal bureaucracy is intrinsic in the American mind-set. Consequently, complaints about OSHA are common. Even supporters occasionally join the ranks of the critics. Often, the criticisms leveled against OSHA are valid.

Criticisms of OSHA take many different forms. Some characterize OSHA as an overbearing bureaucracy with little or no sensitivity to the needs of employers who are struggling to survive in a competitive marketplace. At the same time, others label OSHA as timid and claim it does not do enough. At different times and in different cases, both points of view have probably been at least partially accurate.

Most criticism of OSHA comes in the aftermath of major accidents or a workplace disaster. Such events typically attract a great deal of media attention, which, in turn, draws the attention of politicians. In such cases, the criticism tends to focus on the question, "Why didn't OSHA prevent this disaster?" At congressional hearings, detractors will typically answer this question by claiming that OSHA spends too much time and too many resources dealing with matters of little consequence while ignoring real problems. Supporters of

OSHA will typically answer the question by claiming that a lack of resources prevents the agency from being everywhere at once. There is a measure of validity in both answers.

On one hand, OSHA has made a significant difference in the condition of the workplace in this country. On the other hand, large, centralized bureaucratic agencies rarely achieve a high level of efficiency. This is compounded in OSHA's case by the fact that the agency is subject to the ebb and flow of congressional support, particularly in the area of funding. Consequently, OSHA is likely to continue to be an imperfect organization subject to ongoing criticism.

Other Agencies and Organizations

Although OSHA is the most widely known safety and health organization in the federal government, it is not the only one. Figure 5-5 lists associations, agencies, and organizations (including OSHA) with safety and health as part of their mission.

Of those listed, the most important to modern safety and health professionals are NIOSH and OSHRC. The missions of these organizations are summarized in the following paragraphs.

NIOSH

The **National Institute for Occupational Safety and Health (NIOSH)** is part of the Department of Health and Human Services (HHS). (Recall that OSHA is part of the Department of Labor.) NIOSH has two broad functions: research and education. The main focus of the agency's research is on toxicity levels and human tolerance levels of hazardous substances. NIOSH prepares recommendations along these lines for OSHA standards dealing with hazardous substances. NIOSH studies are also published and made available to employers. Each year, NIOSH publishes updated lists of toxic materials and recommended tolerance levels. These publications represent the educational component of NIOSH's mission.

The Department of Health and Human Services describes NIOSH as follows:

> In 1973, NIOSH became a part of the Centers for Disease Control (CDC), an arm of the Public Health Service in the Department of Health and Human Services. NIOSH is unique among federal research institutions because it has the authority to conduct research in the workplace, and to respond to requests for assistance from employers and employees.

NIOSH also consults with the Department of Labor (DOL) and other federal, state, and local government agencies to promote occupational safety and health and makes recommendations to DOL about worker exposure limits.

NIOSH estimates that over 7,000 people are killed at work each year, and nearly 12 million nonfatal injuries occur in the workplace. In addition to death and injury, it is estimated that over 10 million men and women are exposed to hazardous substances in their jobs that can eventually cause fatal or debilitating diseases. To help establish priorities in developing research and control of these hazards, NIOSH developed a list of the ten leading work-related diseases and injuries:

1. Occupational lung diseases.
2. Musculoskeletal injuries.

FIGURE 5-5 Agencies, organizations, and associations dealing with safety and health.

American Public Health Association
1015 15th St. NW
Washington, DC 20005

Bureau of Labor Statistics
U.S. Department of Labor
Washington, DC 20212

Bureau of National Affairs, Inc.
Occupational Safety and Health Reporter
1231 25th St. NW
Washington, DC 20037

Commerce Clearing House
Employee Safety and Health Guide
4205 W. Peterson Ave.
Chicago, IL 60646

Environmental Protection Agency
401 M St. SW
Washington, DC 20001

Mine Safety and Health Administration
4015 Wilson Blvd.
Rm. 601
Arlington, VA 22203

National Institute for Occupational Safety and Health (NIOSH)
4676 Columbia Pkwy.
Cincinnati, OH 45226

Occupational Safety and Health Administration
U.S. Department of Labor
200 Constitution Ave.
Washington, DC 20210

Occupational Safety and Health Review Committee (OSHRC)
Washington, DC 20210

U.S. Consumer Product Safety Commission
Washington, DC 20207

3. Occupational cancers.
4. Occupational cardiovascular disease.
5. Severe occupational traumatic injuries.
6. Disorders of reproduction.
7. Neurotoxic disorders.
8. Noise-induced hearing loss.
9. Dermatological conditions.
10. Psychological disorders.[30]

Figure 5-6 is an organizational chart showing the major divisions in NIOSH. The four principal divisions are described in the following paragraphs.

FIGURE 5-6 The major divisions of NIOSH.

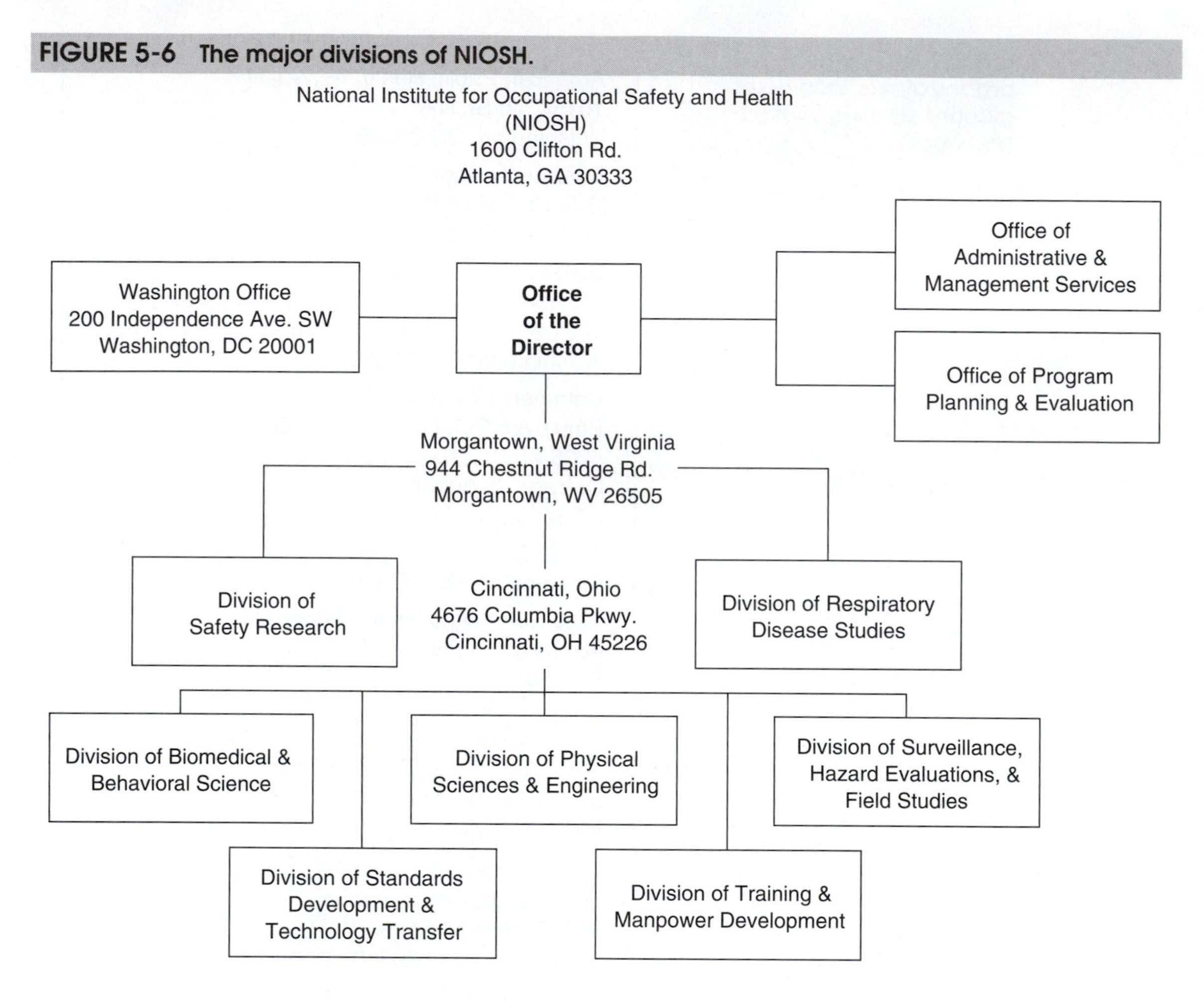

Division of Biomedical and Behavioral Science

The **Division of Biomedical and Behavioral Science (DBBS)** conducts research in the areas of toxicology, behavioral science, ergonomics, and the health consequences of various physical agents.[31] DBBS investigates problems created by new technologies and develops biological monitoring and diagnostic aids to ensure that the workplace is not responsible for diminished health, functional capacity, or life expectancy of workers. Consultation and data are furnished for developing the criteria for workplace exposure standards.

DBBS conducts laboratory and work-site research into the psychological, behavioral, physiological, and motivational factors relating to job stress as well as those induced by chemical and physical agents. The division assesses physical work capacity and tolerance for environmental conditions as influenced by age, gender, body type, and physical fitness. Interventions and control procedures using various approaches are developed and tested for their ability to reduce undue job stress.

The DBBS toxicology research involves dose-response and methods development studies to evaluate the effects of toxic agents and to develop techniques that utilize biomarkers as indicators of toxic exposures and early indicators of pathological change. The studies include applications of cellular biology, immunochemistry, pharmacokinetics, and neurophysiology. Through laboratory analysis of biological samples from animals exposed experimentally and humans exposed occupationally to workplace hazards, DBBS provides clinical and biochemical consultations for ascertaining the extent of exposure and for diagnosing occupational diseases.

DBBS conducts laboratory and work-site research on hazards from physical agents such as noise, vibration, and nonionizing energy sources. Studies seek to identify exposure factors that are significant to the health and well-being of the workforce. Investigations, including instrumentation and methods development, for characterizing and relating exposure factors to biological and performance changes in animal and human populations are also undertaken.

Division of Respiratory Disease Studies

The **Division of Respiratory Disease Studies (DRDS)** is the focal point for the clinical and epidemiological research that NIOSH conducts on occupational respiratory diseases.[32] The division provides legislatively mandated medical and autopsy services and conducts medical research to fulfill NIOSH's responsibilities under the Federal Mine Safety and Health Amendments Act of 1977.

The division conducts field studies of occupational respiratory diseases, in addition to designing and interpreting cross-sectional and prospective morbidity and mortality studies relating to occupational respiratory disease. Field studies are conducted at mines, mills, and other industrial plants where occupational respiratory diseases occur among workers. The division uses epidemiological techniques, including studies of morbidity and mortality, to detect common characteristics related to occupational respiratory diseases.

To formulate and implement programs that will identify factors involved in the early detection and differential rates of susceptibility to occupational respiratory diseases, DRDS conducts cell biology research to determine the role of microorganisms and environmental exposure in these diseases. The division also provides autopsy evaluations and a pathology research program. Research is conducted on immunological mechanisms and cell physiology to determine the effects of environmental exposure as it relates to occupational respiratory diseases.

DRDS provides for planning, coordinating, and processing medical examinations mandated under the Federal Mine Safety and Health Amendments Act of 1977 and operates a certification program for medical facilities and physicians who participate in the examination program. DRDS also evaluates and approves employer programs for the examination of employees in accordance with published regulations as well as arranging for the examination of employees who work at locations that lack an approved examination program. The division conducts the National Coal Workers' Autopsy Program and performs research into the postmortem identification and quantification of occupational respiratory exposures.

Division of Surveillance, Hazard Evaluations, and Field Studies

The **Division of Surveillance, Hazard Evaluations, and Field Studies (DSHEFS)** conducts surveillance of the nation's workforce and workplaces to assess the magnitude and extent of job-related illnesses, exposures, and hazardous agents.[33] DSHEFS conducts legislatively mandated health hazard evaluations and industrywide epidemiological research programs, including longitudinal studies of records and clinical and environmental field studies and surveys. DSHEFS also provides technical assistance on occupational safety and health problems to other federal, state, and local agencies; other technical groups; unions; employers; and employees.

Surveillance efforts are designed for the early detection and continuous assessment of the magnitude and extent of occupational illnesses, disabilities, deaths, and exposures to hazardous agents, using new and existing data sources from federal, state, and local agencies; labor; industry; tumor registries; physicians; and medical centers. DSHEFS also conducts evaluation and validation studies of reporting systems covering occupational illnesses, with the intent of improving methods for measuring the magnitude of occupational health problems nationwide.

Division of Training and Manpower Development

The **Division of Training and Manpower Development (DTMD)** implements Section 21 of the OSH Act, which sets forth training and education requirements.[34] DTMD develops programs to increase the numbers and competence of safety and health professionals. This continuing education program provides short-term technical training courses including seminars, independent study packages, and specialized workshops to federal, state, and local government; private industry; labor unions; and other organizations in the field of safety and health. The curriculum development component develops courseware and other training materials for NIOSH-sponsored training programs, including those represented by in-house faculty as well as those conducted by universities and other outside training organizations.

The educational resource development program continually assesses human resource needs for safety and health practitioners and researchers on a nationwide basis. To help meet the demand, DTMD administers a major training grant program to foster the development of academically based training programs for occupational physicians, occupational health nurses, industrial hygienists, toxicologists, epidemiologists, and safety professionals. DTMD also develops specific criteria for the selection of qualified organizations to conduct research training, graduate education, continuing education, and outreach programs to expand the network of knowledgeable professionals in occupational safety and health.

DTMD initiates special-emphasis projects targeted to physicians, engineers, managers, vocational education students, and science teachers to facilitate the inclusion of occupational safety and health knowledge in their formal program of study. The division establishes a collaborating relationship with the many professional societies and accrediting bodies to formalize the process of long-term commitment through professional networking.

OSHRC

The **Occupational Safety and Health Review Commission (OSHRC)** is not a government agency. Rather, it is an independent board whose members are appointed by the president and given quasi-judicial authority to handle contested OSHA citations. When a citation, proposed penalty, or abatement period issued by an OSHA area director is contested by an employer, OSHRC hears the case. OSHRC is empowered to review the evidence, approve the recommendations of the OSHA area director, reject those recommendations, or revise them by assigning substitute values. For example, if an employer contests the amount of a proposed penalty, OSHRC is empowered to accept the proposed amount, reject it completely, or change it.

Mine Safety and Health Administration (MSHA)

The mining industry is exempt from OSHA regulations. Instead, mining is regulated by the Metal and Non-Metallic Mine Safety Act. OSHA does regulate those aspects of the industry that are not directly involved in actual mining work. There is a formal memorandum of understanding between OSHA and the Mine Safety and Health Administration (MSHA), the agency that enforces the Metal and Non-Metallic Mine Safety Act.

In 1977, Congress passed the Mine Safety and Health Act, which established the Mine Safety and Health Administration (MSHA) as a functional unit within the U.S. Department of Labor. MSHA works with MESA to ensure that the two agencies do not become embroiled in jurisdictional disputes.

Federal Railroad Administration

Railroads, for the most part, fall under the jurisdiction of OSHA. The Federal Railroad Administration (FRA) exercises limited jurisdiction over railroads in situations involving working conditions. Beyond this, railroads must adhere to the standards for *General Industry* in CFR Part 1910. OSHA and FRA personnel coordinate to ensure that jurisdictional disputes do not arise.

OSHA's General Industry Standards

The most widely applicable OSHA standards are the *General Industry Standards.* These standards are found in 29 CFR 1910. Part 1910 consists of 21 subparts, each carrying an uppercase-letter designation. Subparts A and B contain no compliance requirements. The remaining subparts are described in the following subsections.

Subpart C: General Safety and Health Provisions

The only compliance standard in Subpart C is *Access to Employee Exposure and Medical Records.* Employers that are required to keep medical and exposure records must do the following: (1) maintain the records for the duration of employment plus 30 years, and (2) give employees access to their individual personal records.

Subpart D: Walking–Working Surfaces

Subpart D contains the standards for all surfaces on which employees walk or work. Specific sections of Subpart D are as follows:

1910.21	Definitions
1910.22	General requirements
1910.23	Guarding floor and wall openings and holes
1910.24	Fixed industrial stairs
1910.25	Portable wood ladders
1910.26	Portable metal ladders
1910.27	Fixed ladders
1910.28	Safety requirements for scaffolding
1910.29	Manually propelled mobile ladder stands and scaffolds (towers)
1910.30	Other working surfaces

Subpart E: Means of Egress

Subpart E requires employers to ensure that employees have a safe, accessible, and efficient means of escaping a building under emergency circumstances. Specific sections of Subpart E are as follows:

1910.35	Definitions
1910.36	General requirements
1910.37	Maintenance safeguards and operational features for exit routes
1910.38	Emergency action plan
1910.39	Fire protection plan

Emergency circumstances may be caused by fire, explosions, hurricanes, tornadoes, flooding, terrorist acts, earthquakes, nuclear radiation, or other acts of nature not listed here.

Subpart F: Powered Platforms

Subpart F applies to powered platforms, mechanical lifts, and vehicle-mounted work platforms. The requirements of this subpart apply only to employers who use this type of equipment in facility maintenance operations. Specific sections of Subpart F are as follows:

1910.66	Powered platforms for building maintenance
1910.67	Vehicle-mounted elevating and rotating work platforms
1910.68	Manlifts

Subpart G: Health and Environmental Controls

The most widely applicable standard in Subpart G is 1910.95 (occupational noise exposure). Other standards in this subpart pertain to situations where ionizing and/or nonionizing radiation are present. Specific sections of Subpart G are as follows:

1910.94	Ventilation
1910.95	Occupational noise exposure
1910.96	Ionizing radiation
1910.97	Nonionizing radiation
1910.98	Effective dates

Subpart H: Hazardous Materials

Four of the standards in Subpart H are widely applicable. Section 1901.106 is an extensive standard covering the use, handling, and storage of flammable and combustible liquids. Of particular concern are fire and explosions. Section 1910.107 applies to indoor spray-painting processes and processes in which paint (powder coating) is applied in powder form (e.g., electrostatic powder spray).

Section 1910.119 applies to the management of processes involving specifically named chemicals and flammable liquids and gases. Section 1910.120 contains requirements relating to emergency response operations and hazardous waste. All the standards contained in Subpart H are as follows:

1910.101	Compressed gases (general requirements)
1910.102	Acetylene

1910.103	Hydrogen
1910.104	Oxygen
1910.105	Nitrous oxide
1910.106	Flammable and combustible liquids
1910.107	Spray finishing using flammable and combustible materials
1910.108	Dip tanks containing flammable and combustible materials
1910.109	Explosive and blasting agents
1910.110	Storage and handling of liquefied petroleum gases
1910.111	Storage and handling of anhydrous ammonia
1910.119	Process safety management of highly hazardous chemicals
1910.120	Hazardous waste operations and emergency response

Subpart I: Personal Protective Equipment

Subpart I contains three of the most widely applicable standards: 1910.132 General Requirements; 1910.133 Eye and Face Protection; and 1910.134 Respiratory Protection. The most frequently cited OSHA violations relate to these and the other personal protective equipment standards. All the standards in this subpart are as follows:

1910.132	General requirements
1910.133	Eye and face protection
1910.134	Respiratory protection
1910.135	Occupational head protection
1910.136	Occupational foot protection
1910.137	Electrical protective devices
1910.138	Hand protection

Subpart J: General Environment Controls

This subpart contains standards that are widely applicable because they pertain to general housekeeping requirements. An especially important standard contained in this subpart is 1910.146: Permit-Required Confined Spaces. A confined space is one that meets any or all of the following criteria:

- Large enough and so configured that a person can enter it and perform assigned work tasks therein.
- Continuous employee occupancy is not intended.

The *lockout/tagout* standard is also contained in this subpart. All the standards in this subpart are as follows:

1910.141	Sanitation
1910.142	Temporary labor camps
1910.144	Safety color code for marking physical hazards
1910.145	Accident prevention signs and tags
1910.146	Permit-required confined space
1910.147	Control of hazardous energy (lockout/tagout)
1910.148	Standards organizations
1910.149	Effective dates

Subpart K: Medical and First Aid

This is a short subpart, the most important section of which pertains to eyeflushing. If employees are exposed to *injurious corrosive materials,* equipment must be provided for quickly flushing the eyes and showering the body. The standard also requires medical personnel to be readily available. "Readily available" can mean that there is a clinic or hospital nearby. If such a facility is not located nearby, employers must have a person on hand who has had first-aid training. The standard in this subpart is

1910.151	Medical seminars and first aid

Subpart L: Fire Protection

This subpart contains the bulk of OSHA's fire protection standard. These standards detail the employer's responsibilities concerning fire brigades, portable fire-suppression equipment, fixed fire-suppression equipment, and fire-alarm systems. Employers are not required to form fire brigades, but if they choose to, employers must adhere to the standard set forth in 1910.156. The standards in this subpart are as follows:

FIRE PROTECTION

1910.155	Scope, application, and definitions applicable to this subpart
1910.156	Fire brigades

PORTABLE FIRE-SUPPRESSION EQUIPMENT

1910.157	Portable fire extinguishers
1910.158	Standpipe and hose systems

FIXED FIRE-SUPPRESSION EQUIPMENT

1910.159	Automatic sprinkler systems
1910.160	Fixed extinguishing systems, general
1910.161	Fixed extinguishing systems, dry chemical
1910.162	Fixed extinguishing systems, gaseous agent
1910.163	Fixed extinguishing systems, water spray and foam

OTHER FIRE PROTECTION SYSTEMS

1910.164	Fire detection systems
1910.165	Employee alarm systems

Subpart M: Compressed Gas/Air

This subpart contains just three sections and only one standard, 1910.169. This standard applies to compressed-air equipment that is used in drilling, cleaning, chipping, and hoisting. There are many other uses of compressed air in work settings, but 1910.169 applies only to these applications. The standard in this subpart is

1910.169	Air receivers

Subpart N: Materials Handling and Storage

This subpart contains one broad standard—1910.176—and several more specific standards relating to 1910.176. Subpart N is actually limited in scope. It applies only to the handling and storage of materials, changing rim wheels on large vehicles, and the proper use of specific equipment identified in the standards' titles. All the standards in this subpart are as follows:

1910.176	Handling materials—general
1910.177	Servicing multi-piece and single-piece rim wheels
1910.178	Powered industrial trucks
1910.179	Overhead and gantry cranes
1910.180	Crawler locomotive and truck cranes
1910.181	Derricks
1910.183	Helicopters
1910.184	Slings

Subpart O: Machinery and Machine Guarding

This subpart contains standards relating to specific types of machines. The types of machines covered are identified in the titles of the standards contained in Subpart O. These standards are as follows:

1910.211	Definitions
1910.212	General requirements for all machines
1910.213	Woodworking machinery requirements
1910.214	Cooperage machinery
1910.215	Abrasive wheel machinery
1910.216	Mills and calendars in the rubber and plastics industries
1910.217	Mechanical power presses
1910.218	Forging machines
1910.219	Mechanical power-transmission apparatus

Subpart P: Hand Tools/Portable Power Tools

This subpart contains standards relating to the use of hand tools, portable power tools, and compressed-air-powered tools. The types of tools covered in this subpart, in addition to typical hand tools, include jacks, saws, drills, sanders, grinders, planers, power lawnmowers, and other tools. The standards contained in this subpart are as follows:

1910.241	Definitions
1910.242	Hand- and portable-powered tools and equipment, general
1910.243	Guarding of portable tools and equipment
1910.244	Other portable tools and equipment

Subpart Q: Welding, Cutting, and Brazing

Welding, cutting, and brazing are widely used processes. This subpart contains the standards relating to these processes in all their various forms. The primary safety and health concerns are fire protection, employee personal protection, and ventilation. The standards contained in this subpart are as follows:

1910.251	Definitions
1910.252	General requirements
1910.253	Oxygen-fuel gas welding and cutting

1910.254	Arc welding and cutting
1910.255	Resistance welding
1910.256	Sources of standards
1910.257	Standards organizations

Subpart R: Special Industries

This subpart is different from others in Part 1910. Whereas other subparts deal with specific processes, machines, and materials, Subpart R deals with specific industries. Each separate standard relates to a different category of industry. The standards contained in this subpart are as follows:

1910.261	Pulp, paper, and paperboard mills
1910.262	Textiles
1910.263	Bakery equipment
1910.264	Laundry machinery and operations
1910.265	Sawmills
1910.266	Pulpwood logging
1910.268	Telecommunications
1910.272	Grain handling facilities

Subpart S: Electrical

This subpart contains standards divided into the following two categories: (1) design of electrical systems, and (2) safety-related work practices. These standards are excerpted directly from the National Electrical Code. Those included in Subpart S are as follows:

1910.301	Introduction
1910.302	Electric utilization systems
1910.303	General requirements
1910.304	Wiring design and protection
1910.305	Wiring methods, components, and equipment for general use
1910.306	Specific-purpose equipment and installations
1910.307	Hazardous (classified) locations
1910.308	Special systems
1910.331	Scope
1910.332	Training
1910.333	Selection and use of work practices
1910.334	Use of equipment
1910.335	Safeguards for personnel protection
1910.399	Definitions applicable to this subpart

Subpart T: Commercial Diving Operations

This subpart applies only to commercial diving enterprises. The standards contained in Subpart T are divided into six categories: (1) general, (2) personnel requirements, (3) general operations and procedures, (4) specific operations and procedures, (5) equipment procedures and requirements, and (6) record keeping. These standards are as follows:

1910.401	Scope and application
1910.402	Definitions

1910.410	Qualifications of dive teams
1910.420	Safe practices manual
1910.421	Pre-dive procedures
1910.422	Procedures during dive
1910.423	Post-dive procedures
1910.424	SCUBA diving
1910.425	Surface-supplied-air diving
1910.426	Mixed-gas diving
1910.427	Liveboating
1910.430	Equipment
1910.440	Record keeping

Subpart Z: Toxic and Hazardous Substances

This is an extensive subpart containing the standards that establish *permissible exposure limits* (PELs) for selected toxic and hazardous substances (those for which PELs have been established). All such substances have an assigned PEL, which is the amount of a given airborne substance to which employees can be exposed during a specified period of time.

The standards relating to these specific toxic and hazardous substances are contained in 1910.1000 through 1910.1450 and are as follows:

1910.1000	Air contaminants
1910.1001	Asbestos
1910.1002	Coal tar pitch volatiles; interpretation of term
1910.1003	4-Nitrobiphenyl
1910.1004	Alpha-Naphthylamine
1910.1006	Methyl chloromethyl ether
1910.1007	Dichlorobenzidine (and its salts)
1910.1008	Bis-Chloromethyl ether
1910.1009	Beta-Naphthylamine
1910.1010	Benzedrine
1910.1011	4-Aminodiphenyl
1910.1012	Ethyleneimine
1910.1013	Beta-Propiolactone
1910.1014	2-Acetylaminofluorene
1910.1015	4-Dimethylaminoazobenzene
1910.1016	N-Nitrosodimethylamine
1910.1017	Vinyl chloride
1910.1018	Inorganic arsenic
1910.1020	Access to employee exposure and medical records
1910.1025	Lead
1910.1026	Chromium
1910.1027	Cadmium
1910.1028	Benzene
1910.1029	Coke oven emissions
1910.1030	Bloodborne pathogens
1910.1043	Cotton dust

1910.1044	1,2-dibromo-3-chloropropane
1910.1045	Acrylonitrile
1910.1047	Ethylene oxide
1910.1050	Methylenedianiline
1910.1051	Butadiene
1910.1052	Methylene chloride
1910.1096	Ionizing radiation
1910.1200	Hazard communication
1910.1201	Retention of DOT markings, placards, and labels
1910.1450	Occupational exposure to hazardous chemicals in laboratories

OSHA's General Industry Standards were covered in some depth in this section because they have the broadest application for students of workplace safety. The CFR also contains standards for the maritime industry. These standards are described in the sections that follow, but in less detail than was devoted to OSHA's General Standards because they are not as widely applicable.

OSHA's Maritime Standards

OSHA's Maritime Standards apply to shipbuilding, ship-repairing, and ship-breaking operations not already covered by U.S. Coast Guard regulations. It is important to note that Coast Guard regulations do take precedence over OSHA Maritime Standards and supersede those standards in cases of overlap or conflict. Part 1915 of 29 CFR contains the standards relating to shipbuilding, ship repairing, and ship breaking. Part 1917 contains the standards for marine terminals, and Part 1918 contains longshoring standards. Part 1919 contains the gear-certification standards.

Part 1915: Shipyard Employment

Part 1915 is divided into seven subparts covering the various aspects of shipyard operations. Those subparts are as follows:

1915.1	Purpose and authority
1915.2	Scope and application
1915.3	Responsibility
1915.4	Definitions
1915.5	Reference specifications, standards, and codes
1915.6	Commercial diving operations
1915.7	Competent person

Part 1917: Marine Terminals

Part 1917 is divided into seven subparts covering the various operations associated with marine terminals. Those subparts are as follows:

Subpart A	Scope and definitions
Subpart B	Marine terminal operations
Subpart C	Cargo handling gear and equipment
Subpart D	Specialized terminals
Subpart E	Personal protection

Subpart F	Terminal facilities
Subpart G	Related terminal operations and equipment

Part 1918: Longshoring

Longshoring involves loading and unloading maritime vessels. Part 1918 contains the standards relating to longshoring and is subdivided into the following subparts:

Subpart A	General provisions
Subpart B	Gangways and gear certification
Subpart C	Means of access
Subpart D	Working surfaces
Subpart E	Opening and closing hatches
Subpart F	Ship's cargo-handling gear
Subpart G	Cargo-handling gear and equipment other than ship's gear
Subpart H	Handling cargo
Subpart I	General working conditions
Subpart J	Personal protective equipment

Gear Certification

Cargo gear- and material-handling devices must be certified as being safe for use in the handling, loading, unloading, and transport of materials. Persons who certify the gear and devices must be properly accredited. Part 1919 contains the following standards regulating the accreditation process:

Subpart C	General safety and health provisions
Subpart D	Occupational health and environment controls
Subpart E	Personal protective and life-saving equipment
Subpart F	Fire protection and prevention
Subpart G	Signs, signals, and barricades
Subpart H	Materials handling, storage, use, and disposal
Subpart I	Tools (hand and power)
Subpart J	Welding and cutting
Subpart K	Electrical
Subpart L	Scaffolding
Subpart M	Floor and wall openings
Subpart N	Cranes, derricks, hoists, elevators, and conveyors
Subpart O	Motor vehicles, mechanized equipment, and marine vessels
Subpart P	Excavations
Subpart Q	Concrete and masonry construction
Subpart S	Underground construction, caissons, and cofferdams
Subpart T	Demolition
Subpart U	Blasting and the use of explosives
Subpart V	Power transmission and distribution
Subpart W	Rollover protective structures; overhead protection
Subpart X	Stairways and ladders
Subpart Y	Commercial diving operations
Subpart Z	Toxic and hazardous substances

OSHA's Construction Standards

These standards apply to employers involved in construction, alteration, or repair activities. To further identify the scope of the applicability of its construction standards, OSHA took the terms *construction, alteration,* and *repair* directly from the Davis-Bacon Act. This act provides minimum wage protection for employees working on construction projects. The implication is that if the Davis-Bacon Act applies to an employer, OSHA's Construction Standards also apply.[35]

These standards are contained in Part 1926 of the CFR Subparts A–Z. OSHA does not base citations on material contained in Subparts A and B. Consequently, those subparts have no relevance here. The remaining subparts are as follows:

Subpart C	General safety and health provisions
Subpart D	Occupational health and environmental controls
Subpart E	Personal protective and life-saving equipment
Subpart F	Fire protection and prevention
Subpart G	Signs, signals, and barricades
Subpart H	Materials handling, storage, use, and disposal
Subpart I	Tools—hand and power
Subpart J	Welding and cutting
Subpart K	Electrical
Subpart L	Scaffolding
Subpart M	Fall protection
Subpart N	Cranes, derricks, hoists, elevators, and conveyors
Subpart O	Motor vehicles, mechanized equipment, and marine operations
Subpart P	Excavations
Subpart Q	Concrete and masonry construction
Subpart R	Steel erection
Subpart S	Underground construction, caissons, and cofferdams
Subpart T	Demolition
Subpart U	Blasting and the use of explosives
Subpart V	Power transmission and distribution
Subpart W	Rollover protective structures; overhead protection
Subpart X	Ladders
Subpart Y	Commercial diving operations
Subpart Z	Toxic and hazardous substances

Standards and Codes

A **standard** is an operational principle, criterion, or requirement—or a combination of these. A **code** is a set of standards, rules, or regulations relating to a specific area. Standards and codes play an important role in modern safety and health management and engineering. These written procedures detail the safe and healthy way to perform job tasks and, consequently, to make the workplace safer and healthier.

Having written standards and codes that employees carefully follow can also decrease a company's exposure to costly litigation. Courts tend to hand down harsher rulings to companies that fail to develop or adapt, implement, and enforce appropriate standards and codes. Consequently, safety and health professionals should be familiar with the standards and codes relating to their company.

Numerous organizations develop standards for different industries. These organizations can be categorized broadly as follows: the government, professional organizations, and technical/trade associations.

Organizations that fall within these broad categories develop standards and codes in a wide variety of areas including, but not limited to, the following: dust hazards, electricity, emergency electricity systems, fire protection, first aid, hazardous chemicals, instrumentation, insulation, lighting, lubrication, materials, noise/vibration, paint, power, wiring, pressure relief, product storage and handling, piping materials, piping systems, radiation exposure, safety equipment, shutdown systems, and ventilation.

Figure 5-7 is a list of organizations that develop and publish standards and codes covering a wide variety of fields and areas of concern. Figure 5-8 is a reference for applicable standards and codes providers in various areas of concern (see Figure 5-7 to decode the abbreviations).

FIGURE 5-7 Organizations that develop standards and codes.

Organization	Abbreviation
U.S. Government	
Department of Transportation	DOT
Environmental Protection Agency	EPA
Federal Aviation Administration	FAA
Hazardous Materials Regulation Board	HMRB
Mine Safety and Health Administration	MSHA
National Institute for Standards and Technology	NIST
Occupational Safety and Health Administration	OSHA
U.S. Coast Guard	USCG
Professional Associations	
American Conference of Governmental Industrial Hygienists	ACGIH
American Industrial Hygiene Association	AIHA
American Institute of Chemical Engineers	AIChE
American Society of Heating, Refrigeration, and Air Conditioning Engineers	ASHRAE
American Society of Mechanical Engineers	ASME
Illumination Engineers Society	IES
Institute of Electrical and Electronic Engineers	IEEE
Instrument Society of America	ISA
Technical and Trade Associations	
Air Conditioning and Refrigeration Institute	ARI
Air Moving and Conditioning Association	AMCA
American Association of Railroads	AAR
American Gas Association	AGA

FIGURE 5-7 *(continued)*

Organization	Abbreviation
Technical and Trade Associations—cont'd	
American Petroleum Institute	API
American Water Works Association	AWWA
Chlorine Institute	CI
Compressed Gas Association	CGA
Cooling Tower Institute	CTI
Manufacturing Chemists Association	MCA
Manufacturers Standardization Society	MSS
National Electrical Manufacturers Association	NEMA
National Fluid Power Association	NFPA
Pipe Fabrication Institute	PFI
Scientific Apparatus Makers Association	SAMA
Society and Plastics Industry	SPI
Steel Structures Painting Council	SSPC
Tubular Exchanger Manufacturers Association	TEMA
Testing Organizations	
American National Standards Institute	ANSI
American Society for Testing and Materials	ASTM
National Fire Protection Association	NFPA
National Safety Council	NSC
Underwriters Laboratories, Inc.	UL
Insurance Organizations	
American Insurance Association	ATA
Factory Insurance Association	FIA
Factory Mutual Systems	FMS
Industrial Risk Insurers	IRI
Oil Insurance Association	OIA

Laws and Liability

The body of law pertaining to workplace safety and health grows continually as a result of a steady stream of liability litigation. Often a company's safety and health professionals are key players in litigation alleging negligence on the part of the company when an accident or health problem has occurred. Because safety and health litigation has become so prevalent today, professionals in the field need to be familiar with certain fundamental legal principles relating to such litigation. These principles are explained in the following section.

Fundamental Legal Principles

The body of law that governs safety and health litigation evolves continually. However, even though cases that set new precedents and establish new principles continue to occur, a number of fundamental legal principles surface frequently in the court. The most important of these—as well as several related legal terms—are summarized in the following paragraphs.

FIGURE 5-8 Reference table of standards providers.

Area of Concern	Organizations That Develop and Publish Standards
Dust hazards	ANSI, NFPA, UL, NSC, ACGIH, AIHA, BM, USCG
Electricity	ANSI, NFPA, NSC, ADA, FIA, FM, OIA, API, USCG, OSHA, IEEE, NEMA
Emergency electricity	NFPA, AIA, FM, IEEE, AGA, NEMA, USCG
Fire protection	ANSI, NFPA, UL, NSC, AIA, FIA, OIA, AWWA, API, CGA, MCA, NEMA, BM, USCG, OSHA
First aid	ANSI, NFPA, NSC, AIA, ACGIH, AIHA, CI, MCA, DOT, USCG
Flammability of substances	ANSI, NFPA, UL, NSC, FIA, FM, MCA, SPI, DOT, USCG, NIST
Hazardous chemicals	ANSI, NFPA, UL, NSC, AIA, FIA, FM, ACGIH, AIHA, AIChE, CI, CGA, MCA, DOT, USCG, OSHA
Instrumentation	ANSI, ASTM, NFPA, UL, AIA, FIA, FM, OIA, IEEE, ISA, AWWA, ARI, API, CGA, SAMA, NIST
Insulation	ANSI, ASTM, UL, AIA, FM, OIA, ASHRAE, USCG
Lighting	ASTM, NFPA, UL, AIA, FM, OIA, ASHRAE, USCG
Lubrication	ANSI, NFPA, ASME, AMCA
Materials	ANSI, ASTM, NFPA, UL, NSC, AIA, FM, OIA, ISA, AWWA, CI, CGA, CTI, MCA, TEMA, USCG, HMR
Noise—vibration	ANSI, ASTM, NFPA, UL, NSC, ACGIH, AIHA, ASHRAE, ISA, ARI, AMCA, AGA, NFPA, EPA
Paint/coatings	ANSI, ASTM, UL, AIChE, AWWA, SSPC, HMRB
Power writing	ANSI, NFPA, UL, FIA, FM, OIA, IEEE, API, NEMA, USCG, OSHA
Pressure relief	NFPA, FIA, FM, OIA, API, CI, CGA, USCG, HMRB, OSHA
Product storage/handling	ANSI, NFPA, AIA, FIA, FM, OIA, AIChE, AAR, API, CI, CGA, MCA, USCG, OSHA
Piping materials	ANSI, ASTM, NFPA, UL, AIA, FIA, FM, ASME, ASHRAE, AWWA, ARI, AGA, API, CI, CGA, MSS
Piping systems	ANSI, ASTM, NFPA, UL, AIA, FIA, FM, ASME, ASHRAE, AWWA, ARI, AGA, API, CI, CGA, MSS
Radiation exposure	ANSI, NFPA, NSC, AIA, ACGIH, AIHA, ASME, ISA, DOT
Safety equipment	ANSI, UL, NSC, FM, ACGIH, AIHA, CI, CGA, MCA, BM, USCG, OSHA
Shutdown systems	NFPA, UL, AIA, FIA, OIA, API, USCG
Ventilation	ANSI, NFPA, NSC, AIA, FIA, FM, ACGIH, AIHA, ASHRAE, ISA, CI, DOT, HMRB

Note: Refer to Figure 6-7 for an explanation of abbreviations.

Negligence

Negligence means failure to take reasonable care or failure to perform duties in ways that prevent harm to humans or damage to property. The concept of *gross negligence* means failure to exercise even slight care or intentional failure to perform duties properly, regardless of the potential consequences. *Contributory negligence* means that an injured party contributed in some way to his or her own injury. In the past, this concept was used to protect defendants against negligence charges because the courts awarded no damages to plaintiffs who had contributed in any way to their own injury. Modern court cases have rendered this approach outdated with the introduction of *comparative negligence*. This concept distributes the negligence assigned to each party involved in litigation according to the findings of the court.

Liability

Liability is a duty to compensate as a result of being held responsible for an act or omission. A newer, related concept is *strict liability*. This means that a company is liable for damages caused by a product that it produces, regardless of negligence or fault.

Care

Several related concepts fall under the heading of care. *Reasonable care* is the amount that would be taken by a prudent person in exercising his or her legal obligations toward others. *Great care* means the amount of care that would be taken by an extraordinarily prudent person in exercising his or her legal obligations toward others. *Slight care* represents the other extreme: a measure of care less than what a prudent person would take. A final concept in this category is the *exercise of due care*. This means that all people have a legal obligation to exercise the amount of care necessary to avoid, to the extent possible, bringing harm to others or damage to their property.

Ability to Pay

The concept of **ability to pay** applies when there are a number of defendants in a case, but not all have the ability to pay financial damages. It allows the court to assess all damages against the defendant or defendants who have the ability to pay. For this reason, it is sometimes referred to as the "deep pockets" principle.

Damages

Damages are financial awards assigned to injured parties in a lawsuit. *Compensatory damages* are awarded to compensate for injuries suffered and for those that will be suffered. *Punitive damages* are awarded to ensure that a guilty party will be disinclined to engage in negligent behavior in the future.

Proximate Cause

Proximate cause is the cause of an injury or damage to property. It is that action or lack of action that ties one person's injuries to another's lack of reasonable care.

Willful/Reckless Conduct

Behavior that is even worse than gross negligence is **willful/reckless conduct**. It involves intentionally neglecting one's responsibility to exercise reasonable care.

Tort

A **tort** is an action involving a failure to exercise reasonable care that as a result may lead to civil litigation.

Foreseeability

The concept of **foreseeability** holds that a person can be held liable for actions that result in damages or injury only when risks could have been reasonably foreseen.

The types of questions around which safety and health litigation often revolve are these: Does the company keep employees informed of rules and regulations? Does the company enforce its rules and regulations? Does the company provide its employees with the necessary training? The concepts set forth in this section come into play as both sides in the litigation deal with these questions from their respective points of view.

Health and safety professionals can serve their companies best by (1) making sure that a policy and corresponding rules and regulations are in place; (2) keeping employees informed about rules and regulations; (3) encouraging proper enforcement practices; and (4) ensuring that employees get the education and training they need to perform their jobs safely.

Key Terms and Concepts

- Abatement period
- Ability to pay
- Appeals process
- Closing conference
- Code
- Consultation services
- Damages
- De minimis violations
- Demonstration Program
- Division of Biomedical and Behavioral Science (DBBS)
- Division of Respiratory Disease Studies (DRDS)
- Division of Surveillance, Hazard Evaluations, and Field Studies (DSHEFS)
- Division of Training and Manpower Development (DTMD)
- Employee responsibilities
- Employee rights
- Employer appeals
- Employer responsibilities
- Employer rights
- Foreseeability
- Inspection tour
- Liability
- Merit Program
- National Advisory Committee on Occupational Safety and Health (NACOSH)
- National Institute for Occupational Safety and Health (NIOSH)
- Negligence
- Notice of contest
- Notice of proposed rule making
- Occupational diseases
- Occupational Safety and Health Act (OSH Act)
- Occupational Safety and Health Administration (OSHA)
- Occupational Safety and Health Review Commission (OSHRC)
- Opening conference
- OSHA Form 300
- OSHA Form 300A
- OSHA Form 301
- OSHA Poster 2203
- Other-than-serious violation
- Permanent variance
- Petition for Modification of Abatement (PMA)
- Proposed penalty
- Proximate cause
- Record keeping
- Repeat violations
- Reporting
- Serious violation
- Standard
- Star Program
- State-level OSHA program
- Temporary emergency standards
- Temporary variance
- Tort
- Voluntary Protection Programs (VPPs)
- Willful/reckless conduct
- Willful violation
- Workplace accidents
- Workplace inspection

Review Questions

1. Briefly explain the rationale for the OSH Act.
2. What is OSHA's mission or purpose?
3. List those who are exempted from coverage by OSHA.
4. Explain the difference between an OSHA standard and an OSHA regulation.
5. Explain how the following processes relating to OSHA standards are accomplished: passage of a new standard; request for a temporary variance; appealing a standard.
6. Briefly describe OSHA's latest record-keeping requirements.
7. What are OSHA's reporting requirements?
8. Explain what employers are required to do to keep employees informed.
9. Explain the various components of OSHA's enhanced enforcement policy.
10. Describe how a hypothetical OSHA workplace inspection would proceed from the first step to the last.
11. List and explain three different types of OSHA citations and the typical penalties that accompany them.
12. Describe the process for appealing an OSHA citation.
13. List and briefly explain OSHA's voluntary protection programs.
14. List five employer responsibilities.
15. List five employee rights.
16. Describe the purpose and organization of NIOSH.
17. Define the following legal terms as they relate to workplace safety: *negligence; liability; ability to pay; tort.*

Endnotes

1. U.S. Department of Labor, Occupational Safety and Health Administration, All About OSHA, OSHA 2056, rev. ed. (Washington, DC: U.S. Department of Labor, Occupational Safety and Health Administration, 2002), 1.
2. Ibid.
3. Ibid., 2.
4. Ibid., 5.
5. Ibid., 7.
6. Ibid., 9.
7. Retrieved from http://www.osha.gov/standards. 2003.
8. Ibid.
9. Ibid.
10. U.S. Department of Labor, All About OSHA, 12.
11. Ibid.
12. Ibid., 12–13.
13. Ibid., 17.
14. Ibid., 24–25.
15. Facility Manager's ALERT, October 1, 2002, 3.
16. Ibid.
17. Facility Manager's ALERT, March 17, 2003, 3.
18. Ibid.
19. U.S. Department of Labor, All About OSHA, 27.
20. Ibid., 28
21. U.S. Department of Labor, All About OSHA, 28.
22. Ibid.
23. Ibid., 29.
24. Ibid., 32.
25. Ibid., 34–35.
26. Ibid., 35–36.
27. Ibid., 37.
28. Ibid., 39–40.
29. Ibid., 37.
30. U.S. Department of Health and Human Services, Occupational Safety & Health for Fiscal Year 1988 Under Public Law 91–596 (Washington, DC: Public Health Service, Centers for Disease Control and Prevention, National Institute for Occupational Safety and Health).
31. U.S. Department of Health and Human Services, handout (Washington, DC: Public Health Service, Centers for Disease Control and Prevention, National Institute for Occupational Safety and Health, 1998), 1.
32. Ibid., 2
33. Ibid., 3
34. Ibid., 4
35. For more in-depth coverage of OSHA's construction standards, refer to D. L. Goetsch, Construction Safety and Health (Upper Saddle River, NJ: Prentice Hall, 2003

CHAPTER 6

Workers' Compensation

Major Topics

- Overview of Workers' Compensation
- Historical Perspective
- Workers' Compensation Legislation
- Modern Workers' Compensation
- Workers' Compensation Insurance
- Resolution of Workers' Compensation Disputes
- Injuries and Workers' Compensation
- Disabilities and Workers' Compensation
- Monetary Benefits of Workers' Compensation
- Medical Treatment and Rehabilitation
- Medical Management of Workplace Injuries
- Administration and Case Management
- Cost Allocation
- Problems with Workers' Compensation
- Spotting Workers' Compensation Fraud and Abuse
- Future of Workers' Compensation
- Cost-Reduction Strategies

Overview of Workers' Compensation

The concept of **workers' compensation** developed as a way to allow injured employees to be compensated appropriately without having to take their employer to court. The underlying rationale for workers' compensation had two aspects: (1) fairness to injured employees, especially those without the resources to undertake legal actions that are often long, drawn out, and expensive; and (2) reduction of costs to employers associated with workplace injuries (e.g., legal, image, and morale costs). Workers' compensation is intended to be a no-fault approach to resolving workplace accidents by rehabilitating injured employees and minimizing their personal losses because of their reduced ability to perform and compete in the labor market.[1] Since its inception as a concept, workers' compensation has evolved into a system that pays out approximately $70 million in benefits and medical costs annually. The national average net cost of workers' compensation in the manufacturing sector is almost $6 per $100 of payroll.

Workers' compensation represents a compromise between the needs of employees and the needs of employers. Employees give up their right to seek unlimited compensation for pain and suffering through legal action. Employers award the prescribed compensation (typically through insurance premiums) regardless of the employee's negligence. The theory is that in the long run both employees and employers will benefit more than either would through legal action. As you will see later in this chapter, although workers' compensation has reduced the amount of legal action arising out of workplace accidents, it has not completely eliminated legal actions.

Objectives of Workers' Compensation

Workers' compensation laws are not uniform from state to state. In fact, there are extreme variations. However, regardless of the language contained in the enabling legislation in a specific state, workers' compensation as a concept has several widely accepted objectives:

- Replacement of income
- Rehabilitation of the injured employee
- Prevention of accidents
- Cost allocation[2]

The basic premises underlying these objectives are described in the following paragraphs.

Replacement of Income

Employees injured on the job lose income if they are unable to work. For this reason, workers' compensation is intended to replace the lost income adequately and promptly. Adequate **income replacement** is viewed as replacement of current and future income (minus taxes) at a ratio of two-thirds (in most states). Workers' compensation benefits are required to continue even if the employer goes out of business.

Rehabilitation of the Injured Employee

A basic premise of workers' compensation is that the injured worker will return to work in every case possible, although not necessarily in the same job or career field. For this reason, a major objective of workers' compensation is to rehabilitate the injured employee. The **rehabilitation** program is to provide the needed medical care at no cost to the injured employee until he or she is pronounced fit to return to work. The program also provides vocational training or retraining as needed. Both components seek to motivate the employee to return to the labor force as soon as possible.

Accident Prevention

Preventing future accidents is a major objective of workers' compensation. The theory underlying this objective is that employers will invest in **accident prevention** programs to hold down compensation costs. The payoff to employers comes in the form of lower insurance premiums that result from fewer accidents.

Cost Allocation

The potential risks associated with different occupations vary. For example, working as a miner is generally considered more hazardous than working as an architect. The underlying principle of **cost allocation** is to spread the cost of workers' compensation appropriately and proportionately among industries ranging from the most to the least hazardous. The costs of accidents should be allocated in accordance with the accident history of the industry so that high-risk industries pay higher workers' compensation insurance premiums than do low-risk industries.[3]

Who Is Covered by Workers' Compensation?

Workers' compensation laws are written at the state level, and there are many variations among these laws. As a result, it is difficult to make generalizations. Complicating the issue further is the fact that workers' compensation laws are constantly being amended, revised, and rewritten. Additionally, some states make participation in a workers' compensation program voluntary; others excuse employers with fewer than a specified number of employees.

In spite of the differences among workers' compensation laws in the various states, approximately 80 percent of the employees in the United States are covered by workers' compensation. Those employees who are not covered or whose coverage is limited vary as the laws vary. However, they can be categorized in general terms as follows:

- Agricultural employees
- Domestic employees
- Casual employees
- Hazardous work employees
- Charitable or religious employees
- Employees of small organizations
- Railroad and maritime employees
- Contractors and subcontractors
- Minors
- Extraterritorial employees[4]

According to Hammer, coverage in these types of employment, to the extent there is coverage, varies from state to state as follows:

- **Agricultural employees** have limited coverage in 38 states, Puerto Rico, and the Virgin Islands. In 15 states, workers' compensation coverage for agricultural employees is voluntary. In these states, employers are allowed to provide coverage if they wish but are not required to do so.
- **Domestic employees** have coverage available in all 50 states and Puerto Rico. However, coverage tends to be limited and subject to minimum requirements regarding hours worked and earnings.
- **Casual employees** are employed in positions in which the work is occasional, incidental, and scattered at irregular intervals. Such employees are not typically afforded workers' compensation coverage.
- **Hazardous employment** is the only type afforded workers' compensation coverage in some states. To qualify, a particular type of employment must be on an approved list of hazardous or especially hazardous jobs. However, the trend in these states is to broaden the list of approved jobs.

- **Charitable or religious employees** are not afforded workers' compensation in most states when this work is irregular, temporary, or short-term.
- **Small organizations** that employ fewer than a stipulated number of employees do not fall under the umbrella of workers' compensation in 26 states.
- **Railroad and maritime workers** are not typically covered by workers' compensation. However, in most cases, they are covered by the Federal Employer's Liability Act. This act disallows the use of common law defenses by employers if sued by an employee for negligence.
- **Contractors and subcontractors** are those who agree to perform a job or service for an agreed amount of money in a nondirected, nonsupervised format. In essence, contract and subcontract employees are viewed as being self-employed. For this reason, they are not covered by workers' compensation.
- **Minors** are afforded regular workers' compensation coverage as long as they are legally employed. In some states, coverage is significantly higher for minors who are working illegally.
- **Extraterritorial employees** are those who work in one state but live in another. In these cases, the employee is usually on temporary duty. Such employees are typically afforded the workers' compensation coverage of their home state.[5]

Historical Perspective

Before workers' compensation laws were enacted in the United States, injured employees had no way to obtain compensation for their injuries except to take their employer to court. Although common law did require employers to provide a safe and healthy work environment, injured employees bore the burden of proof that negligence in the form of unsafe conditions contributed to these injuries. According to the Society of Manufacturing Engineers, prior to passage of workers' compensation, employees often had to sue their employer to receive compensation for injuries resulting from a workplace accident or occupational disease, even when the following circumstances prevailed:

- The employee was disabled or died as the result of a **workplace accident** or **occupational disease**.
- The injury might have been expected to occur when the risks and hazards of the job were considered.
- **Worker negligence** on the part of a fellow worker or the injured employee clearly caused the injury.[6]

Proving that an injury was the result of employee negligence was typically too costly, too difficult, and too time consuming to be a realistic avenue of redress for most injured employees. According to Somers and Somers, a New York commission determined that it took from six months to six years for an injured worker's case to work its way through the legal system.[7] Typically, injured workers, having lost their ability to generate income, could barely afford to get by, much less pay medical expenses, legal fees, and court costs. Another inhibitor was the ***fear factor***. Injured employees who hoped to return to work after recovering were often afraid to file suit because they feared retribution by their employer. Employers not only might refuse to give them their jobs back but also might **blackball** them with other employers. Add to this that fellow employees were often afraid to testify to the negligence of the employer, and it is easy to see why few injured workers elected to take their employers to court.

Even with all of these inhibitors, some injured employees still chose to seek redress through the courts in the days before workers' compensation. Those who did faced a difficult challenge because the laws at that time made it easy for employers to defend themselves successfully. All an employer had to do to win a decision denying the injured plaintiff compensation was show that at least one of the following conditions existed at the time of the accident:

1. **Contributory negligence was a factor in the accident.** Contributory negligence meant that the injured worker's own negligence contributed to the accident. Even if the employee's negligence was a very minor factor, it was usually enough to deny compensation in the days before workers' compensation.
2. **There was negligence on the part of a fellow worker.** As with contributory negligence, negligence by a fellow employee, no matter how minor a contributing factor it was, could be sufficient to deny compensation. This defense was known as "the fellow servant rule."
3. **There was assumption of risk on the part of the injured employee.** If an employee knew that the job involved risk, he or she could not expect to be compensated when the risks resulted in accidents and injuries.[8] This defense relied on a long-standing principle of tort law known as "assumption of risk."

Because the majority of workplace accidents involve at least some degree of negligence on the part of the injured worker or fellow employees, employers typically won these cases. Because it required little more than a verbal warning by the employer to establish grounds for assumption of risk, the odds against an injured employee being awarded compensation become clear.

In his book *American Social Science,* Gagilardo gives an example of a case that illustrates how difficult it was to win compensation in the days before workers' compensation.[9] He relates the example of an employee who contracted tuberculosis while working under clearly hazardous conditions for a candy-making company. She worked in a wet, drafty basement that admitted no sunlight. Dead rats floated in the overflow of a septic tank that covered the basement floor, and a powerful stench permeated the workplace. Clearly, these were conditions that could contribute to the employee contracting tuberculosis. However, she lost the case and was denied compensation. The ruling judge justified the verdict as follows:

> We think that the plaintiff, as a matter of law, assumed the risk attendant upon her remaining in the employment (*Wager v. White Star Candy Company,* 217 N.Y. Supp. 173).

Situations such as this eventually led to the enactment of workers' compensation laws in the United States.

Workers' Compensation Legislation

Today, all 50 states, the District of Columbia, Guam, and Puerto Rico have workers' compensation laws. However, these laws did not exist prior to 1948. Considering that Prussia passed a workers' compensation law in 1838, the United States was obviously slow to adopt the concept. In fact, the first workers' compensation law enacted in the United States did not pass until 1908, and it applied only to federal employees working in especially hazardous jobs. The driving force behind passage of this law was President Theodore Roosevelt, who as governor of New York had seen the results of workplace accidents firsthand. Montana was the first state to pass a compulsory workers' compensation law. However, it was short-lived. Ruling that the law was unconstitutional, the Montana courts overturned it.

In 1911, the New York Court of Appeals dealt proponents of workers' compensation a serious blow. The New York state legislature had passed a compulsory workers' compensation law in 1910. However, in the case of *Ives v. South Buffalo Railway Company* (201 N.Y. 271, 1911), the New York Court of Appeals declared the law unconstitutional based on the contention that it violated the due process clause in the Fourteenth Amendment to the U.S. Constitution.[10]

This ruling had a far-reaching impact. According to Hammer, "The prestige of the New York court influenced legislators in many of the other states to believe that any compulsory law also would be held unconstitutional."[11] However, even with such precedent-setting cases as *Ives* on the books, pressure for adequate workers' compensation grew as unsafe working conditions continued to result in injuries, diseases, and deaths. In fact, shortly after the New York Court of Appeals released its due process ruling, tragedy struck in a New York City textile factory.

On March 25, 1911, the building that housed the Triangle Shirtwaist Factory on its eighth floor caught fire and burned.[12] As a result of the fire, 149 of the company's 600 workers died, and another 70 were injured. Although the cause of the accident could not be determined, it was clear to investigators and survivors alike that unsafe conditions created by the management of the company prevented those who died or were injured from escaping the fire.

Exit passageways on each floor of the building were unusually narrow (20 inches wide), which made it difficult for employees to carry out bolts of material. A wider exit on each floor was kept locked to force employees to use the narrow exit. The two elevators were slow and able to accommodate only small groups at a time.

As the fire quickly spread, employees jammed into the narrow passageways, crushing each other against the walls and underfoot. With all exits blocked, panic-stricken employees began to jump out of windows and down the elevator shafts. When the pandemonium subsided and the fire was finally brought under control, the harsh realization of why so many had been trapped by the deadly smoke and flames quickly set in.

The owners were brought into court on charges of manslaughter. Although they were not convicted, the tragedy did focus nationwide attention on the need for a safe workplace and adequate workers' compensation. As a result, new, stricter fire codes were adopted in New York, and in spite of the state court's ruling in *Ives,* the state legislature passed a workers' compensation law.

The next several years saw a flurry of legislation in other states relating to workers' compensation. In response to demands from workers and the general public, several states passed limited or noncompulsory workers' compensation laws. Many such states held back out of fear of being overturned by the courts. Others, particularly Washington, publicly disagreed with the New York Court of Appeals and passed compulsory laws. The constitutionality debate continued until 1917 when the U.S. Supreme Court ruled that workers' compensation laws were acceptable.

Modern Workers' Compensation

Since 1948, all states have had workers' compensation laws. However, the controversy surrounding workers' compensation has not died. As medical costs and insurance premiums have skyrocketed, many small businesses have found it difficult to pay the premiums. Unrealistic workers' compensation rates are being cited more and more frequently as contributing to the demise of small business in America.

The problem has even developed into an economic development issue. Business and industrial firms are closing their doors in those states with the highest workers' compensation rates and moving to states with lower rates. States with lower rates are using this as part of their recruiting package to attract new businesses and industry. Where low-rate states border high-rate states, businesses are beginning to move their offices across the border to the low-rate state while still doing business in the high-rate state.

Critics are now saying that workers' compensation has gotten out of hand and is no longer fulfilling its intended purpose. To understand whether this is the case, one must begin with an examination of the purpose of workers' compensation. According to Hammer, the U.S. Chamber of Commerce identified the following six basic objectives of workers' compensation:

1. To provide an appropriate level of income and medical benefits to injured workers or to provide income to the worker's dependents regardless of fault.
2. To provide a vehicle for reducing the amount of personal injury litigation in the court system.
3. To relieve public and private charities of the financial strain created by workplace injuries that go uncompensated.
4. To eliminate time-consuming and expensive trials and appeals.
5. To promote employer interest and involvement in maintaining a safe work environment through the application of an experience-rating system.
6. To prevent accidents by encouraging frank, objective, and open investigations of the causes of accidents.[13]

Early proponents of workers' compensation envisioned a system in which both injured workers and their employers would win. Injured workers would receive prompt compensation, adequate medical benefits, and appropriate rehabilitation to allow them to reenter the workforce and be productive again. Employers would avoid time-consuming, expensive trials and appeals and would improve relations with employees and the public in general.

What proponents of workers' compensation did not anticipate were the following factors: (1) employees who would see workers' compensation as a way to ensure themselves a lifelong income without the necessity of work; (2) enormous increases in the costs of medical care with corresponding increases in workers' compensation insurance premiums; and (3) the radical differences among workers' compensation laws passed by the various states.

Not all employees abide by the spirit of workers' compensation (i.e., rehabilitation in a reasonable amount of time). Attempted abuse of the system was perhaps inevitable. Unfortunately, such attempts result in a return to what workers' compensation was enacted to eliminate: time-consuming, drawn-out, expensive legal battles and the inevitable appeals.

Proponents of workers' compensation **reform** can cite a long list of cases that illustrate their point. The city of Pittsfield, Massachusetts, was once overwhelmed by workers' compensation claims. One of the more remarkable cases concerned a city worker who was receiving workers' compensation benefits as the result of a back injury. While collecting benefits, he was a star player for a local softball team.[14] He eventually agreed to waive his right to compensation for a lump sum settlement of $12,000 plus $3,000 for his lawyer, and city officials considered themselves cheaply rid of him.[15]

Another Pittsfield city employee in the Department of Public Works was injured and began collecting workers' compensation at a rate of $295.50 per week. In addition to his job with the city, this worker owned a small diesel oil company. When his workers' compensation benefits were called into question because he owned a business that produced an income, the injured employee sold the business to his son.[16]

One of the favorite examples of opponents of workers' compensation is that of the "fat deputy sheriff."

> The deputy sheriff was already despondent over the breakup of the extramarital affair he was having with a married colleague, when his supervisor made an expensive mistake. He rated the already overstressed sheriff's job performance as substandard and told him he was too fat. The sheriff promptly filed a claim for workers' compensation benefits to cover job-related pressures stemming from his performance evaluation that contributed to his mental injury.[17]

Medical costs have skyrocketed in the United States since the 1960s. There are many reasons for this. During this same period, the costs associated with other basic human needs including food, clothing, transportation, shelter, and education have also increased markedly. Increases in medical costs can be explained, at least partially, as the normal cost of living increases experienced in other sectors of the economy. However, the costs associated with medical care have increased much faster and much more than the costs in these other areas. The unprecedented increases can be attributed to two factors: (1) technological developments that have resulted in extraordinary but costly advances in medical care; and (2) a proliferation of **litigation** that has driven the cost of malpractice insurance steadily up. Each of these factors has contributed to higher medical costs. For example, X-ray machines that cost thousands of dollars have been replaced by magnetic resonance imaging (MRI) systems that may cost millions. **Malpractice** suits that once might not even have gone to court now result in multimillion-dollar settlements. Such costs are, of course, passed on to whoever pays the medical bill—in this case, employers who must carry workers' compensation insurance. California's workers' compensation system, for example, costs employers $7 billion annually in insurance premiums. Between 1987 and 1998, the cost of workers' compensation premiums increased by more than 60 percent. This trend continues.

In addition to contributing individually to increased medical costs, technology and litigation have interacted in such a way as to increase costs even further. This interaction occurs as follows. First, an expensive new technology is developed that enhances the predictive and/or prescriptive capabilities of the medical profession. Second, expensive malpractice suits force medical practitioners to be increasingly cautious and, accordingly, to order even more tests than a patient's symptoms may suggest. Finally, the tests involve expensive new technologies, adding even more to the cost of medical care.

Early supporters of the concept did not anticipate the radical differences among workers' compensation laws in the various states. The laws themselves differ, as do their interpretations. The differences are primarily in the areas of benefits, penalties, and workers covered. These differences translate into differences in the rates charged for workers' compensation insurance. As a result, the same injury incurred under the same circumstances but in different states can yield radically different benefits for the employee.

The potential for abuse, steadily increasing medical costs that lead to higher insurance premiums, and differences among workers' compensation laws all contribute to the controversy that still surrounds this issue. As business and industry continue to protest that workers' compensation has gotten out of hand, it will continue to be a heated issue in state legislatures. As states try to strike the proper balance between meeting the needs of the workforce while simultaneously maintaining a positive environment for doing business, workers' compensation will be an issue with which they will have to deal.

Workers' Compensation Insurance

The costs associated with workers' compensation must be borne by employers as part of their overhead. In addition, employers must also ensure that the costs will be paid even if they go out of business. The answer for most employers is workers' compensation insurance.

In most states, workers' compensation insurance is compulsory. Exceptions to this are New Jersey, South Carolina, Texas, and Wyoming. New Jersey allows ten or more employers to form a group and self-insure. Texas requires workers' compensation only for *carriers*, as defined in Title 25, Article 911–A, Section II, Texas state statutes. Wyoming requires workers' compensation only for employers involved in specifically identified *extrahazardous occupations*.

A common thread woven through all the various compensation laws is the requirement that employers carry workers' compensation insurance. There are three types: **state funds**, **private insurance**, and **self-insurance**. Figure 6-1 summarizes the methods of insurance coverage allowed in a representative sample of states. Regardless of the method of coverage chosen, rates can vary greatly from company to company and state to state. Rates are affected by a number of different factors including the following:

- Number of employees
- Types of work performed (risk involved)
- Accident experience of the employer
- Potential future losses
- Overhead and profits of the employer
- Quality of the employer's safety program
- Estimates by actuaries

Insurance companies use one of the following six methods in determining the premium rates of employers:

1. **Schedule rating.** Insurance companies establish baseline safety conditions and evaluate the employer's conditions against the baselines. Credits are awarded for conditions that are better than the baseline, and debits are assessed for conditions that are worse. Insurance rates are adjusted accordingly.
2. **Manual rating.** A manual of rates is developed that establishes rates for various occupations. Each occupation may have a different rate based on its perceived

FIGURE 6-1 Workers' compensation coverage methods allowed for selected states.

State	State Fund	Private Insurer	Individual Employer Self-Insurance	Group of Employer's Self-Insurance
Alabama	No	Yes	Yes	Yes
Arkansas	No	Yes	Yes	Yes
California	Competitive	Yes	Yes	No
Florida	No	Yes	Yes	Yes
Indiana	No	Yes	Yes	No
Kansas	No	Yes	Yes	Yes
Maryland	Competitive	Yes	Yes	Yes
Montana	Competitive	Yes	Yes	Yes
Nebraska	No	Yes	Yes	No
New Jersey	No	Yes	Yes	No*
New Mexico	Competitive	Yes	Yes	Yes
Vermont	No	Yes	Yes	No
Washington	Exclusive	No	Yes	Yes
Wisconsin	No	Yes	Yes	No

* Permits ten or more employees licensed by the state to group self-insure.

Source: U.S. Department of Labor, 2005.

level of hazard. The overall rate for the employer is a pro-rata combination of all the individual rates.

3. **Experience rating.** Employers are classified by type. Premium rates are assigned based on predictions of average losses for a given type of employer. Rates are then adjusted either up or down according to the employer's actual experience over the past three years.
4. **Retrospective rating.** Employees pay an established rate for a set period. At the end of the period, the actual experience is assessed, and an appropriate monetary adjustment is made.
5. **Premium discounting.** Large employers receive discounts on their premiums based on their size. The theory behind this method is that it takes the same amount of time to service a small company's account as it does a large company's, but the large company produces significantly more income for the insurer. **Premium discounts** reward the larger company for its size.
6. **Combination method.** The insurer combines two or more of the other methods to arrive at premium rates.[18]

The trend nationwide for the past decade has been for premiums to increase markedly. For example, over the past ten years, some states experienced increases of over 60 percent. This trend will ensure that workers' compensation remains a controversial issue in the state legislatures.

Resolution of Workers' Compensation Disputes

One of the fundamental objectives of workers' compensation is to avoid costly, time-consuming litigation. Whether this objective is being accomplished is questionable. When an injured employee and the employer's insurance company disagree on some aspect of the compensation owed (e.g., weekly pay, length of benefits, degree of disability), the disagreement must be resolved. Most states have an arbitration board for this purpose. Neither the insurance company nor the injured employee is required to hire an attorney. However, many employees do. There are a number of reasons for this. Some don't feel they can adequately represent themselves. Others are fearful of the "big business running over the little guy" syndrome. In any case, workers' compensation litigation is still very common and expensive.

Allowable attorney fees are set by statute, administrative rule, or policy in most states. In some states, attorney fees can be added to the injured employee's award. In others, the fee is a percentage of the award.

Injuries and Workers' Compensation

The original workers' compensation concept envisioned compensation for workers who were injured in on-the-job accidents. What constituted an accident varied from state to state. However, all original definitions had in common the characteristics of being *sudden* and *unexpected*. Over the years, the definition of an accident has undergone continual change. The major change has been a trend toward the elimination of the sudden characteristic. In many states, the gradual onset of a disease as a result of prolonged exposure to harmful substances or a harmful environment can now be considered an accident for workers' compensation purposes.

A **harmful environment** does not have to be limited to its physical components. Psychological factors (such as stress) can also be considered. In fact, the highest rate of growth in workers' compensation claims over the past two decades has been in the area of stress-related injuries.

The National Safety Council maintains statistical records of the numbers and types of injuries suffered in various industries in the United States. Industries are divided into the following categories: agriculture, mining, construction, manufacturing, transportation/public utilities, trade, services, and public sector. Injuries in these industrial sectors are classified according to the type of accident that caused them. Accident types include overexertion, being struck by or against an object, falls, bodily reactions, caught in or between objects, motor vehicle accident, coming in contact with radiation or other caustics, being rubbed or abraded, and coming in contact with temperature extremes.

Over 30 percent of all disabling work injuries are the result of overexertion when all industry categories are viewed in composite. The next most frequent cause of injuries is struck by/struck against objects at 24 percent. Falls account for just over 17 percent. The remainder are fairly evenly distributed among the other accident types just listed.[19]

AOE and COE Injuries

Workers' compensation benefits are owed only when the **injury arises *out of employment (AOE)*** or **occurs in the *course of employment (COE)***. When employees are injured undertaking work prescribed in their job description, work assigned by a supervisor, or work normally expected of employees, they fall into the AOE category. Sometimes, however, different circumstances determine whether the same type of accident is considered to be AOE. For example, say a soldering technician burns her hand while repairing a printed circuit board that had been rejected by a quality control inspector. This injury would be classified as AOE. Now suppose the same technician brings a damaged printed circuit board from her home stereo to work and burns her hand while trying to repair it. This injury would not be covered because the accident did not arise from her employment. Determining whether an injury should be classified as AOE or COE is often a point of contention in workers' compensation litigation.

Who Is an Employee?

Another point of contention in workers' compensation cases is the definition of the term *employee*. This is an important definition because it is used to determine AOE and COE. A person who is on the company's payroll, receives benefits, and has a supervisor is clearly an **employee**. However, a person who accepts a service contract to perform a specific task or set of tasks and is not directly supervised by the company is not considered an employee. Although definitions vary from state to state, there are common characteristics. In all definitions, the workers must receive some form of remuneration for work done, and the employer must benefit from this work. Also, the employer must supervise and direct the work, both process and result. These factors—supervision and direction—are what set **independent contractors** apart from employees and exclude them from coverage. Employers who use independent contractors sometimes require the contractors to show proof of having their own workers' compensation insurance.

Unless an employer provides transportation, employees are not generally covered by workers' compensation when traveling to and from work. However, in some circumstances, they can be covered. Consider the following example:

> Mary works at home three days a week and in the office two days. Wednesday is typically one of her days to telecommute from home. Around ten in the morning, Mary got a call from her supervisor asking her to rush to the office for an unplanned but important meeting and to bring several files that are critical to the subject of the meeting. On the way, Mary had an accident and was injured. Since she was transporting job-specific files at the request of her supervisor, the case was considered work-related, and Mary was eligible for workers' compensation.[20]

Another example shows that Mary's case was special. Tom works in an office in a large city. Because he preferred to drive his own car to work instead of using the public transportation system, Tom had to expose himself to dangerous rush-hour traffic twice every workday. One day, Tom was injured in an accident while driving home from work. He filed a workers' compensation claim, but it was denied because driving to and from work in his own automobile was not considered work related.[21]

Disabilities and Workers' Compensation

Injuries that are compensable typically fall into one of four categories: (1) temporary partial disability, (2) temporary total disability, (3) permanent partial disability, and (4) permanent total disability (Figure 6-2). Determining the extent of disability is often a

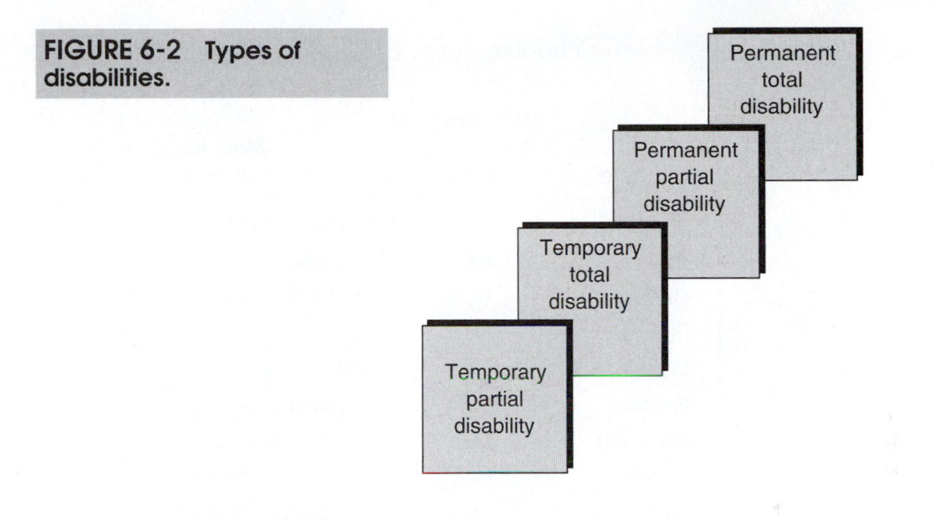

FIGURE 6-2 Types of disabilities.

contentious issue. In fact, it accounts for more workers' compensation litigation than any other issue. Further, when a disability question is litigated, the case tends to be complicated because the evidence is typically subjective, and it requires hearing officers, judges, or juries to determine the future.

Temporary Disability

Temporary disability is the state that exists when it is probable that an injured worker, who is currently unable to work, will be able to resume gainful employment with no or only partial disability. Temporary disability assumes that the employee's condition will substantially improve. Determining whether an employee is temporarily disabled is not normally difficult. Competent professionals can usually determine the extent of the employee's injuries, prescribe the appropriate treatment, and establish a timeline for recovery. They can then determine if the employee will be able to return to work and when the return might take place.

There is an important point to remember when considering a temporary disability case: The ability to return to work relates only to work with the company that employed the worker at the time of the accident.

Temporary disability can be classified as either *temporary total disability* or *temporary partial disability*. A **temporary total disability** classification means the injured worker is incapable of any work for a period of time but is expected to recover fully. Most workers' compensation cases fall in this classification. A **temporary partial disability** means the injured worker is capable of light or part-time duties. Depending on the extent of the injury, temporary partial disabilities sometimes go unreported. This practice is allowable in some states. It helps employers hold down the cost of their workers' compensation premium. This is similar to not reporting a minor fender bender to your automobile insurance agent.

Most states prescribe in law the benefits owed in temporary total disability cases. Factors prescribed typically include a set percentage of an employee's wage that must be paid and a maximum period during which benefits can be collected. Figure 6-3 shows this information for a geographically distributed selection of states. Because workers' compensation legislation changes continually, this figure is provided only as an illustration of how benefits are prescribed in the laws of the various states. Actual rates are subject to change.

FIGURE 6-3 Temporary total disability benefits for selected states.

State	Percentage of Employee's Wage	Maximum Period
Alabama	66⅔	Duration of disability
Arkansas	66⅔	450 weeks
California	66⅔	Duration of disability
Florida	66⅔	104 weeks
Indiana	66⅔	500 weeks
Kansas	66⅔	Duration of disability
Maryland	66⅔	Duration of disability
Montana	66⅔	Duration of disability*
Nebraska	66⅔	Duration of disability
New Jersey	70	400 weeks
New Mexico	66⅔	Duration of disability
Vermont	66⅔	Duration of disability
Washington	60–75	Duration of temporary disability
Wisconsin	60–75	Duration of disability

* Or until worker is released to preinjury job or similar employment.

Source: U.S. Department of Labor, 2005.

Permanent Partial Disability

Permanent partial disability is the condition that exists when an injured employee is not expected to recover fully. In such cases, the employee will be able to work again but not at full capacity. Often employees who are partially disabled must be retrained for another occupation.

Permanent partial disabilities can be classified as *schedule* or *nonscheduled* disabilities. ***Schedule disabilities*** are typically the result of nonambiguous injuries such as the loss of a critical but duplicated body part (e.g., arm, ear, hand, finger, or toe). Because such injuries are relatively straightforward, the amount of compensation that they generate and the period of time that it will be paid can be set forth in a standard schedule. A compilation of information from such schedules for a geographically distributed list of states is shown in Figure 6-4. Workers' compensation legislation changes continually; therefore this figure is provided only as an example. Actual rates are subject to change continually.

Nonschedule injuries are less straightforward and must be dealt with on a case-by-case basis. Disabilities in this category tend to be the result of head injuries, the effects of which can be more difficult to determine. The amount of compensation awarded and the period over which it is awarded must be determined by studying the evidence. Awards are typically made based on a determination of disability percentage. For example, if it is determined that an employee has a 25 percent disability, the employee might be entitled to 25 percent of the income he or she could have earned before the injury with normal career progression factored in.

FIGURE 6-4 Permanent partial disability benefits for selected states in 2005. Based on actual wages lost and not subject to minimums.

State	Percentage of Employee's Wage	Maximum Period
Alabama	66⅔	300 weeks
Arkansas	66⅔	450 weeks
California	66⅔	619.25 weeks
Florida	*	364 weeks
Indiana	66⅔	500 weeks
Kansas	66⅔	415 weeks
Maryland	66⅔	Duration of disability
Montana	66⅔	350 weeks
Nebraska	66⅔	300 weeks
New Jersey	70	600 weeks
New Mexico	66⅔	500 weeks
Vermont	66⅔	330 weeks
Washington	—	—
Wisconsin	66⅔	1,000 weeks

* Shall not exceed 66⅔.

Source: U.S. Department of Labor.

Four approaches to handling permanent partial disability cases have evolved. Three are based on specific theories, and the fourth is based on a combination of two or more of these theories. The three theories are (1) whole-person theory, (2) wage-loss theory, and (3) loss of wage-earning capacity theory.

Whole-Person Theory

The **whole-person theory** is the simplest and most straightforward of the theories for dealing with permanent partial disability cases. Once it has been determined that an injured worker's capabilities have been permanently impaired to some extent, this theory is applied like a subtraction problem. What the worker can do after recuperating from the injury is determined and subtracted from what he or she could do before the accident. Factors such as age, education, and occupation are not considered.

Wage-Loss Theory

The **wage-loss theory** requires a determination of how much the employee could have earned had the injury not occurred. The wages actually being earned are subtracted from what could have been earned, and the employee is awarded a percentage of the difference. No consideration is given to the extent or degree of disability. The only consideration is loss of actual wages.

Loss of Wage-Earning Capacity Theory

The most complex of the theories for handling permanent partial disability cases is the **loss of wage-earning capacity theory**, because it is based not just on what the employee earned at the time of the accident, but also on what he or she might have earned in the future. Making such a determination is obviously a subjective undertaking. Factors considered include past job performance, education, age, gender, and advancement potential at the time of the accident, among others. Once future earning capacity has been determined, the extent to which it has been impaired is estimated, and the employee is awarded a percentage of the difference. Some states prescribe maximum amounts of compensation and maximum periods within which it can be collected. For example, in Figure 6-4, Alabama sets 300 weeks as the maximum period for collecting on a nonschedule injury. Maryland, on the other hand, awards compensation for the duration of the disability. Scheduled disabilities are typically compensated for the duration in all states.

The use of schedules has reduced the amount of litigation and controversy surrounding permanent partial disability cases. This is the good news aspect of schedules. The bad news aspect is that they may be inherently unfair. For example, a surgeon who loses his hand would receive the same compensation as a laborer with the same injury if the loss of a hand is scheduled.

Permanent Total Disability

A **permanent total disability** exists when an injured employee's disability is such that he or she cannot compete in the job market. This does not necessarily mean that the employee is helpless. Rather, it means an inability to compete reasonably. Handling permanent total disability cases is similar to handling permanent partial disability cases except that certain injuries simplify the process. In most states, permanent total disability can be assumed if certain specified injuries have been sustained (i.e., loss of both eyes or both arms). In some states, compensation is awarded for life. In others, a time period is specified. Figure 6-5 shows the maximum period that compensation can be collected for a geographically distributed list of states. Notice in this Figure 6-5 that the time periods range from 401 weeks (Texas) to life (California and Wisconsin).

Monetary Benefits of Workers' Compensation

The **monetary benefits** accruing from workers' compensation vary markedly from state to state. The actual amounts are of less importance than the differences among them. Of course, the amounts set forth in schedules change frequently. However, for the purpose of comparison, consider that at one time the loss of a hand in Pennsylvania resulted in an award of $116,245. The same injury in Colorado brought only $8,736.

When trying to determine a scheduled award for a specific injury, it is best to locate the latest schedule for the state in question. One way to do this is to contact the following agency:

U.S. Department of Labor
Employment Standards Administration
Office of State Liaison and Legislative Analysis
Division of State Workers' Compensation Programs
200 Constitution Ave. NW

FIGURE 6-5 Duration of permanent total disability benefits for selected states.

State	Maximum Paid
Alabama	Duration of disability
Arkansas	Duration of disability
California	Life
Florida	Duration of disability
Indiana	500 weeks
Kansas	Duration of disability
Maryland	Duration of disability
Montana	Duration of disability
Nebraska	Duration of disability
New Jersey	450 weeks (life in some cases)
New Mexico	Life
Vermont	Duration of disability*
Washington	Life
Wisconsin	Life

* Minimum of 330 weeks.

Source: U.S. Department of Labor, 2005.

Washington, DC 20210
http://www.dol.gov

Death and Burial Benefits

Workers' compensation benefits accrue to the families and dependents of workers who are fatally injured. Typically, the remaining spouse receives benefits for life or until remarriage. However, in some cases, a time period is specified. Dependents typically receive benefits until they reach the legal age of maturity unless they have a condition or circumstances that make them unable to support themselves even after attaining that age. Figure 6-6 contains the death benefits accruing to surviving spouses and children for a geographically distributed list of states. Because the actual amounts of benefits are subject to change, these are provided for illustration and comparison only.

Further expenses are provided in addition to death benefits in all states except Oklahoma. As is the case with all types of workers' compensation, the amount of burial benefits varies from state to state and is subject to change. Figure 6-7 contains the maximum burial benefits for a geographically distributed list of states.

Medical Treatment and Rehabilitation

All workers' compensation laws provide for payment of the medical costs associated with injuries. Most states provide full coverage, but some limit the amount and duration of coverage. For example, in Arkansas, employer liability ceases six months after an injury occurs in those cases in which the employee is able to continue working or six

FIGURE 6-6 Death benefits for surviving spouses and children for selected states.

State	Percentage of Employee's Wage: Spouse Only	Percentage of Employee's Wage: Spouse and Children	Maximum Period
Alabama	50	66⅔	500 weeks
Arkansas	35	66⅔	Widow/widowerhood; children until 18 or married
California	66⅔	66⅔	—
Florida	50	66⅔	Widow/widowerhood; children until 18
Indiana	66⅔	66⅔	500 weeks
Kansas	66⅔	66⅔	Widow/widowerhood; children until 18
Maryland	66⅔	66⅔	Widow/widowerhood; children until 18
Montana	66⅔	66⅔	Surviving spouse—10 years; children until 18
Nebraska	66⅔	75	Widow/widowerhood; children until 18
New Jersey	50	70	Widow/widowerhood; children until 18
New Mexico	66⅔	66⅔	700 weeks
Vermont	66⅔	76⅔	Widow/widowerhood until 62; children until 18
Washington	60	70	Widow/widowerhood; children until 18
Wisconsin	66⅔	—	300 weeks

months after he or she returns to work in cases where there is a period of recuperation. In either case, the employer's maximum financial liability is $10,000.[22] In Ohio, medical benefits for silicosis, asbestosis, and coal miner's pneumoconiosis are paid only in the cases of temporary total or permanent total disability.[23]

The laws also specify who is allowed or required to select a physician for the injured employee. The options can be summarized as follows:

- **Employee selects the physician of choice.** This option is available in Alaska, Arizona, Delaware, Hawaii, Illinois, Kentucky, Louisiana, Maine, Massachusetts, Mississippi, Nebraska, New Hampshire, North Dakota, Ohio, Oklahoma, Oregon, Rhode Island, Texas, the Virgin Islands, Washington, West Virginia, Wisconsin, and Wyoming.

FIGURE 6-7 Maximum burial allowances for selected states.

State	Maximum Allowance
Alabama	$3,000
Arkansas	6,000
California	5,000
Florida	2,500
Indiana	6,000
Kansas	5,000
Maryland	5,000
Montana	1,400
Nebraska	6,000
New Jersey	3,500
New Mexico	3,000
Vermont	5,500
Washington	*
Wisconsin	6,000

* 200% of state's average monthly wage

Note: Figures are provided for comparison and illustration only. They are subject to change.
Source: U.S. Department of Labor, 2005.

- **Employee selects the physician from a list provided by the state agency.** This option applies in Connecticut, Nevada, New York, and the District of Columbia.
- **Employee selects the physician from a list provided by the employer.** This option applies in Georgia, Tennessee, and Virginia.
- **Employer selects the physician.** This option applies in Alabama, Florida, Idaho, Indiana, Iowa, Maryland, Montana, New Jersey, New Mexico, North Carolina, South Carolina, and South Dakota.
- **Employer selects the physician, but the selection may be changed by the state agency.** This option applies in Arkansas, Colorado, Kansas, Minnesota, Missouri, Utah, and Vermont.
- **Employer selects the physician, but after a specified period of time, the employee may choose another.** This option applies only in Puerto Rico.

Rehabilitation and Workers' Compensation

Occasionally an injured worker will need rehabilitation before he or she can return to work. There are two types of rehabilitation: medical and vocational. Both are available to workers whose ability to make a living is inhibited by physical and/or mental work-related problems.

Medical rehabilitation consists of providing whatever treatment is required to restore to the extent possible any lost ability to function normally. This may include such services as physical therapy or the provision of prosthetic devices. **Vocational rehabilitation**

involves providing the education and training needed to prepare the worker for a new occupation. Whether the rehabilitation services are medical or vocational in nature or both, the goal is to restore the injured worker's capabilities to the level that existed before the accident.

Medical Management of Workplace Injuries

Out-of-control workers' compensation cases in the 1990s led to the concept of **medical management of workplace injuries**.[24] Through better management of workers' compensation claims, more than 30 states have merged the concepts of workers' compensation and managed care. The goals of these state-level efforts are to (1) speed up the processing of workers' compensation claims; (2) reduce costs; (3) reduce fraud and abuse; and (4) improve medical management of workplace injuries.

Workers' compensation and managed care have been merged through the creation of *Health Partnership Programs (HPPs)*. HPPs are partnerships between employers and their state's Bureau of Workers' Compensation (BWC). Employers who choose to participate (some states mandate participation) are required to have a *managed care organization (MCO)* that provides medical management of workplace injuries and illnesses.

Ohio's HPP is an example of how this concept typically works. According to Clairmonte Cappelle,

> When a workplace injury occurs or an illness manifests itself, the employee reports it to the employer and seeks initial medical treatment. At this early stage, the employer or health care provider informs the MCO, which files a first report of injury electronically with the state and begins medical management of the case. The MCO prefers that an injured worker stay within its health care provider network for care. However, injured workers may choose their own doctors and hospitals from the list of 100,000 BWC-certified providers. Except for emergency situations, workers who select non-BWC-certified health care providers will not have their workers' compensation medical costs covered by the state. Once a claim is filed, MCOs work with employers, injured workers, health care providers, third-party administrators and BWC. This includes authorizing certain medical procedures, processing providers' bills for payment by BWC and driving the return-to-work process.[25]

The HPPs are effective because an MCO coordinates the paperwork generated by the injured employee, the employer, health care providers, and the BWC. As the Ohio example illustrates, the results of this coordination are clear. The average time required to process an injury report prior to implementation of the HPP was more than 66 days. The HPP reduced this to approximately 33 days, effectively cutting reporting time in half. Whether managed medical care will be able to continue this positive trend remains to be seen, but the states that have implemented HPPs have learned the following lessons:

- It is better to mandate HPPs than to make them optional.
- Cost containment is only part of the goal. Managed care programs must also include criteria such as lost wages, ability to return to work, and administrative costs to the employer.
- Employees want choice in selecting health care providers.
- Smart return-to-work programs are critical.

Administration and Case Management

Even though OSHA specifies what constitutes a recordable accident, it is not uncommon for minor injuries to go unreported. The employee may be given the rest of the day off or treated with first aid and returned to work. This is done to avoid time-consuming paperwork and to hold down the cost of workers' compensation insurance. However, if an accident results in a serious injury, several agencies must be notified. What constitutes a serious injury, like many workers' compensation issues, can differ from state to state. However, as a rule, an injury is serious if it requires over 24 hours of active medical treatment (this does not include passive treatment such as observation). Of course, a fatality, a major disfigurement, or the loss of a limb or digit is also considered serious and must be reported.

At a minimum, the company's insurer, the state agency, and the state's federal counterpart must be notified. Individual states may require that additional agencies be notified. All establish a time frame within which notification must be made. Once the notice of injury has been filed, there is typically a short period before the victim or dependents can begin to receive compensation unless inpatient hospital care is required. However, when payments do begin to flow, they are typically retroactive to the date of the injury.

State statutes also provide a maximum time period that can elapse before a compensation claim is filed. The notice of injury does not satisfy the requirement of filing a claim notice. The two are separate processes. The statute of limitations on **claim notices** varies from state to state. However, most limit the period to no more than a year except in cases of work-related diseases in which the exact date of onset cannot be determined.

All such activities—filing injury notices, filing claim notices, arriving at settlements, and handling disputes—fall under the collective heading of administration and case management. Most states have a designated agency that is responsible for administration and case management. In addition, some states have independent boards that conduct hearings and hear appeals when disputes arise.

Once a workers' compensation claim is filed, an appropriate settlement must be reached. Three approaches can be used to settle a claim: (1) **direct settlement**, (2) **agreement settlement**, and (3) **public hearing**. The first two are used in uncontested cases, the third in contested cases.

1. **Direct settlement.** The employer or its insurance company begins making what it thinks are the prescribed payments. The insurer also sets the period over which payments will be made. Both factors are subject to review by the designated state agency. This approach is used in Arkansas, Michigan, Mississippi, New Hampshire, Wisconsin, and the District of Columbia.
2. **Agreement settlement.** The injured employee and the employer or its insurance company work out an agreement on how much compensation will be paid and for how long. Such an agreement must be reached before compensation payments begin. Typically, the agreement is reviewed by the designated state administrative agency. In cases where the agency disapproves the agreement, the worker continues to collect compensation at the agreed-on rate until a new agreement can be reached. If this is not possible, the case becomes a contested case.
3. **Public hearing.** If an injured worker feels he or she has been inadequately compensated or unfairly treated, a hearing can be requested. Such cases are known as *contested cases*. The hearing commission reviews the facts surrounding

the case and renders a judgment concerning the amount and duration of compensation. Should the employee disagree with the decision rendered, civil action through the courts is an option.

Cost Allocation

Workers' compensation is a costly concept. From the outset, one of the basic principles has been cost allocation. **Cost allocation** is the process of spreading the cost of workers' compensation across an industry so that no individual company is overly burdened. The cost of workers' compensation includes the costs of premiums, benefits, and administration. These costs have risen steadily over the years.

When workers' compensation costs by industry are examined, there are significant differences. For example, the cost of workers' compensation for a bank is less than one-half of 1 percent of gross payroll. For a ceramics manufacturer, the percentage might be as high as 3 or 4 percent.

Cost allocation is based on the **experience rating** of the industry. In addition to being the fairest method (theoretically) of allocating costs, this approach is also supposed to give employers an incentive to initiate safety programs. Opinions vary as to the fairness and effectiveness of this approach. Arguments against it include the following: (1) small firms do not have a sufficient number of employees to produce a reliable and accurate picture of the experience rating; (2) firms that are too small to produce an experience rating are rated by class of industry, thereby negating the incentive figure; (3) premium rates are more directly sensitive to experience levels in larger firms but are less so in smaller companies; and (4) in order to hold down experience ratings, employers may put their efforts into fighting claims rather than preventing accidents. Not much hard research has been conducted to determine the real effects of cost allocation. Such research is badly needed to determine if the theoretical construct of workers' compensation is, in reality, valid.

Problems with Workers' Compensation

There are serious problems with workers' compensation in the United States. On one hand, there is evidence of abuse of the system. On the other hand, many injured workers who are legitimately collecting benefits suffer a substantial loss of income. Complaints about workers' compensation are common from all parties involved with the system (employers, employees, and insurance companies).

Earlier in this chapter, the example of the overweight deputy sheriff who applied for benefits because he was distraught over the breakup of his extramarital love affair and a poor performance evaluation was cited as an example of abuse. This individual is just one of thousands who are claiming that job stress has disabled them to the point that workers' compensation is justified. In 1980, there were so few stress claims they were not even recorded as a separate category. Today, they represent a major and costly category.

Stress claims are more burdensome than physical claims because they are typically reviewed in an adversarial environment. This leads to the involvement of expert medical witnesses and attorneys. As a result, even though the benefits awarded for stress-related injuries are typically less than those awarded for physical injuries, the cost of stress claims is often higher because of the litigation.

In addition to abuse, a steadily increasing caseload is also a problem. For example, California saw its caseload increase by 40,000 in one year.[26] As this trend continues, the cost of workers' compensation will increase on a parallel track.

Although the cost of workers' compensation is increasing steadily, the amount of compensation going to injured workers is often disturbingly low. In a given year, if workers' compensation payments in the United States amount to around $27 billion (which is typical), $17 billion of this goes to benefits. Almost $10 billion is taken up by medical costs.[27] The amount of wages paid to injured workers in most states is 66 percent. This phenomenon is not new.

> Almost one-half million families each year are faced with getting by on a drastically reduced income because of a disabling injury suffered by the principal income earner. On-the-job accidents are supposed to be covered by workers' compensation and all states have compensation systems. However, the injured worker rarely receives an income that comes close to what he or she was earning before the accident.[28]

The most fundamental problem with workers' compensation is that it is not fulfilling its objectives. Lost income is not being adequately replaced, the number of accidents has not decreased, and the effectiveness of cost allocation is questionable. Clearly, the final chapter on workers' compensation has not yet been written.

Spotting Workers' Compensation Fraud and Abuse

There is evidence of waste, fraud, and abuse of the system in all states that have passed workers' compensation laws. However, the public outcry against fraudulent claims is making states much less tolerant of, and much more attentive to, abuse. For example, the Ohio legislature passed a statute that allows criminal charges to be brought against employees, physicians, and lawyers who give false information in a workers' compensation case. This is a positive trend. However, even these measures will not completely eliminate abuse.

For this reason, it is important for organizations to know how to spot employees who are trying to abuse the system by filing fraudulent workers' compensation claims. Following are some factors that should cause employers to view claims with suspicion. However, just because one or more of these factors are present does not mean that an employee is attempting to abuse the system. Rather, these are simply factors that should raise cautionary flags:

- The person filing the claim is never home or available by telephone or has an unlisted telephone number.
- The injury in question coincides with a layoff, termination, or plant closing.
- The person filing the claim is active in sports.
- The person filing the claim has another job.
- The person filing the claim is in line for early retirement.
- The rehabilitation report contains evidence that the person filing the claim is maintaining an active lifestyle.
- No organic basis exists for disability. The person filing the claim appears to have made a full recovery.
- The person filing the claim receives all mail at a post office box and will not divulge a home address.

- The person filing the claim is known to have skills such as carpentry, plumbing, or electrical that could be used to work on a cash basis while feigning a disability.
- There are no witnesses to the accident in question.
- The person filing the claim has relocated out of state or out of the country.
- Demands for compensation are excessive.
- The person filing the claim has a history of filing.
- Doctors' reports are contradictory in nature.
- A soft tissue injury is claimed to have produced a long-term disability.
- The injury in question occurred during hunting season.[29]

These factors can help organizations spot employees who may be trying to abuse the workers' compensation system. It is important to do so because legitimate users of the system are hurt, as are their employers, by abusers of the system. If one or more of these factors are present, employers should investigate the claim carefully before proceeding.

Future of Workers' Compensation

The future of workers' compensation will be characterized by an ongoing need for reform as well as higher premiums, higher deductibles, and less coverage in insurance policies. The key to reforming workers' compensation is finding a way to allocate more of the cost to benefits and medical treatment and less to administration and litigation. Key elements in any reform effort are as follows:

- Stabilizing workers' compensation costs over the long term.
- Streamlining administration of the system.
- Reducing the costs associated with the resolution of medical issues.
- Limiting stress-related claims.
- Limiting vocational rehabilitation benefits.
- Increasing benefits paid for temporary and permanent disabilities.
- Reducing the amount that insurers may charge for overhead.
- Providing more public input into the setting of rates.
- Taking litigation out of the process to the extent possible.
- Improving medical treatment management.
- Improving overall case management.
- Streamlining the claim notification and processing system.
- Requiring more sufficient justification from insurance carriers concerning their rates.

Controlling Rising Workers' Compensation Costs

The first and best line of defense against escalating workers' compensation costs is a safe and healthy workplace.[30] However, in today's litigious business environment, even this may not be enough. In order to be an effective member of the team responsible for keeping workers' compensation costs in check, safety and health professionals need to cooperate with risk management professionals in implementing the following strategies:

1. Establish an effective safety and health program, and document it clearly and comprehensively for the workers' compensation underwriters.

2. Review workers' compensation claims to ensure that they are accurate before they are submitted to underwriters.
3. Analyze concentrations of risk by location and have comprehensive, up-to-date plans on hand for preventing and responding to catastrophic events.
4. Advise risk management professionals on potential hazards and related risks so they can make informed decisions concerning levels of coverage and deductibles.
5. Communicate frequently with risk management personnel—you and they are on the same team.
6. Develop strategies for dealing with terrorism—many workers' compensation underwriters exclude terrorism from their coverage.[31]

Cost-Reduction Strategies

Safety and health professionals are responsible for helping their organizations hold down workers' compensation costs. Of course, the best way to accomplish this goal is to maintain a safe and healthy workplace, thereby preventing the injuries that drive up the costs. This section presents numerous other strategies that have proven effective in reducing workers' compensation costs after injuries have occurred, which happen in even the safest environments.

General Strategies

Regardless of the type of organization, there are several rules of thumb that can help reduce workers' compensation claims. These general strategies are as follows:

1. **Stay in touch with the injured employee.** Let injured employees know that they have not been forgotten and that they are not isolated. Answer all their questions and try to maintain their loyalty to the organization.
2. **Have a return-to-work program and use it.** The sooner an injured employee returns to work, even with a reduced workload, the lower workers' compensation costs will be. Reduced costs can, in turn, lower the organization's insurance premium. When using the return-to-work strategy, be cautious. Communicate with the employee and his or her medical treatment team. Have a clear understanding of the tasks that can be done and those that should be avoided, such as how much weight the employee can safely lift.

   Colledge and Johnson recommend using the "SPICE" model for improving the effectiveness of return-to-work programs. It consists of the following components:

   - Simplicity
   - Proximity
   - Immediacy
   - Centrality
   - Expectancy[32]

   Simplicity means that the medical professionals who treat injured employees should work closely with safety professionals to prevent "system"-induced complications. Such complications occur when employees become convinced their injuries are more serious than they really are because of ominous-sounding diagnostic terminology and complicated tests and treatments. Medical professionals and safety personnel should work together to keep the terminology simple and to explain tests and treatments in easily understood lay terms.

Proximity means keeping the injured employee as close to the job as possible. Employees who are physically separated from their place of employment and their fellow employees also become mentally separated. Within a short time, what used to be "us" can become "them." Giving as much injury care as possible at the work site, providing light-duty assignments, and communicating regularly with employees whose injuries preclude on-site assignments or treatment will keep employees connected and maintain the advantages of proximity.

Immediacy means that the faster an employee's injury claims can be handled, the less likely he or she will be to develop psychosocial issues that can complicate the recovery process. The longer it takes to process a claim, conduct a diagnosis, and begin treatments the more likely the employee is going to worry about the injury and to become accustomed to being off work. Immediate diagnosis, processing, and treatment can decrease the amount of time that elapses before the employee can begin a return-to-work program.

Centrality means getting the employee, his or her family, the medical professionals handling the case, insurance personnel, and the employer to agree on a common vision for successfully returning the injured party to work as soon as possible. It is important for injured employees and their families to know that everyone involved has the same goal and that everyone is working in good faith to achieve that goal.

Expectancy means creating the expectation that getting the employee well and back to work is the goal of all parties involved. It is achieved by communicating this message clearly to all parties and reinforcing it by establishing short-term goals and timelines for actually being back on the job. Achievement of each respective short-term goal should move the employee one step closer to recovery and return to work.

3. **Determine the cause of the accident.** The key to preventing future accidents and incidents is determining the cause of the accident in question, and the key to holding down workers' compensation costs is preventing accidents. Eliminating the root cause of every accident is fundamental to any cost-containment effort.

Specific Strategies

In addition to the general strategies just presented, there are numerous specific cost-containment strategies that have proven to be effective. These specific strategies are presented in this section.

1. **Cultivate job satisfaction.** According to Jon Gice, increasing job satisfaction is just as important as eliminating physical hazards in the workplace.[33] High stress, aggression, alienation, and social maladjustment—all factors associated with, and aggravated by, a lack of job satisfaction—can make employees less attentive while working. An inattentive employee is an accident waiting to happen. Gice recommends the following strategies for improving job satisfaction:
   - Recognize and reward employees.
   - Communicate frequently and openly with employees about job-related problems.

- Give employees as much control over their work as possible.
- Encourage employees to talk freely among themselves.
- Practice conflict management.
- Provide adequate staffing and expense budgets.
- Encourage employees to use employee assistance programs.[34]

2. **Make safety part of the culture.** The Ohio Division of Safety and Hygiene recommends the following steps for making safety part of the organizational culture as a way to reduce workers' compensation costs:

   - Ensure visible, active leadership, involvement, and commitment from senior management.
   - Involve employees at all levels in the safety program and recognize them for their efforts.
   - Provide comprehensive medical care, part of which is a return-to-work program.
   - Ensure effective communication throughout the organization.
   - Coordinate all safety and health processes.
   - Provide orientation and training for all employees.
   - Have written safe work practices and procedures.
   - Have a comprehensive written safety policy.
   - Keep comprehensive safety records and analyze the data contained in those records.[35]

3. **Have a systematic cost-reduction program.** To reduce costs, an organization should have a systematic program that can be applied continually and consistently. Alpha Meat Packing Company of South Gate, California, has had success using the following strategies:
   - Insert safety notes and reminders in employees' paycheck envelopes.
   - Call injured employees at home to reassure them that they will have a job when they return.
   - Keep supervisors trained on all applicable safety and health issues, procedures, rules, and so on.
   - Hold monthly meetings to review safety procedures, strategies, and techniques.
   - Reward employees who give suggestions for making the workplace safer.[36]

4. **Use integrated managed care.** Managed care is credited by many with reducing workers' compensation costs nationwide. Others claim that cost reduction has occurred because managed care dangerously restricts the types and amount of health care provided to injured employees. GE Aircraft Engines is an advocate of managed care that is fully integrated as follows:

   - A comprehensive safety and hazard-prevention program to keep employees safe and healthy.
   - On-site medical clinics.
   - Plant compensation teams that include nurses, rehabilitation technicians, and third-party adjudicators.
   - Claims management.
   - Companywide safety and health database.
   - Return-to-work program.[37]

Key Terms and Concepts

- Accident prevention
- Agreement settlement
- Assumption of risk
- Blackball
- Claim notice
- Contributory negligence
- Cost allocation
- Direct settlement
- Employee
- Experience rating
- Extraterritorial employees
- Fear factor
- Harmful environment
- Income replacement
- Independent contractor
- Injury arises out of employment (AOE)
- Injury occurs in the course of employment (COE)
- Litigation
- Loss of wage-earning capacity theory
- Malpractice
- Manual rating
- Medical management of workplace injuries
- Medical rehabilitation
- Monetary benefits
- Occupational disease
- Permanent partial disability
- Permanent total disability
- Premium discounting
- Private insurance
- Public hearing
- Reform
- Rehabilitation
- Retrospective rating
- Schedule disabilities
- Schedule rating
- Self-insurance
- State funds
- Stress claims
- Temporary partial disability
- Temporary total disability
- Vocational rehabilitation
- Wage-loss theory
- Whole-person theory
- Worker negligence
- Workers' compensation
- Workplace accident

Review Questions

1. Explain the underlying rationale of workers' compensation as a concept.
2. List four objectives of workers' compensation.
3. List five types of employees who may not be covered by workers' compensation.
4. What is meant by the term *contributory negligence*?
5. What is meant by the term *assumption of risk*?
6. Explain the reasons for the unprecedented increases in medical costs in the United States.
7. What are the three types of workers' compensation insurance?
8. Insurance companies use one of six methods for determining the premium rates of employers. Select three and explain them.
9. How can one determine if an injury should be considered serious?
10. Explain the concepts of AOE and COE.
11. Distinguish between an employee and an independent contractor.
12. Define the following terms: *temporary disability* and *permanent disability*.
13. Explain the following theories of handling permanent partial disability cases: *whole-person*, *wage-loss*, and *loss of wage-earning capacity*.
14. Distinguish between medical and vocational rehabilitation.
15. What are the three approaches for settling workers' compensation claims?
16. Explain the concept of medical management of workplace injuries.
17. Explain the theory of cost allocation.
18. Summarize briefly the problems most widely associated with workers' compensation.
19. What types of actions are workers' compensation reform movements likely to recommend in the future?
20. Explain the most common workers' compensation cost-reduction strategies.

Endnotes

1. "Workers' Compensation Beginner's Guide." Retrieved from http://beginnersguide.com/small_business/workers_compensation/ April 4, 2006.
2. Ibid.
3. Ibid.
4. Ibid.
5. Ibid.
6. Ibid.
7. H. M. Somers and A. R. Somers, *Workmen's Compensation* (New York: Wiley, 1945), 29.
8. Ibid.

9. D. Gagilardo, *American Social Science* (New York: Harper & Row, 1949), 149.
10. "Workers' Compensation: Beginner's Guide."
11. Ibid.
12. Ibid.
13. Ibid.
14. D. Goetsch, *Workers' Comp Cases,* Report 2006 (Niceville, FL: The Development Institute, April 2006), 12–13.
15. Ibid.
16. Ibid.
17. "Workers' Comp Update," *Occupational Hazards* 60, no. 7: 93.
18. Ibid.
19. Society of Manufacturing Engineers, *Accident Facts* (Chicago: Society of Manufacturing Engineers, 1998), 36.
20. D. Goetsch, "Workers' Comp Cases," 14.
21. Ibid.
22. U.S. Department of Labor, *State Workers' Compensation Laws* (Washington, DC: U.S. Department of Labor, January 2005).
23. Ibid.
24. C. Cappelle, "Making a Strong Case for Managed Care," *Occupational Hazards* 61, no. 4: 67–71.
25. Ibid., 68.
26. National Safety Council, *Accident Facts* (Chicago: National Safety Council 2004), 43.
27. Ibid.
28. U.S. Department of Labor, *Federal Worker 2000* (Washington, DC: U.S. Department of Labor, March 2000), 2.
29. D. Goetsch, "Workers' Comp Fraud Detection Checklist," Report 2005–6 (Niceville, FL: The Development Institute, April 2005), 8.
30. Christine Fuge, "Tough Times Ahead for Workers' Compensation," *Occupational Hazards* 64, no. 12: 39–42.
31. Ibid., 40.
32. A. Colledge and H. Johnson, "The S.P.I.C.E. Model for Return To Work," *Occupational Health & Safety* 69, no. 2: 64–69.
33. Jon Gice, *The Relationship between Job Satisfaction and Workers Compensation Claims* (Malvern, PA: Chartered Property Casualty Underwriters Society, 720 Providence Rd., 19355–0709).
34. Ibid.
35. Ohio Division of Safety and Hygiene, "Ohio Prompts 10 Steps to Reduced Comp Costs," *Occupational Hazards,* August 1996, 59.
36. "Workers' Comp Update: Meatpacking Industry Cuts Comp Claims," *Occupational Hazards*, May 1996, 103.
37. Candace Goforth, "Workers' Comp: Is Managed Care the Answer?" *Occupational Hazards,* October 1996, 126.

CHAPTER 7

Accident Investigation and Reporting

Major Topics

- Types of Accident Investigations
- When to Investigate
- What to Investigate
- Who Should Investigate
- Conducting the Investigation
- Interviewing Witnesses
- Reporting Accidents
- Ten Accident Investigation Mistakes to Avoid

Dan Hartshorn defines an accident as "any unplanned event that causes injury, illness, property damage or harmful disruption of work process."[1]

When an accident occurs, it is important that it be investigated thoroughly. The results of a comprehensive accident report can help safety and health professionals pinpoint the cause of the accident. This information can then be used to prevent future accidents, which is the primary purpose of accident investigation.

The Society of Manufacturing Engineers describes the importance of thoroughly investigating accidents:

> The primary reason for investigating an accident is not to identify a scapegoat, but to determine the cause of the accident. The investigation concentrates on gathering factual information about the details that led to the accident. If investigations are conducted properly, there is the added benefit of uncovering problems that did not directly lead to the accident. This information benefits the ongoing effort of reducing the likelihood of accidents. As problems are revealed during the investigation, action items and improvements that can prevent similar accidents from happening in the future will be easier to identify than at any time.[2]

This chapter gives prospective and practicing safety and health professionals the information they need to conduct thorough, effective accident investigations and prepare comprehensive accident reports.

Types of Accident Investigations

There are *accident reports* and there are *accident-analysis reports*. An **accident report** is completed when the accident in question represents only a minor incident. It answers the following questions: *who*, *what*, *where*, and *when*. However, it does not answer the *why* question.[3] An accident report can be completed by a person with very little formal investigation and reporting training or experience. Supervisors often complete accident reports which, in turn, might be used later as part one of a more in-depth accident report. OSHA's Form 301 (see Chapter 6) can be used for accident reports.

An **accident-analysis report** is completed when the accident in question is serious. This level of report should answer the same questions as the regular accident report plus one more—*why*. Consequently it involves a formal accident analysis. The analysis is undertaken for the purpose of determining the root cause of the accident. Accident analysis requires special skills and should be undertaken only by an individual with those skills. There are two reasons for this. First, the accident analysis must identify the actual root cause or the company will expend resources treating only symptoms or, even worse, solving the wrong problem. Second, serious accidents are always accompanied by the potential for litigation. If there might be legal action as a result of an accident, it is important to have a professional conduct the investigation even if it means bringing in an outside consultant.

How can safety and health professionals determine when an accident report is sufficient and when an accident-analysis report is called for? Accident reports are called for when the accident in question is a minor incident that did *not* result in any of the following circumstances: death, loss of consciousness, medical treatment beyond first aid, more than one additional day of lost work beyond the day of the accident, or any kind of modifications to the injured employee's work duties beyond those that might occur on the day of the accident.

Accident-analysis reports are called for when any of the following circumstances result from the accident in question: death, loss of consciousness, professional medical treatment beyond first aid, one or more days of lost work over and above any time lost the day of the accident, or any modifications to the injured employee's work duties beyond those that might occur on the day of the injury.

When to Investigate

Of course, the first thing to do when an accident takes place is to implement **emergency procedures**. This involves bringing the situation under control and caring for the injured worker. As soon as all emergency procedures have been accomplished, the accident investigation should begin. Waiting too long to complete an investigation can harm the results. This is an important rule of thumb to remember. Another is that *all* accidents, no matter how small, should be investigated. **Evidence** suggests that the same factors that cause minor accidents cause major accidents.[4] Further, a near miss should be treated like an accident and investigated thoroughly.

There are several reasons why it is important to conduct investigations immediately. First, immediate investigations are more likely to produce accurate information. Conversely, the longer the time span between an accident and an investigation, the greater the likelihood of important facts becoming blurred as memories fade. Second, it is important to collect information before the accident scene is changed and before witnesses begin comparing notes. Human nature encourages people to change their stories to agree with those of other witnesses.[5] Finally, an immediate investigation is evidence of management's commitment to preventing future accidents. An immediate response shows that management cares.[6]

What to Investigate

The purpose of an **accident investigation** is to collect facts. It is not to find fault. It is important that safety and health professionals make this distinction known to all involved. **Fault finding** can cause reticence among witnesses who have valuable information to share. **Causes** of the accident should be the primary focus. The investigation should be guided by the following words: **who**, **what**, **when**, **where**, **why**, and **how**.

This does not mean that mistakes and breaches of precautionary procedures by workers go unnoted. Rather, when these things are noted, they are recorded as facts instead of faults. If fault must be assigned, that should come later, after all the facts are in. The distinction is a matter of emphasis. The National Safety Council summarizes this approach:

> As you investigate, don't put the emphasis on identifying who could be blamed for the accident. This approach can damage your credibility and generally reduce the amount and accuracy of information you receive from workers. This does not mean you ignore oversights or mistakes on the part of employees nor does it mean that personal responsibility should not be determined when appropriate. It means that the investigation should be concerned with only the facts. In order to do a quality job of investigating accidents you must be objective and analytical.[7]

In attempting to find the facts and identify causes, certain questions should be asked, regardless of the nature of the accident. The Society of Manufacturing Engineers recommends using the following questions when conducting accident investigations:

- What type of work was the injured person doing?
- Exactly what was the injured person doing or trying to do at the time of the accident?
- Was the injured person proficient in the task being performed at the time of the accident? Had the worker received proper training?
- Was the injured person authorized to use the equipment or perform the process involved in the accident?
- Were there other workers present at the time of the accident? If so, who are they, and what were they doing?
- Was the task in question being performed according to properly approved procedures?
- Was the proper equipment being used, including personal protective equipment?
- Was the injured employee new to the job?
- Was the process, equipment, or system involved new?
- Was the injured person being supervised at the time of the accident?

- Are there any established safety rules or procedures that were clearly not being followed?
- Where did the accident take place?
- What was the condition of the accident site at the time of the accident?
- Has a similar accident occurred before? If so, were corrective measures recommended? Were they implemented?
- Are there obvious solutions that would have prevented the accident?[8]

The answers to these questions should be carefully and copiously recorded. You may find it helpful to dictate your findings into a microcassette recorder. This approach allows you to focus more time and energy on investigating and less on taking written notes.

Regardless of how the findings are recorded, it is important to be thorough. What may seem like a minor unrelated fact at the moment could turn out to be a valuable fact later when all the evidence has been collected and is being analyzed.

Common Causes of Accidents

Hartshorn places many of the common causes of accidents in the following categories: **personal beliefs and feelings**, **decision to work unsafely**, **mismatch or overload**, **systems failures**, **traps**, **unsafe conditions**, and **unsafe acts**.[9] The common causes in each of these categories can help investigators determine the root cause of an accident.

Personal beliefs and feelings. Causes in this category include the following: individual did not believe the accident would happen to him or her; individual was working too fast, showing off, or being a know-it-all; individual ignored the rules out of contempt for authority and rules in general; individual gave in to peer pressure; and individual had personal problems that clouded his or her judgment.

Decision to work unsafely. Some people, for a variety of reasons, feel it is in their best interests or to their benefit to work unsafely. Hence, they make a conscious decision to do so.

Mismatch or overload. Causes in this category include the following: individual is in poor physical condition; individual is fatigued; individual has a high stress level; individual is mentally unfocused or distracted; the task required is too complex or difficult; the task required is boring; the physical environment is stressful (e.g., excessive noise, heat, dust, or other factors); the work in question is very demanding—even for an individual in good physical condition; and individual has a negative attitude (e.g., hostile, uncooperative, apathetic, etc.).

Systems failure. Causes in this category consist of the various errors management makes that are not grossly negligent or *serious and willful*. Common causes in this category include lack of a clear policy; lack of rules, regulations, and procedures; poor hiring procedures; inadequate monitoring and inspections; failure to correct known hazards; insufficient training for employees; rules that are in place but are not enforced; no rewarding or reinforcement of safe behavior; inadequate tools and equipment provided; production requirements set too high; inadequate communication to employees of safety concerns, statistics, and rules; poor safety management; no or insufficient job safety analysis; and insufficient management support for safety.

Traps. Poor design of workstations and processes can create *traps* that, in turn, lead to unsafe behavior. Common causes in this category include defective equipment; failure to provide, maintain, and replace proper personal protective equipment; failure to train employees in the proper use of their personal protective equipment; overly complicated and confusing controls; poorly laid out work area; mechanical lifting equipment that is inadequate for the jobs required of it; uncontrolled hazards that might lead to slips and falls; excessive reaching, bending, stooping, and twisting; excessive contact pressure, vibration, or force; awkward postures that result from poor workstation or tool design; excessive temperature extremes; insufficient lighting; and insufficient ventilation.

Unsafe conditions. Common causes in this category include the following: unsafe condition created by the person injured in the accident; unsafe condition created by a fellow employee; unsafe condition created by a third party; unsafe condition created by management; unsafe condition knowingly overlooked by management; and unsafe condition created by the elements (e.g., rain, sun, snow, ice, wind, darkness, etc.).

Unsafe acts. Common causes in this category include the following: individual chooses to ignore the rules; people are involved in horseplay or fighting; individual uses drugs or alcohol; individual uses unauthorized tools or equipment; individual chooses an improper work method; individual fails to ask for information or other resources needed to do the job safely; individual forgets a rule, regulation, or procedure; individual does not pay proper attention; and individual uses improper body mechanics.

Who Should Investigate

Who should conduct the accident investigation? Should it be the responsible supervisor? The safety and health professional? A higher-level manager? An outside specialist? There is no simple answer to this question, and there is disagreement among professional people of goodwill.

In some companies, the supervisor of the injured worker conducts the investigation. In others, a safety and health professional performs the job. Some companies form an investigative team; others bring in outside specialists. There are several reasons for the various approaches used. Factors considered in deciding how to approach accident investigations include:

- Size of the company
- Structure of the company's safety and health program
- Type of accident
- Seriousness of the accident
- Technical complexity
- Number of times that similar accidents have occurred
- Company's management philosophy
- Company's commitment to safety and health

After considering all the variables just listed, it is difficult to envision a scenario in which the safety and health professional would not be involved in conducting an accident investigation. If the accident in question is very minor, the injured employee's supervisor may conduct the investigation, but the safety and health professional should at least study the accident report and be consulted regarding recommendations for corrective action.

If the accident is so serious that it has widespread negative implications in the community and beyond, responsibility for the investigation may be given to a high-level manager or corporate executive. In such cases, the safety and health professional should assist in conducting the investigation. If a company prefers the team approach, the safety and health professional should be a member of the team and, in most cases, should chair it. Regardless of the approach preferred by a given company, the safety and health professional should play a leadership role in collecting and analyzing the facts and developing recommendations.

Conducting the Investigation

The questions in the previous section summarize what to look for when conducting accident investigations. Figure 7-1 lists five steps to follow in conducting an accident investigation.[10] These steps are explained in the following paragraphs.

Isolate the Accident Scene

You may have seen a crime scene that was sealed off by the police. The entire area surrounding such a scene is typically blocked off by barriers or heavy yellow tape. This is done to keep curious onlookers from removing, disturbing, or unknowingly destroying vital evidence. This same approach should be used when conducting an accident investigation. As soon as emergency procedures have been completed and the injured worker has been removed, the accident scene should be *isolated* until all pertinent evidence has been collected or observed and recorded. Further, nothing but the injured worker should be removed from the scene. If necessary, a security guard should be posted to

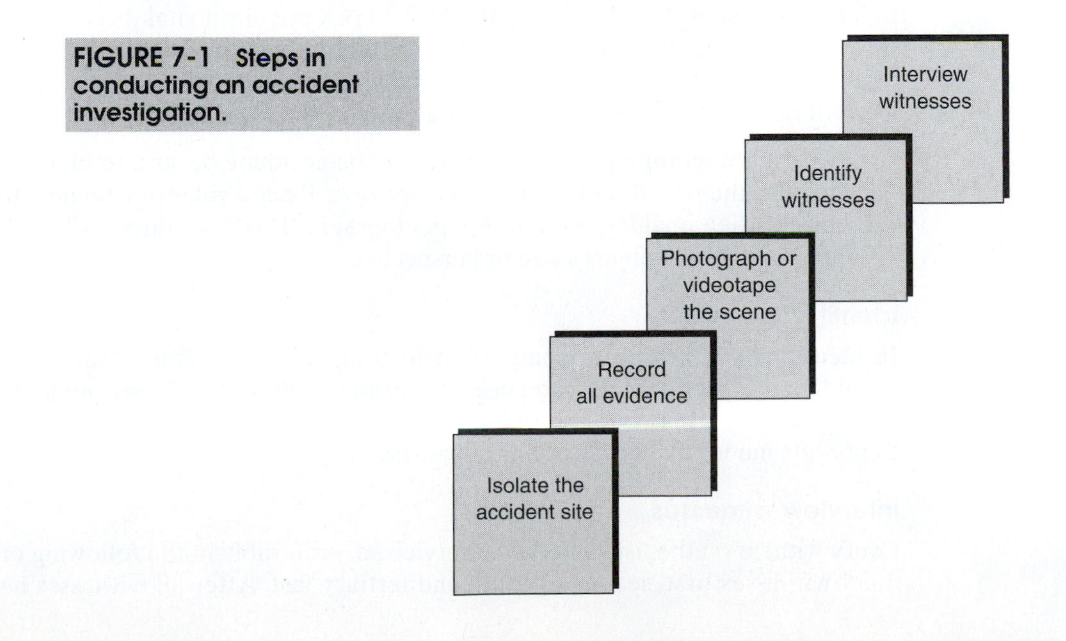

FIGURE 7-1 Steps in conducting an accident investigation.

maintain the integrity of the **accident scene**. The purpose of **isolating the scene** is to maintain as closely as possible the conditions that existed at the time of the accident.

Record All Evidence

It is important to make a permanent record of all *pertinent evidence* as quickly as possible. There are three reasons for this: (1) certain types of evidence may be perishable; (2) the longer an accident scene must be isolated, the more likely it is that evidence will be disturbed, knowingly or unknowingly; and (3) if the isolated scene contains a critical piece of equipment or a critical component in a larger process, pressure will quickly mount to get it back in operation. Evidence can be recorded in a variety of ways including written notes, sketches, photography, videotape, dictated observations, and diagrams. In deciding what to record, a good rule of thumb is *if in doubt, record it.* It is better to record too much than to skip evidence that may be needed later after the accident scene has been disturbed.

Photograph or Videotape the Scene

This step is actually an extension of the previous step. Modern photographic and videotaping technology has simplified the task of observing and recording evidence. Safety and health professionals should be proficient in the operation of a camera, even if it is just an instant camera, and a videotaping camera.

The advent of the digital camera has introduced a new meaning for the concept of "instant photographs." Using a digital camera in conjunction with a computer, photographs of accident scenes can be viewed immediately and transmitted instantly to numerous different locations. Digital camera equipment is especially useful when photographs of accident scenes in remote locations are needed.

Both still and video cameras should be on hand, loaded, and ready to use immediately should an accident occur. As with the previous step, a good rule of thumb in photographing and videotaping is *if in doubt, shoot it.* When recording evidence, it is better to have more shots than necessary than it is to risk missing a vital piece of evidence.

A problem with photographs is that, by themselves, they don't always reveal objects in their proper perspective. To overcome this shortcoming, the National Safety Council recommends the following technique:

> When photographing objects involved in the accident, be sure to identify and measure them to show the proper perspective. Place a ruler or coin next to the object when making a close-up photograph. This technique will help to demonstrate the object's size or perspective.[11]

Identify Witnesses

In **identifying witnesses**, it is important to compile a witness list. Names on the list should be recorded in three categories: (1) **primary witnesses**, (2) **secondary witnesses**, and (3) **tertiary witnesses** (Figure 7-2). When compiling the witness list, ask employees to provide names of all three types of witnesses.

Interview Witnesses

Every witness on the list should be interviewed, preferably in the following order: primary witnesses first, secondary next, and tertiary last. After all witnesses have been

FIGURE 7-2 Categories of accident witnesses.

- Primary witnesses are eyewitnesses to the accident.
- Secondary witnesses are witnesses who did not actually see the accident happen, but were in the vicinity and arrived on the scene immediately or very shortly after the accident.
- Tertiary witnesses are witnesses who were not present at the time of the accident nor afterward but may still have relevant evidence to present (e.g., an employee who had complained earlier about a problem with the machine involved in the accident).

interviewed, it may be necessary to reinterview witnesses for clarification or corroboration. Interviewing witnesses is such a specialized process that the next major section is devoted to it.

Interviewing Witnesses

The techniques used for interviewing accident witnesses are designed to ensure that the information is objective, accurate, as untainted by the personal opinions and feelings of witnesses as possible, and able to be corroborated. For this reason, it is important to understand the *when, where,* and *how* of interviewing the accident witnesses.

When to Interview

Immediacy is important. Interviews should begin as soon as the witness list has been compiled and, once begun, should proceed expeditiously. There are two main reasons for this. First, a witness's recollections will be best right after the accident. The more time that elapses between the accident and the interview, the more blurred the witness's memory will become. Second, immediacy avoids the possibility of witnesses comparing notes and, as a result, changing their stories. This is just human nature, but it is a tendency that can undermine the value of testimony given and, in turn, the facts collected. Recommendations based on questionable facts are not likely to be valid. Also, witnesses should be interviewed individually and separately, preferably before they have talked to each other.

Where to Interview

The best place to interview is at the accident scene. If this is not possible, interviews should take place in a private setting elsewhere. It is important to ensure that all distractions are removed, interruptions are guarded against, and the witness is not accompanied by other witnesses. All persons interviewed should be allowed to relate their recollections without fear of contradiction or influence by other witnesses or employees. It is also important to select a neutral location in which witnesses will feel comfortable. Avoid the "**principal's office syndrome**" by selecting a location that is not likely to be intimidating to witnesses.

How to Interview

The key to getting at the facts is to put the witness at ease and to listen. Listen to what is said, how it is said, and what is not said. Ask questions that will get at the information listed earlier in this chapter, but phrase them in an **open-ended** format. For example, instead of asking "Did you see the victim pull the red lever?" phrase your question as follows: "Tell me what you saw." Don't lead witnesses with your questions or influence them with gestures, facial expressions, tone of voice, or any other form of nonverbal communication. Interrupt only if absolutely necessary to seek clarification on a critical point. Remain nonjudgmental and objective.

The information being sought in an accident investigation can be summarized as *who, what, when, where, why,* and *how* (Figure 7-3). As information is given, it may be necessary to take notes. If you can keep your notetaking to a minimum during the interview, your chances of getting uninhibited information are increased. Notetaking can distract and even frighten a witness.

An effective technique is to listen during the interview and make mental notes of critical information. At the end of the interview, summarize what you have heard and have the witness verify your summary. After the witness leaves, develop your notes immediately.

A question that sometimes arises is, "Why not tape the interview?" Safety and health professionals disagree on the effectiveness and advisability of taping. Those who favor taping claim it allows the interviewer to concentrate on listening without having to worry about forgetting a key point or having to interrupt the witnesses to jot down

FIGURE 7-3 Questions to ask when interviewing witnesses.

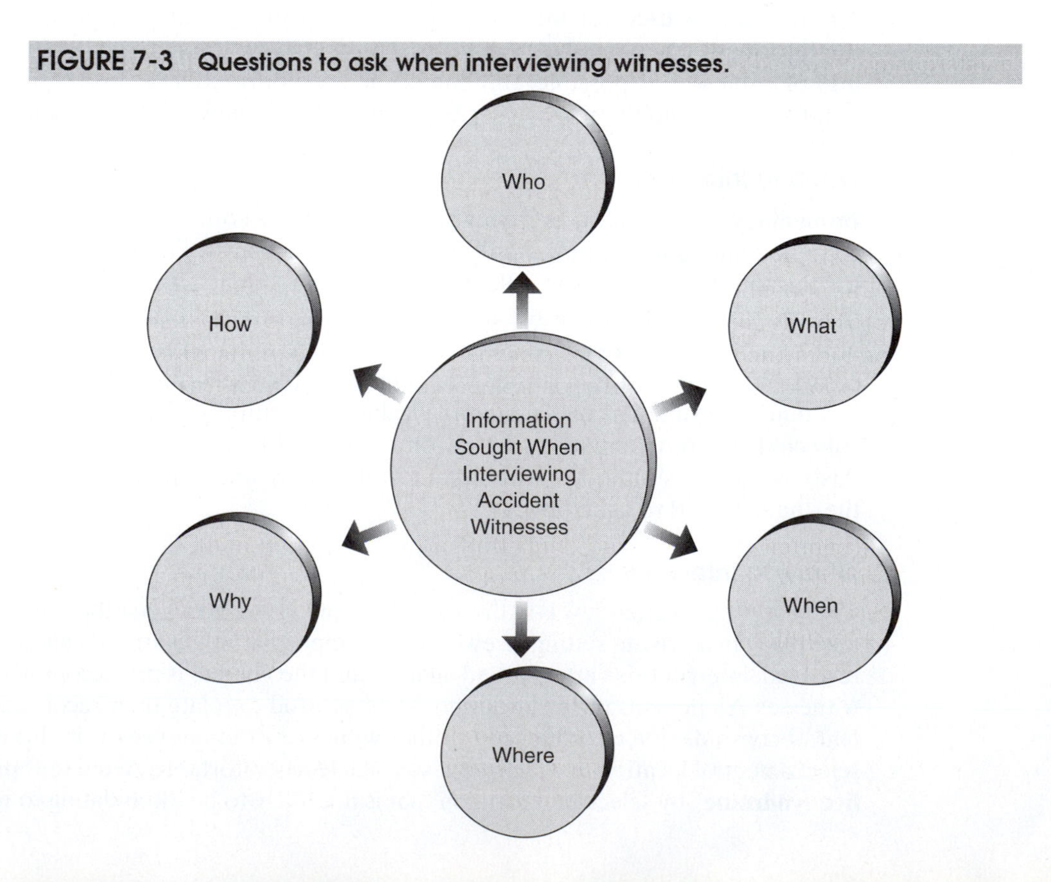

critical information. It also preserves everything that is said for the record as well as the tone of voice in which it is said. A complete transcript of the interview also ensures that information is not taken out of context.

Those opposed to taping say that taping devices tend to inhibit witnesses so that they are not as forthcoming as they would be without taping. Taping also slows down the investigation while the taped interview is transcribed and the interviewer wades through voluminous testimony trying to separate critical information from irrelevant information.

In any case, if the interview is to be taped, the following rules of thumb should be applied:

- Use the smallest, most unobtrusive taping device available, such as a microcassette recorder.
- Inform the witness that the interview will be taped.
- Make sure the taping device is working properly and that the tape it contains can run long enough so that you don't have to interrupt the witness to change it.
- Take time at the beginning of the interview to discuss unrelated matters long enough to put the witness at ease and overcome the presence of the taping device.
- Make sure that personnel are available to transcribe the tapes immediately.
- Read the transcripts as soon as they are available and highlight critical information.

An effective technique to use with eyewitnesses is to ask them to reenact the accident for you. Of course, the effectiveness of this technique is enhanced if the reenactment can take place at the accident site. However, even when this is not possible, an eyewitness reenactment can yield valuable information.

In using the **reenactment** technique, a word of caution is in order. If an **eyewitness** does exactly what the victim did, there may be another accident. Have the eyewitnesses explain what they are going to do before letting them do it. Then, have them *simulate* rather than actually perform the steps that led up to the accident.

Reporting Accidents

An accident investigation should culminate in a comprehensive *accident report*. The purpose of the report is to record the findings of the accident investigation, the cause or causes of the accident, and recommendations for corrective action.

OSHA has established requirements for reporting and record keeping. According to OSHA document 2056,

> Employers of 11 or more employees must maintain records of occupational injuries and illnesses as they occur. Employers with 10 or fewer employees are exempt from keeping such records unless they are selected by the Bureau of Labor Statistics (BLS) to participate in the Annual Survey of Occupational Injuries and Illnesses.[12]

All injuries and illnesses are supposed to be recorded, regardless of severity, if they result in any of the outcomes shown in Figure 7-4. If an accident results in the death of an employee or hospitalization of five or more employees, a report must be submitted to the nearest OSHA office within 48 hours. This rule applies regardless of the size of the company. Reporting locally within an organization for insurance, legal, prevention, and management purposes and reporting for OSHA purposes can be two different

FIGURE 7-4 OSHA record-keeping requirements.

Injuries and illnesses must be recorded if they result in any of the following:

- Death
- One or more lost work days
- Restriction of motion or work
- Loss of consciousness
- Transfer to another job
- Medical treatment (more than first aid)

tasks. OSHA's reporting requirements were covered in Chapter 5. Reporting procedures in this section pertain to local in-house reporting.

Accident report forms vary from company to company. However, the information contained in them is fairly standard. Regardless of the type of form used, an accident report should contain at least the information needed to meet the record-keeping requirements set forth by OSHA. This information includes at least the following, according to the National Safety Council:

- Case number of the accident
- Victim's department or unit
- Location and date of the accident or date that an illness was first diagnosed
- Victim's name, social security number, gender, age, home address, and telephone number
- Victim's normal job assignment and length of employment with the company
- Victim's employment status at the time of the accident (i.e., temporary, permanent, full-time, part-time)
- Case numbers and names of others injured in the accident
- Type of injury and body part(s) injured (e.g., burn to right hand; broken bone, lower right leg) and severity of injury (i.e., fatal, first aid only required, hospitalization required)
- Name, address, and telephone number of the physician called
- Name, address, and telephone number of the hospital to which the victim was taken
- Phase of the victim's work day when the accident occurred (e.g., beginning of shift, during break, end of shift, and so on)
- Description of the accident and how it took place, including a step-by-step sequence of events leading up to the accident
- Specific tasks and activities with which the victim was involved at the time of the accident (e.g., task: mixing cleaning solvent; activity: adding detergent to the mixture)
- Employee's posture or proximity related to his or her surroundings at the time of the accident (e.g., standing on a ladder; bent over at the waist inside the robot's work envelope)
- Supervision status at the time of the accident (i.e., unsupervised, directly supervised, indirectly supervised)
- Causes of the accident
- Corrective actions that have been taken so far
- Recommendations for additional corrective action[13]

In addition to these items, you may want to record such additional information as the list of witnesses; dates, times, and places of interviews; historical data relating to similar accidents; information about related corrective actions that were made previously but had not yet been followed up on; and any other information that might be relevant. Figure 7-5 is an example of an accident report form that meets the OSHA record-keeping specifications.

FIGURE 7-5 Sample accident report form.

ACCIDENT REPORT FORM

Fairmont Manufacturing Company
1501 Industrial Park Road
Fort Walton Beach, FL 32548
904-725-4041

Victim-Related Information

Person completing report ______________________ Case no. ______________

Victim's name ____

Gender ______________________________________ Age ______________

Date of accident/illness ____

Victim's home address/telephone ______________________________________

Victim's assignment at the time of the accident and length of time in that assignment:

____

____

____

Victim's normal job and length of time in that job: ______________________

____

____

Time of injury/illness and phase of victim's work day: ____________________

____

____

Severity of the injury (e.g., hospitalization required, first aid required, etc.): ________

____

____

Type of injury and body part(s) injured: ______________________________

____

Exact location of the accident (which facility, department, place within the department):

____

Physician and hospital: ____

____

Note: Complete one form for each injured worker.

FIGURE 7-5 ***(continued)***

Accident-Related Information

Accident description with step-by-step sequence of events: ____________________

____

____

Task and specific activity at the time of the accident: ____________________

____

____

Posture/proximity of employee at the time of the accident: ____________________

____

____

Supervision status at the time of the accident: ____________________

____

____

Apparent causes including conditions, actions, events, and activities and other contributing factors: ____________________

____

____

Recommendations for corrective action: ____________________

____

____

Case numbers and names of other persons injured in the accident: ____________

____

____

Witnesses to the accident, and dates/places of their interviews:

______________________________ ______________________________

______________________________ ______________________________

______________________________ ______________________________

______________________________ ______________________________

______________________________ ______________________________
Reporter name Date

______________________________ ______________________________
Employee name Date

Why Some Accidents Are Not Reported

In spite of OSHA's reporting specifications, some accidents still go unreported. According to Cunningham and Kane,

> The majority of accidents are not being reported! Articles in the Wall Street Journal testify to this fact. Many firms failed to report OSHA recordable incidents, presumably either to avoid OSHA inspections that result from poor

incident rates, or to achieve statistical goals. The saddest part of non-reporting of accidents is that they are not investigated to determine and eliminate the causes.[14]

There are several reasons why accidents go unreported. Be familiar with these reasons so that you can do your part to overcome them. Cunningham and Kane list the main reasons as follows:

1. **Red tape.** Some people see the paperwork involved in accident reporting as red tape and, therefore, don't report accidents just to avoid paperwork.
2. **Ignorance.** Not all managers and supervisors are as knowledgeable as they should be about the reasons for accident reporting. Many are not familiar with OSHA's reporting specifications.
3. **Embarrassment.** Occasionally, people do not report an accident because they are embarrassed by their part in it. A supervisor who did not properly supervise or a manager who has not provided the proper training for employees may be embarrassed to file a report.
4. **Record-spoiling.** Some accidents go unreported just to preserve a safety record, such as the record for days worked without an accident.
5. **Fear of repercussions.** Some accidents go unreported because the people involved are afraid of being found at fault, being labeled accident prone, and being subjected to other negative repercussions.
6. **No feedback.** Some accidents go unreported because those involved feel filing a report is a waste of time. This typically happens when management does not respond to recommendations made in earlier accident reports.[15]

Clearly, these reasons for not reporting accidents present safety and health professionals with a challenge. To overcome these inhibitors, it is necessary to develop a simple reporting system that will not be viewed as too much bureaucratic paperwork to have to do. Safety and health professionals must educate personnel at all levels concerning the purpose of accident reporting and why it is important. An important step is to communicate the fact that fault finding is not the purpose. Another important step is to follow up to ensure that recommendations are acted on or that employees are made aware of why they aren't. This helps ensure the integrity of the process.

Discipline and Accident Reporting

Fault finding is not the purpose of an accident investigation. However, an investigation sometimes reveals that an employee has violated or simply overlooked safety regulations. Should such violations be condoned? According to Kane and Cunningham,

> Many companies condone nonconformance to safety rules as long as no injury results. However, if the nonconformance results in an accident involving an injury, the disciplinary boom is promptly lowered. This inconsistency inevitably leads to resentment and failure to report accidents and a hiding of accident problems.[16]

There is a built-in dilemma here that modern safety and health professionals must be prepared to handle. On one hand, it is important that fault finding not be seen as the purpose of an accident investigation. Such a perception limits the amount of information that can be collected. On the other hand, if those workers whose behavior leads to accidents are not disciplined, the credibility of the safety program is undermined. Kane and Cunningham recommend the following procedures for handling this dilemma: Never discipline an employee because he or she had an accident. Always discipline employees for noncompliance with safety regulations.[17]

Such an approach applied with consistency will help maintain the integrity of both the accident investigation process and the overall safety program.

Ten Accident Investigation Mistakes To Avoid

The amount of information you collect and how you collect it will go a long way toward determining how effective your resultant corrective actions will be following a workplace accident. According to William R. Coffee Jr., safety and health professionals should avoid the following commonly made mistakes when investigating accidents.

- **Failing to investigate near misses.** A near miss is simply an accident that did not happen because of luck. Consequently, investigating near misses can reveal critical accident prevention information.
- **Taking ineffective corrective action.** Ineffective corrective action is often the result of a cursory accident investigation. When investigating, look for the root cause, not the symptoms. Corrective action based on symptoms will not prevent future accidents.
- **Allowing your biases to color the results of the investigation.** Look for facts and be objective when investigating an accident. Do not make assumptions or jump to conclusions. One of the best ways to eliminate bias in accident investigations is to use a standard, structured routine and to skip no steps in the routine.
- **Failing to investigate in a timely manner.** Accident investigations should begin as soon as possible after the accident. The longer you wait to begin, the more likely it is that evidence will be lost, corrupted, or compromised. For example, once people start talking to each other about what they saw, invariably their memories will be shaped by the opinions of their fellow workers and witnesses. People walking through an accident scene can compromise the integrity of the scene by unwittingly destroying evidence.
- **Failing to account for human nature when conducting interviews.** Often what those involved in an accident as well as witnesses to an accident will say during an interview will be shaped by their desire to escape blame, deflect blame to someone else, or protect a friend. This is why it is important to interview witnesses and others involved privately and individually, and to look for corroborating evidence to support (or refute) their input.
- **Failing to learn investigation techniques.** Before investigating an accident, safety and health professionals should complete specialized training or undertake self-study to learn investigation techniques such as those presented in this chapter. An unskilled investigator is not likely to conduct a valid investigation.
- **Allowing politics to enter into an investigation.** The goal of an investigation is and must be to identify the root cause so that appropriate corrective action can be

taken. Personal likes, dislikes, favoritism, and office politics will corrupt an investigation from the outset.

- **Failing to conduct an in-depth investigation.** Everyone is in a hurry and investigating an accident was not on your agenda for the day. In addition, there is sometimes pressure from higher management to "get this thing behind us." Such pressures and circumstances can lead to a rushed investigation in which the goal is to get it over with, not to find the root cause of the accident. Surface-level investigations almost ensure that the same type of accident will happen again.
- **Allowing conflicting goals to enter into an investigation.** The ultimate goal of an accident investigation is to prevent future accidents and injuries. However, even when that is your goal, there may be other people who have different goals. Some may see the investigation as an opportunity to deflect blame, others may see it as an opportunity to protect the organization from litigation, and some may see it as a way to explain not meeting production quotas or performance standards. Safety and health professionals should be aware that other agendas may be in play every time an accident investigation is conducted. For this reason, objectivity, structure, and routine are critical.
- **Failing to account for the effects of uncooperative people.** One would think that employees and management personnel would automatically want to cooperate in accident investigation to ensure that similar accidents are prevented in the future. Unfortunately, this is not always the case. People will not always cooperate for a variety of reasons—all growing out of the concept of perceived self-interest. Further, the lack of cooperation will not always be overt. In fact, often it will be covert (e.g., a person you need to interview keeps putting you off or canceling meetings). Safety and health professionals need to understand that self-interest is one of the most powerful motivators of human beings and factor this into their planning for accident investigations.[18]

These 10 mistakes will probably never be completely eliminated from every accident investigation. However, if safety and health professionals are aware of them, they can at least ensure that such mistakes are minimized. The fewer of these mistakes that are made during an accident investigation, the better the quality of the investigation and the more likely that it will lead to effective corrective action.

KEY TERMS AND CONCEPTS

- Accident-analysis report
- Accident investigation
- Accident report
- Accident scene
- Causes
- Decision to work unsafely
- Emergency procedures
- Evidence
- Eyewitness
- Fault finding
- How
- Identify witnesses
- Immediacy
- Isolate the scene
- Mismatch or overload
- Open-ended
- Personal beliefs and feelings
- Primary witness
- Principal's office syndrome
- Reenactment
- Secondary witness
- Systems failure
- Tertiary witness
- Traps
- Unsafe acts
- Unsafe conditions
- What
- When
- Where
- Who
- Why

Review Questions

1. Explain the rationale for investigating accidents.
2. When should an investigation be reported? Why?
3. Explain the difference between an accident report and an accident-analysis report.
4. List the categories of the most common causes of accidents.
5. What are the terms that should guide the conduct of an accident investigation?
6. What role should the safety and health professional play in the conduct of an accident investigation?
7. List and explain the steps for conducting an accident investigation.
8. Why is it important to record all pertinent evidence relating to an accident immediately after an accident has occurred?
9. What can you do when taking close-up photographs to put them in the proper perspective?
10. List and differentiate among the three categories of witnesses to an accident.
11. Briefly explain the when and where of interviewing witnesses.
12. Briefly explain the how of interviewing witnesses.
13. What is the purpose of an accident report?

Endnotes

1. D. Hartshorn, "Solving Accident Investigation Problems," *Occupational Hazards* 65, no. 1: 57.
2. Society of Manufacturing Engineers (SME), *Tool and Manufacturing Engineers Handbook* 6 (Dearborn, MI: Society of Manufacturing Engineers, 1998), 12–21.
3. Hartshorn, "Solving Problems," 57–58.
4. SME, *Tool and Manufacturing,* 12–21.
5. Ibid.
6. Ibid.
7. National Safety Council, *Supervisor's Safety Manual* (Chicago: National Safety Council, 1991), 69–70.
8. SME, *Tool and Manufacturing,* 12–21.
9. Hartshorn, "Solving Problems," 58–59.
10. National Safety Council, *Supervisor's Safety Manual,* 71.
11. Ibid.
12. U.S. Department of Labor, OSHA 2056, 2003 (Revised), 11.
13. National Safety Council, *Supervisor's Safety Manual,* 76–77.
14. J. Cunningham and A. Kane, "Accident Reporting—Part I: Key to Prevention," *Safety & Health* 139, no. 4, 70.
15. Ibid., 70–71.
16. A. Kane and J. Cunningham, "Accident Reporting—Part II: Consistent Discipline Is Vital," *Safety & Health* 139, no. 5, 78.
17. Ibid.
18. W. R. Coffee Jr., "Avoid These 10 Mistakes," *Occupational Health & Safety* 74, no. 5, 44–47.

CHAPTER 8

Ergonomic Hazards: Musculoskeletal Disorders (MSDs) and Cumulative Trauma Disorders (CTDs)

Major Topics

- Ergonomics Defined
- Human Factors and Ergonomic Hazards
- Factors Associated with Physical Stress
- Ergonomics: A Political Football
- OSHA's Voluntary Ergonomics Guidelines
- Worksite Analysis Program for Ergonomics
- Hazard Prevention and Control
- Medical Management Program
- Training and Education
- Common Indicators of Problems
- Identifying Specific Ergonomic Problems
- Ergonomic Problem-Solving Strategies
- Economics of Ergonomics
- Cumulative Trauma Disorders (CTDs)
- Participatory Ergonomics

The history of workplace development in the Western world is characterized by jobs and technologies designed to improve processes and productivity. All too often in the past, little or no concern was given to the impact of the job process or technology on workers. As a result, work processes and machines have sometimes been unnecessarily dangerous. Another result has been that new technologies have sometimes failed to live up to expectations. This is because, even in the age of high technology, human involvement in work processes is still the key to the most significant and enduring productivity improvements. If a machine or system is uncomfortable, difficult, overly complicated, or dangerous to use, human workers will not be able to derive its full benefit.

The proliferation of uncomfortable and dangerous workplace conditions, whether created by job design or unfriendly technologies, is now widely recognized as harmful to productivity, quality, and worker safety and health. The advent of the science of ergonomics is making the workplace more physically friendly. This, in turn, is making the workplace a safer and healthier place.

Ergonomics Defined

Minimizing the amount of physical stress in the workplace requires continuous study of the ways in which people and technology interact. The insight learned from this study must then be used to improve the interaction. This is a description of the science of **ergonomics**. For the purpose of this book, ergonomics is defined as follows:

> Ergonomics is a multidisciplinary science that seeks to conform the workplace and all of its physiological aspects to the worker. Ergonomics involves the following:
>
> > Using special design and evaluation techniques to make tasks, objects, and environments more compatible with human abilities and limitations.
> >
> > Seeking to improve productivity and quality by reducing workplace stressors, reducing the risk of injuries and illnesses, and increasing efficiency.

The word *ergonomics* is derived from the Greek language. *Ergon* is Greek for *work; nomos* means *laws*. Therefore, in a literal sense, ergonomics means work laws. In practice, it consists of the scientific principles (laws) applied in minimizing the physical stress associated with the workplace (work). Figure 8-1 summarizes some of the widely accepted benefits of ergonomics.

There are benefits to be derived from ergonomics (as listed in Figure 8-1). There are also problems, both financial and health related, that can result from giving too little attention to ergonomics. The matter is complicated further because health problems tend to multiply a company's financial problems. Consequently, modern safety and health professionals need to be well versed in ergonomics.

FIGURE 8-1 Benefits of ergonomics.

- Improved health and safety for workers
- Higher morale throughout the workplace
- Improved quality
- Improved productivity
- Improved competitiveness
- Decreased absenteeism and turnover
- Fewer workplace injuries/health problems

Human Factors and Ergonomic Hazards

When the topic of ergonomics is discussed, the term *human factors* will usually find its way into the conversation.[1] But what is meant by the term? Tillman and Tillman define it as follows: Consumers are demanding safe and effective products. However, not all people have control over products they use. Therefore, all products must be carefully designed. For example, if a child car seat fails because it does not fit the child or is difficult to install, everyone will lose: the child, the parent, the designer, and the manufacturer. Human factors is a profession to help ensure that equipment and systems are safe and easy to operate by human beings. A human factors researcher gathers and analyzes data on human beings (how they work, their size, their capabilities and limitations). A human factors engineer works with designers as a team to incorporate data into designs to make sure people can operate and maintain the product or system. Human factors professionals then determine the skills needed to operate or maintain a finished product. Human factors is difficult to define because it is a compilation of many sciences dealing with both humans and machines. Some of the disciplines human factors experts are trained in include the following: psychology, anthropology, engineering, biology, medicine, education, and physiology.[2]

Human Factors Defined

Tillman and Tillman further define human factors as a "science combining research and application of human data."[3] The concept can also be viewed as a science that bridges research about human beings and the application of that research in designing products and systems for human beings.

Human Factors in Action

Tillman and Tillman provide several examples of how the science of human factors fits into the systems design process. Perhaps the best way to get a feel for the concept of human factors is to consider several of these examples:

1. **Predesign analysis.** In this stage of the design process, human factors professionals conduct research to answer such questions as: What is the best way for humans to interact with computers? What factors contribute to fatigue and stress in an office environment? How can designers overcome these factors?
2. **Preliminary design.** In this stage, human factors professionals study machine and human capabilities to determine which tasks should be undertaken manually and which should be automated.
3. **Detail design and development.** In design and development, human factors professionals define the environment required for operator safety, enhanced operator performance, and the reduction or prevention of operator stress and fatigue.
4. **Test and evaluation.** In this stage of the process, human factors professionals test actual humans in using the prototype equipment or system.[4]

Human Factors and Safety

Human factors can play an important role in both product safety and workplace safety (where many products are used). What follows is how the science of human factors can help reduce both product and workplace hazards:

1. **Hazard elimination by design.** Human error is frequently the root cause or a contributing cause in accidents on the job. Intelligent design can reduce human errors by providing controls that are simple to understand and operate and by proving human–machine interaction that is not boring or overly demanding physically.
2. **Provision and location of safety devices.** The design and location of safety devices such as emergency cutoff switches can reduce human error on the job, correspondingly reducing the chances of an accident.
3. **Provision of warning devices.** The color, location, and wording of warning devices; the pitch and volume of warning signals; and the design of caution markings on gauges and video displays are all important factors in reducing the likelihood of human error that might lead to an accident. The science of human factors can help determine the appropriate way to apply all of these factors in a given setting.
4. **Establishment of procedures/provision of training.** When hazards cannot be realistically designed out of a system, administrative procedures for hazard reduction must be established and training relating to those procedures must be provided. Human factors professionals can help establish appropriate administrative procedures and help develop the necessary training.[5]

Factors Associated with Physical Stress

Eight variables that can influence the amount of *physical stress* experienced on the job are as follows:

Sitting versus standing

Stationary versus moveable/mobile

Large demand for strength/power versus small demand for strength/power

Good horizontal work area versus bad horizontal work area

Good vertical work area versus bad vertical work area

Nonrepetitive motion versus repetitive motion

Low surface versus high surface contact

No negative environmental factors versus negative environmental factors

The following paragraphs summarize the extent of this influence and the forms that it can take.[6]

Sitting versus Standing

Generally speaking, sitting is less stressful than standing. Standing for extended periods, particularly in one place, can produce unsafe levels of stress on the back, legs, and feet. Although less so than standing, sitting can be stressful unless the appropriate precautions are taken. These precautions include proper posture, a supportive back rest, and frequent standing/stretching movement.

Stationary versus Mobile

Stationary jobs are those done primarily at one workstation. Of course, even these jobs involve movement at the primary workstation and occasional movement to other areas. Mobile jobs, on the other hand, require continual movement from one station to another. The potential for physical stress increases with stationary jobs when workers fail to take such precautions as periodically standing/stretching/moving. The potential for physical stress increases with mobile jobs when workers carry materials as they move from station to station.

Large versus Small Demand for Strength/Power

In classifying jobs by these two criteria, it is important to understand that repeatedly moving small amounts of weight over a period of time can have a cumulative effect equal to the amount of stress generated by moving a few heavy weights. Regardless of whether the stress results from lifting a few heavy objects or repeated lifting of lighter objects, jobs that demand larger amounts of strength/power are generally more stressful than those requiring less.

Good versus Bad Horizontal Work Area

A good **horizontal work area** is one that is designed and positioned so that it does not require the worker to bend forward or to twist the body from side to side. Horizontal work areas that do require these movements are bad. Bad horizontal work surfaces increase the likelihood of physical stress.

Good versus Bad Vertical Work Area

Good **vertical work areas** are designed and positioned so that workers are not required to lift their hands above their shoulders or bend down in order to perform any task. Vertical work areas that do require these movements are bad. Bad vertical work areas increase the likelihood of physical stress.

Nonrepetitive versus Repetitive Motion

Repetitive motion jobs involve short-cycle motion that is repeated continually. Nonrepetitive jobs involve a variety of tasks that are not, or only infrequently, repeated. Repetition can lead to monotony and boredom. When this happens, the potential for physical stress increases.

Low versus High Surface Contact

Surface stress can result from contact with hard surfaces such as tools, machines, and equipment. High-surface-contact jobs tend to be more stressful in a physical sense than are low-surface-contact jobs.

SAFETY FACT

High Cost of Pain

When employees are in pain, they respond by taking sick leave from work. Sick leave due to unspecified pain costs employers more than $3 billion in lost workdays. Approximately 17 million employees in the United States take an average of three days of sick leave per year to deal with unspecified pain (e.g., headaches, neck pain, back pain, and menstrual pain). Often the pain is not work related.

Absence versus Presence of Environmental Factors

Generally, the more **environmental factors** with which a worker has to contend on the job, the more stressful the job. For example, personal protective equipment, although conducive to reducing environmental hazards, can increase the amount of physical stress associated with the job.

Ergonomics: A Political Football

OSHA's attempts to develop an ergonomics standard have been the subject of bitter disputes between labor and management in the private sector. These disputes, in turn, have found their way into the political arenas of both the U.S. Congress and the executive branch. OSHA established its voluntary ergonomics guidelines in 1989. Soon thereafter, organized labor began its campaign to have the guidelines made mandatory through passage of an ergonomics standard. During the Clinton administration (1993–2000), Congress steadfastly refused to approve OSHA's proposed ergonomics standard.

Consequently, just before leaving office, President Clinton signed an executive order putting the elements of the proposed ergonomics standard in place as a rule. However, with the election of George W. Bush as president, Congress used the Congressional Review Act (CRA) to overturn former President Clinton's executive order. Organized labor and other supporters of OSHA vowed to make congressional opponents of an ergonomics standard and President Bush pay for overturning the executive order that had in essence made the voluntary guidelines mandatory. However, the most ardent opponents in Congress of an ergonomics standard all won their elections and were returned to Congress, now even more determined to squelch any attempts by supporters to pass a standard.

OSHA has responded to all of this in two ways. First, OSHA continues to develop voluntary guidelines beyond those that apply to industry in general for businesses in specific industrial classifications. Second, OSHA has begun to claim that it will use its **general duty clause** to enforce ergonomic safety. However, there is little evidence that this is actually happening. After vowing to make enforcement through the general duty clause part of its four-pronged attack on musculoskeletal hazards, OSHA is, in reality, focusing most of its efforts on the other three prongs: voluntary guidelines, outreach and assistance, and continuing research.

OSHA's Voluntary Ergonomics Guidelines

OSHA first published guidelines for general safety and health program management in 1989. OSHA's **ergonomics guidelines** are voluntary and are designed to provide employers with the information and guidance needed to meet their obligations under the OSH Act regarding ergonomics.

OSHA followed these guidelines with another set designed specifically for the meat-packing industry. In June 2002, OSHA announced voluntary guidelines for retail grocery stores and the poultry-processing industry. These specific guidelines represent a model for guidelines that are likely to be developed for other specific industries. Meat packing and poultry processing were singled out because of the high incidence of **cumulative trauma disorders (CTDs)** associated with meat packing and processing.

CTDs are injuries that result from an accumulation of repetitive motion stress. For example, using scissors continually over time can cause a CTD in the hand and wrist.

OSHA's Ergonomics Standard (Voluntary Guidelines)

OSHA's voluntary guidelines were well received by employees and labor organizations with one major caveat: the "voluntary" nature of the guidelines. Many people interested in occupational safety and health advocated a mandatory standard for ergonomics. This led to discussions concerning the development by OSHA of a new ergonomics standard. The standard has not yet been developed.

However, OSHA's proposed ergonomics standard failed to win approval from Congress. Consequently, after much political wrangling, OSHA officials decided to convert the proposed ergonomics standard into voluntary ergonomics guidelines. OSHA's current plan for reducing ergonomic hazards in the workplace has four elements:

1. Voluntary guidelines for specific industries.
2. Enforcement of the guidelines under the general duty clause of the OSH Act 5(a)(1). This is a controversial element of the four-part plan because some employers think that OSHA is using the general duty clause to make the "voluntary" guidelines mandatory against the stated will of Congress.
3. Compliance assistance to help employers reduce ergonomic hazards.
4. Research into ergonomic issues to help identify gaps in the body of knowledge surrounding this topic.

Enforcement by OSHA

Although the ergonomics guidelines developed by OSHA are voluntary, at least for the foreseeable future, the agency does claim that it will use the general duty clause of the OSH Act to enforce the guidelines in certain situations. OSHA's criteria for applying the general duty clause are as follows:

1. Is there currently an ergonomic hazard that is causing injuries?
2. Does the employer in question know about the hazard (or should the employer know)?
3. Are the injuries caused by the ergonomic hazard resulting in serious physical harm?
4. Are there feasible alternatives available to the employer for reducing, abating, or minimizing the hazard?

Although the use of the general duty clause to enforce voluntary ergonomics guidelines is controversial and has been questioned by many employers in theory, in practice this is an approach that is rarely used by OSHA.

Rationale for the Voluntary Guidelines

Musculoskeletal disorders (MSDs) and cumulative trauma disorders (CTDs) account for more than 30 percent of all occupational injuries and illnesses in the United States. Such injuries and illnesses result in more than 600,000 lost workdays annually and account for approximately one of every three dollars spent on workers' compensation every year. A conservative estimate of these costs is more than $20 billion annually.

Organizations that are losing valuable employee time to CTDs and MSDs are well advised to implement the type of ergonomics program set forth in OSHA's standard on a voluntary, self-enforced basis.

Application of the Voluntary Guidelines

OSHA's ergonomics guidelines are geared toward manufacturing and materials handling in the general industry sector. They do not apply to construction, maritime operations, agriculture, or employers that operate a railroad, although they can be applied to other jobs in which the type of work that is fundamental and necessary to perform the job result in MSDs. The ergonomics guidelines apply to more than 1.5 million employees nationwide. Examples of jobs to which the guidelines apply are as follows:

- Patient-handling jobs (nurse assistants and orderlies)
- Shipping, receiving, and delivery (package sorting, handling, delivery, etc.)
- Baggage handlers
- Warehouse work (manual tasks)
- Beverage and water handling and delivery
- Grocery/retail store stocking and bagging
- Garbage and trash collecting
- Assembly-line work
- Piecework assembly
- Product inspection (involving manual tasks such as weighing objectives)
- Meat, poultry, and fish processing
- Machine loading, unloading, and operation
- Textile manufacturing
- Food preparation assembly-line work
- Commercial banking
- Cabinet making
- Tire making

Proposed Requirements of the Voluntary Guidelines

Organizations that fall into the general industry classifications of manufacturing and manual material handling are asked by the guidelines to implement a "basic ergonomics program." This amounts to assigning responsibility for ergonomics to one individual and informing employees about the risks of MSD-related injuries, symptoms of such injuries, and why early reporting of symptoms is important. In addition, the basic program requires employers to establish a system that employees can use to report symptoms of MSD injuries.

The so-called full ergonomics program set forth in the guidelines is not required unless and until an employee's job is determined, by use of a basic screening tool (such as the one shown in Figure 8-2) to have met the "action trigger." The full program consists of the following components:

- Management leadership and employee participation
- Employee participation
- Training
- Record keeping
- Job hazard analysis and control

FIGURE 8-2 Basic screening tool for VDT workstations.

Workstation Checklist

for

Video Display Terminals

The workstation should be designed to ensure the following conditions:

A. *Head* and *neck* should be about upright.

B. *Head, neck,* and *trunk* should face forward.

C. *Trunk* should be about perpendicular to floor.

D. *Shoulders* and *upper arms* should be about perpendicular to floor and relaxed.

E. *Upper arms* and *elbows* should be close to body.

F. *Forearms, wrists,* and *hands* should be straight and parallel to floor.

G. Wrists and hands should be straight.

H. *Thighs* should be about parallel to floor, and lower legs should be about perpendicular to floor.

I. *Feet* should rest flat on floor or be supported by a stable footrest.

J. *VDT tasks* should be organized in a way that allows employee to vary them with other work activities or to take short breaks or recovery pauses while at the VDT workstation.

SEATING (The Chair)

1. *Backrest* should provide support for employee's lower back.
2. *Seat width* and *depth* should accommodate specific employee.
3. *Seat front* should not press against the back of employee's knees and lower legs.
4. *Seat* should have cushioning and be rounded with a "waterfall" front.
5. *Armrests* should support both forearms while the employee performs VDT tasks and not interfere with movement.

KEYBOARD/INPUT DEVICE

The keyboard/input device should be designed for doing VDT tasks so that. . . .

6. *Keyboard/input device platforms* are stable and large enough to hold keyboard and input device.
7. *Input device* (mouse or trackball) is located right next to keyboard so it can be operated without reaching.
8. *Input device* is easy to activate, and shape/size fits hand of specific employee (not too big/small).
9. *Wrists* and *hands* do not rest on sharp or hard edge.

FIGURE 8-2 *(continued)*

MONITOR

The monitor should be designed for VDT tasks so that. . . .

10. *Top line* of screen is at or below eye level so the employee is able to read it without bending head or neck down/back.
11. *Employees with bifocals/trifocals* are able to read the screen without bending their heads or necks backward.
12. *Monitor distance* allows employees to read the screen without leaning their heads, necks, or trunks forward/backward.
13. *Monitor position* is directly in front of the employee so he or she does not have to twist the head or neck.
14. *No glare* (e.g., from windows, lights) is present on the screen that may cause employees to assume an awkward position to read screen.

WORK AREA

The work area is designed for doing VDT tasks so that. . . .

15. *Thighs* have clearance space between chair and the VDT table/keyboard platform.
16. *Legs* and *feet* have clearance space under the VDT table so the employee is able to get close enough to the keyboard/input device.

ACCESSORIES

17. *Document holder,* if provided, should be stable and large enough to hold documents that are used.
18. *Document holder,* if provided, should be placed at about the same height and distance as the monitor screen so there is little head movement while the employee looks from document to screen.
19. *Wrist rest,* if provided, should be padded and free of sharp and square edges.
20. *Wrist rest,* if provided, should allow employees to keep forearms, wrists, and hands straight and parallel to ground when using the keyboard/input device.
21. *Telephone* can be used with head upright and shoulders relaxed if the employee does VDT tasks at the same time.

GENERAL

22. Workstation and equipment should have sufficient adjustability so that the employee is able to be in a safe working posture and to make occasional changes in posture while performing VDT tasks.
23. VDT workstation, equipment, and accessories should be maintained in serviceable condition and function properly.

Note to Employers:
Workstation must meet ALL of criteria A–J and all but two of criteria 1–23.

Source: Adapted from Appendix D-2 of OSHA Standard 1910.900.

- Work restriction protection
- MSD management
- Program evaluation

Worksite Analysis Program for Ergonomics

Although complex analyses are best performed by a professional ergonomist, the "ergonomics team"—or any qualified person—can use this program to conduct a *worksite analysis* and identify stressors in the workplace. The purpose of the following information is to give a starting point for finding and eliminating those tools, techniques, and conditions that may be the source of ergonomic problems.[7]

In addition to analyzing current workplace conditions, planned changes to existing and new facilities, processes, materials, and equipment should be analyzed to ensure that changes made to enhance production will also reduce or eliminate **ergonomic risk factors**. As emphasized before, this program should be adapted to each individual workplace.

The discussion of the recommended program for worksite analysis is divided into four main parts: (1) gathering information from available sources; (2) conducting baseline screening surveys to determine which jobs need closer analysis; (3) performing ergonomic job hazard analyses of those workstations with identified risk factors; and (4) after implementing control measures, conducting periodic surveys and follow-up studies to evaluate changes.

Information Sources

Records Analysis and Tracking

The essential first step in worksite analysis is **records analysis and tracking** to develop the information necessary to identify ergonomic hazards in the workplace. Existing medical, safety, and insurance records, including OSHA 300 logs, should be analyzed for evidence of injuries or disorders associated with CTDs. Health care providers should participate in this process to ensure confidentiality of patient records.

Incidence Rates

Incidence rates for upper extremity disorders and/or back injuries should be calculated by counting the incidence of CTDs and reporting the number for each 100 full-time workers per year by facility:

$$\text{Incidence rate} = \frac{(\text{Number of new cases/yr}) \times (200{,}000 \text{ work hrs}) \text{ per facility}}{\text{Number of hours worded/facility/yr}}{}^{*}$$

Screening Surveys

The second step in worksite analysis is to conduct baseline screening surveys. Detailed baseline screening surveys identify jobs that put employees at risk of developing CTDs. If the job places employees at risk, an effective program will then require the ergonomic job hazard analysis.**

*Adapted from OSHA 3123.

**The same method should be applied to departments, production lines, or job types within the facility.

Checklist

The survey is performed with an ergonomic checklist. This checklist should include components such as posture, materials handling, and upper extremity factors. (The checklist should be tailored to the specific needs and conditions of the workplace.)

Ergonomic Risk Factors

Identification of **ergonomic hazards** is based on ergonomic risk factors: conditions of a job process, workstation, or work method that contribute to the risk of developing CTDs. Not all these risk factors will be present in every CTD producing job, nor is the existence of one of these factors necessarily sufficient to cause a CTD.

CTD Risk Factors

Some of the risk factors for CTDs of the upper extremities include the following:

- Repetitive and/or prolonged activities
- Forceful exertions, usually with the hands (including pinch grips)
- Prolonged static postures
- Awkward postures of the upper body, including reaching above the shoulders or behind the back, and twisting the wrists and other joints to perform tasks
- Continued physical contact with work surfaces (soft tissue compression)
- Excessive vibration from power tools
- Cold temperatures
- Inappropriate or inadequate tool design
- High wrist acceleration
- Fatigue (inadequate recovery time)
- Use of gloves

Back Disorder Risk Factors

Risk factors for back disorders include:

- Bad body mechanics such as continued bending over at the waist, continued lifting from below the knees or above the shoulders, and twisting at the waist, especially while lifting
- Lifting or moving objects of excessive weight or asymmetric size
- Prolonged sitting, especially with poor posture
- Lack of adjustable chairs, footrests, body supports, and work surfaces at workstations
- Poor grips on handles
- Slippery footing

Multiple Risk Factors

Jobs, operations, or workstations that have **multiple risk factors** have a higher probability of causing CTDs. The combined effect of several risk factors in the development of CTDs is sometimes referred to as *multiple causation*.

Ergonomic Job Hazard Analyses

At this point, the employer has identified—through the information sources and screening surveys just discussed—jobs that place employees at risk of developing

CTDs. As an essential third step in the worksite analysis, an effective ergonomics program requires a job hazard analysis for each job so identified.

The job hazard analysis should be routinely performed by a qualified safety and health professional, preferably an ergonomist, for jobs that put workers at risk of developing CTDs. This type of analysis helps verify lower risk factors at light-duty or restricted activity work positions and to determine if risk factors for a work position have been reduced or eliminated to the extent feasible.

Workstation Analysis

An adequate **workstation analysis** would be expected to identify all risk factors present in each studied job or workstation. For upper extremities, three measures of repetitiveness are the total hand manipulations per cycle, the cycle time, and the total manipulations or cycles per work shift. Force measurements may be noted as an estimated average effort and a peak force. They may be recorded as light, moderate, or heavy.

Tools should be checked for excessive vibration. The tools, personal protective equipment, and dimensions and adjustability of the workstation should be noted for each job hazard analysis. Finally, hand, arm, and shoulder postures and movements should be assessed for levels of risk.

Lifting Hazards

For manual materials handling, the maximum weight-lifting values should be calculated.

DVD/Videotape Method

The use of DVD/videotape, where feasible, is suggested as a method for analysis of the work process. Slow-motion videotape or equivalent visual records of workers performing their routine job tasks should be analyzed to determine the demands of the task on the worker and how each worker actually performs each task.

Periodic Ergonomic Surveys

The fourth step in worksite analysis is to conduct periodic review. Periodic surveys should be conducted to identify previously unnoticed factors or failures or deficiencies in work practices or engineering controls. The periodic review process should also include feedback, follow-up, and trend analysis.

Feedback and Follow-Up

A reliable system should be provided for employees to notify management about conditions that appear to be hazardous and to utilize their insight and experience to determine work practice and engineering controls. This may be initiated by an ergonomics questionnaire and maintained through an active safety and health committee or by employee participation with the ergonomics team. Reports of ergonomic hazards or signs and symptoms of potential CTDs should be investigated by ergonomics screening surveys and appropriate ergonomic hazard analysis in order to identify risk factors and controls.

Trend Analysis

Trends of injuries and illnesses related to actual or potential CTDs should be calculated, using several years of data where possible. Trends should be calculated for several departments, process units, job titles, or workstations. These trends may also be used to determine which work positions are most hazardous and need to be analyzed by the qualified person.

Using standardized job descriptions, incidence rates may be calculated for work positions in successive years to identify trends. Using trend information can help determine the priority of screening surveys and/or ergonomic hazard analyses.

Hazard Prevention and Control

Engineering solutions, where feasible, are the preferred method for ergonomic **hazard prevention and control**. The focus of an ergonomics program is to make the job fit the person—not to make the person fit the job. This is accomplished by redesigning the workstation, work methods, or tool to reduce the demands of the job, including high force, repetitive motion, and awkward postures. A program with this goal requires research into currently available controls and technology. It should also include provisions for utilizing new technologies as they become available and for in-house research and testing. Following are some examples of engineering controls that have proven to be effective and achievable.

Workstation Design

Workstations should be designed to accommodate the persons who actually use them; it is not sufficient to design for the average or typical worker. Workstations should be easily adjustable and should be either designed or selected to fit a specific task, so that they are comfortable for the workers who use them. The work space should be large enough to allow for the full range of required movements, especially where knives, saws, hooks, and similar tools are used.

Design of Work Methods

Traditional work method analysis considers static postures and repetition rates. This should be supplemented by addressing the force levels and the hand and arm postures involved. The tasks should be altered to reduce these and the other stresses associated with CTDs. The results of such analyses should be shared with the health care providers to assist in compiling lists of light-duty and high-risk jobs.

Tool Design and Handles

Tools should be selected and designed to minimize the risks of upper extremity CTDs and back injuries. In any tool design, a variety of sizes should be available. Examples of criteria for selecting tools include the following:

- Designing tools to be used by either hand, or providing tools for both left- and right-handed workers.
- Using tools with triggers that depress easily and are activated by two or more fingers.
- Using handles and grips that distribute the pressure over the fleshy part of the palm, so that the tool does not dig into the palm.
- Designing and selecting tools for minimum weight; counterbalancing tools heavier than one or two pounds.

- Selecting pneumatic and power tools that exhibit minimal vibration and maintaining them in accordance with manufacturer's specifications or with an adequate vibration-monitoring program. Wrapping handles and grips with insulation material (other than wraps provided by the manufacturer for this purpose) is normally *not* recommended, as it may interfere with a proper grip and increase stress.

Medical Management Program

An effective **medical management program** for cumulative trauma disorders is essential to the success of an employer's ergonomic program in industries with a high incidence of CTDs.[8] It is not the purpose of these guidelines to dictate medical practice for an employer's **health care providers**. Rather, they describe the elements of a medical management program for CTDs to ensure early identification, evaluation, and treatment of signs and symptoms; to prevent their recurrence; and to aid in their prevention. Medical management of CTDs is a developing field, and health care providers should monitor developments on the subject. These guidelines represent the best information currently available.

A physician or occupational health nurse (OHN) with training in the prevention and treatment of CTDs should supervise the program. Each work shift should have access to health care providers in order to facilitate treatment, surveillance activities, and recording of information. Where such personnel are not employed full-time, the part-time employment of appropriately trained health care providers is recommended.

In an effective ergonomics program, health care providers should be part of the ergonomics team interacting and exchanging information routinely to prevent and treat CTDs properly. The major components of a medical management program for the prevention and treatment of CTDs are trained first-level health care providers, health surveillance, employee training and education, early reporting of symptoms, appropriate medical care, accurate record keeping, and quantitative evaluation of CTD trends throughout the plant.

Trained and Available Health Care Providers

Appropriately trained health care providers should be available at all times and on an ongoing basis as part of the ergonomics program. In an effective medical management program, first-level health care providers should be knowledgeable in the prevention, early recognition, evaluation, treatment, and rehabilitation of CTDs, as well as in the principles of ergonomics, physical assessment of employees, and OSHA record-keeping requirements.

Periodic Workplace Walk-Through

In an effective program, health care providers should conduct periodic, systematic workplace walk-throughs to remain knowledgeable of operations and work practices, to identify potential light-duty jobs, and to maintain close contact with employees. Health care providers should also be involved in identifying risk factors for CTDs in the workplace as part of the ergonomics team.

These walk-through surveys should be conducted every month or whenever a particular job task changes. A record should be kept documenting the date of the walk-through, areas visited, risk factors recognized, and action initiated to correct identified problems. Follow-up should be initiated to correct problems identified and should be documented to ensure that corrective action is taken when indicated.

Symptoms Survey

Those responsible for the medical management program should develop a standardized measurement to determine the extent of work-related disorder symptoms in each area of the plant. This measurement will help determine which jobs are exhibiting problems and to measure progress of the ergonomics program.

Institute a Survey

A **symptoms survey** of employees should be conducted to measure employee awareness of work-related disorders and to report the location, frequency, and duration of discomfort. Body diagrams should be used to facilitate gathering this information. Surveys normally should not include employee's personal identifiers to encourage employee participation.

The survey is one method for identifying areas or jobs where potential CTD problems exist. The major strength of the survey approach is in collecting data on the number of workers who may be experiencing some form of CTD. Reported pain symptoms by several workers on a specific job would indicate the need for further investigation of that job.

Conduct the Survey Annually

Conducting the survey annually should help detect any major change in the prevalence, incidence, and/or location of reported symptoms.

Keep List of Light-Duty Jobs

The ergonomist or other qualified person should analyze the physical procedures used in the performance of each job, including lifting requirements, postures, hand grips, and frequency of repetitive motion.

The ergonomist and health care providers should develop a list of jobs with the lowest ergonomic risk. For such jobs, the ergonomic risk should be described. This information will assist health care providers in recommending assignments to light- or restricted-duty jobs. The light-duty job should, therefore, not increase ergonomic stress on the same muscle-tendon groups. Health care providers should likewise develop a list of known high-risk jobs. Supervisors should periodically review and update the lists.

Health Surveillance

Baseline

The purpose of baseline health surveillance is to establish a base against which changes in health care status can be evaluated, not to prevent people from performing work. Prior to assignment, all new and transferred workers who are to be assigned to positions involving exposure of a particular body part to ergonomic stress should receive a baseline health surveillance.

Conditioning Period Follow-Up

New and transferred employees should be given the opportunity during a four- to six-week break-in period to condition their muscle-tendon groups prior to working at full capacity. Unfortunately, this is not always possible; however, when it is possible, conditioning should be done. Health care providers should perform a follow-up assessment of these workers after the break-in period (or after one month, if the break-in period is longer than a month) to determine if conditioning of the muscle-tendon groups has been successful; whether any reported soreness or stiffness is transient and consistent with normal adaptation to the job or whether it indicates the onset of a CTD; and if problems are identified, what appropriate action and further follow-up are required.

Periodic Health Surveillance

Periodic health surveillance—every two to three years—should be conducted on all workers who are assigned to positions involving exposure of a particular body part to ergonomic stress. The content of this assessment should be similar to that outlined for the baseline. The worker's medical and occupational history should be updated.

Employee Training and Education

Health care providers should participate in the training and education of all employees, including supervisors and other plant management personnel, on the different types of CTDs and means of prevention, causes, early symptoms, and treatment of CTDs. This information should be reinforced during workplace walk-throughs and the individual health surveillance appointments. All new employees should be given such education during orientation. This demonstration of concern and the distribution of information should facilitate the early recognition of CTDs prior to the development of more severe and disabling conditions and increase the likelihood of compliance with prevention and treatment.

Encourage Early Report of Symptoms

Employees should be encouraged by health care providers and supervisors to report early signs and symptoms of CTDs to the in-plant health facility. This allows for timely and appropriate evaluation and treatment without fear of discrimination or reprisal by employers. It is important to avoid any potential disincentives for employee reporting, such as limits on the number of times that an employee may visit the health unit.

Protocols for Health Care Providers

Health care providers should use written protocols for health surveillance and the evaluation, treatment, and follow-up of workers with signs or symptoms of CTDs. A qualified health care provider should prepare the protocols. These protocols should be available in the plant health facility. Additionally, the protocols should be reviewed and updated annually or as state-of-the-art evaluation and treatment of these conditions changes.

Evaluation, Treatment, and Follow-Up of CTDs

If CTDs are recognized and treated appropriately early in their development, a more serious condition can likely be prevented. Therefore, a good medical management program that seeks to identify and treat these disorders early is important.

OSHA Record-Keeping Forms

The Occupational Safety and Health Act and record-keeping regulations in Title 29, Code of Federal Regulations (CFR) 1904 provide specific recording requirements that comprise the framework of the occupational safety and health recording system. The Bureau of Labor Statistics (BLS) has issued guidelines that provide official agency interpretations concerning the record keeping and reporting of occupational injuries and illnesses. These guidelines—U.S. Department of Labor, BLS: *Record-Keeping Guidelines for Occupational Injuries and Illnesses,* September 1986 (or later editions as published)— provide supplemental instructions for the OSHA record-keeping forms and should be available in every plant health care facility. Because health care providers often provide information for OSHA logs, they should be aware of record-keeping requirements and participate in fulfilling them.

Monitor Trends

Health care providers should periodically (e.g., quarterly) review health care facility sign-in logs, OSHA Form 300, and individual employee medical records to monitor trends for CTDs in the plant. This ongoing analysis should be made in addition to the symptoms survey to monitor trends continuously and to substantiate the information obtained in the annual symptoms survey. The analysis should be done by department, job title, work area, and so on.

The information gathered from the annual symptoms survey will help identify areas or jobs where potential CTD problems exist. This information may be shared with anyone in the plant because employees' personal identifiers are not solicited. The analysis of medical records (e.g., sign-in logs and individual employee medical records) may reveal areas or jobs of concern, but it may also identify individual workers who require further follow-up. The information gathered while analyzing medical records is confidential; thus, care must be exercised to protect the individual employee's privacy. The information gained from the CTD trend analysis and symptoms survey will help determine the effectiveness of the various programs initiated to decrease CTDs in the plant.

Training and Education

The fourth major program element for an effective ergonomics program is training and education.[9] The purpose of training and education is to ensure that employees are sufficiently informed about the ergonomic hazards to which they may be exposed and thus able to participate actively in their own protection.

Training and education allow managers, supervisors, and employees to understand the hazards associated with a job or process, their prevention and control, and their medical consequences. A training program should include all affected employees, engineers and maintenance personnel, supervisors, and health care providers.

The program should be designed and implemented by qualified persons. Appropriate special training should be provided for personnel responsible for administering the program. The program should be presented in language and at a level of understanding appropriate for the individuals being trained. It should provide an overview of the potential risk of illnesses and injuries, their causes and early symptoms, the means of prevention, and treatment.

The program should also include a means for adequately evaluating its effectiveness. This may be achieved by using employee interviews, testing, and observing work practices to determine if those who received the training understand the material and the work practices to be followed.

Common Indicators of Problems

Does my company have ergonomic problems? Are injuries and illnesses occurring because too little attention is paid to ergonomic factors? These are questions that modern safety and health professionals should ask themselves. But how you answer such questions? According to the National Institute for Occupational Safety and Health (NIOSH), the factors discussed in the following paragraphs can be examined to determine if ergonomic problems exist in a given company.[10]

Apparent Trends in Accidents and Injuries

By examining accident reports, record-keeping documents such as OSHA Form 300 (Log of Work-Related Illnesses and Injuries), first-aid logs, insurance forms, and other available records of illnesses or injuries, safety and health professionals can identify trends if they exist. A pattern or a high incidence rate of a specific type of injury typically indicates that an ergonomic problem exists.

Incidence of Cumulative Trauma Disorders

Factors associated with CTDs include a high level of repetitive work, greater than normal levels of hand force, awkward posture, high levels of vibration, high levels of mechanical stress, extreme temperatures, and repeated hand-grasping or pinch-gripping. By observing the workplace and people at work, safety and health professionals can determine the amount of exposure that employees have to these factors and the potential for ergonomics-related problems.

Absenteeism and High Turnover Rates

High absentee rates and high turnover rates can be indicators of ergonomic problems. People who are uncomfortable on the job to the point of physical stress are more likely to miss work or leave for less stressful conditions.

Employee Complaints

A high incidence of **employee complaints** about physical stress or poor workplace design can indicate the presence of ergonomic problems.

Employee-Generated Changes

Employees tend to adapt the workplace to their needs. The presence of many workplace adaptations, particularly those intended to decrease physical stress, can indicate the presence of ergonomic problems. Have employees added padding, modified personal protective equipment, brought in extra lighting, or made other modifications? Such **employee-generated changes** may be evidence of ergonomic problems.

Poor Quality

Poor quality, although not necessarily caused by ergonomic problems, can be the result of such problems. Poor quality is at least an indicator that there may be ergonomic problems. Certainly, poor quality is an indicator of a need for closer inspection.

Manual Material Handling

The incidence of musculoskeletal injuries is typically higher in situations that involve a lot of **manual material handling**. Musculoskeletal injuries increase significantly when the job involves one or more of the following: lifting large objects, lifting bulky objects, lifting objects from the floor, and lifting frequently. When such conditions exist, the company has ergonomic problems.

Identifying Specific Ergonomic Problems

A task analysis of the job in question can identify specific ergonomic problems. Figure 8-3 lists the types of problems that can be identified by a thorough task analysis.[11]

General Observation

General observation of a worker or workers performing the task(s) in question can be an effective task analysis technique. The effectiveness is usually enhanced if the workers are not aware that they are being observed. When observing employees at work, be especially attentive to tasks requiring manual material handling and repetitive movements.

Questionnaires and Interviews

Questionnaires and **interviews** can be used for identifying ergonomic problems. Questionnaires are easier to distribute, tabulate, and analyze, but interviews generally provide more in-depth information.

Videotaping and Photography

Videotaping technology has simplified the process of task analysis considerably. Videotaping records the work being observed as it is done, it is silent so it is not intrusive, and

FIGURE 8-3 Problems that can be pinpointed by a task analysis.

- Tasks that involve potentially hazardous movements
- Tasks that involve frequent manual lifting
- Tasks that involve excessive wasted motion or energy
- Tasks that are part of a poor operations flow
- Tasks that require unnatural or uncomfortable posture
- Tasks with high potential for psychological stress
- Tasks with a high fatigue factor
- Tasks that could or should be automated
- Tasks that involve or lead to quality control problems

such capabilities as freeze and playback enhance the observer's analysis capabilities significantly. Photography can also enhance the observer's analysis capabilities by recording each motion or step involved in performing a task. If photography is used, be aware that flashes can be disruptive. High-speed film will allow you to make photographs without using a flash.

Drawing or Sketching

Making a neat sketch of a workstation or a drawing showing workflow can help identify problems. Before using a drawing or sketch as part of a task analysis, make sure that it is accurate. Ask an employee who is familiar with the area or process sketched to check the drawing.

Measuring the Work Environment

Measurements can help identify specific ergonomic problems. How far must a worker carry the material manually? How high does a worker have to lift an object? How much does an object weigh? How often is a given motion repeated? Answers to these and similar questions can enhance the effectiveness of the analysis process.

Understanding the Ergonomics of Aging

When identifying specific ergonomic problems in the workplace, don't overlook the special challenges presented by aging workers. A good ergonomics program adapts the job to the person. Because nearly 30 percent of the workforce is 45 years of age or older, organizations must be prepared to adapt workstations to employees whose physical needs are different from those of their younger counterparts.

In adapting workstations and processes for employees who are 45 or older, keep in mind the following rules of thumb:

- Nerve conduction velocity, hand-grip strength, muscle mass, range of motion, and flexibility all begin to diminish about age 45.
- Weight and mass tend to increase through about the early fifties.
- Height begins to diminish beginning around age 30.
- Lower back pain is more common in people 45 years of age and older.
- Visual acuity at close range diminishes with age.[12]

These rules of thumb mean that safety and health professionals cannot take a "one-size-fits-all" approach to ergonomics. Adaptations for older workers must be individualized and should take aging factors into account.

Ergonomic Problem-Solving Strategies

The factors that influence stress were explained earlier in this chapter. These factors can be combined in different ways, and the ways in which they are combined determine the type and amount of stress experienced. For the purpose of recommending ergonomic problem-solving strategies, the following combinations can be used:

- Seated repetitive work with light parts
- Seated work with larger parts
- Seated control work
- Standing work

FIGURE 8-4 Adjust the angle of work to reduce stress.

Source: From a drawing by Robin J. Miller.

- Standing for heavy lifting and/or carrying work in one place or in motion
- Work with hands above chest height
- Work with hand tools
- Work with VDTs[13]

Seated Repetitive Work with Light Parts

This type of work can produce more physical stress than you may suspect. Back, neck, shoulder, and lower leg pain are commonly associated with this type of work. Some of the problems associated with seated repetitive work are the result of the nature of the job. The fixed work position and repetitive motion can contribute to ergonomic problems. To solve these problems, it may be necessary to modify both the job and the workstation. Improvement strategies include the following:

- Include other work tasks to break the monotony of repetition.
- Use job rotation, with workers rotating from one or more different jobs.
- Adjust the height of the work surface and/or position.
- Use an adjustable chair equipped with hand, wrist, or arm supports as appropriate.
- Make sure that there is sufficient legroom (height, width, and depth).
- Use ergonomic devices to adjust the height and angle of work (see Figure 8-4).[14]

Seated Work with Larger Parts

This type of work, which involves interacting with objects that may be too large to manipulate manually, is associated with assembly and welding jobs. Problems associated

FIGURE 8-5 Use adjustable work surfaces to reduce stress.

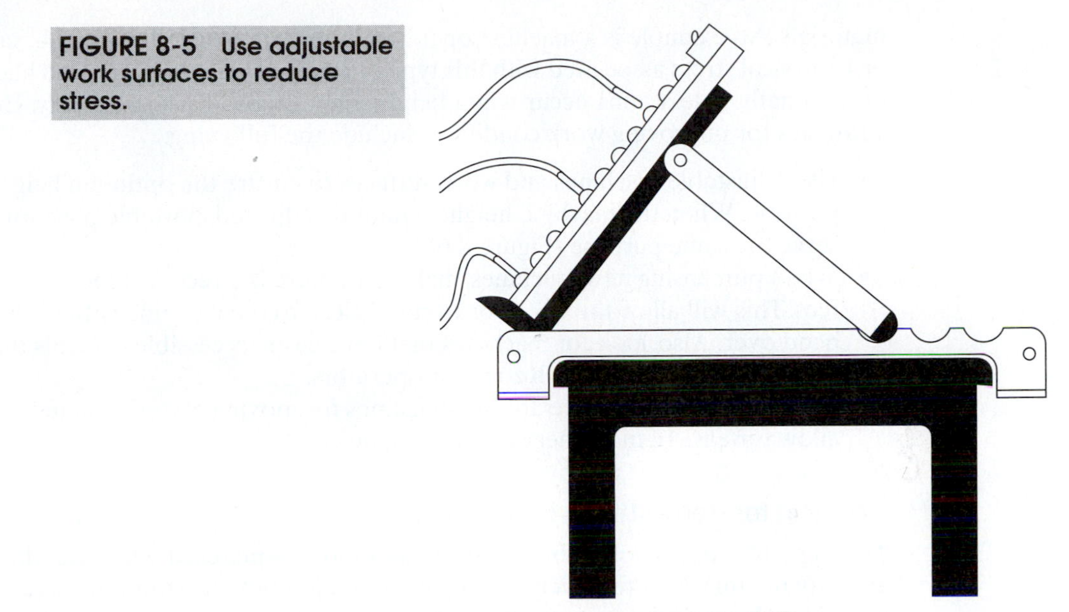

Source: From a drawing by Robin J. Miller.

with this type of work are typically related to posture, illumination, reach, and lifting. Ergonomic strategies for improving work conditions include the following:

- Use technology to lift and position the work for easy access that does not require bending, twisting, and reaching.
- Use supplemental lighting at the worksite.
- Use adjustable chairs and work surfaces as appropriate (see Figure 8-5).[15]

Seated Control Work

This type of work involves sitting in one location and using wheels, levers, knobs, handles, and buttons to control a process, system, or piece of equipment. The physical stress associated with seated control work is typically the result of excessive vibration or bending and twisting to achieve better visibility. Ergonomic strategies for improving work conditions include:

- Use an adjustable swivel chair with inflatable back and seat support.
- Provide comfortable and convenient locations for control devices.
- Use control devices that meet two standards: finger control systems that do not require more than five newtons (1.1 pounds); hand levers that do not exceed 20 newtons (4.5 pounds).
- Position the control seat so that a clear line of sight exists between the work and the person controlling it.
- Provide a ladder if the workstation is more than 14 inches above ground.[16]

Standing Work

This category includes most jobs that are performed while standing. Such jobs do not involve a great deal of repetitive motion but do involve handling medium to heavy materials. An example is a machine operator's job (lathe, mill, drill, punch, saw, and so on). Physical stress associated with this type of work includes leg, arm, and back strains. Occasionally, side strains occur when bending and twisting are necessary. Ergonomic strategies for improving work conditions include the following:

- Use adjustable machines and work surfaces to ensure the optimum height and position. When the machine height cannot be adjusted, portable platforms can serve the same purpose (Figure 8-6).
- When purchasing new machines, make sure there is a recess at the bottom for feet. This will allow the operator to stand close to the machine without having to bend over. Also, look for machines that have easily accessible controls that fall within a comfortable reach zone for operators.
- Provide ample free space around machines for moving material in and out and to allow for ease of movement in servicing machines.[17]

Standing for Heavy Lifting and Carrying

This type of work involves heavy lifting and moving material while standing. Lifting and moving may be a relatively small part of the job but are required somewhat regularly. The physical stress most commonly associated with this type of work is back and muscle strains resulting from improper lifting. Falls can also be a problem. Ergonomic strategies for improving work conditions include the following:

- Eliminate manual lifting to the extent possible using various lifting and hoisting technologies.
- Where manual lifting is necessary, train workers in proper lifting techniques.
- Provide sufficient room around all objects to allow lifting without twisting.

FIGURE 8-6 Adjust machines or work surfaces to reduce stress.

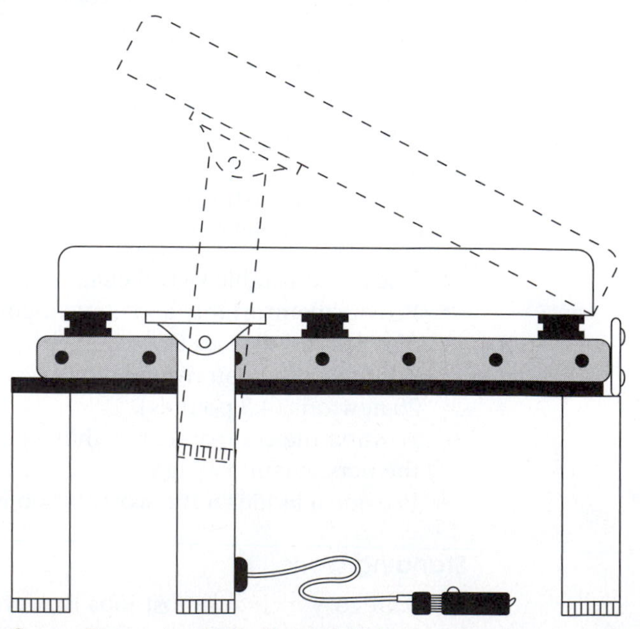

Source: From a drawing by Robin J. Miller.

- Supply the appropriate personal protection equipment such as sure-grip shoes and gloves.
- Keep floors around materials to be lifted clean and dry to prevent slips.
- Do not allow manual carrying of heavy objects upstairs. Stairs increase the physical stress of carrying and, in turn, the potential for injury.[18]

Work with Hands above the Chest

This type of work can be done in either a standing or sitting position. It may or may not involve material handling. Physical stress associated with this type of work includes neck, upper body, and heart strain. Of these, the most potentially dangerous is heart strain. Prolonged work with the arms above the shoulder level requires the heart to work harder to pump blood to the elevated areas. Ergonomic strategies for improving work conditions include:

- Eliminate manual lifting to the extent possible by raising the work floor using lifts and various other technologies.
- Use extension arms or poles when the work floor cannot be raised.
- When purchasing new machines, look for machines with controls that are easily accessible below the horizontal plane of a worker's shoulders.[19]

Work with Hand Tools

All the types of work presented in this section may involve the use of hand tools to some extent. However, because hand tools introduce a variety of potential hazards that are indigenous to their use, they are best examined as a separate work type. Physical stress associated most commonly with the use of hand tools includes carpal tunnel syndrome (CTS) and muscle strains of the lower arm, hands, and wrist. Ergonomic strategies for improving the work conditions focus primarily on improving hand positions during the use of tools, enhancing the worker's grip on tools, and minimizing the amount of twisting involved. Following are some of these strategies:

- Select tools that are designed to keep hands in the rest position (palm down, gently curved, thumb outstretched, and knuckle of the index finger higher than that of the little finger).
- Reduce stress on the hand by selecting tools that have thick, rather than thin, handles (a good range for the diameter is 0.8 to 1.2 inches).
- Select tools that have enhanced gripping surfaces on handles such as knurling, filing, or other enhancements.
- To the extent possible, eliminate twisting by selecting tools designed so that the direction of movement or function is the same as the direction in which force is applied or by using technology (e.g., power screwdriver).
- For tools that do not involve twisting, select handles that have an oval-shaped cross section.
- Select tools with handles made of hard, nonpermeable materials that will not absorb toxic liquids that could be harmful to the skin.[20]

Work with VDTs

The video display terminal, primarily because of the all-pervasive integration of personal computers in the workplace, is now the most widely used piece of office equipment. This

FIGURE 8-7 Ergonomics of VDTs. The left of the diagram highlights optimal postures and positions for the computer user.

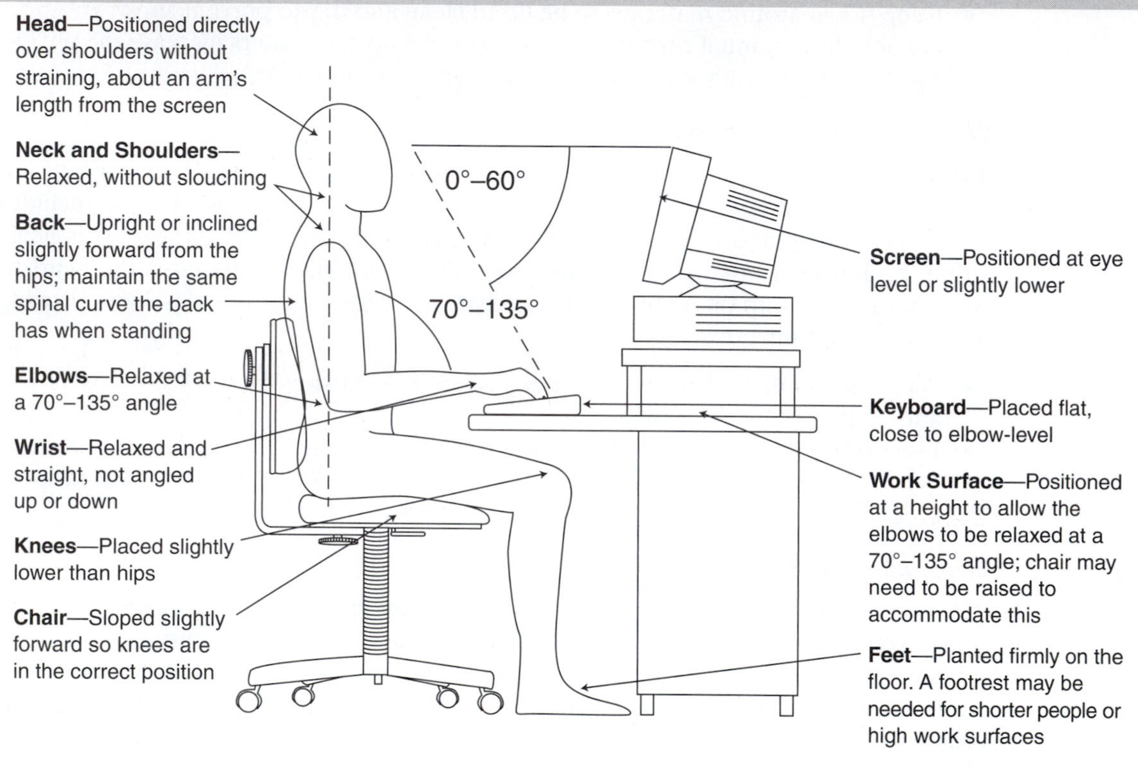

fact coupled with the ergonomic hazards associated with VDTs have created a whole new range of concerns for safety and health professionals. Using ergonomics to design a workspace will make it easier, safer, more comfortable, and more efficient to use. Following are some strategies that can be used to reduce the hazards associated with VDTs:

- **Arrange the keyboard properly.** It should be located in front of the user, not to the side. Body posture and the angle formed by the arms are critical factors (see Figure 8-7).
- **Adjust the height of the desk.** Taller employees often have trouble working at average height desks. Raising the desk with wooden blocks can solve this problem.
- **Adjust the tilt of the keyboard.** The rear portion of the keyboard should be lower than the front.
- **Encourage employees to use a soft touch on the keyboard and when clicking a mouse.** A hard touch increases the likelihood of injury.
- **Encourage employees to avoid wrist resting.** Resting the wrist on any type of edge can increase pressure on the wrist.
- **Place the mouse within easy reach.** Extending the arm to its full reach increases the likelihood of injury.
- **Remove dust from the mouse ball cavity.** Dust can collect, making it difficult to move the mouse. Blowing out accumulated dust once a week will keep the mouse easy to manipulate.

- **Locate the VDT at a proper height and distance.** The VDT's height should be such that the top line on the screen is slightly below eye level. The optimum distance between the VDT and user will vary from employee to employee, but it will usually be between 16 and 32 inches.
- **Minimize glare.** Glare from a VDT can cause employees to adopt harmful postures. Changing the location of the VDT, using a screen hood, and closing or adjusting blinds and shades can minimize glare.
- **Reduce lighting levels.** Reducing the lighting level in the area immediately around the VDT can eliminate vision strain.
- **Dust the VDT screen.** VDT screens are magnets to dust. Built-up dust can make the screen difficult to read, contributing to eye strain.
- **Eliminate telephone cradling.** Cradling a telephone receiver between an uplifted shoulder and the neck while typing can cause a painful disorder called **cervical radiculopathy** (compression of the cervical vertebrae in the neck). Employees who need to talk on the telephone while typing should wear a headset.
- **Require typing breaks.** Continuous typing for extended periods should be avoided. **Repetitive strain injuries** are cumulative. Breaking up the repetitive motion in question (typing and clicking) can help prevent the accumulation of strain.[21]

Economics of Ergonomics

Perhaps the most underresearched subject relating to ergonomics is the cost-effectiveness of safety and health. There are two main reasons for this: (1) such research is often more complex and extensive than the safety and health measures that have been undertaken; and (2) many decision makers think such studies are irrelevant because promoting safety and health is the right thing to do, regardless of costs. As with safety and health in general, there are no in-depth studies available that conclusively pin down the cost benefits of specific ergonomic measures.

According to Bedal,

> Does the application of ergonomic principles make good business sense? The good news is most industry experts believe ergonomics does make good business sense. But the bad news is that if you're looking for studies to prove ergonomics is worth the investment, you'll be hard pressed to find them. Very few true cost-benefit analyses on applying ergonomic principles in an industrial setting have been done.[22]

There is disagreement among well-informed professionals on whether ergonomic improvements should be expected to meet the test of cost–benefit analysis. However, there is growing support for research that will produce reliable data. This point of view is beginning to characterize the outlook of safety and health professionals in government and academe.[23] Consequently, it is important to understand the problems in attempting to undertake or participate in cost–benefit studies.

Inhibitors of Cost–Benefit Studies

A number of factors inhibit hard research into the economics of ergonomics. Bedal summarizes the most problematic of these as follows:

- Record-keeping systems in industry are not sufficient to support such studies. As a result, a base of comparison from which to work has not been established.
- Industry does not track injuries and illnesses in ways that provide the controls necessary for true hard research. There are no control groups against which to compare groups of injured workers.
- It is difficult and sometimes impossible to determine what improvements can be attributed directly to specific ergonomic strategies and what improvements should be attributed to other factors.
- Undertaking hard research studies requires a commitment of both time and money. A longitudinal study of the effects of a given ergonomic improvement may take three to five years to produce reliable data.
- Follow-up evaluations of injuries that summarize the direct and indirect costs do not exist to the extent necessary to contribute to hard research.[24]

These inhibitors cannot be overcome unless industrial firms are willing to invest enormous sums of money and time. Although the data produced by such investments would be valuable, they are generally not perceived as adding to a company's bottom line. Therefore, industry is likely to continue to rely on academe and governmental agencies for research. In today's intensely competitive international marketplace, industry is likely to invest all available funds in efforts that are perceived as improving productivity and quality. Consequently, hard research into the economics of ergonomics may remain a low priority for some time to come.

Cumulative Trauma Disorders (CTDs)

The personal computer has become an all-pervasive and universal work tool. Jobs from the shop floor to the executive office now involve frequent, repetitive computer use. This means that people in the workplace are typing and clicking at an unprecedented pace. Frequent and, for some, constant computer use have led to an explosion of injuries that until now were seen mostly in the meat-packing industry. Collectively, these injuries are known as **cumulative trauma disorders (CTDs)**.

Definition

CTD is an umbrella term that covers a number of injuries caused by forceful or awkward movements repeated frequently over time. Other aggravating factors include poor posture, an improperly designed workstation, poor tool design, and job stress. CTDs occur to the muscles, nerves, and tendons of the hands, arms, shoulders, and neck.

Classifications of CTDs

For years, CTDs have been incorrectly referred to as *carpal tunnel syndrome,* which is actually one type of CTD. This is like referring to all trees as oaks. Figure 8-8 is a checklist of the most common CTDs organized into four broad classifications.

Muscle and Tendon Disorders

Tendons connect muscles to bones. They can accommodate very little in the way of stretching and are prone to injury if overused. Overworking a tendon can cause small

FIGURE 8-8 CTD checklist—types of injury by classification.

Muscle and Tendon Disorders

✓ Tendinitis
✓ Muscle damage (myofacial)
✓ Tenosynovitis
✓ Stenosing tenosynovitis
- DeQuervain's disease
- Trigger finger (flexor tenosynovitis)

✓ Shoulder tendinitis
✓ Bicipital tendinitis
✓ Rotator cuff tendinitis
✓ Forearm tendinitis
- Flexor carpi radialis tendinitis
- Extensor tendinitis
- Flexor tendinitis

✓ Epicondylitis
✓ Ganglion cysts

Cervical Radiculopathy

Tunnel Syndromes

✓ Carpal tunnel syndrome
✓ Radial tunnel syndrome
✓ Sulcus ulnaris syndrome
✓ Cubital tunnel syndrome
✓ Guyon's canal syndrome

Nerve and Circulation Disorders

✓ Thoracic outlet syndrome
✓ Raynaud's disease

tears in it. These tears can become inflamed and cause intense pain. This condition is known as **tendinitis**.

Myofacial muscle damage can also be caused by overexertion. It manifests itself in soreness that persists even when resting. Muscles may burn and be sensitive to the touch. When sore muscles become inflamed and swell, the symptoms are aggravated even further by nerve compression. Tendons that curve around bones are encased in protective coverings called *sheathes*. Sheathes contain a lubricating substance known as synovial fluid. When tendons rub against the sheath too frequently, friction is produced. The body responds by producing additional synovial fluid. Excess build-up of this fluid can cause swelling that, in turn, causes pressure on the surrounding nerves, causing a condition known as **tenosynovitis**.

Chronic tenosynovitis is known as *stenosing tenosynovitis* of which there are two types: DeQuervain's disease and flexor tenosynovitis (trigger finger). DeQuervain's disease affects the tendon at the junction of the wrist and thumb. It causes pain when the thumb is moved or when the wrist is twisted. Flexor tenosynovitis involves the locking of a digit in a bent position, hence the term *trigger finger*. However, it can occur in any finger.

Shoulder tendinitis is of two types: bicipital and rotator cuff tendinitis. Bicipital tendinitis occurs at the shoulder joint where the bicep muscle attaches. The rotator cuff is a group of muscles and tendons in the shoulder that move the arm away from the body and turn it in and out. Pitchers in baseball and quarterbacks in football often experience rotator cuff tendinitis.

Forearm tendinitis is of three types: flexor carpi radialis tendinitis, extensor tendinitis, and flexor tendinitis. Flexor carpi radialis tendinitis causes pain in the wrist at the base of the thumb. Extensor tendinitis causes pain in the muscles in the top of the hand, making it difficult to straighten the hands. Flexor tendinitis causes pain in the fingers, making them difficult to bend.

Epicondylitis and *ganglion cysts* are two muscle and tendon disorders. Epicondylitis (lateral) affects the outside of the elbow, whereas epicondylitis (medial) affects the inside. The common term for this disorder is "tennis elbow." Ganglion cysts grow on the tendon, tendon sheath, or synovial lining, typically on top of the hand, on the nail bed, above the wrist, or on the inside of the wrist.

Cervical Radiculopathy

This disorder is most commonly associated with holding a telephone receiver on an upraised shoulder while typing. This widely practiced act can cause compression of the cervical discs in the neck, making it painful to turn the head. Putting the body in an unnatural posture while using the hands is always dangerous.

Tunnel Syndromes

Tunnels are conduits for nerves that are formed by ligaments and other soft tissues. Damage to the soft tissues can cause swelling that compresses the nerves that pass through the tunnel. These nerves are the median, radial, and ulnar nerves that pass through a tunnel in the forearm and wrist. Pain experienced with tunnel injuries can be constant and intense. In addition to pain, people with a tunnel injury might experience numbness, tingling, and a loss of gripping power. The most common ***tunnel syndromes*** are carpal tunnel syndrome, radial tunnel syndrome, sulcus ulnaris syndrome, cubital tunnel syndrome, and Guyon's canal syndrome.

Nerve and Circulation Disorders

When friction or inflammation cause swelling, both nerves and arteries can be compressed, restricting the flow of blood to muscles. This can cause a disorder known as *thoracic outlet syndrome*. The symptoms of this disorder are pain in the entire arm, numbness, coldness, and weakness in the arm, hand, and fingers.

If the blood vessels in the hands are constricted, *Raynaud's disease* can result. Symptoms include painful sensitivity, tingling, numbness, coldness, and paleness in the fingers. It can affect one or both hands. This disorder is also known as *vibration syndrome* because it is associated with vibrating tools.

Preventing CTDs

The best way to prevent CTDs is proper work design. In addition, it also helps make employees aware of the hazards that can cause it. These hazards include poor posture at the workstation, inappropriate positioning of the hands and arms, a heavy hand on a keyboard or mouse, and any other act that repeatedly puts the body in an unnatural posture while using the hands. Ergonomically sound workstations can help prevent CTDs, especially when they can be modified to fit the individual employee. However, even the best ergonomic design cannot prevent a heavy hand on the keyboard or mouse. Consequently, ergonomics is only part of the answer. Following are some preventive strategies that can be applied in any organization:

1. **Teach employees the warning signs.** CTDs occur cumulatively over time; they sneak up on people. Employees should be aware of the following warning signs: weakness in the hands or forearms, tingling, numbness, heaviness in the hands, clumsiness, stiffness, lack of control over the fingers, cold hands, and tenderness to the touch.
2. **Teach employees how to stretch.** Employees whose jobs involve repetitive motion work such as typing may help prevent CTDs by using stretching exercises. Limbering up the hands and forearms each day before starting work and again after long breaks such as the lunch hour may help eliminate the stress on muscles and tendons that can lead to CTDs. The term "may" is used because the jury is still out as to the efficacy of stretching. There is no consensus in the ergonomics community on this preventive measure.
3. **Teach employees to start slowly.** Long-distance runners typically start slowly, letting their bodies adjust and their breathing find its rhythm. They pick up the pace steadily, until eventually settling in at a competitive pace. This approach is an excellent example of how employees in CTD-prone jobs should work. Teach employees to limber up, then begin slowly and increase their pace gradually.
4. **Avoid the use of wrist splints.** Teach employees to position their hands properly without using wrist splints. Splints can cause the muscles that they support to atrophy, thereby actually increasing the likelihood of problems.
5. **Start an exercise group.** Exercises that strengthen the hands and forearms coupled with exercises that gently stretch hand and forearm muscles may be a preventive measure. Exercises that strengthen the back can help improve posture, and good posture helps prevent CTDs. As with stretching, the jury is still out on the efficacy of exercise as a preventive measure.
6. **Select tools wisely.** CTS and other cumulative trauma disorders (CTDs) are most frequently associated with the repetitive use of VDTs and hand tools. Selecting and using hand tools properly can help prevent CTDs. Figure 8-9 is a checklist for the proper selection and use of hand tools. Note that ergonomically designed hand tools will not overcome poor job design. Good job design and proper tool selection together are the best strategy.

FIGURE 8-9 Checklist for safe selection and use of hand tools.

Use Anthropometric Data

Anthropometric data has to do with human body dimensions. Such data can be used to determine the proper handle length, grip span, tool weight, and trigger length when selecting tools.

Reduce Repetition

Repetition is a hazard that can and should be reduced using such strategies as the following:

- Limit overtime.
- Change the process.
- Provide mechanical assists.
- Require breaks.
- Encourage stretching and strengthening exercises.
- Automate where possible.
- Rotate employees regularly.
- Distribute work among more employees.

Reduce the Force Required

The more force required, the more potential for damage to soft tissue. Required force can be reduced using the following strategies:

- Use power tools wherever possible.
- Use the power grip instead of the pinch grip.
- Spread the force over the widest possible area.
- Eliminate slippery, hard, and sharp gripping surfaces.
- Use jigs and fixtures to eliminate the pinch grip.

Minimize Awkward Postures

Awkward postures contribute to CTDs. The following strategies can reduce posture hazards:

- Keep the wrist in a neutral position.
- Keep elbows close to the body (90°–110° where bent).
- Avoid work that requires overhead reaching.
- Minimize forearm rotation.

Participatory Ergonomics

Participatory ergonomics is the involvement of people at work in planning for and controlling the ergonomic aspects of their work environment. There is one caveat to be observed here: Before people are involved in planning and controlling the ergonomic aspects of their work environment, they must first be equipped with prerequisite knowledge. This means that employees will typically require training before they can become involved in participatory ergonomics.[25]

Participatory ergonomics is an approach to intervention that combines the best of two worlds: outside expertise and inside experience. When undertaking an ergonomics intervention in the workplace, the most common approach is to form a participatory ergonomics (or PE) team. This team should consist of the outside expert(s) in the field of ergonomics, an internal safety and health professional, management personnel, and employees. Before beginning the intervention process, the team is trained—typically by the outside expert(s) in the field of ergonomics. The goal of the PE team is to design ergonomic interventions that are tailored specifically to the workplace in question.

Nine of ten studies reviewed by the Institute for Work & Health showed that PE interventions can have a positive effect on safety and health in the workplace. These studies showed that PE interventions had a positive effect on the musculoskeletal symptoms of workers and on reducing workplace injuries as well as workers' compensation claims.[26]

Making PE Interventions Effective

The Institute for Work & Health found that there are several factors that, when taken together, serve to make PE interventions effective. These factors are:

- Acceptance of the PE team members by other employees and unions (where applicable)
- Acceptance of the PE team members by management
- Availability of an ergonomics expert to lead the team and provide the prerequisite training
- Sufficient support from management in the form of resources

When these factors are present, an organization can reduce injuries and workers' compensation claims by applying the concept of participatory ergonomics when making workplace interventions.

Key Terms and Concepts

- Cervical radiculopathy
- Cumulative trauma disorders (CTDs)
- Employee complaints
- Employee-generated changes
- Environmental factors
- Ergonomic hazards
- Ergonomic risk factors
- Ergonomics
- General duty clause
- General observation
- Hazard prevention and control
- Health care providers
- High wrist acceleration
- Horizontal work area
- Incidence rates
- Interviews
- Manual material handling
- Medical management program
- Multiple risk factors
- OSHA's voluntary ergonomics guidelines
- Participatory ergonomics
- Physical stress
- Questionnaires
- Records analysis and tracking
- Repetitive motion
- Repetitive strain injury
- Symptoms survey
- Tendinitis
- Tenosynovitis
- Tunnel syndromes
- Vertical work area
- Worksite analysis
- Workstation analysis

Review Questions

1. Define the term *ergonomics*. Explain its origins.
2. Explain how the following opposing factors can influence the amount of physical stress associated with a job: sitting versus standing, large versus small demand for strength/power, nonrepetitive versus repetitive motion.

3. Explain the concept of human factors and how it relates to ergonomics.
4. Name the elements of OSHA's voluntary ergonomics guidelines.
5. What are the four main parts of OSHA's recommended worksite analysis program for ergonomics?
6. List five risk factors associated with CTDs.
7. Briefly explain the steps in conducting an ergonomic job hazard analysis.
8. Briefly explain the components of a hazard prevention and control program.
9. Who should participate in an ergonomics training program?
10. List and briefly explain three common indicators of the existence of ergonomic problems.
11. Describe three approaches that can be used to pinpoint specific ergonomic problems.
12. Describe an ergonomic problem-solving strategy for each of the following types of work: seated repetitive work with light parts, work with hands above the chest height, and work with hand tools.
13. Explain why so little hard research has been done concerning the economics of ergonomics.
14. Define the term *CTD*. Describe the most common types.

ENDNOTES

1. P. Tillman and B. Tillman, *Human Factors Essentials* (New York: McGraw-Hill, 1991), 3–22.
2. Ibid., 3–4.
3. Ibid., 9.
4. Ibid., 12–13.
5. Ibid., 21–22.
6. *Ergonomics Report,* April 7, 2006. Retrieved from http://www.ergonomicsreport.com/publish.
7. Adapted from OSHA 3123.
8. Ibid.
9. Ibid.
10. National Institute for Occupational Safety and Health (NIOSH), "Elements of Ergonomics Programs." Retrieved from http://www.cdc.gov/niosh.
11. Ibid., 12.
12. S. Minter, "Ergonomic Challenge: The Aging Workforce," *Occupational Hazards* 64, no. 9: 6.
13. NIOSH, 13–14.
14. Ibid., 14.
15. Ibid., 15.
16. Ibid., 16.
17. Ibid., 17.
18. Ibid., 18.
19. Ibid., 19.
20. Ibid., 20.
21. Roberta Carson, "Ergonomic Innovations: Free to a Good Company," *Occupational Hazards,* January 1996, 61–64.
22. T. Bedal, "The Economics of Ergonomics: What Are the Paybacks?", *Safety & Health* 142, no. 4: 34.
23. Ibid., 38.
24. Ibid., 39.
25. D. Cole, I. Rivilis, D. Van Eerd, K. Cullen, E. Irvin, and D. Kramer, "Effectiveness of Participatory Ergonomic Interventions: A Systematic Review," Institute for Work & Health, January 2005. Retrieved from http://www.iwh.on.ca.
26. Ibid.

CHAPTER 9

Stress and Safety

Major Topics

- Workplace Stress Defined
- Sources of Workplace Stress
- Human Reactions to Workplace Stress
- Measurement of Workplace Stress
- Shift Work, Stress, and Safety
- Improving Safety by Reducing Workplace Stress
- Stress in Safety Managers
- Stress and Workers' Compensation

Workplace Stress Defined

Our emotions are affected in the workplace by social, occupational, environmental, and psychological factors that we perceive as threats. These perceived threats are external stimuli and **stress** is the reaction of the human body to these stimuli. The amount of stress felt depends as much on the individual's ability to deal with the external stimuli as the relative intensity of the stimuli. For example, the threat of falling out of a boat into the water would cause more stress for a person who cannot swim than it would for a person who is an accomplished swimmer.

In this example, the water is just as deep and the boat is just as far from shore for both individuals, but the threat and the amount of stress it causes are less for the one who can swim than for the nonswimmer. This is because of the good swimmer's ability to deal with the threatening situation. This same rule of thumb can be applied to all perceived threats in the workplace.

As they often do, management personnel and line employees often view stress from two different perspectives. Managers tend to view stress as an individual problem tied to the personality and emotional makeup of the employee. Employees often view stress as a problem induced by poor supervision, unrealistic expectations, and other management shortcomings. Practically speaking, both managers and employees are partially right in their opposing perspectives. On one hand, the personality and emotional makeup of employees are both factors in how they respond to the stimuli that produce stress. On the other hand, management shortcomings are also factors.

However it is defined, stress is a serious problem in the modern workplace. Stress-related medical bills and the corresponding absentee rates cost employers over $150 billion annually. Almost 15 percent of all occupational disease claims are stress related.

Workplace stress involves the emotional state resulting from a perceived difference between the level of occupational demand and a person's ability to cope with this demand. Because preparations and emotions are involved, workplace stress is considered a subjective state. An environment that a worker finds stressful may generate feelings of tension, anger, fatigue, confusion, and anxiety.

Workplace stress is primarily a matter of person–workload fit. The status of the person–workload fit can influence the acceptance of the work and the level of acceptable performance of that work. The perception of workload may be affected by the worker's needs and his or her level of job satisfaction. The relation between job demands and the worker's ability to meet those demands further influence workplace stress. Because workplace stress may be felt differently by people in similar situations, it must be concluded that there are many causes of workplace stress.

Sources of Workplace Stress

The sources of on-the-job stress may involve physical working conditions, work overload, **role ambiguity**, lack of **feedback**, personality, personal and family problems, or role conflict. Other sources of workplace stress are discussed in the following paragraphs.

- **Task complexity** relates to the number of different demands made on the worker. A job perceived as being too complex may cause feelings of inadequacy and result in emotional stress. Repetitive and monotonous work may lack complexity so that the worker becomes bored and dissatisfied with the job and possibly experiences some stress associated with the boredom.
- **Control** of employees over their job responsibilities can also be a source of workplace stress. Being responsible for work without being in control of it is a formula for stress. The more work can be organized and structured to allow for maximum control by those who do it, the less stressful the work will be. In addition, even when workers do feel as if they control the work they are responsible for, if they are monitored electronically their stress levels tend to rise. In the eyes of a line employee, electronic monitoring is the same as having a supervisor standing over your shoulder constantly. Job control has also been tied to cardiovascular disease and heart attacks. Workers with high stress related to a lack of job control experience elevated levels of plasma fibrinogen—a blood-clotting factor that has been linked to cardiovascular events such as heart attacks. A study by Belgian scientists at Ghent University tied not just job stress, but also the specific type of stress associated with a lack of job control to plasma fibrinogen—a specific medical marker for cardiovascular problems.[1]
- A **feeling of responsibility** for the welfare or safety of family members may produce on-the-job stress. Being responsible for the welfare of his or her family may cause a worker to feel that options to take employment risks are limited. A worker may then perceive that he or she is "trapped in the job." Overly constrained employment options may lead to anxiety and stress. The feeling of being responsible for the safety of the general public has also been shown to be a

stressor. Air traffic controllers are known to experience intense stress when their responsibility for public safety is tested by a near-accident event. A feeling of great responsibility associated with a job can transform a routine activity into a stress-inducing task.

- **Job security** involves the risk of unemployment. A worker who believes that his or her job is in jeopardy will experience anxiety and stress. The ready availability of other rewarding employment and a feeling that one's professional skills are needed reduce the stress associated with job security issues.
- An **organizational culture** leaves the employee feeling left out, out of the loop, and ill-informed. Organizations in which managers fail to communicate frequently and effectively with employees are creating high-stress environments for workers.
- **Work schedules** that are unpredictable, never-changing, and ever-changing can induce stress in employees. Employees have lives outside of their jobs. Consequently, the ability to predict their work schedules is important. When work schedules are unpredictable, stress increases. It also increases when work schedules are inflexible—they cannot be changed no matter what other obligation the employee might have. On the other hand, ever-changing work schedules such as those associated with shift work can also increase the level of employee stress. The big-picture issue with regard to work schedules is the level of control employees have over their lives. The less control, the more stress.
- **Home and family problems** can create added stress for workers. There was a time when employees were expected to "leave their problems at the front door" when they came to work. This, of course, is a practical impossibility. The demands of raising children, working out home and job schedules, dealing with the conflicting agendas inherent in dual-career households, and otherwise handling the everyday work and home conflicts that inevitably arise can markedly increase an employee's stress level.
- **Work relationships** can lead to on-the-job stress. People are social beings by nature. They like to get along with the people they spend a lot of time with. However, office politics, turf battles, and internal competition for recognition and rewards can be hard on work relationships. When employees do not get along with their fellow workers, stress levels increase.
- **Human resource management (HRM) issues** can be a source of workplace stress. People who work have a vested interest in their wages, salaries, working conditions, and benefits. If even one of these factors is a negative, employee stress levels can increase significantly. In addition to these factors, other stress-inducing HRM issues include being underemployed, failing to get promoted, and working in a position that is clearly not valued by management.
- **Workload demands** can stimulate stress when they are perceived as being overwhelming. These demands may involve time constraints and cognitive constraints such as speed of decision making and mandates for attention. Workload demands may also be physically overwhelming if the worker is poorly matched to the physical requirements of the job or is fatigued. Whenever the worker believes the workload to be too demanding, stress can result.
- **Psychological support** from managers and coworkers gives a feeling of acceptance and belonging and helps defuse stress. A lack of such support may increase the perception of a burdensome workload and result in stress.

- The lack of **environmental safety**, including the potential for workplace violence, can also be a cause of stress. Feeling that one is in danger can be a stressor. Workers need to feel safe from environmental hazards such as extreme temperatures, pressure, electricity, fire, explosives, toxic materials, ionizing radiation, noises, and dangerous machinery. To reduce the potential for stress due to environmental hazards, workers should feel that their managers are committed to safety and that their company has an effective safety program.

Common Causes of Stress in the Workplace

- The company was recently purchased by another company.
- Downsizing or layoffs have occurred in the past year.
- Employee benefits were significantly cut recently.
- Mandatory overtime is frequently required.
- Employees have little control over how they do their work.
- The consequences of making a mistake on the job are severe.
- Workloads vary greatly.
- Most work is machine paced or fast paced.
- Workers must react quickly and accurately to changing conditions.
- Personal conflicts on the job are common.
- Few opportunities for advancement are available.
- Workers cope with a great deal of bureaucracy in getting work done.
- Staffing, money, or technology is inadequate.
- Pay is below the going rate.
- Employees are rotated among shifts.[2]

It has long been known among safety and health professionals and practitioners of occupational medicine that stress can have a detrimental effect on attendance, productivity, employee retention, and morale. The sources of stress contained in this list are just some of the more common ones. There are many others. For example, technological developments have increased stress levels on the job.

The fact that cellular telephones can be obnoxiously intrusive and distracting has tended to increase stress levels at work. Continual change in the field of computer technology—change that forces workers to upgrade their skills almost as soon as they become comfortable with a given software package—has become a source of stress in the workplace. Stress has become such an all-pervasive source of safety and health–related problems in the workplace that professionals in the field must acknowledge it and work to reduce its harmful effects like any other workplace hazard.

Human Reactions to Workplace Stress

Human reactions to workplace stress may be grouped into the following categories: subjective or emotional (anxiety, aggression, guilt); behavioral (accident proneness, trembling); cognitive (inability to concentrate or make decisions); physiological (increased heart rate and blood pressure); and organizational (absenteeism and poor productivity). Continual or persistent stress has been linked to many physiological problems. Initially, the effects may be psychosomatic, but with continued stress, the symptoms show up as actual organic dysfunction. The most common forms of stress-related diseases are gastrointestinal, particularly gastric or duodenal ulcers. Research has linked some autoimmune diseases with increased long-term workplace stress.[3]

The human response to workplace stress can be compared to a rubber band being stretched. As the stress continues to be applied, the rubber band stretches until a limit is reached when the rubber band breaks. For humans, various physical and psychological changes are observed with the repetitive stimuli of stress. Until the limit is reached, the harmful effects can be reversed. With an increase in intensity or duration of the stress beyond the individual's limit, the effects on the human become pathological.[4]

There are three stages of the human stress response: (1) alarm, (2) resistance, and (3) exhaustion.[5] The alarm reaction occurs when the stress of a threat is sensed. The **stage of alarm** is characterized by pallor, sweating, and an increased heart rate. This stage is usually short. It prepares the body for whatever action is necessary.

When the stress is maintained, the **stage of resistance** initiates a greater physical response. The alarm symptoms dissipate, and the body develops an adaptation to the stress. The capacity for adaptation during this stage is limited.

Eventually, with sustained stress, the **stage of exhaustion** is reached. This stage is demonstrated by the body's failure to adapt to the continued stress. Psychosomatic diseases such as gastric ulcers, colitis, rashes, and autoimmune disorders may begin during this stage. The tendency to develop a specific stress-related disease may be partially predetermined by heredity, personal habits such as smoking, and personality.

From an evolutionary viewpoint, the adverse effects of stress on health may be considered a maladaptation of humans to stress. What does this tell us? Either we (1) learn to do away with all stress (unlikely); (2) avoid all stressful situations (equally unlikely); (3) learn to adapt to being sick because of stress (undesirable); or (4) learn to adapt to workplace stress (the optimal choice). The first step in learning to adapt to stress is understanding the amount of stress to which we are subjected.

Measurement of Workplace Stress

Workplace stress can be seen as an individual's psychological reaction to the work environment. Although psychological response cannot be directly measured in physical terms, one method commonly employed uses a measurement of mental workload. Mental workload can be measured in one of three ways:

1. With **subjective ratings**, the workers are asked to rate their perceived level of workload. The perceived workload is then viewed as a direct reflection of workplace stress. The workers may be asked to rate their mood in relation to the work situation. The data gathered by this method are obviously subjective and state dependent. **State-dependent** data are directly related to the circumstances or state under which they are collected and, therefore, have a built-in state bias.
2. **Behavioral time sharing** techniques require the simultaneous performance of two tasks. One of the tasks is considered the primary or most important; the other is of secondary importance. The decrease in performance efficiency of the secondary task is considered an index of workload for behavioral time sharing or human multitasking. Workplace stress is thought to increase as behavioral time sharing increases.

3. **Psychophysiological techniques** require simultaneous measurement of heart rate and brain waves, which are then interpreted as indexes of mental workload and workplace stress. Behavioral time sharing and psychophysiological techniques are related to theoretical models, making data easier to interpret. These two techniques also require sophisticated equipment and data collection methods.

Subjective ratings may be collected using questionnaires or survey instruments. These instruments may ask about the physical working conditions, the individual's health and mental well-being, and perceived overall satisfaction with the job. The data may then be compared to standardized scales developed by various researchers.

Psychosocial questionnaires evaluate workers' emotions about their jobs. Workers may be asked about job satisfaction, workload, pace, opportunities for advancement, management style, and organizational climate. Psychosocial questionnaires are another form of subjective rating and are also subject to state-dependent bias in the data. Regardless of the measurement method, because workplace stress is dependent on personal awareness, no direct means of measuring workplace stress are now available.

Shift Work, Stress, and Safety

Shift work can require some employees to work when the majority of people are resting. In some cases, shift work requires rotating between two or three different starting times, which may vary by eight hours or more. Shift work has traditionally been required by the medical community, the transportation industry, utilities, security, and, increasingly, by retail sales.[6]

Basic physiological functions are scheduled by the biological clock called the **circadian rhythm**. Most children in the United States grow up on the day shift, going to school during the day and sleeping at night. After a life of being on the day shift, the body perceives a change in work shift as being stressful. If the person takes a job starting at midnight, his or her body will still expect to be sleeping at night and active during the day.

Many physical and psychological functions are affected by circadian rhythm. Blood pressure, heart rate, body temperature, and urine production are measurably slower at night. These same functions are normally faster during the day (active time).

Behavioral patterns also follow the circadian slower-at-night and more-active-during-the-day pattern. Sleep demand and sleep capacity for people age 14 or older are greatest at night and least during daylight hours. Alertness has been determined to be decreased at night.

Workers surveyed have consistently reported lower job satisfaction with rotating shifts. Day-shift workers with the same task definitions report higher job satisfaction and less stress than their second- or third-shift counterparts. Rotating shifts over several weeks can result in desensitization to the circadian rhythms. With this desensitization comes a measurable loss in productivity, increased numbers of accidents, and reported subjective discomfort. After returning to a predictable shift, workers regained their biological clock and circadian rhythm.

Not working the normal day-shift hours results in an increase in workplace stress, with rotating shifts being the most stressful. From a safety viewpoint, shift workers are subjected to more workplace stress in terms of weariness, irritability, depression, and a lack of interest in work. Shift work increases workplace stress and may lead to a less safe worker.

Reducing the Stress Associated with Shift Work

Shift work is and will probably always be a fact of life for employees in certain occupations. More than 15 million people in the United States have jobs that require shift work. To reduce the stress associated with shift work, safety and health professionals can apply the following strategies:

1. Encourage shift workers to exercise regularly. Regular exercise can have the double benefit of improving the quality of an individual's sleep and relieving pent-up stress.
2. Encourage shift workers to avoid caffeine, alcohol, or other drugs that can inhibit their ability to sleep.
3. If shift workers cannot sleep without some type of sleep aid, the food supplement melatonin or other natural sleep inducers should be recommended rather than sleeping pills that contain synthetic chemicals and can have side effects that might contribute to other stress-inducing effects.[7]

Improving Safety by Reducing Workplace Stress

Not all sources of stress on the job can be eliminated, and employment screening is unlikely to identify all those who are sensitive to stress. People can learn to adapt to stress, however. **Training** can help people recognize and deal with stress effectively. Employees need to know what is expected of them at any given time and to receive recognition when it is deserved. Managers can reduce role ambiguity and stress caused by lack of feedback by providing frequent feedback.

Stress can result from low participation or lack of **job autonomy**. A manager can help employees realize their full potential by helping them match their career goals with the company's goals and giving them more control over their jobs.

Managers can help design jobs in ways that lead to worker satisfaction, thereby lessening work stress. **Physical stress** can be reduced by improving the work environment and establishing a sound safety and health program. Managers can also assist in the effort to provide varied and independent work with good possibilities for contact and collaboration with fellow workers and for personal development.

Organizational approaches to coping with work stress include avoiding a monotonous, mechanically **controlled pace**, standardized motion patterns, and constant repetition of **short-cycle operations**. Other stress-inducing work design features to avoid include jobs that do not make use of a worker's knowledge and initiative, that lack human contact, and that have authoritarian-type supervision.

There are also several individual approaches to coping with stress. One of the most important factors in dealing with stress is learning to recognize its symptoms and taking them seriously. Handling stress effectively should be a lifelong activity that gets easier with practice. Keeping a positive mental attitude can help defuse some otherwise stressful situations.

People can analyze stress-producing situations and decide what is worth worrying about. Individuals can effectively respond to a stressful workload by **delegating responsibility** instead of carrying the entire load. Relaxation techniques can also help reduce the effects of stress. Some common relaxation methods include meditation, biofeedback, music, and exercise.

The following strategies are recommended for reducing workplace stress.

- Management recognizes workplace stress and takes steps regularly to reduce this stress.
- Mental health benefits are provided in the employee's health insurance coverage.
- The employer has a formal **employee communications** program.
- Employees are given information on how to cope with stress.
- Workers have current, accurate, and clear **job descriptions**.
- Management and employees talk openly with one another.
- Employees are free to talk with each other during work.
- Employers offer exercise and other stress-reduction classes.
- Employees are recognized and rewarded with nonmonetary prizes for their contributions.
- Work rules are published and are the same for everyone.
- Child care programs are available.
- Employees can work **flexible hours**.
- Perks are granted fairly based on a person's level in the organization.
- Workers have the training and **technology access** that they need.
- Employers encourage work and personal support groups.
- Workers have a place and time to relax during the workday.
- Elder care programs are available.
- Employees' work spaces are not crowded.
- Workers can put up personal items in their work areas.
- Management appreciates humor in the workplace.[8]

Additional ways to reduce stress in the workplace include the following:

- Match workload and pace to the training and abilities of employees.
- Make an effort to match work schedules with the personal lives of employees.
- Clearly define work roles.
- Before giving employees additional duties beyond their normal work roles, make sure they receive the necessary training.
- Enforce work rules equitably.
- Promote teamwork among employees and encourage it throughout the organization.
- Involve employees in making decisions that affect them.
- Inform employees in a timely manner of organizational changes that might affect them.[9]

There is no one clear answer to workplace stress. The suggestions given here are a good starting place for management and employees to begin the process of being aware of and dealing effectively with workplace stress.

Stress in Safety Managers

Safety and health management can be a stressful profession. Robert Scherer of Wright State University identified the following four conditions that frequently trigger stress among modern safety and health professionals:

- Role overload
- Coping with regulatory breakdown

- Communication breakdown
- Competing loyalties[10]

According to Scherer, safety and health professionals are sometimes overloaded when corporate *downsizing* results and they are delegated more and more responsibilities. Trying to keep up with the ever-changing multitude of regulations is a stress-inducing challenge. Communication relating to safety and health is always a challenge. However, when economic forces focus an organization's attention on other matters, it can be even more difficult than usual to get the safety and health message across. This increased difficulty can lead to increased stress. Line managers who are more concerned with meeting production quotas than with the safety of their employees sometimes try to influence safety and health managers—their colleagues and sometimes their friends—to look the other way. This subjects safety and health managers to the pressures of competing loyalties.[11]

Scherer recommends that safety and health managers cope with these four common triggers of stress by applying the following strategies: (1) prioritize activities by focusing on those that present the most risk to the organization; (2) work closely with the organization's legal staff and subscribe to an online CD-ROM updating service; (3) formalize communication and hold regularly scheduled safety and health meetings for all operating employees; and (4) focus on the risks to the organization and refuse to take sides.[12]

Stress and Workers' Compensation

There are serious problems with workers' compensation in the United States. On one hand, there is evidence of abuse of the system. On the other hand, many injured workers who are legitimately collecting benefits suffer a substantial loss of income. Complaints about workers' compensation are common from all parties involved with the system (employers, employees, and insurance companies).

Stress claims are more burdensome than physical claims because they are typically reviewed in an adversarial environment. This leads to the involvement of expert medical witnesses and attorneys. As a result, even though the benefits awarded for stress-related injuries are typically less than those awarded for physical injuries, the cost of stress claims is often higher because of litigation.

Key Terms and Concepts

- Behavioral time sharing
- Circadian rhythm
- Control
- Controlled pace
- Delegate responsibility
- Employee communications
- Environmental safety
- Feedback
- Feeling of responsibility
- Flexible hours
- Human reaction
- Job autonomy
- Job description
- Job security
- Physical stress
- Psychological support
- Psychophysiological techniques
- Psychosocial questionnaires
- Role ambiguity
- Shift work
- Short-cycle operations
- Stage of alarm
- Stage of exhaustion
- Stage of resistance
- State dependent
- Stress
- Subjective rating
- Task complexity
- Technology access
- Training
- Workload demands
- Workplace stress

Review Questions

1. Define *stress*.
2. How is workplace stress different from general stress?
3. List five sources of workplace stress and give an on-the-job example for each source.
4. Explain why lack of job autonomy may cause workplace stress.
5. Give five categories of human reaction to workplace stress.
6. How are psychosomatic reactions to stress and actual physiological illness related?
7. How are some autoimmune diseases and workplace stress related?
8. Explain three stages of human reaction to stress.
9. Explain three ways in which mental workload can be measured.
10. Discuss efforts to rid the workplace of all causes of workplace stress.
11. What type of data do psychosocial questionnaires provide? Discuss the bias in this type of data.
12. Discuss how shift work causes workplace stress. Give suggestions for minimizing workplace stress from shift work.
13. Give specific steps that can be taken by managers to help reduce workplace stress.
14. Discuss at least five methods to reduce workplace stress, according to the life insurance company research.
15. Explain how individuals can reduce workplace stress.

Endnotes

1. Occupational Health & Safety Online, "Study: Job Stress Linked to Increased Inflammation," September 2005. Retrieved from http://www.ohsonline.com/stevens/ohspub.nsf/d3d5b4f93862266e8625670c006dbc58/c16.
2. B. Farms, *Occupational Medicine Practical Guidelines: Evaluation and Management of Common Health Problems and Functional Recovery in Workers,* 2nd ed. (Elk Grove Village, IL: OEM Press, 2004), 492.
3. Ibid., 494.
4. Ibid.
5. Ibid.
6. D. Morshead, "Stress and Shiftwork," *Occupational Health & Safety* 71, no. 4: 36–37.
7. Ibid.
8. Farms, 498.
9. Ibid.
10. R. Scherer. "Stress Homes in on Our Safety's Ranks," *Occupational Hazards,* January 1994, 156–157.
11. Ibid., 157.
12. Ibid.

CHAPTER 10

Mechanical Hazards and Machine Safeguarding

Major Topics

- Common Mechanical Injuries
- Safeguarding Defined
- OSHA's Requirements for Machine Guarding
- Risk Assessment in Machine Operation
- Requirements for All Safeguards
- Point-of-Operation Guards
- Point-of-Operation Devices
- Machine Guarding Self-Assessment
- Feeding and Ejection Systems
- Robot Safeguards
- Control of Hazardous Energy (Lockout/Tagout Systems)
- General Precautions
- Basic Program Content
- Taking Corrective Action

Failure to provide proper machine guards and enforce their use can be costly for companies. A manufacturing firm in Syracuse, New York, learned this fact the hard way when it was cited by OSHA for failure to provide appropriate machine guards and require their proper use by employees. Because the company had been issued a similar citation earlier, it was fined $119,000. Mechanical hazards that are not properly guarded are implicated in thousands of workplace injuries every year.

As another example, a windshield manufacturing firm was fined more than $105,000 by OSHA when it allowed employees to be exposed to electrical hazards and knowingly failed to meet lockout/tagout standards. The small gains in productivity that might be obtained by willfully bypassing mechanical safeguards on machines can cost companies huge fines and even expensive medical bills.

Mechanical hazards are those associated with power-driven machines, whether automated or manually operated. Concerns about mechanical hazards date back to the Industrial Revolution and the earliest days of mechanization. Machines driven by steam, hydraulic, or electric power introduced new hazards into the workplace. In spite of advances in safeguarding technologies and techniques, mechanical hazards are still a major concern today. In addition, automated machines have introduced new concerns.

Common Mechanical Injuries

In an industrial setting, people interact with machines that are designed to drill, cut, shear, punch, chip, staple, stitch, abrade, shape, stamp, and slit such materials as metals, composites, plastics, and elastomers. If appropriate safeguards are not in place or if workers fail to follow safety precautions, these machines can apply the same procedures to humans. When this happens, the types of **mechanical injuries** that result are typically the result of cutting, tearing, shearing, crushing, **breaking**, straining, or puncturing (see Figure 10-1). Information about each of these hazards is provided in the following paragraphs.

Cutting and Tearing

A cut occurs when a body part comes in contact with a sharp edge. The human body's outer layer consists of the following, starting from the outside: *epidermis*, the tough

FIGURE 10-1 Some common mechanical hazards.

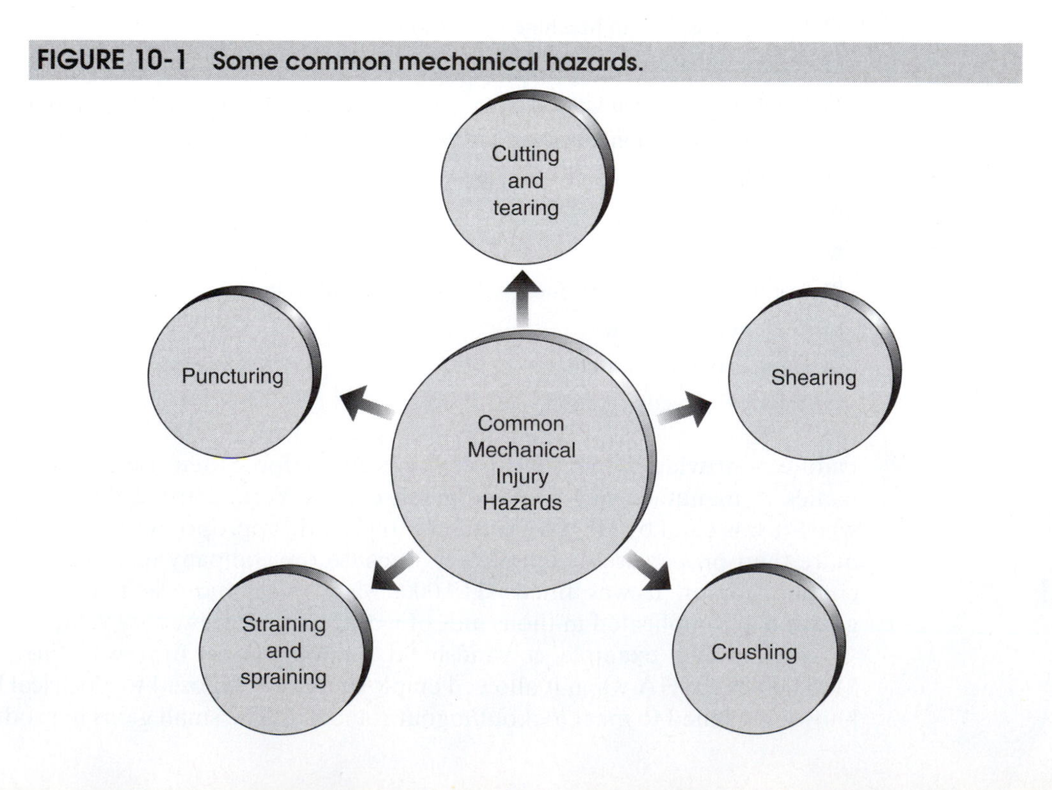

outer covering of the skin; *dermis,* the greatest part of the skin's thickness; *capillaries*, the tiny blood vessels that branch off the small arteries and veins in the dermis; *veins*, the blood vessels that collect blood from the capillaries and return it to the heart; and *arteries*, the larger vessels that carry blood from the heart to the capillaries in the skin. The seriousness of **cutting** or **tearing** the skin depends on how much damage is done to the skin, veins, arteries, muscles, and even bones.

Shearing

To understand what **shearing** is, think of a paper cutter. It shears the paper. Power-driven shears for severing paper, metal, plastic, elastomers, and composite materials are widely used in manufacturing. In times past, such machines often amputated fingers and hands. These tragedies typically occurred when operators reached under the shearing blade to make an adjustment or placed materials there and activated the blade before fully removing their hand. Safeguards against shearing accidents are explained later in this chapter.

Crushing

Injuries from **crushing** can be particularly debilitating, painful, and difficult to heal. They occur when a part of the body is caught between two hard surfaces that progressively move together, thereby crushing anything between them. Crushing hazards can be divided into two categories: *squeeze-point* types and *run-in points*.

Squeeze-point hazards exist where two hard surfaces, at least one of which must be in motion, push close enough together to crush any object that may be between them. The process can be slow, as in a manually operated vise, or fast, as with a metal-stamping machine.

Run-in point hazards exist where two objects, at least one of which is rotating, come progressively closer together. Any gap between them need not become completely closed. It need only be smaller than the object or body part lodged in it. Meshing gears and belt pulleys are examples of run-in point hazards (see Figures 10-2, 10-3, and 10-4).

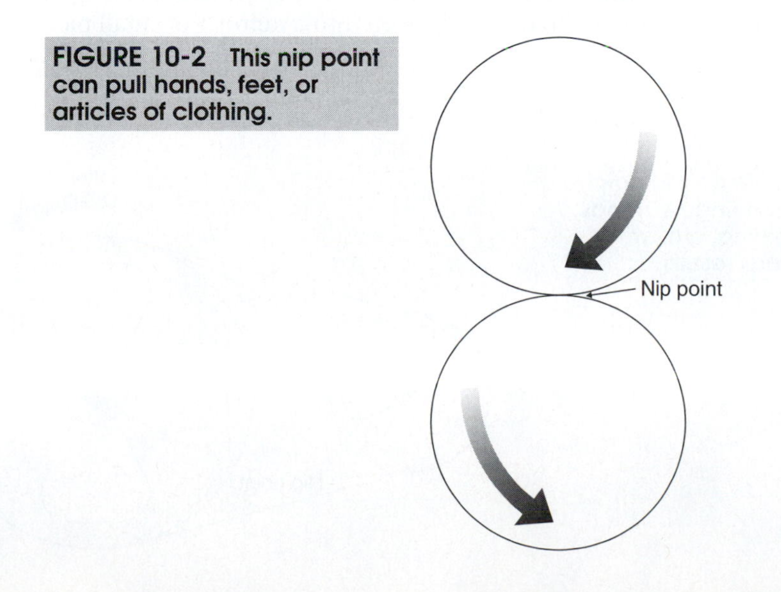

FIGURE 10-2 This nip point can pull hands, feet, or articles of clothing.

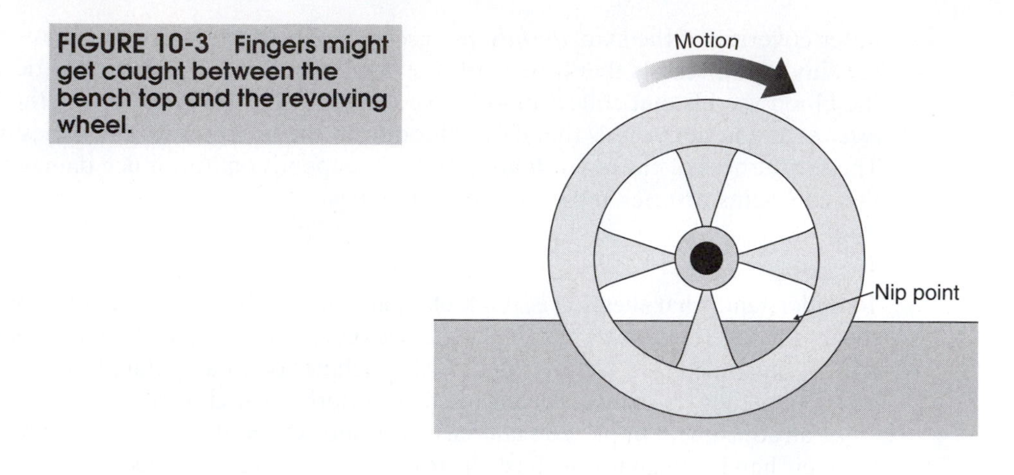

FIGURE 10-3 Fingers might get caught between the bench top and the revolving wheel.

Body parts can also be crushed in other ways—for example, a heavy object falling on a foot or a hammer hitting a finger. However, these are impact hazards, which are covered in Chapter 11.

Breaking

Machines used to deform engineering materials in a variety of ways can also cause broken bones. A break in a bone is known as a *fracture*. Fractures are classified as simple, compound, complete, and incomplete.

A simple fracture is a break in a bone that does not pierce the skin. A compound fracture is a break that has broken through the surrounding tissue and skin. A complete fracture divides the affected bone into two or more separate pieces. An incomplete fracture leaves the affected bone in one piece but cracked.

Fractures are also classified as transverse, oblique, and comminuted. A transverse fracture is a break straight across the bone. An oblique fracture is diagonal. A comminuted fracture exists when the bone is broken into a number of small pieces at the point of fracture.

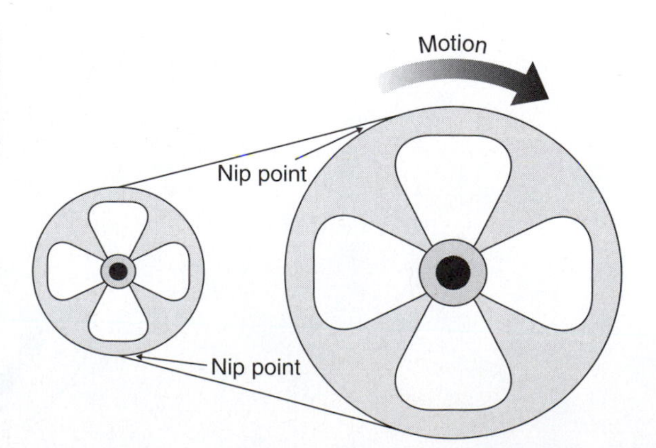

FIGURE 10-4 Nip points can catch fingers, hands, hair, clothing, etc., with dangerous results.

Straining and Spraining

There are numerous situations in an industrial setting when **straining** of muscles or **spraining** of ligaments is possible. A strain results when muscles are overstretched or torn. A sprain is the result of torn ligaments in a joint. Strains and sprains can cause swelling and intense pain.

Puncturing

Punching machines that have sharp tools can puncture a body part if safety precautions are not observed or if appropriate safeguards are not in place. **Puncturing** results when an object penetrates straight into the body and pulls straight out, creating a wound in the shape of the penetrating object. The greatest hazard with puncture wounds is the potential for damage to internal organs.

Safeguarding Defined

All the hazards explained in the previous section can be reduced by the application of appropriate safeguards. CFR 1910 Subpart O contains the OSHA standards for machinery and machine guarding (1910.211–1910.222). The National Safety Council defines **safeguarding** as follows:

> Machine safeguarding is to minimize the risk of accidents of machine–operator contact. The contact can be:
>
> 1. An individual making the contact with the machine—usually the moving part—because of inattention caused by fatigue, distraction, curiosity, or deliberate chance taking;
> 2. From the machine via flying metal chips, chemical and hot metal splashes, and circular saw kickbacks, to name a few;
> 3. Caused by the direct result of a machine malfunction, including mechanical and electrical failure.[1]

Safeguards can be broadly categorized as point-of-operation guards, point-of-operation devices, and feeding/ejection methods. The various types of safeguards in these categories are explained later in this chapter.

OSHA's Requirements for Machine Guarding

The OSHA standard containing the general requirements for machine guarding is 29 CFR 1910.212. A more specific standard (29 CFR 1926.300) exists for the construction industry. This section focuses on 29 CFR 1910.212—OSHA's requirements for all industries. Those requirements are summarized as follows:

> **Types of guarding.** One or more methods of machine guarding must be provided to protect people from such point of operation hazards as nip points, rotating parts, flying chips, and sparks. "Point of operation" refers to the area on the machine where work is performed on the material being processed. Examples of point-of-operation machine guards are barriers, two-hand switches and tripping devices, and electronic sensors.

General requirements for machine guards. Where possible, guards should be affixed to the machine in question. When this is not possible, guards should be secured in the most feasible location and method away from the machine. Guards must be affixed in such a way that they do not create a hazard themselves.

Guarding the point of operation. Any point of operation that might expose a person to injury must be guarded. Guarding devices must comply with all applicable standards. In the absence of applicable standards, the guard must be designed, constructed, and installed in such a way as to prevent the machine operator from having any part of his body (including clothing, hair, etc.) in the danger zone during the operating cycle of the machine.

Machines requiring point of operation guards. The following are examples of machines that require point of operation guards: guillotine cutters, shears, alligator shears, power presses, milling machines, power saws, jointers, portable power tools, forming rolls, and calendars.

Exposure of blades. Fans must be guarded in any case in which the periphery of the fan blades is less than seven feet above the floor or working level. Guards for fans shall have no openings that exceed one-half inch.

Anchoring fixed machinery. Machines that are designed to be fixed in one location must be securely anchored to prevent movement.[2]

Risk Assessment in Machine Operation

Risk assessment in this context is the process of quantifying the level of risk associated with the operation of a given machine. It should be a structured and systematic process that answers the following four specific questions:

- How *severe* are potential injuries?
- How *frequently* are employees exposed to the potential hazards?
- What is the *possibility* of avoiding the hazard if it does occur?
- What is the *likelihood* of an injury should a safety control system fail?[3]

The most widely used risk-assessment technique is the decision tree, coupled with codes representing these four questions and defined levels of risk. Figure 10-5 is an

FIGURE 10-5 Risk-assessment decision tree.

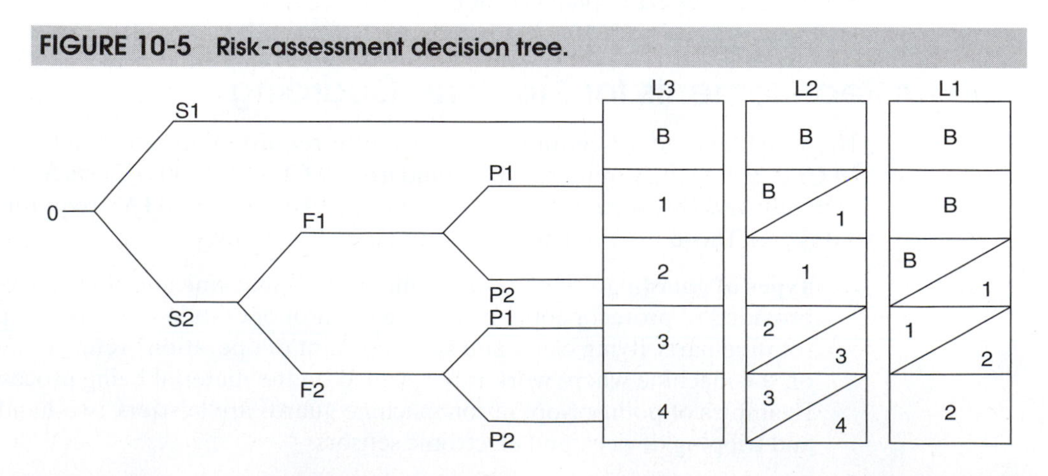

example of a risk-assessment decision tree. In this example, the codes and their associated levels of risk are as follows:

S = Severity

QUESTION 1: SEVERITY OF POTENTIAL INJURIES

S1 Slight injury (bruise, abrasion)
S2 Severe injury (amputation or death)
F = Frequency

QUESTION 2: FREQUENCY OF EXPOSURE TO POTENTIAL HAZARDS

F1 Infrequent exposure
F2 From frequent to continuous exposure
P = Possibility

QUESTION 3: POSSIBILITY OF AVOIDING THE HAZARD IF IT DOES OCCUR

P1 Possible
P2 Less possible to not impossible
L = Likelihood

QUESTION 4: LIKELIHOOD THAT THE HAZARD WILL OCCUR

L1 Highly unlikely
L2 Unlikely
L3 Highly likely
Risk Levels

Associated risk factors ranging from lowest (B) to highest (4) By applying the decision tree in Figure 10-5 or a similar device, the risk associated with the operation of a given machine can be quantified. This allows safety personnel to assign logical priorities for machine safeguarding and hazard prevention.

Requirements for All Safeguards

The various machine motions present in modern industry involve mechanisms that rotate, reciprocate, or do both. This equipment includes tools, bits, chucks, blades, spokes, screws, gears, shafts, belts, and a variety of different types of stock. Safeguards can be devised to protect workers from harmful contact with such mechanisms while at the same time allowing work to progress at a productive rate. The National Safety Council has established the following requirements for safeguards.

1. **Prevent contact.** Safeguards should prevent human contact with any potentially harmful machine part. The prevention extends to machine operators and any other person who might come in contact with the hazard.
2. **Be secure and durable.** Safeguards should be attached so that they are secure. This means that workers cannot render them ineffective by tampering with or disabling them. This is critical because removing safeguards in an attempt to speed production is a common practice. Safeguards must also be durable

enough to withstand the rigors of the workplace. Worn-out safeguards won't protect workers properly.

3. **Protect against falling objects.** Objects falling onto moving machine mechanisms increase the risk of accidents, property damage, and injury. Objects that fall on a moving part can be quickly hurled out, creating a dangerous projectile. Therefore, safeguards must do more than just prevent human contact. They must also shield the moving parts of machines from falling objects.
4. **Create no new hazard.** Safeguards should overcome the hazards in question without creating new ones. For example, a safeguard with a sharp edge, unfinished surface, or protruding bolts introduces new hazards while protecting against the old.
5. **Create no interference.** Safeguards can interfere with the progress of work if they are not properly designed. Such safeguards are likely to be disregarded or disabled by workers feeling the pressure of production deadlines.
6. **Allow safe maintenance.** Safeguards should be designed to allow the more frequently performed maintenance tasks (e.g., lubrication) to be accomplished without the removal of guards. For example, locating the oil reservoir outside the guard with a line running to the lubrication point will allow for daily maintenance without removing the guard.[4]

Design and construction of safeguards are highly specialized activities requiring a strong working knowledge of machines, production techniques, and safety. However, it is critical that all the factors explained in this section be considered and accommodated during the design process.

Point-of-Operation Guards

Guards are most effective when used at the point of operation, which is where hazards to humans exist. Point-of-operation hazards are those caused by the shearing, cutting, or bending motions of a machine. Pinch-point hazards result from guiding material into a machine or transferring motion (e.g., from gears, pressure rollers, or chains and sprockets). Single-purpose safeguards, because they guard against only one hazard, typically are permanently fixed and nonadjustable. Multiple-purpose safeguards, which guard against more than one hazard, typically are adjustable.[5]

Point-of-operation guards are of three types, each with its own advantages and limitations: fixed, interlocked, and adjustable (Figure 10-6).

- **Fixed guards** provide a permanent barrier between workers and the point of operation. They offer the following advantages: They are suitable for many specific applications, can be constructed in-plant, require little maintenance and are suitable for high-production, repetitive operations. Limitations include the following: They sometimes limit visibility, are often limited to specific operations, and sometimes inhibit normal cleaning and maintenance.
- **Interlocked guards** shut down the machine when the guard is not securely in place or is disengaged. The main advantage of this type of guard is that it allows safe access to the machine for removing jams or conducting routine maintenance without the need for taking off the guard. There are also limitations. Interlocked guards require careful adjustment and maintenance and, in some cases, can be easily disengaged.

FIGURE 10-6 Point-of-operation guards.

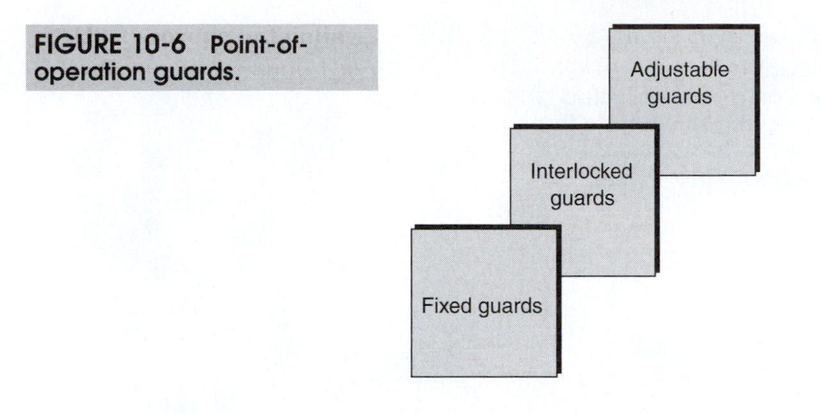

- **Adjustable guards** provide a barrier against a variety of different hazards associated with different production operations. They have the advantage of flexibility. However, they do not provide as dependable a barrier as other guards do, and they require frequent maintenance and careful adjustment.

Figures 10-7 through 10-12 show various guards used in modern manufacturing settings.

FIGURE 10-7 Series 12 PRO-TECH-TOR GATE GUARD used on an open-back power press.

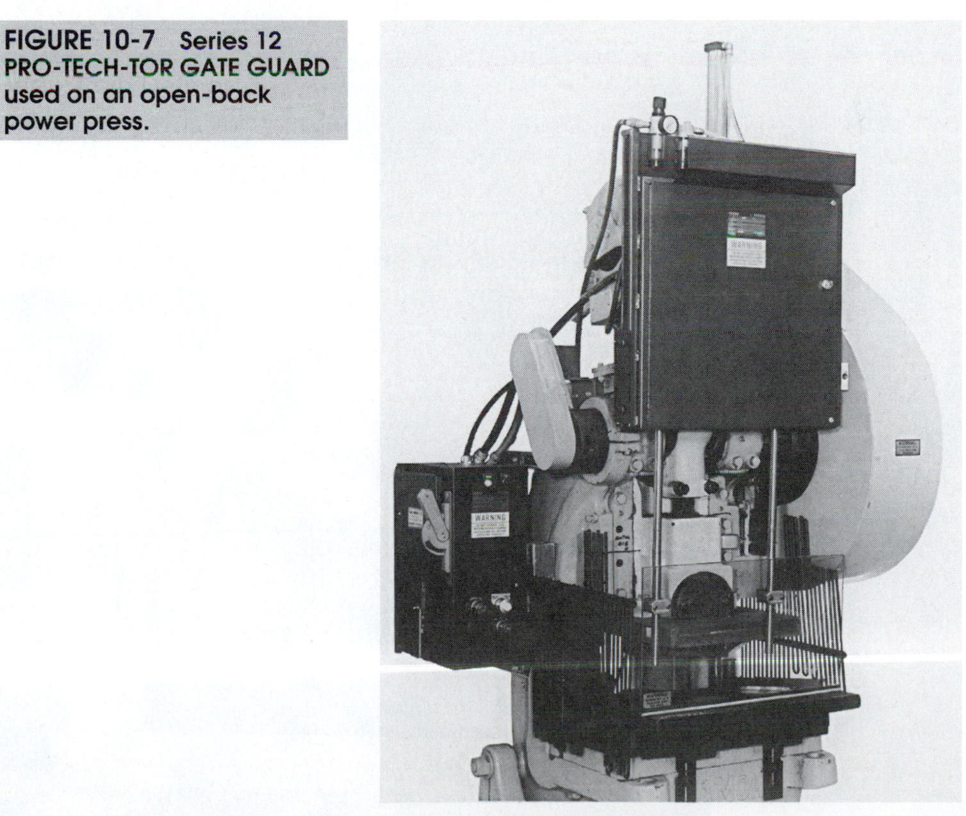

Courtesy of PROTECH SYSTEMS.

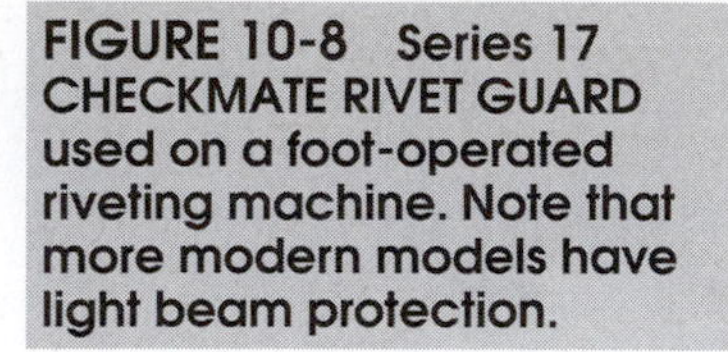

FIGURE 10-8 Series 17 CHECKMATE RIVET GUARD used on a foot-operated riveting machine. Note that more modern models have light beam protection.

Courtesy of PROTECH SYSTEMS.

FIGURE 10-9 When the doors are opened, the milling tool stops automatically.

FIGURE 10-10 In order for this shearing machine to cut, both the foot pedal and the hand button must be engaged.

FIGURE 10-11 This door protects the operator in the event of an exploding or shattering grinding wheel.

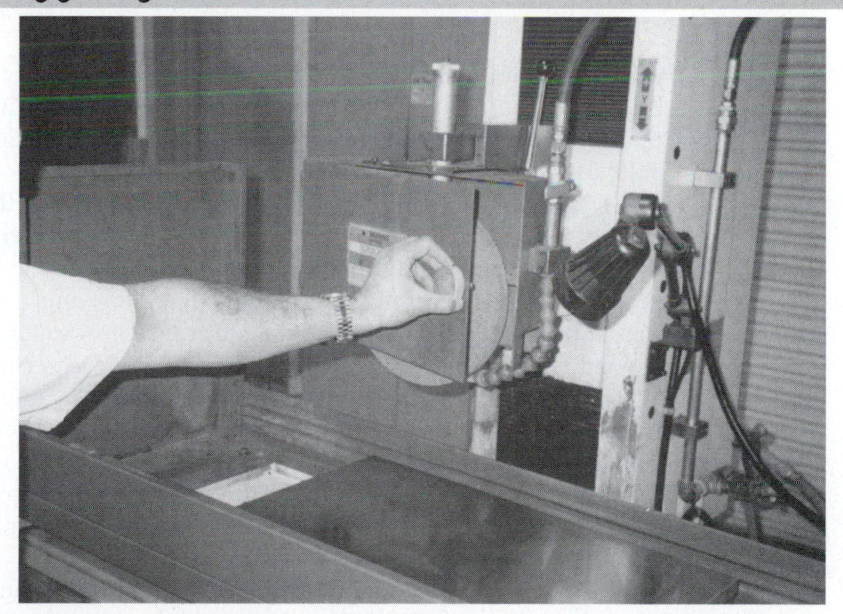

Point-of-Operation Devices

A number of different **point-of-operation devices** can be used to protect workers. The most widely used are explained in the following paragraphs.

- **Photoelectric devices** are optical devices that shut down the machine whenever the light field is broken. These devices allow operators relatively free movement.

FIGURE 10-12 The safety door on this drilling machine must be closed or the drill will not operate.

They do have limitations, including the following: They do not protect against mechanical failure, they require frequent calibration, they can be used only with machines that can be stopped, and they do not protect workers from parts that might fly out of the point-of-operation area.

- **Radio-frequency devices** are capacitance devices that brake the machine if the capacitance field is interrupted by a worker's body or another object. These devices have the same limitations as photoelectric devices.
- **Electromechanical devices** are contact bars that allow only a specified amount of movement between the worker and the hazard. If the worker moves the contact bar beyond the specified point, the machine will not cycle. These devices have the limitation of requiring frequent maintenance and careful adjustment.
- **Pullback devices** pull the operator's hands out of the danger zone when the machine starts to cycle. These devices eliminate the need for auxiliary barriers. However, they also have limitations. They limit operator movement, must be adjusted for each individual operator, and require close supervision to ensure proper use.
- **Restraint devices** hold the operator back from the danger zone. They work well, with little risk of mechanical failure. However, they do limit the operator's movement, must be adjusted for each individual operator, and require close supervision to ensure proper use.
- **Safety trip devices** include trip wires, trip rods, and body bars. All these devices stop the machine when tripped. They have the advantage of simplicity. However, they are limited in that all controls must be activated manually. They protect only the operator and may require the machine to be fitted with special fixtures for holding work.

FIGURE 10-13 Series 25 EAGLE EYE INFRA-RED LIGHT BARRIER. A point-of-operation guarding system on a roller press machine.

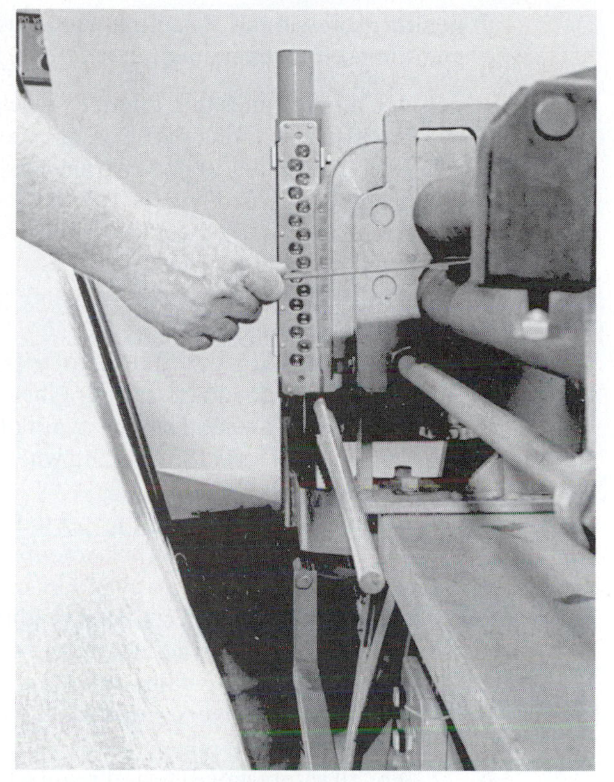

Courtesy of PROTECH SYSTEMS.

- **Two-hand controls** require the operator to use both hands concurrently to activate the machine (e.g., a paper cutter or metal-shearing machine). This ensures that hands cannot stray into the danger zone. Although these controls do an excellent job of protecting the operator, they do not protect onlookers or passers-by. In addition, some two-hand controls can be tampered with and made operable using only one hand.
- **Gates** provide a barrier between the danger zone and workers. Although they are effective at protecting operators from machine hazards, they can obscure the work, making it difficult for the operator to see.[6]

Figure 10-13 is a point-of-operation device that shuts down the machine when an operator's hand breaks a light beam. How quickly will the machine shut down? The stopping distance equation is as follows:

$$SD = \text{Stoptime}\,[\text{Seconds} \times \text{Hand speed constant } (63''/\text{second})]$$

Machine Guarding Self-Assessment

One of the most effective ways to ensure that machines are properly guarded is to conduct periodic self-assessments. These self-assessments can be conducted by safety personnel, supervisors, or employees. Developing self-assessment criteria and encouraging supervisors and employees to use them daily is an excellent strategy for safety and health professionals. The following questions can be used for conducting machine guarding self-assessments:

1. Are all machines that might expose people to rotating parts, nip points, flying chips, sparks, flying particles, or other similar hazards properly guarded?
2. Are all mechanical power transmission belts and the nip points they create properly guarded?
3. Are all exposed power shafts located less than seven feet above the working level properly guarded?
4. Are all hand tools and other hand-operated equipment inspected regularly for hazardous conditions?
5. Is compressed air used to clean tools, machines, equipment, and parts reduced to less than 30 pounds per square inch (psi)?
6. Are power saws and similar types of equipment properly guarded?
7. Are the tool rests for grinding wheels set to within one-eighth or less of the grinding wheel?
8. Are hand tools regularly inspected on a systematic basis for burred ends, cracked handles, and other potentially hazardous conditions?
9. Are all compressed gas cylinders inspected regularly and systematically for obvious signs of defects, deep rusting, or leakage?
10. Do all employees who handle and store gas cylinders and valves know how to do so without causing damage?
11. Are all air receivers periodically and systematically inspected, including safety valves?
12. Are all safety valves tested regularly, systematically, and frequently?[7]

Feeding and Ejection Systems

Feeding and ejection systems can be effective safeguards if properly designed and used. The various types of feeding and ejection systems available for use with modern industrial machines are summarized as follows:

- **Automatic feed** systems feed stock to the machine from rolls. Automatic feeds eliminate the need for operators to enter the danger zone. Such systems are limited in the types and variations of stock that they can feed. They also typically require an auxiliary barrier guard and frequent maintenance.
- **Semiautomatic feed** systems use a variety of approaches for feeding stock to the machine. Prominent among these are chutes, moveable dies, dial feeds, plungers, and sliding bolsters. They have the same advantages and limitations as automatic feed systems.
- **Automatic ejection** systems eject the work pneumatically or mechanically. The advantage of either approach is that operators don't have to reach into the danger zone to retrieve workpieces. However, these systems are restricted to use with relatively small stock. Potential hazards include blown chips or debris and noise. Pneumatic ejectors can be quite loud.[8]
- **Semiautomatic ejection** systems eject the work using mechanisms that are activated by the operator. Consequently, the operator does not have to reach into the danger zone to retrieve workpieces. These systems do require auxiliary barriers and can be used with a limited variety of stock.

Robot Safeguards

Robots have become commonplace in modern industry. The safety and health concerns relating to robots are covered in Chapter 19. Only the guarding aspects of robot safety are covered in this section. The main hazards associated with robots are (1) entrapment of a worker between a robot and a solid surface, (2) impact with a moving robot arm, and (3) impact with objects ejected or dropped by the robot.

The best guard against these hazards is to erect a physical barrier around the entire perimeter of a robot's **work envelope** (the three-dimensional area established by the robot's full range of motion). This physical barrier should be able to withstand the force of the heaviest object that a robot could eject.

Various types of shutdown guards can also be used. A guard containing a sensing device that automatically shuts down the robot if any person or object enters its work envelope can be effective. Another approach is to put sensitized doors or gates in the perimeter barrier that automatically shut down the robot as soon as they are opened.

These types of safeguards are especially important because robots can be deceptive. A robot that is not moving at the moment may simply be at a stage between cycles. Without warning, it might make sudden and rapid movements that could endanger any person inside the work envelope.

Control of Hazardous Energy (Lockout/Tagout Systems)

OSHA's standard for the control of hazardous energy, often referred to as the "lockout/tagout" standard, is 29 CFR 1910.147. The purpose of this standard is to protect people in the workplace from hazardous energy while they are performing service or maintenance on machines, tools, and equipment. A key element of the standard is to prevent the accidental or inadvertent activation of a machine while it is being serviced or repaired. The lockout/tagout standard identifies the proper procedures for shutting down machines and equipment and locking or tagging it out so that accidental or inadvertent activation does not occur. The standard also calls for employee training and periodic inspections. The overall requirement of 29 CFR 1910.147 is that before service or maintenance are performed, the machines or equipment in question must be disconnected from their energy source, and the energy source must be either locked out or tagged out to prevent accidental or inadvertent activation.[9]

Lockout/Tagout Language

The following terms and phrases are frequently used in the language of lockout/tagout. Safety and health professionals should be knowledgeable of these terms:

- **Affected employee.** Employees who perform their jobs in areas in which the procedure in question is implemented and in which service or maintenance operations are performed. Affected employees do not implement energy control procedures unless they are *authorized*.
- **Authorized employee.** Employees who perform service or maintenance on a machine and use lockout/tagout procedures for their own protection.
- **Energized.** Machines, equipment, and tools are energized if they are connected to an energy source or when they still contain stored or residual energy even after being disconnected.

- **Capable of being locked out.** A device is considered to be capable of being locked out if it meets one of the following requirements: (1) it has a hasp to which a lock can be attached; (2) it has another appropriate integral part through which a lock can be attached; (3) it has a built-in locking mechanism; or (4) it can be locked without permanently dismantling, rebuilding, or replacing the energy-isolating device.
- **Energy-isolating device.** Any mechanical device that physically prevents the release or transmission of energy (e.g., circuit breakers, disconnect switches, blocks, etc.).
- **Energy source.** Any source of power that can activate a machine or piece of equipment (e.g., electrical, mechanical, hydraulic, pneumatic, chemical, thermal, etc.).
- **Energy control procedure.** A written document containing all the information an authorized person needs to know in order to properly control hazardous energy when shutting down a machine or equipment for maintenance or service.
- **Energy control program.** A systematic program for preventing the accidental or inadvertent energizing of machines or equipment during maintenance or servicing. This is sometimes called the organization's lockout/tagout program.
- **Lockout.** Placing a lockout device such as a padlock on an energy-isolating device to prevent the accidental or inadvertent energizing of a machine during maintenance or servicing.
- **Lockout device.** Any device (see Figure 10-14) that uses a positive means to keep an energy-isolation device in the *safe* position to prevent the accidental or inadvertent energizing of a machine or piece of equipment.
- **Tagout.** Placing a tag (see Figure 10-15) on an energy-isolation device to warn people so that they do not accidentally or inadvertently energize a machine or piece of equipment.

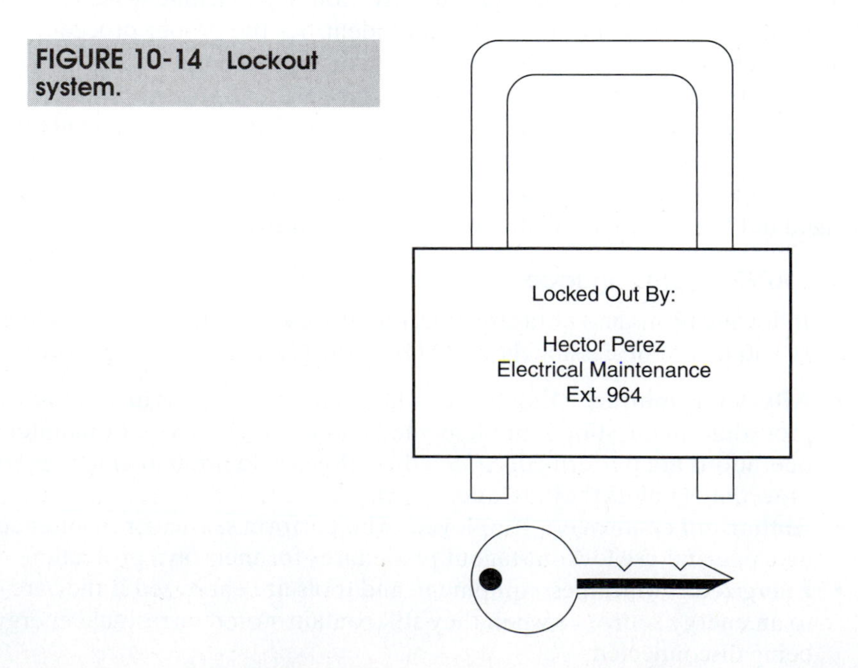

FIGURE 10-14 Lockout system.

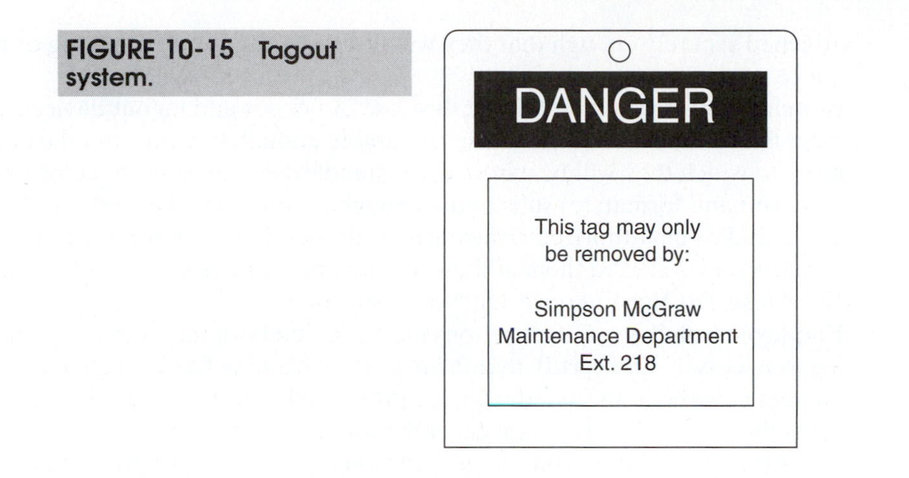

FIGURE 10-15 Tagout system.

- **Tagout device.** Any prominent warning device such as a tag that can be affixed to an energy-isolation device to prevent the accidental or inadvertent energizing of a machine or piece of equipment.

Provisions of the Standard

OSHA's standard for control of hazardous energy contains provisions in the following areas: energy control program, energy control procedure, energy-isolating devices, lockout/tagout devices, periodic inspections, application of controls and lockout/tagout devices, removal of locks and tags, testing or positioning of machines, outside personnel, group lockout or tagout, and shift or personnel changes. Those provisions are as follows:

- **Energy control program.** Organizations must establish energy control programs that have fully documented energy control procedures, provide employee training, and ensure periodic inspections.
- **Energy control procedure.** Organizations must develop, document, and use energy control procedures that contain at least the following elements: (a) a statement on how the procedure will be used; (b) procedural steps used to shut down, isolate, block, and secure machines or equipment; (c) steps designating the safe placement, removal, and transfer of lockout/tagout devices and who has responsibility for them; and (d) specific requirements for testing machines or equipment to verify the effectiveness of energy control measures.
- **Energy-isolating devices.** Organizations must have appropriate energy-isolating devices for preventing the accidental or inadvertent release of energy on all machines and equipment. The preferred type of device is one that can be locked out. However, when this type of device is not feasible, tagout procedures may be used. When a tagout system is used, employees must receive training on the following limitations of tags: (a) tags are just warning devices and do not provide the safety margin of locks; (b) tags may be removed only by the person who affixes them and should never be bypassed, ignored, or otherwise overcome; (c) tags must be legible and understandable by all employees (this can mean providing labels in more than one language); (d) tags and their means of attachment must be made of material that is durable enough to withstand the environment to which they will be subjected; (e) tags can evoke a false sense of security; and (f) tags must be

attached securely enough that they will not come off during servicing or maintenance operations.

- **Requirements for lockout/tagout devices.** Lockout and tagout devices must meet the following requirements: (a) durable enough to withstand the environment to which they will be exposed; (b) standardized in terms of color or size and print and format; (c) substantial enough to minimize the likelihood of accidental or unauthorized removal; and (d) identifiable in terms of the employee who affixed them and the warning message (e.g., Do Not Start, Do Not Close, Do Not Energize, Do Not Open, etc.).
- **Employee training.** Organizations must provide both initial training and retraining as necessary and certify that the necessary training has been given to all employees covered in the standard (e.g., authorized, affected, and other). Training for authorized employees must cover the following topics at a minimum: (a) details about the types and magnitude of hazardous energy sources present in the workplace; and (b) methods, means, and procedures for isolating and controlling these sources. Training for affected employees (usually machine operators or users) and other employees must cover the following topics at a minimum: (a) how to recognize when the energy control procedure is implemented; and (b) the importance of never attempting to start up a locked out or tagged out machine. All training must ensure that employees understand the purpose, function, and restrictions of the energy control program, and that authorized employees have the knowledge and skills necessary to properly apply, use, and remove energy controls.
- **Periodic inspections.** Inspections must be performed at least annually to ensure that the energy control program is up-to-date and being properly implemented. In addition, the organization must certify that the periodic inspections have actually taken place.
- **Application of controls and lockout/tagout devices.** Controls and lockout/tagout devices must be applied properly. The appropriate procedure consists of the following steps that must be implemented in sequence: (a) prepare for shutdown; (b) shut down the machine or equipment in question; (c) affix the lockout or tagout device; (d) render safe any residual or stored energy that might remain in the machine; and (e) verify that the energy source has been effectively isolated and that the machine or equipment has been effectively deenergized.
- **Removal of locks or tags.** Before locks or tags are removed, the following procedures must be completed: (a) inspect the work area to ensure that nonessential items have been removed and that the machine is capable of operating properly; (b) check the area around the machine to ensure that all employees are safely back or removed from the area; (c) notify affected employees immediately after removing the energy control devices and before energizing the machine; and (d) ensure that energy control devices are removed by the individual who affixed them (if this is not possible, make sure that the person who does remove them follows the proper procedures).
- **Additional safety requirements.** OSHA allows for special circumstances as set forth in this subsection. When a machine must be energized in order to test or position it, energy control devices may be removed only as follows: (a) clear the machine of tools and materials; (b) remove employees from the area; (c) remove the devices as set forth in the standard; (d) energize the machine and conduct the

test or proceed with the positioning procedure; and (e) deenergize the machine, isolate the energy source, and reapply the energy control devices. Organizations must ensure that outside personnel such as contractors are fully informed about energy control procedures. In group lockout or tagout situations, each individual employee performing maintenance or service tasks must be protected by his or her own personal energy control device. Organizations must have specific procedures for ensuring continuity in spite of personnel and shift changes.

Evaluating Lockout/Tagout Programs

Lockout/tagout violations are frequently cited by OSHA during on-site inspections. The following questions developed by Linda F. Johnson can be used to evaluate an organization's lockout/tagout program:

- Are all machinery or equipment capable of movement required to be deenergized or disengaged and blocked or locked out during cleaning, servicing, adjusting, or setting up operations?
- Where the power disconnect equipment does not disconnect the electrical control circuit, are the appropriate electrical enclosures identified?
- If the power disconnect for equipment does not disconnect the electrical control circuit, is a means provided to ensure that the control circuit can be disconnected and locked out?
- Is it required to lock out main power disconnects instead of locking out control circuits?
- Are all equipment control valve handles equipped with a means for locking out?
- Does the lockout procedure require that stored energy—whether it is mechanical, hydraulic, or air—be released or blocked before the equipment is locked out for repairs?
- Are appropriate employees provided with individually keyed personal safety locks?
- Are these employees required to keep personal control of their keys while they have safety locks in use?
- Is only the employee exposed to the hazard required to install or remove the safety lock?
- Are employees required to check the safety lockout by attempting a startup after making sure no one is exposed?
- After the safety is checked, does the employee again place the switch in the "off" position?
- Are employees instructed always to push the control circuit stop button before reenergizing the main power switch?
- Are all employees who are working on locked-out equipment identified by their locks or accompanying tags?
- Are enough accident prevention signs, tags, and safety padlocks provided for any reasonably foreseeable repair emergency?
- When machine operations, configuration, or size require the operator to leave his or her control station to install tools or perform other operations, is he or she required to lock or tag out separately any parts of the machine that could move if accidentally activated?
- If the equipment or lines cannot be shut down, locked out, and tagged, is a safe job procedure established and rigidly followed?

- Have employees been trained not to start machinery or equipment if it is locked out or tagged out?
- Are all workers notified when the machinery or equipment they usually use is shut down and locked out for maintenance or servicing purposes?
- After maintenance is completed, is the machinery checked to ensure that nonessential items have been removed and the machine is operationally intact?
- Before the machinery is activated, are employees removed from possible danger?
- When the machinery is fully operational, are employees notified?[10]

General Precautions

The types of safeguards explained in this chapter are critical. In addition to these specific safeguards, there are also a number of general precautions that apply across the board in settings where machines are used. Some of the more important general precautions are as follows:

- All operators should be trained in the safe operation and maintenance of their machines.
- All machine operators should be trained in the emergency procedures to take when accidents occur.
- All employees should know how to activate emergency shutdown controls. This means knowing where the controls are and how to activate them.
- Inspection, maintenance, adjustment, repair, and calibration of safeguards should be carried out regularly.
- Supervisors should ensure that safeguards are properly in place when machines are in use. Employees who disable or remove safeguards should be disciplined appropriately.
- Operator teams (two or more operators) of the same system should be trained in coordination techniques and proper use of devices that prevent premature activation by a team member.
- Operators should be trained and supervised to ensure that they dress properly for the job. Long hair, loose clothing, neckties, rings, watches, necklaces, chains, and earrings can become caught in equipment and, in turn, pull the employee into the hazard zone.
- Shortcuts that violate safety principles and practices should be avoided. The pressures of deadlines should never be the cause of unsafe work practices.
- Other employees who work around machines but do not operate them should be made aware of the emergency procedures to take when an accident occurs.

Basic Program Content

Machine safeguarding should be organized, systematic, and comprehensive. A company's safeguarding program should have at least the following elements:

- Safeguarding policy that is part of a broader companywide safety and health policy
- Machine hazard analysis
- Lockout/tagout (materials and procedures)
- Employee training
- Comprehensive documentation
- Periodic safeguarding audits (at least annually)

FIGURE 10-16 Selected examples of problems and corresponding actions.

Problem	Action
Machine is operating without the safety guard.	Stop machine immediately and activate the safety guard.
Maintenance worker is cleaning a machine that is operating.	Stop machine immediately and lock or tag it out.
Visitor to the shop is wearing a necktie as he observes a lathe in operation.	Immediately pull the visitor back and have him remove the tie.
An operator is observed disabling a guard.	Stop the operator, secure the guard, and take disciplinary action.
A robot is operating without a protective barrier.	Stop the robot and erect a barrier immediately.
A machine guard has a sharp, ragged edge.	Stop the machine and eliminate the sharp edge and ragged burrs by rounding it off.

Taking Corrective Action

What should be done when a mechanical hazard is observed? The only acceptable answer to this question is, take *immediate corrective action*. The specific action indicated will depend on what the problem is. Figure 10-16 shows selected examples of problems and corresponding corrective actions.

These are only a few of the many different types of problems that require corresponding corrective action. Regardless of the type of problem, the key to responding is immediacy. As shown in the examples given earlier in this chapter, waiting to take corrective action can be fatal. It is important to note that it is often prudent to exceed some OSHA guarding requirements (i.e., the 7-foot rule).

Key Terms and Concepts

- Adjustable guards
- Affected employee
- Authorized employee
- Automatic ejection
- Automatic feed
- Breaking
- Crushing
- Cutting
- Electromechanical devices
- Energized
- Energy control procedure
- Energy control program
- Energy source
- Energy-isolating device
- Fixed guards
- Gates
- Interlocked guards
- Lockout device
- Lockout/tagout systems
- Mechanical hazards
- Mechanical injuries
- Photoelectric devices
- Point-of-operation devices
- Point-of-operation guards
- Pullback devices
- Puncturing
- Radio-frequency devices
- Restraint devices
- Risk assessment
- Safeguarding
- Safety trip devices
- Semiautomatic ejection
- Semiautomatic feed
- Shearing
- Spraining
- Straining
- Tagout
- Tagout device
- Tearing
- Two-hand controls
- Work envelope

Review Questions

1. List and briefly explain the common types of mechanical injury hazards.
2. Explain the concept of safeguarding.
3. Summarize OSHA's requirements for machine guarding.
4. What are the requirements all safeguards should meet?
5. Describe the three types of point-of-operation guards.
6. Describe four types of point-of-operation devices.
7. What are the relative advantages and disadvantages of feeding and ejection systems?
8. Describe the primary hazards associated with robots.
9. Explain how to guard against the hazards associated with robots.
10. What is a lockout system?
11. What is a tagout system?
12. What impact may a lockout/tagout system have if carefully followed nationwide?
13. Summarize the main provisions of OSHA's lockout/tagout standard.
14. Explain the concept of risk assessment as it relates to machine operation.
15. Explain how to evaluate lockout/tagout programs.

Endnotes

1. National Safety Council, Guards: Safeguarding Concepts Illustrated, 7th ed. (Chicago: National Safety Council, 2002), 1.
2. Retrieved from http://www.osha.gov/pls/oshaweb.
3. EN 954, Part I, "Safety of Machinery—Principles of Safety Related to Control Systems," European Union, 1997.
4. National Safety Council, Guards: Safeguarding Concepts Illustrated, 2–3.
5. Ibid., 36.
6. Ibid., 38–39.
7. Retrieved from http://online.misu.nodak.edu/19577/BADM309checklist.htm.
8. Ibid., 44.
9. Retrieved from http://www.osha.gov/SLTC/smallbusiness/sec11.html.
10. L. Johnson, "The 'Red Flags' of LOTC," Occupational Health & Safety 68, no. 3:55.

Falling, Impact, Acceleration, Lifting, and Vision Hazards

Major Topics

- Causes of Falls
- Kinds of Falls
- Walking and Slipping
- Slip and Fall Prevention Programs
- OSHA Fall Protection Standards
- Ladder Safety
- Impact and Acceleration Hazards (head, eyes, face, and feet protection)
- Lifting Hazards
- Standing Hazards
- Hand Protection
- Personal Protective Equipment
- Forklift Safety (Powered Industrial Trucks)

Some of the most common accidents in the workplace happen as the result of slipping, falling, and improper lifting. Impact from a falling object is also a common cause of accidents. This chapter provides the information needed by modern safety and health professionals to prevent such accidents. It also provides specific information about head, hand, back, eye, face, and foot protection as well as forklift safety.

Causes of Falls

More than 16 percent of all disabling work-related injuries are the result of falls. Clearly, falls are a major concern of safety and health professionals. The primary causes of falls are as follows:

- A foreign object on the walking surface
- A design flaw in the walking surface

- Slippery surfaces
- An individual's impaired physical condition[1]

A **foreign object** is any object that is out of place or in a position to trip someone or to cause a slip. There is an almost limitless number of **design flaws** that may cause a fall. A poorly designed floor covering, a ladder that does not seat properly, or a catwalk that gives way are all examples of design flaws that may cause falls. Slippery surfaces are particularly prevalent in industrial plants where numerous different lubricants and cleaning solvents are used.

Automobile accidents are often caused when a driver's attention is temporarily drawn away from the road by a visual distraction. This is also true in the workplace. Anything that distracts workers visually can cause a fall. When a person's physical condition is impaired for any reason, the potential for falls increases. This is a particularly common problem among aging workers. Understanding these causes is the first step in developing fall prevention techniques.

Kinds of Falls

Falling from ladders and other elevated situations is covered later in this chapter. This section deals with the more common surface falls. Such falls can be divided into the following four categories:

- **Trip and fall** accidents occur when workers encounter an unseen foreign object in their path. When the employee's foot strikes the object, he or she trips and falls.
- **Stump and fall** accidents occur when a worker's foot suddenly meets a sticky surface or a defect in the walking surface. Expecting to continue at the established pace, the worker falls when his or her foot is unable to respond properly.
- **Step and fall** accidents occur when a person's foot encounters an unexpected step down (e.g., a hole in the floor or a floorboard that gives way). This can also happen when an employee thinks he or she has reached the bottom of the stairs when, in reality, there is one more step.
- **Slip and fall** accidents occur when the worker's center of gravity is suddenly thrown out of balance (e.g., an oily spot causes a foot to shoot out from under the worker). This is the most common type of fall.

Walking and Slipping

Judging by the number of injuries that occur each year as the result of slipping, it is clear that walking can be hazardous to a worker's health. This is, in fact, the case when walking on an unstable platform. A stable platform for walking is any surface with a high degree of traction that is free of obstructions. It follows that an unstable platform is one lacking traction, one on which there are obstructions, or both.

Measuring Surface Traction

In order to understand *surface traction, you* must have a basis for comparison. An effective way for comparing the relative traction of a given surface is to use the **coefficient of friction**, which is a numerical comparison of the resistance of one surface (shoe or boot) against another surface (the floor).

FIGURE 11-1 Coefficients of friction and relative traction ratings.

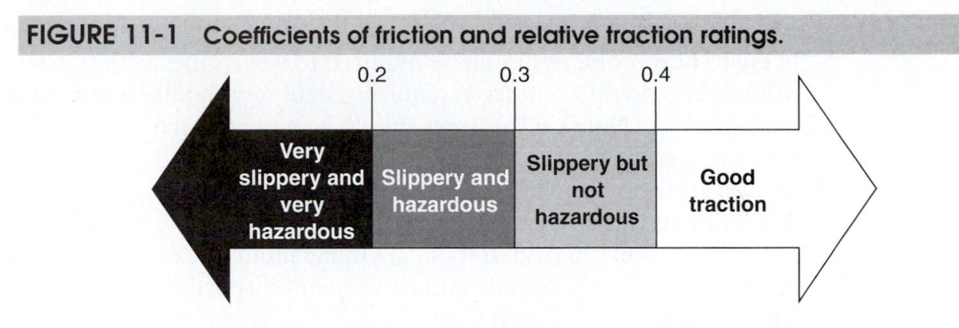

Figure 11-1 is a continuum showing coefficients of friction ratings from very slippery to good traction. Surfaces with a coefficient of friction of 0.2 or less are very slippery and very hazardous. At the other end of the continuum, surfaces with a coefficient of friction of 0.4 or higher have good traction.

To gain a feel for what different coefficients actually mean, consider the following: (1) ice has a coefficient of friction of 0.10; (2) concrete has a coefficient of 0.43; (3) linoleum has a coefficient of 0.33; and (4) waxed white oak has a coefficient of 0.24. Compare these coefficients with Figure 11-1 to determine the degree of hazard and how the surfaces compare.

Factors That Decrease Traction

Good housekeeping can be a major factor in reducing slip and fall hazards. Water, oil, soap, coolant, and cleaning solvents left on a floor can decrease traction and turn an otherwise safe surface into a danger zone. For example, the friction coefficient of concrete (0.43) is reduced by almost 15 percent if the concrete is wet. Rubber-soled shoes can decrease slipping hazards somewhat, but changing the type of shoe is not enough to ensure safety. Additional precautions are needed.

General Strategies for Preventing Slips

Modern safety and health professionals are concerned with preventing slips and falls. Slip prevention should be a part of the company's larger safety and health program. Here are some strategies that can be used to help prevent slipping:

1. **Choose the right material from the outset.** Where the walking surface is to be newly constructed or an existing surface is to be replaced, safety and health professionals should encourage the selection of surface materials that have the highest possible coefficient of friction. Getting it right from the start is the best way to prevent slipping accidents.
2. **Retrofit an existing surface.** If it is too disruptive or too expensive to replace a slippery surface completely, **retrofit** it with friction enhancement devices or materials. Such devices or materials include runners, skid strips, carpet, grooves, abrasive coatings, grills, and textured coverings.
3. **Practice good housekeeping.** Regardless of the type of surface, keep it clean and dry. Spilled water, grease, oil, solvents, and other liquids should be removed immediately. When the surface is wet intentionally, as when cleaning or mopping, rope off the area and erect warning signs.

4. **Require nonskid footwear.** Employees who work in areas where slipping is likely to be a problem should be required to wear shoes with special nonskid soles. This is no different from requiring steel-toed boots to protect against falling objects. **Nonskid footwear** should be a normal part of a worker's personal protective equipment.
5. **Inspect surfaces frequently.** Employees who are working to meet production deadlines may be so distracted that they don't notice a wet surface, or they may notice it but feel too rushed to do anything about it. Consequently, safety and health professionals should conduct frequent inspections and act immediately when a hazard is identified.[2]

Effective strategies for preventing slips and falls include the following:

1. Review and analyze accident statistics to determine where slip and fall accidents are happening and why; then take the appropriate corrective measures.
2. Monitor the condition of walking surfaces continually and make appropriate preventive corrections immediately.
3. Make sure that ramps and sloped floors have high-friction surfaces.
4. Use safety mats, nonslip flooring, and slip-resistant safety shoes.
5. Make sure that stairs have handrails.
6. Make sure that visibility is good in potentially hazardous areas. Add extra lighting if necessary. Also make sure that the color of paint in these areas is bright and helpful in calling attention to potential hazards.
7. Make sure that spills are cleaned up immediately and that the underlying cause of the spill is corrected.
8. Make sure that employees who work in potentially hazardous areas select and wear the right slip-resistant footwear.
9. Use appropriate technologies such as **vertical incidence tribometers (VITs)** to measure the slip resistance of floors and take appropriate action based on the results.
10. Conduct periodic audits of walking surfaces throughout the facilities in question, document carefully the findings, and take appropriate action in a systematic way.[3]

Specific Strategies for Preventing Slips

In addition to the general strategies already explained there are a number of specific strategies that will be helpful for preventing slips and falls. These specific strategies are presented in this section.[4]

Building Lobbies

The lobbies to buildings often have slick, highly polished floor surfaces designed to project a certain image. Unfortunately, such floors represent a real hazard for slipping and falling. The hazard level is often increased when outside moisture from rain, sleet, or snow is brought in by pedestrian traffic. The following prevention strategies will help decrease the hazard level: (1) use large welcome mats that are wide enough to allow several "drying steps" to be taken before reaching the slick floor; (2) provide umbrella holders so that dripping umbrellas are not brought onto the slick floor; (3) monitor the floor surface continually and dry any moisture that makes its way onto the floor immediately; and (4) substitute nonslip surfaces for the slick, highly polished flooring.

Restrooms

Because of the nature of their use, restrooms are likely to have tiled flooring. Certain types of tiles become slippery when water from the sinks, toilets, or urinals splash onto the floor or overflow. The hazards in this situation are multiplied when soap is added in restrooms. The following prevention strategies will help decrease the hazard level in restrooms: (1) monitor restrooms continually and clean up spills immediately; (2) use "wet floors" warning signs; (3) block off any wet areas until they dry; and (4) conduct periodic inspections of public restrooms on a systematic basis.

Kitchens

Some business and industrial firms have commercial kitchens for providing meals to their personnel. When this is the case, the most common hazard is polymerized grease. Grease on almost any floor surface, but especially on tile, concrete, and linoleum floors, creates a serious slip and fall hazard. The following strategies will help decrease the level of hazard in commercial kitchens: (1) use a nonslip floor surface; (2) require kitchen employees to wear slip-resistant footwear; and (3) frequently dry clean the floors after-hours using a method other than wet mopping (which typically just distributes the grease).

Processing Areas

Companies that process materials typically experience high hazard levels in the processing areas. Processing often requires the use of fluids that can spill over and onto the floor. For example, companies that process meat have to contend with blood, fats, and meat juices on the floor. Companies that process chemicals must contend with spillage of those chemicals. The following strategies will help decrease the hazard level in processing areas: (1) use nonslip flooring; (2) monitor floor surfaces continually and take immediate action to clean up spills; (3) require processing employees to wear slip-resistant footwear; and (4) inspect and clean floor surfaces on a regular basis.

Slip and Fall Prevention Programs

Every year slips, trips, and falls cause more than one million workplace injuries and approximately 16,000 deaths. A company's overall safety and health program should include a slip and fall prevention component. Such a component should have the following elements:

1. **A policy statement/commitment.** Statement to convey management's commitment. Areas that should be included in the policy statement are management's intent, scope of activity, responsibility, accountability, the safety professional's role, authority, and standards.
2. **Review and acceptance of walkways.** Establish the criteria that will be used for reviewing all walking surfaces and determining if they are acceptable. For example, a criterion may be a minimum coefficient of friction value. Regardless of the criteria, the methodology that will be used for applying them to the review and acceptance of walkways should also be explained.
3. **Reconditioning and retrofitting.** Include recommendations and timetables for reconditioning or retrofitting existing walking surfaces that do not meet review and acceptance criteria.

4. **Maintenance standards and procedures.** State the maintenance standards for walking surfaces (e.g., how often surfaces should be cleaned, resurfaced, replaced, and so on). In addition, this section should contain procedures for meeting the standards.
5. **Inspections, audits, tests, and records.** Provide a comprehensive list of inspections, audits, and tests (including the types of tests) that will be done, how frequently, and where. Maintain records of the results.
6. **Employee footwear program.** Specify the type of footwear required of employees who work on different types of walking surfaces.
7. **Defense methods for legal claims.** Outline the company's legal defenses so that aggressive action can be taken immediately should a lawsuit be filed against the company. In such cases, it is important to be able to show that the company has not been negligent (e.g., the company has a slip and fall prevention program that is in effect).
8. **Measurement of results.** Contain the following two parts: (a) an explanation of how the program will be evaluated and how often (e.g., comparison of yearly, quarterly, or monthly slip and fall data); (b) records of the results of these evaluations.[5]

OSHA Fall Protection Standards

The OSH Act mentions fall protection in several places. Although the General Industry Standards are silent on fall protection, the problem is covered in the following subparts:

Subpart D	Walking/working surfaces
Subpart F	Powered platforms, manlifts, and vehicle-mounted work platforms
Subpart R	Special industries

In addition to these OSHA standards, the American National Standards Institute (ANSI) publishes a Fall Protection Standard (ANSI Z359.1: *Safety Requirements for Personal Fall Arrest Systems, Subsystems, and Components*). The most comprehensive and most controversial fall protection standard is OSHA's Fall Protection Standard for the construction industry (Subpart M of 29 CFR 1926).

OSHA's Fall Protection Standard for Construction

OSHA's current Fall Protection Standard sets the *trigger height* at 6 feet. This means that any construction employee working higher than 6 feet off the ground must use a fall protection device such as a safety harness and line (see Figure 11-2).

This trigger height means that virtually every small residential builder and roofing contractor is subject to the standard. Because most residential builders and roofing contractors are small, Subpart M of 29 CFR 1926 is a source of much controversy.

OSHA officials argue that the 6-foot trigger height saves up to 80 lives per year and prevents more than 56,000 injuries. The rationale is that 6 percent of all lost-time fall injuries in the construction industry are caused by falls from less than 10 feet. Opponents counter that the cost of complying with the standard is almost $300 million annually. Commercial contractors, whose employees typically work much higher than the 6- to 16-foot range, are not concerned about the height controversy.

FIGURE 11–2 Personal fall arrest harness.

Courtesy of Dalloz Fall Protection.

Items from OSHA Regulation 1926 that apply specifically to fall protection in scaffolding work are as follows:

- 1926.451(g)(2) reads: "The employer shall have a competent person determine the feasibility and safety of providing fall protection for employees erecting or dismantling supported scaffolds. Employers are required to provide fall protection for employees erecting or dismantling supported scaffolds where the installation and use of such protection is feasible and does not create a greater hazard."

- 1926.502(d)(15) reads: "Anchorages used for attachment of personal fall arrest equipment shall be independent of any **anchorage** being used to support or suspend platforms and capable of supporting at least 5,000 pounds (22.2 kilograms) per employee attached, or shall be designed, installed and used as follows:
  i. as part of a complete personal fall arrest system which maintains a safety factor of at least two; and
  ii. under the supervision of a qualified person."
- 1926.451(d)(16) "Scaffold."
  i. limit maximum arresting force on an employee to 1500 pounds (4 kg) when used with a **body belt**;
  ii. limit maximum arresting force on an employee to 1,800 pounds (8 kg) when used with a **body harness**;
  iii. be rigged such that an employee can neither free-fall more than 6 feet (1.8 m), nor contact any lower level;
  iv. bring an employee to a complete stop and limit maximum deceleration distance an employee travels to 3.5 feet (1.07 m); and
  v. have sufficient strength to withstand twice the potential impact energy of an employee free-falling a distance of 6 feet (1.8 m), or the free fall distance permitted by the system, whichever is less.[6]

OSHA's Recommendations for Effective Fall Protection

Because slip and fall accidents account for approximately one million workplace injuries every year, organizations obviously need to have a strong fall prevention program in place. But what does it take to have an effective fall prevention program? OSHA recommends the following strategies:

- **Have a plan.** An organization should develop a written plan that is part of its larger safety and health plan. The fall protection plan should contain a statement of commitment from both management and employees, rules and regulations relating to fall protection, and an explanation of the training program and training requirements.
- **Establish proper fall protection requirements.** Require the use of fall protection equipment any time an employee works more than 4 feet above the floor in general industry, 6 feet or more in construction, and 10 feet or more when on scaffolding.
- Provide proper fall protection equipment and procedures and require their use. Organizations should determine what types of fall protection equipment and procedures are needed, provide them to employees, and require their proper use. As examples, this might include personal fall arrest systems, guardrails, safety nets, positioning devices, warning lines, controlled access zones, and safety monitoring.
- **Ensure fall protection device replacement.** Organizations should replace fall protection devices on a regular schedule even if there are no significant signs of wear.
- **Ensure proper use and type of equipment.** Ensure that the fall protection equipment provided to employees is the proper type for the situation in question and that employees inspect it before putting it on, that it fits properly, and that it is properly attached to anchorage points.

- **Provide training.** Provide fall protection training for supervisors and employees including how to recognize fall-related hazards and how to properly use all applicable fall protection equipment.[7]

Ladder Safety

Jobs that involve the use of ladders (see Figure 11-3) introduce their own set of safety problems, one of which is an increased potential for falls. The National Safety Council recommends that ladders be inspected before every use and that employees who use them follow a set of standard rules.[8]

Inspecting Ladders

Taking a few moments to look over a ladder carefully before using it can prevent a fall. The NSC recommends the following when inspecting a ladder:

- See if the ladder has the manufacturer's instruction label on it.
- Determine whether the ladder is strong enough.
- Read the label specifications about weight capacity and applications.
- Look for the following conditions: cracks on side rails; loose rungs, rails, or braces; or damaged **connections** between rungs and rails.
- Check for heat damage and corrosion.
- Check wooden ladders for moisture that may cause them to conduct electricity.
- Check metal ladders for burrs and sharp edges.
- Check fiberglass ladders for signs of *blooming,* deterioration of exposed fiberglass.[9]

Do's and Don'ts of Ladder Use

Many accidents involving ladders result from improper use. Following a simple set of rules for the proper use of ladders can reduce the risk of falls and other ladder-related accidents. The NSC recommends the following do's and don'ts of ladder use:

- Check for slipperiness on shoes and ladder rungs.
- Secure the ladder firmly at the top and bottom.
- Set the ladder's base on a firm, level surface.
- Apply the **four-to-one ratio** (base one foot away from the wall for every 4 feet between the base and the support point).
- Face the ladder when climbing up or down.
- Barricade the base of the ladder when working near an entrance.
- Don't lean a ladder against a fragile, slippery, or unstable surface.
- Don't lean too far to either side while working (stop and move the ladder).
- Don't rig a makeshift ladder; use the real thing.
- Don't allow more than one person at a time on a ladder.
- Don't allow your waist to go any higher than the last rung when reaching upward on a ladder.
- Don't separate the individual sections of extension ladders and use them individually.
- Don't carry tools in your hands while climbing a ladder.
- Don't place a ladder on a box, table, or bench to make it reach higher.[10]

FIGURE 11-3 Portable ladder/stair.

Courtesy of Lapeyre Stair, Inc.

OSHA standards for walking and working surfaces and ladder safety are set forth in 29 CFR Part 1910 (Subpart D). The standards contained in Subpart D are as follows:

1910.21	Definitions
1910.22	General requirements
1910.23	Guarding floor and wall openings and holes
1910.24	Fixed industrial stairs
1910.25	Portable wood ladders
1910.26	Portable metal ladders
1910.27	Fixed ladders
1910.28	Safety requirements for scaffolding
1910.29	Manually propelled mobile ladder stands and scaffolds (towers)
1910.30	Other working surfaces
1910.31	Sources of standards
1910.32	Standards organizations

What to Do after a Fall

If, in spite of your best efforts, a fall occurs on the job, what employees do in the immediate aftermath can mean the difference between life and death for the victim. First, make sure your organization has a fall rescue plan in place that includes the following: (1) training for all personal in how to carry out a rescue—what to do and what not to do; (2) proper equipment on-site and readily available; (3) coordination with local emergency authorities; and (4) assigned responsibilities. Then, in addition, make sure all employees who work at heights or with others who work at heights understand the following basics:[11]

- **Never work alone.** These should always be two or more people working in close proximity when working at heights. It is important to have someone available to intiate the fall rescue plan.
- **Keep legs moving.** When a worker is dangling from his fall arrest gear, it is important for him to keep his legs moving—not frantically, but just rhythmically and regularly. This will help prevent the venous pooling of blood that can lead to shock. If it is possible, the suspended worker should try to move into an upright position.
- **Raise the worker to a seated position.** Once the suspended worker has been brought to the ground, the tendency is to lie him down in a horizontal position. This is a mistake because it can suddenly release pooled blood that can strain the heart and cause death. Instead, move the victim into a seated position.

Monitor Fall Protection Equipment and Know Why It Fails

Although more people than ever are using fall protection equipment on the job, the number of injuries and deaths from falls continues to increase. There are several reasons for this, including poor training, deterioration of equipment over time, and selection of the wrong equipment for the job. However, the reason that should concern safety and health professionals most is failure due to lack of monitoring. In many cases, equipment failures could have been prevented by a systematic monitoring process in which all equipment is inspected before being used. Consequently, it is important for safety and health professionals to understand why fall protection equipment fails, and to ensure that equipment is properly monitored to detect potential failure points before it is used.[12]

When inspecting fall protection equipment, look for the following types of potential problems: weld splatter; webbing cuts and abrasions; broken stitching; frayed or burned webbing; chemical damage; discoloration; deformed hardware; loose, distorted, or broken grommets; knotted webbing; and malfunctioning snap hooks. Just making sure that employees know how to monitor their equipment and that they follow through and do it—every time—will save lives.

Impact and Acceleration Hazards

An employee working on a catwalk drops a wrench. The falling wrench accelerates over the 20-foot drop and strikes an employee below. Had the victim not been wearing a hard hat he might have sustained serious injuries from the impact. A robot loses its grip on a part, slinging it across the plant and striking an employee. The impact from the part breaks one of the employee's ribs. These are examples of accidents involving **acceleration** and **impact**. So is any type of fall because, having fallen, a person's rate of fall accelerates (increases) until striking a surface (impact). Motor vehicle accidents are also acceleration and impact instances.

Because falls were covered in the previous section, this section will focus on hazards relating to the acceleration and impact of objects. Approximately 25 percent of the workplace accidents that occur each year as the result of acceleration and impact involve objects that become projectiles.

Protection from Falling or Accelerating Objects

Objects that fall, are slung from a machine, or otherwise become projectiles pose a serious hazard to the heads, faces, feet, and eyes of workers. Consequently, protecting workers from projectiles requires the use of appropriate personal protective equipment and strict adherence to safety rules by all employees.

Head Protection

Approximately 120,000 people sustain head injuries on the job each year.[13] Falling objects are involved in many of these accidents. These injuries occur in spite of the fact that many of the victims were wearing hard hats. Such statistics have been the driving force behind the development of tougher, more durable hard hats.

Originally introduced in 1919, the hard hats first used for head protection in an industrial setting were inspired by the helmets worn by soldiers in World War I. Such early versions were made of varnished resin-impregnated canvas. As material technology evolved, hard hats were made of vulcanized fiber, then aluminum, and then fiberglass. Today's hard hats are typically made from the thermoplastic material polyethylene, using the injection-molding process.[14] Basic hard hat design has not changed radically since before World War II. They are designed to provide limited protection from impact primarily to the top of the head, and thereby reduce the amount of impact transmitted to the head, neck, and spine.[15]

It is important not only to wear hard hats, but to wear them properly (e.g. never wear them backwards to make a fashion statement). Some companies adhere double-stick tape or flat magnets to the upper visor area to minimize the amount of dust or iron filings that fall into workers' eyes.

The American National Standards Institute (ANSI) standard for hard hats is Z89–1986. OSHA subsequently adopted this standard as its hard hat standard (29 CFR 1010.135).

> This standard calls for testing hard hats for impact attenuation and penetration resistance as well as electrical insulation. Specifically, hard hats are tested to withstand a 40-foot-pound impact, which is equivalent to a two-pound hammer falling about 20 feet. Hard hats are also designed to limit penetration of sharp objects that may hit the top of the hard-hat shell and to provide some lateral penetration protection.[16]

Hard hats can help reduce the risk associated with falling or projected objects, but only if they are worn. The use of hard hats in industrial settings in which falling objects are likely has been mandated by federal law since 1971.[17] In addition to making the use of hard hats mandatory when appropriate and supervising to ensure compliance, Feuerstein recommends the use of incentives.[18] According to Feuerstein,

> It would seem that the sweetest offer a head-injury prevention program makes is a work environment free of injuries from falling objects. But sometimes this ultimate reward is too abstract to excite employees. They need to be led into safety for its own sake by concrete incentives, such as intra-department competition, monetary rewards for good suggestions, points toward prizes, and peer recognition for the most improved behavior.[19]

Resources expended promoting the use of hard hats are resources wisely invested. "Work accidents resulting in head injuries cost employers and workers an estimated $2.5 billion per year in workers' compensation insurance, medical expenses and accident investigation as well as associated costs due to lost time on the job and substitute workers. That is an average cost of $22,500 for each worker who received a head injury."[20]

Eye and Face Protection

Eye and face protection are critical in the modern workplace. Eye injuries are a common and costly phenomenon.

Every day, an estimated 1,000 eye injuries occur in American workplaces, according to the Bureau of Labor Statistics (BLS). The result? More than $300 million a year in lost production time, medical expenses, and workers' compensation. Why is this continuing to happen?

First, too many people are not wearing eye protection. BLS found three out of every five workers with eye injuries were not wearing eye protection. Second, they were not wearing the right kind of eye protection. Flying particles, according to the bureau, cause most eye injuries. Almost 70 percent of injuries resulted from flying or falling objects or sparks striking the eye. About 20 percent of the injuries were caused by contact with chemicals. Nearly half of the accidents occurred in manufacturing; just over 20 percent were in construction.[21]

FIGURE 11-4 ANSI Standard Z87.1–2003.

High-Mass Impact Test—Purpose

This test is intended to ensure the level of mechanical integrity of a protective device and a level of protection from relatively heavy, pointed objects traveling at low speeds. Frames shall be capable of resisting impact from a 500-gram (17.6-ounce) missile with a 30-degree conical heat-treated tip and a 1-mm (.039-inch) radius dropped from a height of 130 cm (51.2 inches). No parts or fragments shall be ejected from the protector that could contact an eye.

High-Velocity Impact Test—Purpose

This test is intended to ensure a level of protection from high-velocity, low-mass projectiles. Frames shall be capable of resisting impact from 6.35-mm (¼-inch) steel balls weighing 1.06 grams (.04 ounce) at 150 feet per second (fps) from 0 degrees to 90 degrees for frames; 250 fps for goggles, 300 fps for face shields.

Impact Test—Drop Ball

A 25.4-mm (1-inch) steel ball, weighing 68 grams (24 ounces), free fall from 127 cm (50 inches).

Lens Thickness

Thickness is 3.0 mm (.118 inch) except lenses that withstand high velocity impact, then 2.0-mm (.079-inch) thickness is acceptable.

Impact Test—Penetration

Lens shall be capable of resisting penetration from a Singer needle on a holder weighing 1.56 ounces dropped freely from 50 inches.

Eye and face protection typically consist of safety glasses, safety goggles, or face shields. The ANSI standard for face and eye protective devices is Z87.1–2003. OSHA has also adopted this standard. It requires that nonprescription eye and face protective devices pass two impact tests: a high-mass, low-speed test and a low-mass, high-speed test. Figure 11-4 summarizes the purpose of the tests and their individual requirements concerning impact and penetration. Figures 11-5,11-6, and 11-7 show examples of the types of devices available for eye and face protection.

FIGURE 11-5 Safety glasses that wrap around for lateral protection.

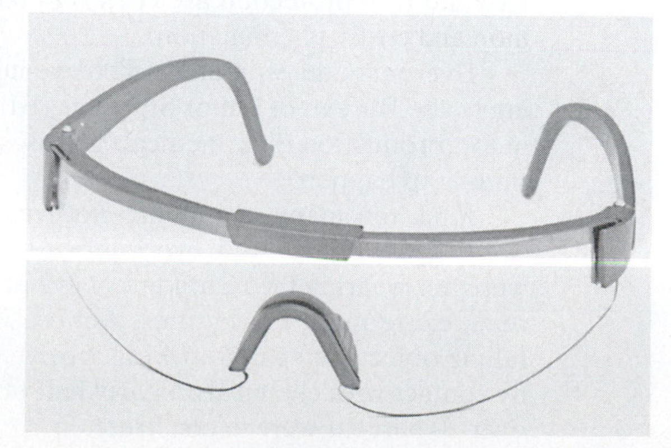

Courtesy of ELVEX Corporation, Bethel, CT.

FIGURE 11-6 Eye safety shield combined with ear and head protection.

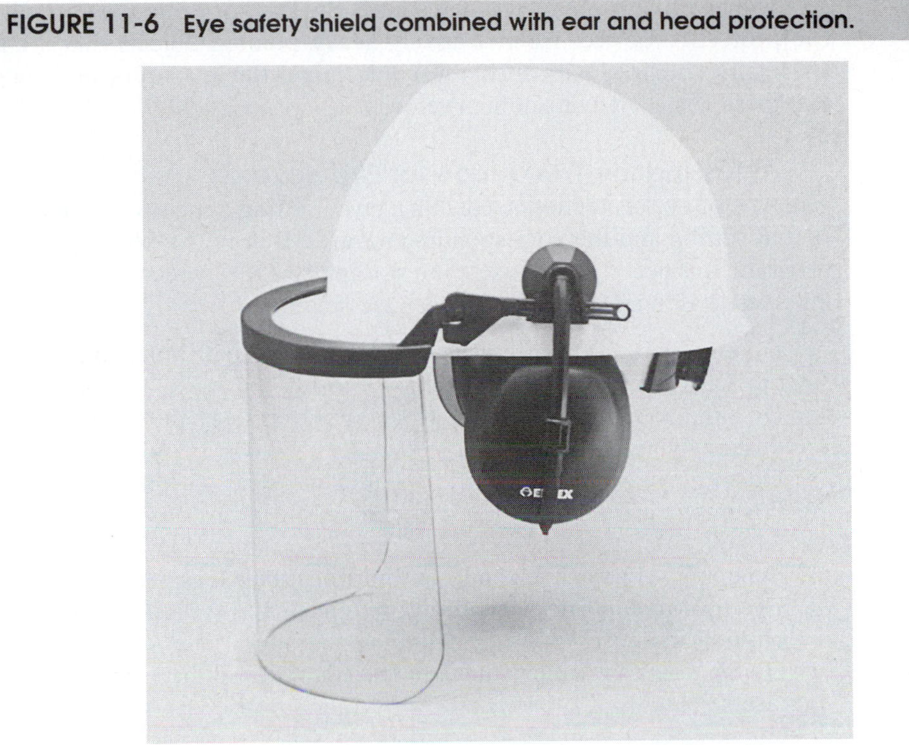

Courtesy of ELVEX Corporation, Bethel, CT.

FIGURE 11-7 Face shield (Huntsman® Model K Facesaver® with 8154L window attached).

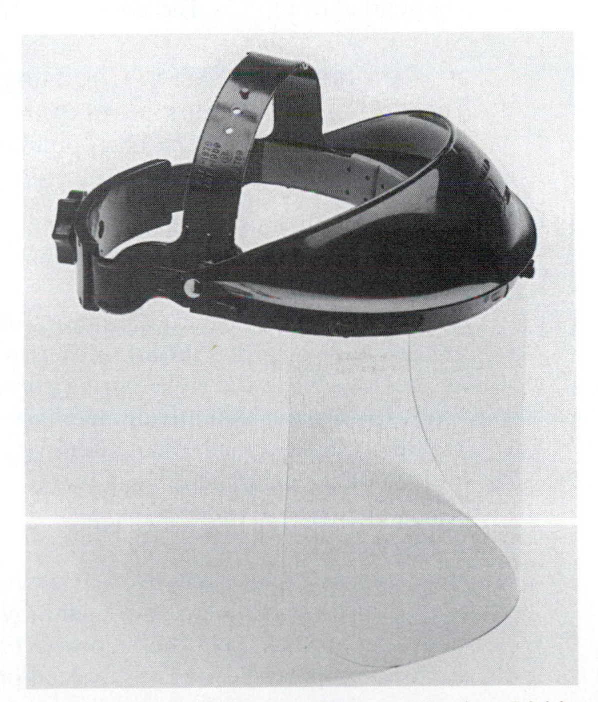

Courtesy of Kedman Company, Huntsman Products Division.

The high-mass impact test determines the level of protection provided by face and eye protective devices from relatively heavy, pointed objects that are moving at low speeds. The high-velocity impact test determines the level of protection provided from low-mass objects moving at high velocity.

Assessing the Workplace for Eye Hazards

The type of eye protection needed in a given setting depends on the type of work done in that setting and the corresponding hazards. Before establishing a vision protection program, it is necessary to assess the workplace. OSHA recommends using the following questions in making a workplace assessment:

- Do employees perform tasks that may produce airborne dust or flying particles?
- Do employees work near others who perform tasks that may produce airborne dust or flying particles?
- Do employees handle hazardous liquid chemicals or blood?
- Do employees work near others who handle hazardous liquid chemicals or blood?
- Do employees work in conditions in which their lens may become fogged?
- Do employees work in situations that may expose their eyes to chemical or physical irritants?
- Do employees work in situations that may expose their eyes to intense light or lasers?[22]

Based on the answers to these questions, a vision protection program can be developed to protect employees. That program should meet certain requirements as recommended by OSHA. These requirements are summarized in the next section.

Requirements When Choosing Vision Protection Devices

There are many different types of eye protection devices available that vary in terms of function, style, fit, lens, and other options. OSHA recommends applying the following criteria when selecting vision protection devices:

- Select only those that meet the standards set forth in ANSI Z87.1–2003.
- Select devices that protect against the specific hazard(s) identified in that assessment.
- Select devices that are as comfortable as possible to wear.
- Select devices that do not restrict vision in any way.
- Select devices the have fogging prevention capabilities built in.
- Select devices that are durable, easy to clean, and easy to disinfect.
- Select devices that do not interfere with the functioning of other personal protective equipment.[23]

Training

Once the workplace has been assessed and eye protection devices have been selected, it is important to provide employees with training in the proper use of the devices. This accomplishes the following: First, it ensures that the eye protection devices are used properly. Second, it shows employees that they have a critical role to play in the protection of their eyes. OSHA recommends training that covers the following topics:

- Why it is important to use the eye protection devices
- How the devices protect the eyes
- Limitations of the devices
- When the devices should be used
- How the devices are properly worn
- How straps are adjusted for both effectiveness and comfort
- How the employee can identify signs of wear that may lessen the effectiveness of the devices
- How the devices are cleaned and disinfected and how often[24]

First Aid for Eye Injuries

Even with proper eye protection, there is still the risk that an employee may sustain an injury. Even the best vision protection program is not perfect. When this happens, the following guidelines for first aid apply:

- Be gentle with the employee. Don't add to the injury with rough treatment.
- Do not attempt to remove objects embedded in the eyeball.
- Rinse the eyes with a copious amount of water for 15 to 30 minutes to remove the chemicals. Call for professional help. Cover both eyes after the rinsing has been completed.
- Never press on an injured eye or put any pressure on it (as when covering the eyes).
- Do not allow the employee to rub his or her eyes.[25]

Contact Lenses in a Chemical Enviornment

For years it was commonly thought among safety and health professionals that workers should not wear contact lenses in a chemical environment. In fact, until 2003 NIOSH recommended that workers in chemical environments not wear contact lenses. However, over time much has been learned about this issue. Contact lenses may, in fact, be worn in chemical environments and NIOSH has published an "intelligence bulletin" (number 59) explaining how to safely wear contact lenses in chemical environments.[26]

There are still environments in which contacts lens should not be worn. These are environments in which certain chemicals such as the following are present:

- 1,2-dibromo-3-chloropropane (DBCP)
- 4,4'-methylene dianiline (MDA)
- Ethyl alcohol
- Ethylene oxide
- Isopropyl alcohol
- Methylene chloride

This list is neither exhaustive nor comprehensive. Rather, it contains a partial list of the types of chemicals that can make a work environment especially hazardous for contact lens wearers. Before allowing employees who wear contact lenses to work in a chemical environment, it is best to conduct a comprehensive hazard assessment in which the types of chemicals that will be present are identified. NIOSH's Intelligence Bulletin 59: "Contact Lens Use in a Chemical Environment" is a good source of information when conducting hazard assessments of chemical environments. Once the hazard assessment has been completed, provided none of the "contact-lens-prohibited" chemicals are present in the work environment, workers who wear contact lenses should use the same eye protection recommended for other workers.

Foot Protection

The OSHA regulations for foot protection are found in 29 CFR 1910.132 and 126. Foot and toe injuries account for almost 20 percent of all disabling workplace injuries in the United States.[27] There are over 180,000 foot and toe injuries in the workplace each year.[28] The major kinds of injuries to the foot and toes are from the following:

- Falls or impact from sharp or heavy objects (this type accounts for 60 percent of all injuries)
- Compression when rolled over by or pressed between heavy objects
- Punctures through the sole of the foot
- Conductivity of electricity or heat
- Electrocution from contact with an energized, conducting material
- Slips on unstable walking surfaces
- Hot liquid or metal splashed into shoes or boots
- Temperature extremes[29]

The key to protecting workers' feet and toes is to match the protective measure with the hazard. This involves the following steps: (1) identify the various types of hazards present in the workplace, (2) identify the types of footwear available to counter the hazards, and (3) require that proper footwear be worn. Shoes selected should meet all applicable ANSI standards and have a corresponding ANSI rating. For example, "a typical ANSI rating is Z41PT83M1–75C–25. This rating means that the footwear meets the 1983 ANSI standard and the steel toe cap will withstand 75 foot pounds of impact and 2,500 pounds of compression."[30]

Modern safety boots are available that provide comprehensive foot and toe protection. The best safety boots provide all of the following types of protection:

- *Steel* toe for impact protection
- *Rubber* or *vinyl* for chemical protection
- *Puncture-resistant* soles for protection against sharp objects
- *Slip-resistant* soles for protection against slippery surfaces
- *Electricity-resistant* material for protection from electric shock

Employers are not required to provide footwear for employees, but they are required (29 CFR 1910.132 and 136) to provide training on foot protection. The training must cover the following topics as a minimum:

- Conditions when protective footwear should be worn
- Type of footwear needed in a given situation
- Limitations of protective footwear
- Proper use of protective footwear

OSHA Regulations Relating to Footwear

Foot protection is a high priority with OSHA. This can be seen in the number of regulations OSHA has developed and now mandates relating to footwear. The most prominent of these and a summary of the respective requirements follow:

1. **29 CFR 1910.132(d):** Requires hazard assessment in the workplace.
2. **29 CFR 1910.136:** Lists the general requirements for foot protection in the workplace. Requires that employers ensure that employees use the appropriate foot protection when working in areas where there is a danger of foot injuries due to falling or rolling objects, or objects piercing the sole, and where such employees' feet are exposed to electrical hazards.
3. **29 CFR 1910.132(f)(a,iv,v):** Specifies the training mandated for employees and fitting of footwear.[31]

In addition to these regulations, the following additional regulations also deal with specific aspects of footwear and foot protection: 29 CFR 1910.94 (ventilation), 29 CFR 1910.156 (foot, hand, eye, face, head, and body protection for employees who serve in fire brigades), 29 CFR 1910.269 (foot protection when working in electric power generation, transmission, and distribution), and 29 CFR 1910.1029 (foot protection when working with coke ovens).

Lifting Hazards

Back injuries that result from improper lifting are among the most common in an industrial setting. In fact, back injuries account for approximately $12 billion in workers' compensation costs annually. The following statistics concerning workplace back injuries illustrate the scope and seriousness of this problem:

- Lower back injuries account for 20 to 25 percent of all workers' compensation claims.
- Thirty-three to 40 percent of all workers' compensation costs are related to lower back injuries.
- Each year, there are approximately 46,000 back injuries in the workplace.
- Back injuries cause 100 million lost workdays each year.
- Approximately 80 percent of the population will experience lower back pain at some point in their lives.[32]

Typical cause of back injuries in the workplace include: improper lifting, reaching, sitting, and bending. **Lifting hazards** such as poor posture, ergonomic factors, and personal lifestyles also contribute to back problems. Consequently, a company's overall safety and health program should have a **back safety/lifting** component.

Back Safety/Lifting Program

Prevention is critical in back safety. Consequently, safety and health professionals need to know how to establish back safety programs that overcome the hazards of lifting and other activities. Dr. Alex Kaliokin recommends the following six-step program:

1. **Display poster illustrations.** Posters that illustrate proper lifting, reaching, sitting, and bending techniques should be displayed strategically throughout the workplace. This is as important in offices as in the plant. Clerical and office personnel

actually sustain a higher proportion of back injuries than employees in general. Sitting too long without standing, stretching, and walking can put as much pressure on the back as lifting.

2. **Preemployment screening.** Preemployment screening can identify people who already have back problems when they apply. This is important because more than 40 percent of back injuries occur in the first year of employment and the majority of these injuries are related to preexisting back problems.
3. **Regular safety inspections.** Periodic inspections of the workplace can identify potential problem areas so that corrective action can be taken immediately. Occasionally bringing a workers' compensation consultant in to assist with an inspection can help identify hazards that company personnel may miss.
4. **Education and training.** Education and training designed to help employees understand how to lift, bend, reach, stand, walk, and sit safely can be the most effective preventive measure undertaken. Companies that provide back safety training report a significant decrease in back injuries.
5. **Use external services.** A variety of external health care agencies can help companies extend their programs. Identify local health care providing agencies and organizations, what services they can provide, and a contact person in each. Maintaining a positive relationship with these external service contact people can increase the services available to employers.
6. **Map out the prevention program.** The first five steps should be written down and incorporated in the company's overall safety and health program. The written plan should be reviewed periodically and updated as needed.[33]

In spite of a company's best efforts, back injuries will still occur. Consequently, safety and health professionals should be familiar with the treatment and therapy that injured employees are likely to receive. Treatment for reconditioning addresses five goals: restoring function, reducing pain, minimizing deficits in strength, reducing lost time, and returning the body to preinjury fitness levels.[34]

A concept that is gaining acceptance in bridging the gap between treatment or therapy and a safe return to work is known as **work hardening**.[35] Work hardening and its objectives are explained as follows: In specially designed "work centers," various work stations, exercise equipment, and aggressive protocols are used for work reconditioning. The objectives are:

- A return to maximum physical abilities as soon as possible
- Improvement of general body fitness
- Reducing the likelihood of reinjury
- Work simulation that duplicates real work conditions[36]

The work centers referred to above replicate in as much detail as possible the injured employee's actual work environment. In addition to undergoing carefully controlled and monitored therapy in the work center, the employee is encouraged to use exercise equipment. Employees who undergo work center therapy should have already completed a program of acute physical therapy and pain management, and they should be medically stable.[37]

Health and safety managers can help facilitate the fastest possible safe resumption of duties by injured employees by identifying local health care providers that use the work-hardening approach. Such services and local providers of them should be made known to higher management so that the company can take advantage of them.

Proper Lifting Techniques

One of the most effective ways to prevent back injuries is to teach employees proper lifting techniques. Following are lifting techniques that should be taught as part of an organization's safety program.

PLAN AHEAD

- Determine if you can lift the load. Is it too heavy or too awkward?
- Decide if you need assistance.
- Check your route to see whether it has obstructions and slippery surfaces.

LIFT WITH YOUR LEGS, NOT YOUR BACK

- Bend at your knees, keeping your back straight.
- Position your feet close to the object.
- Center your body over the load.
- Lift straight up smoothly; don't jerk.
- Keep your torso straight; don't twist while lifting or after the load is lifted.
- Set the load down slowly and smoothly with a straight back and bent knees; don't let go until the object is on the floor.

PUSH, DON'T PULL

- Pushing puts less strain on your back; don't pull objects.
- Use rollers under the object whenever possible.

NIOSH and the Ergonomic Guidelines for Manual Material Handling (EGMMH)

The National Institute for Occupational Safety and Health (NIOSH) originally developed guidelines for lifting and lowering in 1981. The guidelines include a formula for calculating the recommended weight limit (RWL) for a given lifting job. The 1981 formula was simple and easy to use because it considered only a few factors that affect a lifting task. In 1993, the NIOSH guidelines were revised. The formula now takes into account nonsymmetrical lifting and lifting of items that don't have handles. Another important aspect of the guidelines was the new **multitask analysis strategy**. This strategy gives safety professionals a method for considering a variety of related lifting variables and how they interact. This is a much more complicated method than the original, but it is also much more accurate and realistic.

The multitask analysis strategy is particularly useful when dealing with tasks in which the lifting variables change throughout the task. For example, consider the task of a stacking job in which each successive item takes a different vertical location in the stack. The ergonomics of the task change with each successive item added to the stack (e.g., reach span required, height of lift) as do the corresponding hazards.

Because of its complexity, the NIOSH *lifting equation* is now more difficult to use than the original version. Safety professionals should know the formulas to understand and identify the various risk factors that contribute to back injuries. Software that allows safety professionals simply to plug selected values into the formulas is becoming readily available and is recommended. These values may be easily collected using nothing more than a stopwatch and a tape measure.

To apply the new lifting equation, safety professionals need to understand the types of information that they must collect and either turn it over to mathematicians or plug it into a computer program. This information is as follows:

LC: Load constant (always use 51 pounds or 23 kilograms)
HM: Horizontal line measured from the midpoint between the ankles forward to the midpoint between the hands, at both the origin and destination of lift
VM: Vertical line from the floor to the hands (also measured at the origin and destination of the lift)
DM: Vertical distance between the origin and destination of the lift
AM: Turning or twisting angle of asymmetry
FM: Average frequency rate of lifting measured in lifts per minute
CM: Coupling value (Does the item to be lifted have a good, fair, or poor grasping mechanism?)

More recently, NIOSH teamed with the California Occupational Safety and Health Administration (CAL/OSHA) and the Ergonomic Assist and Systems Equipment (EASE) Council of the Material Handling Industry of America to develop the **Ergonomic Guidelines for Manual Material Handling (EGMMH)**. The guidelines were finalized in 2006 and were developed primarily for smaller companies that cannot afford to hire a staff of safety and health professionals. They are divided into four parts: (1) the ergonomic process; (2) a matrix of common material handling tasks; (3) several chapters on how to approach common material handling tasks; and (4) a resource index.

The guidelines provide more than one safe way to approach the various material handling tasks covered, and they incorporate photographs and pictures to illustrate these approaches. Organizations or safety and health professionals who are interested in obtaining a copy of the EGMMH should visit the NIOSH Web site: http://www.cde.goc/niosh

Standing Hazards

Many jobs require that workers stand or walk for prolonged periods. Prolonged walking and/or standing can cause lower back pain, sore feet, varicose veins, and a variety of other related problems. The following hazard mitigation strategies can help minimize these problems for workers who must stand and walk for prolonged periods.

Antifatigue Mats

Antifatigue mats provide cushioning between the feet and hard working surfaces such as concrete floors (see Figure 11-8). This cushioning effect can reduce muscle fatigue and lower back pain. However, too much cushioning can be just as bad as too little. Consequently, it is important to test mats on a trial basis before buying a large quantity. Mats that become slippery when wet should be avoided. In areas where chemicals are used, be sure to select mats that will hold up to the degrading effects of chemicals.

Shoe Inserts

When antifatigue mats are not feasible because employees must move from area to area and, correspondingly, from surface to surface, shoe inserts may be the answer.

FIGURE 11-8 Padded mat.

Courtesy of Tennessee Mat Company, Inc.

Such inserts are worn inside the shoe and provide the same type of cushioning the mats provide. Shoe inserts can help reduce lower back, foot, and leg pain. It is important to ensure proper fit. If inserts make an employee's shoes too tight, they will do more harm than good. In such cases, employees may need to wear a slightly larger shoe size.

Foot Rails

Foot rails added to work stations can help relieve the hazards of prolonged standing. Foot rails allow employees to elevate one foot at a time four or five inches. The elevated foot rounds out the lower back, thereby relieving some of the pressure on the spinal column. Placement of a rail is important. It should not be placed in a position that inhibits movement or becomes a tripping hazard.

Workplace Design

A well-designed workstation can help relieve the hazards of prolonged standing. The key is to design workstations so that employees can move about while they work and can adjust the height of the workstation to match their physical needs.

Sit/Stand Chairs

Sit/stand chairs are higher-than-normal chairs that allow employees who typically stand while working to take quick mini-breaks and return to work without the hazards associated with getting out of lower chairs. They have the advantage of giving the employee's feet, legs, and back an occasional rest without introducing the hazards associated with lower chairs.

Proper Footwear

Proper footwear is critical for employees who stand for prolonged periods. Well-fitting, comfortable shoes that grip the work surface and allow free movement of the toes are best.

Hand Protection

In the United States there are more than 500,000 hand injuries every year. Hand injuries are both serious and costly for employers and for employees. Section 138 of OSHA standard 29 CFR 1910.132 covers personal protective equipment for hands. This standard requires employers to base the selection of hand protection (gloves) on a comprehensive assessment of the tasks performed for a given job, hazards present, and the duration of exposure to the hazards. The assessment must be documented in writing.[38]

With the assessment completed, employers are required to review specification information from manufacturers of safety gloves and select the gloves that are best suited for the individual situation. Selecting just the right gloves for the job has historically been one of employers' greatest difficulties in complying with the standard. It is not a simple task. For example, take the issue of fit. A poorly fitted set of gloves cannot offer the degree of protection that a responsible employer or employee wants. Yet, because manufacturers have not developed a consistent set of metrics for sizing gloves, the only way to determine whether a pair fits properly is for the employee to try them on.

Fit is just one of the problems faced when selecting gloves. Other critical features include the protection capability, comfort, and tactile sensitivity of the gloves. Often, greater comfort and tactile sensitivity can mean less protection. Correspondingly, greater protection can mean less comfort and tactile sensitivity.

In an attempt to simplify the process of selecting the right gloves, the American National Standards Institute (ANSI) and the Industrial Safety Equipment Association (ISEA) developed a joint hand-protection standard, ANSI/ISEA 105–1999. This standard simplifies glove selection by (1) defining characteristics of protection in a variety of critical areas including cuts, puncture resistance, abrasion, protection from cold and heat, chemical resistance (including both permeation and degradation), viral penetration, dexterity, liquid-tightness, and flame/heat resistance; and (2) standardizing the tests used to measure all of these various characteristics.

Common Glove Materials

Depending on the individual hazards available in a given situation, the right gloves for the application may be made of a variety of different materials (see Figure 11-9, 11-10, and11-11). The most widely used materials in manufacturing gloves are as follows:

- **Leather.** Offers comfort, excellent abrasion resistance, and minimum cut resistance.
- **Cotton.** Offers comfort, minimal abrasion resistance, and minimum cut resistance.
- **Aramids.** Offer comfort, good abrasion resistance, excellent cut resistance, and excellent heat resistance.
- **Polyethylene.** Offers comfort, excellent abrasion resistance, and minimal cut resistance. Gloves made of this material should not be subjected to high temperatures.

FIGURE 11-9 Cut-resistant work gloves.

Courtesy of Best Manufacturing Company.

FIGURE 11-10 Synthetic fiber gloves reinforced with stainless steel.

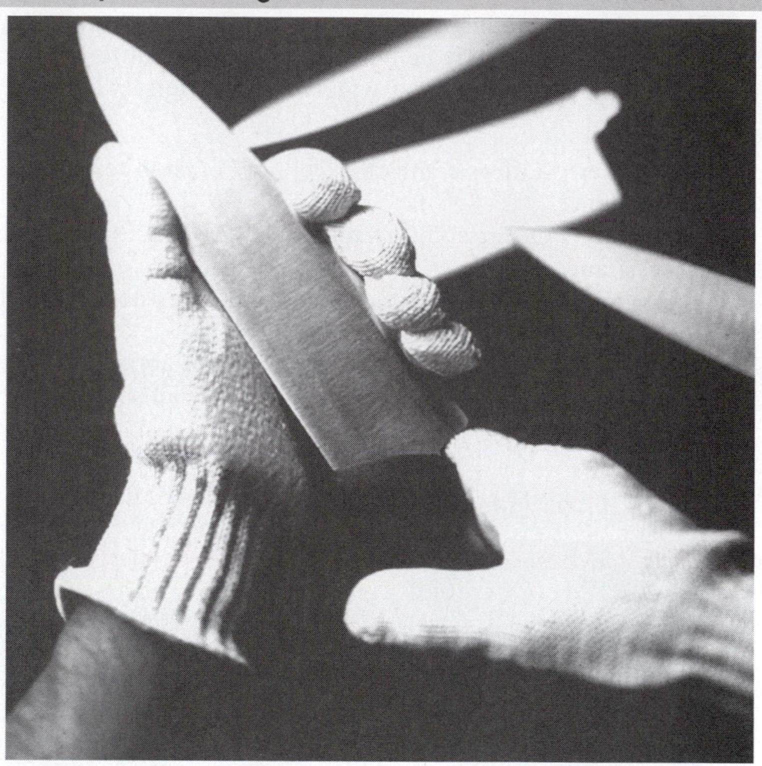

Courtesy of Best Manufacturing Company.

- **Stainless steel cord (wrapped in synthetic fiber).** Offers comfort, good abrasion resistance, and optimal cut resistance.
- **Chain link or metal mesh.** Offers very little comfort, but maximum abrasion and cut resistance.
- **Butyl rubber.** Offers little comfort, but has excellent resistance to heat, ozone, tearing, and certain chemicals including alcohols, aldehydes, ketones, esters, nitriles, gases, amides, acids, and nitro compounds.
- **Nitrile-based material.** Offers greater comfort and protection. Consequently, there is increased use of this type of material for the substrate coating of glasses.
- **Viton rubber.** Offers little comfort, but performs well with chemicals that butyl rubber cannot protect against, including aliphatics, halogenated, and aromatics. Like butyl gloves, viton gloves also perform well in handling alcohols, gases, and acids.[39]

Personal Protective Equipment

Personal protective equipment (PPE) is a critical component in the safety program of most organizations. Head, hand, back, eye, face, foot, skin, and breathing protection all involve the use of PPE. Unfortunately, it can be difficult to convince employees to wear PPE properly or, sometimes, to wear it at all. Employees often balk at the perceived

FIGURE 11-11 Natural rubber gloves.

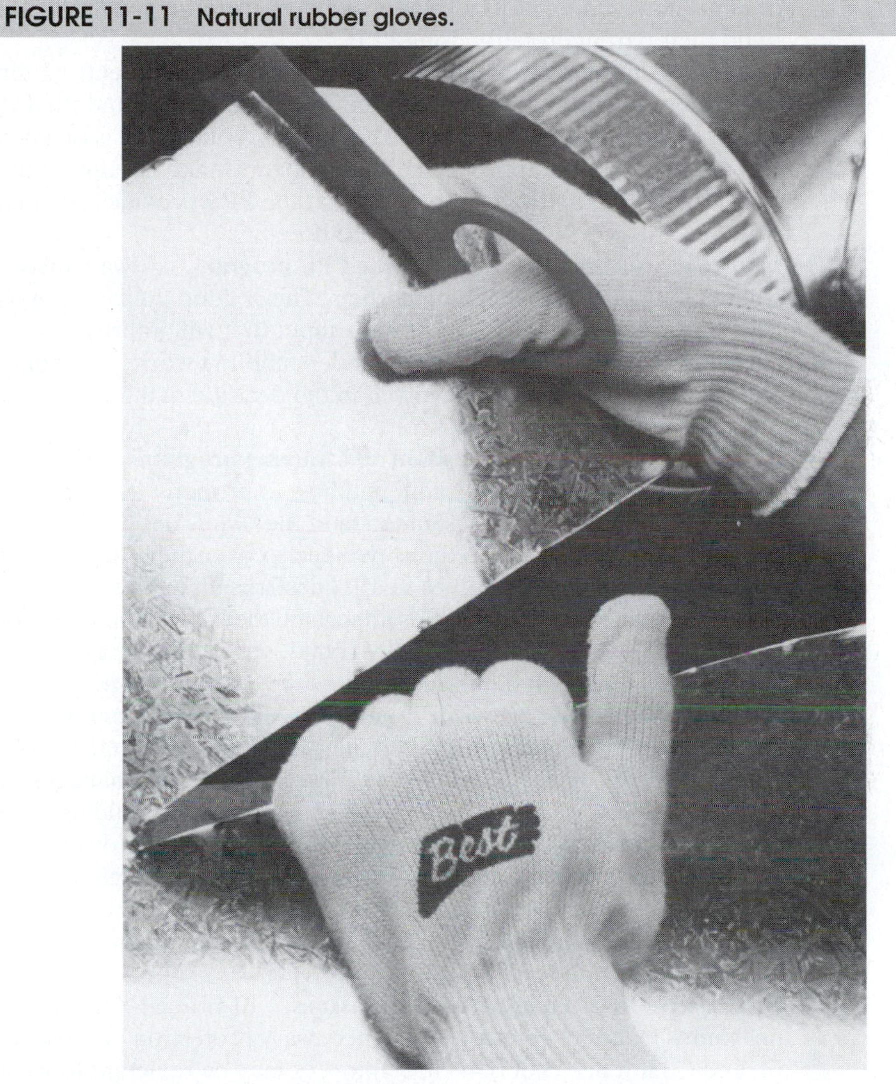

Courtesy of Best Manufacturing Company.

"inconvenience" of PPE. They don't like the way it looks or how it feels. They think it is cumbersome in which to work in or time consuming to put on and take off. Sometimes, they just forget.

Making employees comfortable with PPE is a serious and sometimes difficult challenge for safety and health professionals. The following strategies can be used to meet this challenge:

- **Make maximum use of engineering and administrative controls.** PPE should be the last line of defense in protecting employees from hazardous conditions. Before adopting PPE, organizations should first use every engineering and administrative control available to minimize potential hazards. If employees see that the organization is doing its part by applying these controls, they will be less reluctant to do their part in properly using PPE.

- **Ensure the optimum choice of PPE by using risk assessment.** Employees know when there is a mismatch between the hazards they face and the PPE they are provided. OSHA requires that PPE be selected on the basis of a comprehensive risk assessment. This approach helps employers make the optimum choice when selecting PPE. Employees who know that the PPE provides adequate protection from hazards will be more likely to use it.
- **Involve employees in all aspects of the PPE program.** Always involve employees when making decisions that affect them. This is good policy for two reasons: (1) Employees may be able to provide input that will improve the quality of the decisions being made because they understand the work tasks being discussed; and (2) employees who are involved in the decision making are more likely to buy into and support that decision.
- **Provide comprehensive education and training programs.** Employees need to understand why PPE is important and how to properly use it. Employers should never assume that employees understand the "why" or "how" of PPE. Training programs should begin with the why aspects, cover them thoroughly, and give employees ample opportunities to ask questions and voice concerns. Once employees understand why PPE is important, they should be given comprehensive training on how to use it properly. No employee should be expected to use PPE without first understanding why and how they should use it.
- **Reinforce the proper use of PPE and challenge its improper use.** Employers should never fall into the trap of taking PPE use for granted. Proper behavior relating to PPE should be reinforced by supervisors and managers. Correspondingly, improper use should be challenged. Employers should use PPE properly themselves and reward employees who follow suit. Rewards need not be formal. Publicly complimenting an employee can be reward enough. Correspondingly, when an employee is seen failing to use the required PPE or using it improperly, that employee should be corrected. However, whereas compliments are given publicly, correction should be done in private.
- **Be sensitive to fit, comfort, and style issues.** Ill-fitting PPE poses a double problem: (1) it may not provide the necessary protection because of the improper fit; and (2) if it does not fit well, employees may be reluctant to wear it because it is uncomfortable. Style can also be a problem in that employees are often self-conscious about their appearance. All these factors should be considered when choosing PPE.
- **Work to make PPE a normal part of the uniform.** By applying these strategies, employers can make PPE a normal part of the uniform. When this happens, using PPE will become standard operating procedure, and its proper use will cease to be an issue.[40]

Forklift Safety (Powered Industrial Trucks)

Powered industrial truck or forklift safety is included here because forklift-related injuries often result from impact or acceleration hazards. OSHA issues its standards for forklift safety in 29 CFR 1910.138 under the heading "Powered Industrial Truck."

These regulations apply to forklifts, platform lift trucks, and motorized hand trucks. The latest edition of OSHA's forklift standard is based primarily on a standard produced by the American Society of Mechanical Engineers titled, "Safety for Low Lift and High Lift Trucks" (ASME B56.1).

Forklifts are different from cars and trucks in several ways. Employees who drive forklifts should understand how they are different. The primary differences are as follows:

- Forklifts are typically steered by the rear wheels.
- An empty forklift can be more difficult to steer than one with a load.
- Forklifts are frequently driven in reverse.
- Forklifts have three-point suspension so that the center of gravity can move from the rear of the vehicle closer to the front when it is loaded.

Because of these differences, it is important to ensure that only properly trained employees drive forklifts and that these employees follow some basic rules of accident prevention. The rules fall into four categories: (1) general, (2) lifting, (3) traveling, and (4) placing.

GENERAL RULES

The rules in this section are general and apply to all phases of forklift operation:

- Keep arms, hands, and legs inside the vehicle at all times.
- Face in the direction of travel at all times.
- If the load blocks your view, drive backward.
- Allow plenty of room for braking—at least three vehicle lengths.
- Make sure there is sufficient overhead clearance before moving a load.

RULES FOR PICKING UP A LOAD

- Make sure the load is within the capacity of the forklift.
- Make sure forks are positioned properly.
- Make sure the load is properly balanced.
- Make sure the load is secure.
- Raise the load to the proper height.
- Run the forks all the way into the pallet, and tilt the mast back to stabilize the load before moving.
- Back out and stop completely before lowering the load.

RULES FOR TRAVELING WITH A LOAD

- Always give pedestrians the right-of-way.
- Never allow passengers on the forklift.
- Keep the forks low while moving.
- Keep the load tilted back slightly while moving.
- Drive slowly; a forklift is not a car.
- Slow down at all intersections; stop and sound the horn at blind intersections.
- Drive up and back down ramps and inclines.
- Never lift or lower the load when traveling.
- Keep to the right just as you do when driving a car.
- Watch for oil, grease, and wet spots, which could inhibit traction.

- Cross railroad tracks at a skewed angle, never at a right angle.
- Watch for edges on loading docks and other changes in elevation.
- Maintain at least four seconds of spacing between your forklift and the one in front of you.

RULES FOR PLACING A LOAD

- Stop the forklift completely before raising or lowering the load.
- Move slowly and cautiously with the load raised.
- Never walk or stand under a raised load or allow anyone else to do so.
- Be certain the forks have cleared the pallet before turning and before changing height.
- Stack the load square and straight.
- Check behind and on both sides before backing up.

OSHA's Training Requirements

OSHA estimates that 155,000 injuries and 100 fatalities occur every year as a result of accidents involving powered industrial trucks. Because of this, OSHA revised its training requirements for operators of powered industrial trucks (29 CFR 1910.178).[41] OSHA estimates that employers can save as much as $135 million annually by following these training requirements. Of this sum, $83 million could be saved in direct costs such as medical treatment, and $52 million could be saved in indirect costs.

Initial Training Program Content

Before they are allowed to operate a powered industrial truck (forklift), employees are required by OSHA's 29 CFR 1910.178 to complete initial training in the following topics:

1. Operating instruction, warnings, and precautions for the types of trucks the operator will be authorized to operate.
2. Differences between the truck and the automobile.
3. Truck controls and instrumentation: where they are located, what they do, and how they work.
4. Engine or motor operation.
5. Steering and maneuvering.
6. Visibility (including restrictions due to loading).
7. Fork and attachment adaptation, operation, and use limitations.
8. Vehicle capacity.
9. Vehicle stability.
10. Any vehicle inspection and maintenance the operator will be required to perform.
11. Refueling and/or charging and recharging of batteries.
12. Operating limitations.
13. Any other operating instructions, warnings, or precautions listed in the operator's manual for the types of vehicle the employee is being trained to operate.
14. Workplace-related topics including the following: surface conditions where the vehicle will be operated; composition of loads to be carried and load stability; load manipulation, stacking, and unstacking; pedestrian traffic in areas where the vehicle will be operated; narrow aisles and other restricted places where the vehicle will be operated; hazardous (classified) locations where the vehicle will

be operated; ramps and other sloped surfaces that affect the vehicle's stability; closed environments and other areas where insufficient ventilation or poor vehicle maintenance could cause a buildup of carbon monoxide or diesel exhaust; other unique or potentially hazardous environmental conditions in the workplace that could affect safe operation.

Refresher Training Program Content

Refresher training, including an evaluation of the effectiveness of that training, must be conducted to ensure that operators of powered industrial trucks have the up-to-date knowledge and skills to operate them safely. Refresher training is required whenever any one of the following conditions exists: the operator has been observed operating a vehicle in an unsafe manner; the operator has been involved in an accident or a near-hit accident; the operator has received an evaluation that indicates he or she is not operating the powered truck safely; the operator is assigned to drive a different type of truck; or any condition in the workplace changes in such a manner that it could affect the safe operation of the truck. In addition to these requirements, employers are required to conduct an evaluation of the performance of all industrial powered truck operators at least once every three years.

Key Terms and Concepts

- Acceleration
- Anchorage
- Back safety/lifting
- Body belt/body harness
- Coefficient of friction
- Connector
- Design flaw
- Ergonomic Guidelines for Manual
- Material Handling (EGMMH) Electrical hazard guard
- External services
- Eye and face protection
- Foot protection
- Foreign object
- Four-to-one ratio
- Good housekeeping
- Head protection
- Impact
- Initial training program content
- Ladder safety
- Lanyard/lifeline
- Lifting hazards
- Metatarsal guard
- Multitask analysis strategy
- Nonskid footwear
- OSHA Fall Protection Standard
- Personal fall arrest system
- Preemployment screening
- Puncture-resistant guard
- Reconditioning
- Refresher training program content
- Retrofit
- Review and acceptance
- Self-retracting lifeline/lanyard
- Slip and fall
- Slip-resistant guard
- Slippery surface
- Steel toe
- Step and fall
- Stump and fall
- Surface traction
- Trip and fall
- Vertical incidence tribometer (VIT)
- Work hardening

Review Questions

1. List the primary causes of falls.
2. Explain briefly the most common kinds of falls.
3. Explain how surface traction is measured.
4. List and briefly explain five strategies for preventing slips.
5. Describe the various components of a slip and fall prevention program.
6. Explain the "trigger height" controversy over OSHA's Fall Protection Standard for construction.
7. What are the requirements for personal fall arrest systems in OSHA regulation 1926.451 (d)(16)?
8. What is a lanyard?
9. Explain OSHA's recommendations for effective fall protection, and what action to take when a worker is dangling from his fall arrest gear.
10. Explain how to assess the workplace for eye hazards.

11. What topics should be covered in eye protection training?
12. What should an employee look for when inspecting a ladder?
13. Briefly summarize the evolution of hard hats in this country.
14. List six major kinds of injuries to the foot and toes that occur in the workplace each year, and explain the most widely used footwear terms.
15. What are the typical causes of back injuries in the United States?
16. Describe the six-step back safety/lifting program, and explain what the EGMMH are.
17. List four ways to minimize standing hazards.
18. Explain the strategies for proper lifting that should be taught as part of the safety program.
19. What are the critical factors to consider when selecting gloves?
20. Explain how to get employees more comfortable with using PPE.
21. Explain the training requirement contained in OSHA's standard 29 CFR 1910.178 (powered industrial trucks).

ENDNOTES

1. J. Rhodes, "A Risk Manager's Roadmap," Occupational Health & Safety Online, May 11, 2006, 1–4. Retrived from http://www.stevenspublishing.com/stevens/ohspub.nsf/pubhome/of108dOc21adf8doc21adf8136257.
2. Ibid., 2.
3. Ibid., 2–3.
4. Ibid., 3–4.
5. T. Christensen and S. Rupard, "After the Fall: Why Fall Protection Isn't Always Enough," *Occupational Health & Safety* 75, no 3: 48–49.
6. OSHA Regulation 24 CFR 1926.
7. James L. Nash, "OSHA's Fight against Fatal Falls," *Occupational Hazards* 65, no. 4: 37.
8. National Safety Council, "Ladder Safety Tips." Retrieved from http://www.nsc.org/pubs/sh/clip1099.htm.
9. Ibid.
10. Ibid.
11. T. Christensen and S. Rupard, 49.
12. T. Cox, "Is Your Fall Protection Equipment a Silent Hazard?" *Occupational Health & Safety* 75, no. 5: 69–70.
13. Retrieved from http://public.ansi.org/ansionline/portal/search.
14. Ibid.
15. Ibid.
16. Ibid.
17. P. Feuerstein, "Head Protection Looks Up," *Safety & Health* 144, no. 3: 38.
18. Ibid., 39.
19. Ibid.
20. Bross, "Advances Lead to Tougher, More Durable Hard Hat," *Occupational Health & Safety* 74, no. 10: iv.
21. A. Chambers, "Safety Goggles at a Glance," *Occupational Health & Safety* 71, no. 10: 58.
22. J. Hensel, "Setting Up a Vision Program," *Occupational Health & Safety* 68, no. 10: 36.
23. Ibid.
24. Ibid.
25. Ibid.
26. B. Weissman, "Contact Lenses in a Chemical Enviornment," *Occupational Health & Safety* 74, no. 10: 56–58.
27. J. Goodwin, "A Cure for Common Foot Hazards," *Occupational Health & Safety* 74, no. 7: 84.
28. Ibid.
29. Ibid., 84–86.
30. Ibid., 86.
31. Retrieved from http://www.osha.gov/pls/oshaweb.
32. Retrieved from http://www.nsc.org/ergorisk/.
33. A. Kaliokin, "Six Steps Can Help Prevent Back Injuries and Reduce Compensation Costs," *Safety & Health* 138, no. 4: 50.
34. B. Urborg, "How to Comply with OSHA's Ergonomic Standard." Retrieved from" http://www.nsc.org/news/nr11601.htm.
35. Ibid.
36. Ibid.
37. Ibid.
38. 29 CFR 1910.132, Section 38.
39. T. Busshart, "A Cut Above," *Occupational Safety & Health* 67, no. 1: 36.
40. T. Andrews, "Getting Employees Comfortable with PPE," *Occupational Hazards* 62, no. 1: 35–38.
41. 29 CFR 1910.178—Powered Industrial Trucks.

CHAPTER 12

Hazards of Temperature Extremes

Major Topics

- Thermal Comfort
- Heat Stress and Strain
- Cold Stress
- Burns and Their Effects
- Chemical Burns

Part of providing a safe and healthy workplace is appropriately controlling the temperature, humidity, and air distribution in work areas. A work environment in which the temperature is not properly controlled can be uncomfortable. Extremes of either heat or cold can be more than uncomfortable—they can be dangerous. Heat stress, cold stress, and burns are major concerns of modern safety and health professionals. This chapter provides the information that professionals need to know to overcome the hazards associated with extreme temperatures.

Thermal Comfort

Thermal comfort in the workplace is a function of a number of different factors.[1] Temperature, humidity, air distribution, personal preference, and acclimatization are all determinants of comfort in the workplace. However, determining optimum conditions is not a simple process.

To understand fully the hazards posed by temperature extremes, safety and health professionals must be familiar with several basic concepts related to thermal energy. The most important of these are summarized here:

- **Conduction** is the transfer of heat between two bodies that are touching, or from one location to another within a body. For example, if an employee touches a workpiece that has just been welded and is still hot, heat will be conducted from the workpiece to the hand. Of course, the result of this heat transfer is a burn.

- **Convection** is the transfer of heat from one location to another by way of a moving medium (a gas or a liquid). Convection ovens use this principle to transfer heat from an electrode by way of gases in the air to whatever is being baked.
- **Metabolic heat** is produced within a body as a result of activity that burns energy. All humans produce metabolic heat. This is why a room that is comfortable when occupied by just a few people may become uncomfortable when it is crowded. Unless the thermostat is lowered to compensate, the metabolic heat of a crowd will cause the temperature of a room to rise to an uncomfortable level.
- **Environmental heat** is produced by external sources. Gas or electric heating systems produce environmental heat as do sources of electricity and a number of industrial processes.
- **Radiant heat** is the result of electromagnetic nonionizing energy that is transmitted through space without the movement of matter within that space.

Heat Stress and Strain

The key question that must be answered by safety and health professionals concerning employees whose work may subject them to heat stress is as follows:

> "What are the conditions to which most adequately hydrated, unmedicated, healthy employees may be exposed without experiencing heat strain or any other adverse effects?"

The American Conference of Governmental Industrial Hygienists (ACGIH) publishes a comprehensive manual to help safety and health professionals answer this question for the specific situations and conditions that they face. This manual, titled *TLVs and BEIs: Threshold Limit Values for Chemical Substances and Physical Agents and Biological Exposure Indices,* provides reliable guidance and should be in every safety and health professional's library. In addition to using the information contained in this manual, all safety and health professionals should have a comprehensive heat stress management program in place and apply sound professional judgment.[2]

Heat Stress Defined

Heat stress is the net heat load to which a worker may be exposed from the combined contributions of metabolic effect of work, environmental factors (i.e., air temperature, humidity, air movement, and radiant heat exchange) and clothing requirements. A mild or moderate heat stress may cause discomfort and may adversely affect performance and safety, but it is not harmful to health. As the heat stress approaches human tolerance limits, the risk of heat-related disorders increases.[3]

What follows are some widely used heat stress–related terms safety and health professionals should be familiar with:

Heat exhaustion. This is physical state in which the worker's skin becomes clammy and moist and his or her body temperature is still normal or slightly higher than normal. Heat exhaustion results from loss through sweating of fluid and salt that are not properly replaced during exertion.

Heat cramps. Heat cramps are muscle cramps that can occur when workers exert themselves sufficiently to lose fluids and salt through sweating, but replace only the fluids by drinking large amounts of water containing no salt.

Heat syncope or fainting. Workers who exert themselves in a hot environment will sometimes faint. This is especially the case with workers who are not accustomed to working in such an environment.

Heat rash. Workers who exert themselves in a hot environment in which sweat does not evaporate can develop a prickly rash known as heat rash. Before air conditioning was widely used in the hot and humid summer months in the southeastern United States, children often developed heat rash. Periodic rest breaks in a cool environment that allows sweat to evaporate will prevent heat rash.

Work tolerance time (WTT). WTT is a formula safety and health professionals can use to determine what steps can be taken to allow a worker to safely perform his or her required tasks in the environment in question for the time required. The formula takes into account such factors as temperature, humidity, level of energy that will be expended in performing the task, rest periods, and personal protective equipment (PPE).

Moisture vapor transfer rate (MVTR). The MVTR is a measure of the ability of the fabric used in making personal protective equipment (PPE) to dissipate heat. The best MVTR occurs on an unclothed body. Even the lightest cotton fabric is less capable of dissipating heat when the unclothed body is used for baseline comparisons. The MVTR of impermeable fabric is zero (because the fabric does not allow the skin to "breathe"). The higher the MVTR the better in hot environments.

Heat Strain Defined

Heat strain is the overall physiological response resulting from heat stress. The physiological adjustments are dedicated to dissipating excess heat from the body. **Acclimatization** is a gradual physiological adaptation that improves an individual's ability to tolerate heat stress.

Recognizing Heat Strain

Safety and health professionals, supervisors, and coworkers should know how to recognize heat strain. The following factors are signs of excessive heat strain. Exposure to heat stress should be stopped immediately for any employee experiencing any of these symptoms:

- A sustained rapid heart rate (180 beats per minutes minus the employee's age in years). For example, a 40-year-old employee has a sustained heart rate of 150 beats per minutes. This is a problem because the heart rate exceeds 140 (180 minus 40) beats per minute.
- Core body temperature is greater than 38.5°C.
- Recovery rate one minute after a peak work effort is greater than 110 beats per minute.
- Sudden and severe fatigue, nausea, dizziness, or light-headedness.

These symptoms can be assessed on the spot in real-time. In addition, other symptoms can be monitored only over time. Employees are at greater risk of excessive heat strain if they experience any of the following:

- Profuse sweating that continues for hours.

- Weight loss of more than 1.5 percent of body weight during one work shift.
- Urinary sodium excretion of less than 50 moles (24-hour period).

Clothing

Heat is best removed from the body when there is free movement of cool dry air over the skin's surface. This promotes the evaporation of sweat from the skin, which is the body's principal cooling mechanism. Clothing impedes this process, some types more than others. Encapsulating suits and clothing that is impermeable or highly resistant to the flow of air and water vapor multiply the potential for heat strain.

When assessing heat stress hazards in the workplace, safety and health professionals should consider the added effect of clothing. For example, the *wet bulb globe temperature* (WBGT) of working conditions should be increased by 3.5°C for employees wearing cloth overalls. This factor increases to 5°C with double cloth overalls.

Because the WBGT is influenced by air temperature, radiant heat, and humidity, it can be helpful in establishing a threshold for making judgments about working conditions. WBGT values can be calculated using the following formula:

EXPOSED TO DIRECT SUNLIGHT

$$WBGT = 0.7\,T_{nwb} \text{ to } 0.2\,T_{g} + 0.1\,T_{db}$$

Tnwb = Natural wet bulb temperature
Tg = Globe temperature
Tdb = Dry bulb (air) temperature

NOT EXPOSED TO DIRECT SUNLIGHT

$$WBGT = 0.7\,T_{nwb} + 0.3\,T_{g}$$

These formulas for WBGT give safety and health professionals a beginning point for making judgments. The WBGT must be adjusted for clothing, work demands, and the employee's acclimatization state. The key is to ensure that employees never experience a core body temperature of 38°C or higher. Figures 12-1 and 12-2 provide screening criteria for heat stress exposure. Once the WBGT has been calculated and adjusted for clothing, these figures may be used for factoring in work demands and acclimatization. To use Figures 12-1 and 12-2, apply the following example:

> Several acclimatized employees have a job to do that has a work demand of 75 percent work and 25 percent rest. The WBGT has been computed as 26. The work is considered "heavy." Because the employees will wear long-sleeved shirts and long trousers made of woven material an additional 3.5° must be added: 26 + 3.5 = 29.5°C. Using the proper column and row of Figure 16-1, a WBGT of 26 can be determined. Because the calculated and adjusted WBGT is 29.5, there is a problem. In order to work in these conditions, the employees should adjust the work demand to 25 percent work and 75 percent rest.

Heat Stress Management

Safety and health professionals should continually emphasize the importance of paying attention to recognizable symptoms of heat stress. In addition, they should ensure

FIGURE 12-1 Criteria for determining the allowable work periods for acclimatized employees.

Screening Criteria (°C)

Acclimatized Employees

Work Demands	Light Work	Moderate Work	Heavy Work	Very Heavy Work
100% Work	29.5	27.5	26.0	—
75% Work 25% Rest	30.5	28.5	27.5	—
50% Work 50% Rest	31.5	29.5	28.5	27.5
25% Work 75% Rest	32.5	31.0	30.0	29.5

Source: American Conference of Governmental Industrial Hygienists (ACGIH).

that a comprehensive heat stress management program is in place. Such a program should consist of both general and specific controls.

General Controls

The ACGIH recommends the following general controls:

- Provide accurate verbal and written instructions, training programs, and other information about heat stress and strain.
- Encourage drinking small volumes (approximately 1 cup) of cool water about every 20 minutes.

FIGURE 12-2 Criteria for determining the allowable work periods for employees who are not acclimatized.

Screening Criteria (°C)

Not-Acclimatized Employees

Work Demands	Light Work	Moderate Work	Heavy Work	Very Heavy Work
100% Work	27.5	25.0	22.5	—
75% Work 25% Rest	29.0	26.5	24.5	—
50% Work 50% Rest	30.0	28.0	26.5	25.0
25% Work 75% Rest	31.0	29.0	28.0	26.5

Source: American Conference of Governmental Industrial Hygienists (ACGIH).

- Permit self-limitation of exposure. Encourage coworker observation to detect signs and symptoms of heat strain in others.
- Counsel and monitor those employees who take medications that may compromise normal cardiovascular, blood pressure, body temperature regulation, renal, or sweat gland functions, as well as those who abuse or who are recovering from the abuse of alcohol and other intoxicants.
- Encourage healthy lifestyles, ideal body weight, and electrolyte balance.
- Adjust expectations of those returning to work after absence from heat stress situations and encourage consumption of salty foods (with approval of the employee's physician if on a salt-restricted diet).
- Consider replacement medical screening to identify those susceptible to systemic heat injury.[4]

Specific Controls

The ACGIH recommends the following specific controls:

- Establish engineering controls that reduce the metabolic rate, provide general air movement, reduce process heat and water-vapor release, and shield radiant heat sources, among others.
- Consider administrative controls that set acceptable exposure times, allow sufficient recovery, and limit physiological strain.
- Consider personal protection that has been demonstrated to be effective for the specific work practices and conditions at the location.[5]

Cold Stress

Excessive exposure to cold can lead to hypothermia, which can be fatal. The goal of safety and health professionals in protecting employees from acute cold stress is to prevent the deep body temperature from falling below 36°C (98.6°F) and to prevent cold injuries to body extremities, especially the hands, feet, and head. A fatal exposure to cold typically results from failure to remove the employee from a cold air environment or immersion in cold water.[6]

Excessive exposure to cold stress, even when not fatal, can result in impaired judgment, reduced alertness, and poor decision making. Acute cold stress can cause reduced muscular function, decreased tactile sensitivity, reduced blood flow, and thickening of the synovial fluid. Chronic cold stress can lead to reduced functioning of the peripheral nervous system. All these factors increase the likelihood of accidents and injuries. Figure 12-3 shows the effects of allowing the core body temperature to fall to selected levels.

Whether employees are exposed to cold air or are immersed in cold water, wind can magnify the level of cold stress. This phenomenon is often referred to as windchill. Figure 12-4 shows the effect of wind on selected temperatures. To read this chart, locate the actual temperature (50, 40, 30 . . . 0). Then, find the applicable wind speed. Reading across that row to the right, find the equivalent temperature. For example, if employees are working in an environment that is 30°F and has a wind speed of 15 miles per hour (mph), the equivalent temperature is 9°F.

Preventing Cold Stress

When the equivalent air temperature reaches 232°C (225.6°F), continuous exposure of exposed skin should not be allowed. At equivalent air temperatures of 2°C (35.6°F),

FIGURE 12-3 The body's response to reducing its core temperature.

Effects of Reducing the Core Body Temperature

Core Temperature °C	°F	Body's Response
37.6	99.6	Normal rectal temperature
36.0	96.8	Metabolic rate increases
35.0	95.0	Pronounced shivering
33.0	91.4	Severe hypothermia
30.0	86.0	Progressive loss of consciousness begins
24.0	75.2	Pulmonary edema
20.0	68.0	Cardiac standstill

employees who are immersed in water or whose clothing gets wet should be treated for hypothermia immediately. Figure 12-5 shows selected threshold limit values (TLVs) for employees who work in environments with temperature below freezing.

To use Figure 12-5, locate the applicable temperature in the leftmost column. Reading to the right, locate the applicable wind speed. For example, employees working a four-hour shift in an environment with an air temperature of 232°C and a 5-mph wind should be exposed no longer than 55 minutes at a time and should warm up at least three times during the shift.

When work is to be performed in an environment with an air temperature of 4°C (39.2°F) or less, total body protective clothing is advisable. What follows are several strategies that can be used to decrease the hazards of cold stress:

- When working in a setting in which wind is a factor, reduce the effect of the wind by (1) erecting a windscreen; or (2) wearing wind-breaking clothing.

FIGURE 12-4 Effect of wind on the actual temperature.

Cooling Effect of Wind

	Actual Temperature (°F) and Equivalent Temperatures (°F)					
Wind Speed (in mph)	50	40	30	20	10	0
5	48	37	27	16	6	–5
15	36	22	9	–5	–18	–32
25	30	16	0	–15	–29	–44
35	27	11	–4	–20	–35	–51

Source: U.S. Army Research Institute of Environmental Medicine.

FIGURE 12-5 Partial table for determining TLVs in selected circumstances.

TLVs for a Four-Hour Shift							
Air Temperature		No Wind		5 mph Wind		10 mph Wind	
C°	F°	Max. Work Time	No. of Breaks	Max. Work Time	No. of Breaks	Max. Work Time	No. of Breaks
–32 to –34	–25 to –29	75 Min.	2	55 Min.	3	40 Min.	4
–35 to –37	–30 to –34	55 Min.	3	40 Min.	4	30 Min.	5
–38 to –39	–35 to –39	40 Min.	4	30 Min.	5	Nonemergency Work Prohibited	

Note: This applies to workers properly dressed in dry clothing.

Source: American Conference of Governmental Industrial Hygienists (ACGIH).

- When working in a setting in which clothing may get wet, apply one or more of the following strategies: (1) with light work, wear an outer layer of impermeable clothing; (2) with heavier work, wear an outer layer that is water repellent, but not impermeable (change outerwear as it becomes wet); (3) select outer garments that are ventilated to prevent internal wetting from sweat; (4) if clothing gets wet before going into the cold environment, change first; (5) change socks daily or more often to keep them dry; and (6) use vapor barrier boots to help keep the feet dry.
- When working in a cold setting, use auxiliary heat applied directly to the hands and feet.
- When working in a cold setting, use facial protection to prevent cold stress to the face and lungs.
- If adequate protective clothing that is appropriate for the conditions in question is not available, the work should be modified or suspended until conditions change or the clothing is available.

When work is to be performed in an environment with an air temperature of 212°C (10.4°F), the following additional strategies should be applied:

- Employees should be under continuous observation using either direct supervision or the buddy system.
- The work rate should be paced to avoid sweating. When heavy work is necessary, employees should take frequent warming breaks in heated shelters. If clothing becomes wet—internally or externally—it should be changed during a break.
- Do not allow new employees to work full time in these conditions until they have several days to become accustomed to the conditions and the necessary protective clothing.
- When determining the required work level for employees (light, heavy, or very heavy), consider the weight and bulkiness of protective clothing.

FIGURE 12-6 Checklist for training employees who will work in a cold environment.

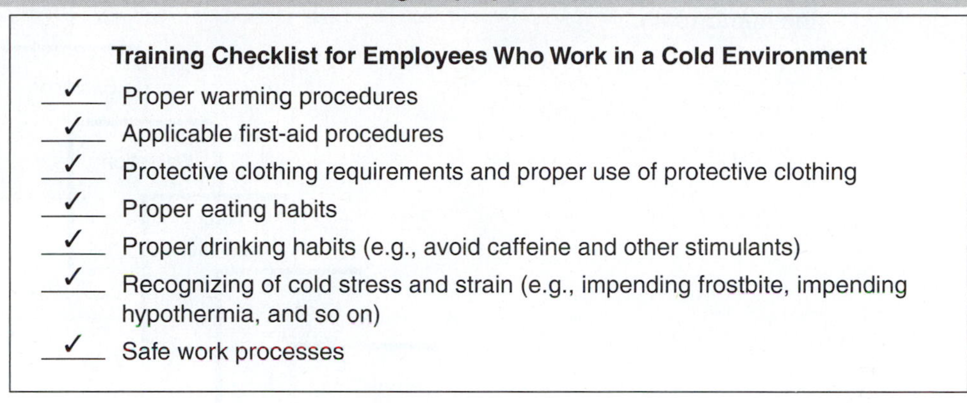
Training Checklist for Employees Who Work in a Cold Environment

- ✓ Proper warming procedures
- ✓ Applicable first-aid procedures
- ✓ Protective clothing requirements and proper use of protective clothing
- ✓ Proper eating habits
- ✓ Proper drinking habits (e.g., avoid caffeine and other stimulants)
- ✓ Recognizing of cold stress and strain (e.g., impending frostbite, impending hypothermia, and so on)
- ✓ Safe work processes

- Organize work in cold environments to minimize long periods of sitting or standing still. Never use unprotected metal chairs or seats.
- Before allowing employees to work in a cold environment, make sure they have been trained in safety and health procedures. Figure 12-6 is a checklist of topics that should be covered as a minimum during employee training.
- When work in a refrigerated room is required, the air velocity should be minimized and maintained at one meter per second (200 feet per minute) or less.
- When work outdoors in snow is required, employees should be provided special safety goggles that protect the eyes from ultraviolet light, glare, and blowing ice crystals.
- Employees who suffer from diseases or take medications that inhibit normal body functions or that reduce normal body tolerances should be prohibited from working in environments where temperatures are at 21°C (30.2°F) or less.
- Employees who are routinely exposed to the following conditions should be medically certified as being suitable for work in such conditions: (1) air temperatures of less than 224°C (211.2°F) with wind speeds less than 5 mph; and (2) air temperatures of less than 218°C (0°F) with wind speeds greater than 5 mph.

Burns and Their Effects

One of the most common hazards associated with heat in the workplace is the burn. Burns can be especially dangerous because they disrupt the normal functioning of the skin, which is the body's largest organ and the most important in terms of protecting other organs. It is necessary first to understand the composition of, and purpose served by, the skin to understand the hazards that burns can represent.

Human Skin

Human skin is the tough, continuous outer covering of the body. It consists of the following two main layers: (1) the outer layer, which is known as the **epidermis**; and (2) the inner layer, which is known as the **dermis**, **cutis**, or **corium**. The dermis is connected to the underlying subcutaneous tissue.

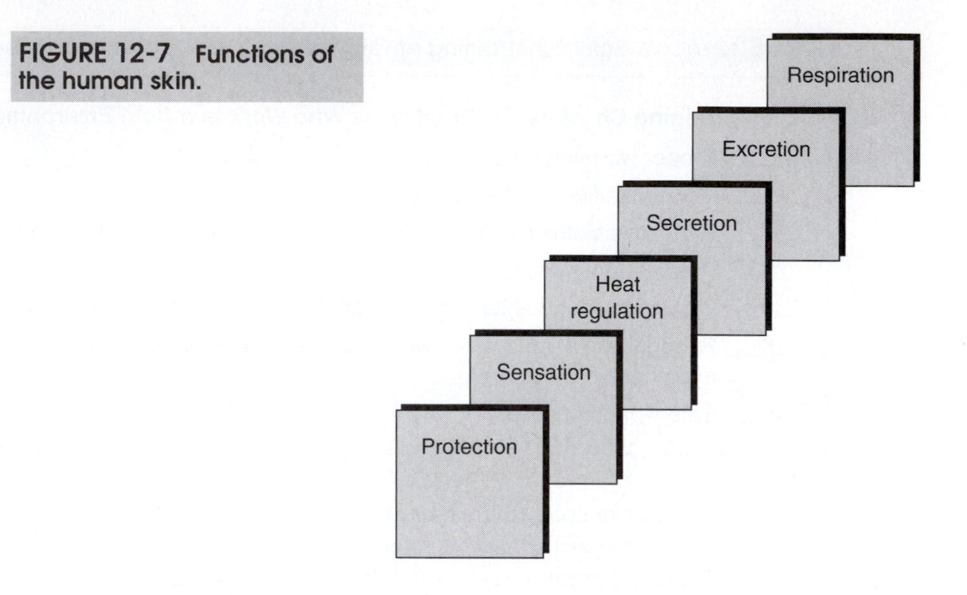

FIGURE 12-7 Functions of the human skin.

The skin serves several important purposes including the following: protection of body tissue, sensation, secretion, excretion, and respiration (see Figure 12-7). Protection from fluid loss, water penetration, ultraviolet radiation, and infestation by microorganisms is a major function of the skin. The sensory functions of touching, sensing cold, feeling pain, and sensing heat involve the skin.

The skin helps regulate body heat through the sweating process. It excretes sweat that takes with it electrolytes and certain toxins. This helps keep the body's fluid level in balance. By giving off minute amounts of carbon dioxide and absorbing small amounts of oxygen, the skin also aids slightly in respiration.

What makes burns particularly dangerous is that they can disrupt any or all of these functions depending on their severity. The deeper the penetration, the more severe the burn.

Severity of Burns

The severity of a burn depends on several factors. The most important of these is the depth to which the burn penetrates. Other determining factors include location of the burn, age of the victim, and amount of burned area.

The most widely used method of classifying burns is by degree (i.e., first-, second-, or third-degree burns). Modern safety and health professionals should be familiar with these classifications and what they mean.

First-degree burns are minor and result only in a mild inflammation of the skin, known as *erythema*. Sunburn is a common form of first-degree burn. It is easily recognizable as a redness of the skin that makes the skin sensitive and moderately painful to the touch.

Second-degree burns are easily recognizable from the blisters that form on the skin. If a second-degree burn is superficial, the skin will heal with little or no scarring. A deeper second-degree burn will form a thin layer of coagulated, dead cells that feels leathery to the touch. A temperature of approximately 98.9°C can cause a second-degree burn in as little as 15 seconds of contact.

FIGURE 12-8 Estimating percentage of body surface area (BSA) burned.

Body area	Percentage
Right arm	9% of BSA
Left arm	9% of BSA
Head/neck	9% of BSA
Right leg	18% of BSA
Left leg	18% of BSA
Back	18% of BSA
Chest/stomach	18% of BSA
Perineum	1% of BSA

Third-degree burns are very dangerous and can be fatal depending on the amount of body surface affected. A third-degree burn penetrates through both the epidermis and the dermis. A deep third-degree burn will penetrate body tissue. Third-degree burns can be caused by both moist and dry hazards. Moist hazards include steam and hot liquids; these cause burns that appear white. Dry hazards include fire and hot objects or surfaces; these cause burns that appear black and charred.

In addition to the depth of penetration of a burn, the amount of surface area covered is also a critical concern. This amount is expressed as a percentage of **body surface area (BSA)**. Figure 12-8 shows how the percentage of BSA can be estimated. Burns covering over 75 percent of BSA are usually fatal.

Using the first-, second-, and third-degree burn classifications in conjunction with BSA percentages, burns can be classified further as minor, moderate, or critical. According to Mertz, these classifications can be summarized as described in the following paragraphs.[7]

Minor Burns

All first-degree burns are considered **minor**. Second-degree burns covering less than 15 percent of the body are considered minor. Third-degree burns can be considered minor provided they cover only 2 percent or less of BSA.

Moderate Burns

Second-degree burns that penetrate the epidermis and cover 15 percent or more of BSA are considered **moderate**. Second-degree burns that penetrate the dermis and cover from 15 to 30 percent of BSA are considered moderate. Third-degree burns can be considered moderate provided they cover less than 10 percent of BSA and are not on the hands, face, or feet.

Critical Burns

Second-degree burns covering more than 30 percent of BSA or third-degree burns covering over 10 percent of BSA are considered **critical**. Even small-area third-degree burns to the hands, face, or feet are considered critical because of the greater potential for infection to these areas by their nature. In addition, burns that are complicated by other injuries (fractures, soft tissue damage, and so on) are considered critical.

Chemical Burns

Chemicals are widely used in modern industry even by companies that do not produce them as part of their product base. Many of the chemicals produced, handled, stored, transported, or otherwise used in industry can cause burns similar to those caused by heat (i.e., first-, second-, and third-degree burns). The hazards of chemical burns are very similar to those of thermal burns.

Chemical burns, like thermal burns, destroy body tissue; the extent of destruction depends on the severity of the burn. However, chemical burns continue to destroy body tissue until the chemicals are washed away completely.

The National Safety Council describes the physiological process in cases of chemical burns:

> Many concentrated chemical solutions have an affinity for water. When they come in contact with body tissue, they withdraw water from it so rapidly that the original chemical composition of the tissue (and hence the tissue itself) is destroyed. In fact, a strong caustic may dissolve even dehydrated animal tissue. The more concentrated the solution, the more rapid is the destruction.[8]

The severity of the burn produced by a given chemical depends on the following factors:

- Corrosive capability of the chemical
- Concentration of the chemical
- Temperature of the chemical or the solution in which it is dissolved
- Duration of contact with the chemical[9]

Effects of Chemical Burns

Different chemicals have different effects on the human body. The harmful effects of selected widely used chemicals are summarized in Figure 12-9.[10] These are only a few

FIGURE 12-9 Harmful effects of selected widely used chemicals.

Chemical	Potential Harmful Effect
Acetic acid	Tissue damage
Liquid bromide	Corrosive effect on the respiratory system and tissue damage
Formaldehyde	Tissue hardening
Lime	Dermatitis and eye burns
Methylbromide	Blisters
Nitric/sulfuric acid mixture	Severe burns and tissue damage
Oxalic acid	Ulceration and tissue damage
White phosphorus	Ignites in air causing thermal burns
Silver nitrate	Corrosive/caustic effect on the skin
Sodium (metal)	Ignites with moisture causing thermal burns
Trichloracetic acid	Tissue damage

of the many chemicals widely used in industry today. All serve an important purpose; however, all carry the potential for serious injury.

The primary hazardous effects of chemical burns are infection, loss of body fluids, and shock, and are summarized in the following paragraphs.[11]

Infection

The risk of **infection** is high with chemical burns—as is it with heat-induced burns—because the body's primary defense against infection-causing microorganisms (the skin) is penetrated. This is why it is so important to keep burns clean. Infection in a burn wound can cause *septicemia* (blood poisoning).

Fluid Loss

Body **fluid loss** in second- and third-degree burns can be serious. With second-degree burns, the blisters that form on the skin often fill with fluid that seeps out of damaged tissue under the blister. With third-degree burns, fluids are lost internally and, as a result, can cause the same complications as a hemorrhage. If these fluids are not replaced properly, the burns can be fatal.

Shock

Shock is a depression of the nervous system. It can be caused by both physical and psychological trauma. In cases of serious burns, it may be caused by the intense pain that can occur when skin is burned away, leaving sensitive nerve endings exposed. Shock from burns can come in the following two forms: (1) primary shock, which is the first stage and results from physical pain or psychological trauma; and (2) secondary shock, which comes later and is caused by a loss of fluids and plasma proteins as a result of the burns.

First Aid for Chemical Burns

There is a definite course of action that should be taken when chemical burns occur, and the need for immediacy cannot be overemphasized. According to the NSC, the proper response in cases of chemical burns is to "wash off the chemical by flooding the burned areas with copious amounts of water as quickly as possible. This is the only method for limiting the severity of the burn, and the loss of even a few seconds can be vital."[12]

In the case of chemical burns to the eyes, the continuous flooding should continue for at least 15 minutes. The eyelids should be held open to ensure that chemicals are not trapped under them.

According to the NSC,

> The Committee on Industrial Ophthalmology, Council of Industrial Health of the American Medical Association has noted the tremendous saving of eyesight among industrial employees brought about by immediate and thorough flushing of harmful chemicals from the eyes by copious amounts of water. It is the belief of the committee that this is the most effective and practical emergency first aid treatment of eyes injured by chemicals.[13]

Clothing is another consideration when an employee comes in contact with a caustic chemical. If chemicals have saturated the employee's clothes, they must be removed quickly. The best approach is to remove the clothes while flooding the body or the affected area. If necessary for quick removal, clothing should be ripped or cut off.

The critical need to apply water immediately in cases of chemical burns means that water must be readily available. Health and safety professionals should ensure that special eye wash and shower facilities are available wherever employees handle chemicals.

Key Terms and Concepts

- Acclimatization
- Body surface area (BSA)
- Chemical burn
- Cold stress
- Conduction
- Convection
- Corium
- Critical burn
- Cutis
- Dermis
- Electrolytes
- Environmental heat
- Epidermis
- First-degree burn
- Fluid loss
- Heat cramps
- Heat exhaustion
- Heat rash
- Heat strain
- Heat stress
- Heat stress management
- Heat syncope
- Hypothermia
- Infection
- Metabolic heat
- Minor burn
- Moderate burn
- Moisture vapor transfer rate (MVTR)
- Radiant heat
- Second-degree burn
- Shock
- Third-degree burn
- Work tolerance time (WTT)

Review Questions

1. Define the following thermal comfort–related terms: *conduction, convection,* and *metabolic heat.*
2. What is heat stress?
3. What is heat strain?
4. Define the following terms: *heat exhaustion, heat cramps, heat syncope, heat rash, work tolerance time,* and *moisture vapor transfer rate.*
5. What are the symptoms of heat exhaustion?
6. How can heat strain be recognized?
7. How does clothing affect the cooling process?
8. What factors influence the WBGT?
9. Describe the various general controls in heat stress management.
10. How can cold stress be prevented?
11. How does wind movement affect the way the body perceives temperature?
12. Describe the symptoms of cold stress and hypothermia.
13. Describe the various components of a cold stress prevention program.
14. Describe the various purposes served by the skin.
15. Describe and differentiate among first-, second-, and third-degree burns.
16. Describe and differentiate among minor, moderate, and critical burns.
17. List the factors that determine the severity of a chemical burn.
18. Explain the hazards of chemical burns besides tissue damage.
19. What should you do if an employee accidentally splashes a caustic chemical on himself or herself?

Endnotes

1. American Conference of Governmental Industrial Hygienists (ACGIH), 2006, TLVs and BEIs (Cincinnati, OH: 2006), 172–181.
2. Ibid., 180–188.
3. Ibid., 181.
4. Ibid., 187.
5. Ibid.
6. Ibid., 171–179.
7. Patricia M. Mertz, Burn Study, an unpublished paper, University of Miami School of Medicine, Department of Dermatology and Cutaneous Surgery, Miami, Florida, October 2000, 2.
8. National Safety Council, "Chemical Burns," Data Sheet 1–523 Rev. 87 (Chicago: National Safety Council): 1.
9. Ibid.
10. Ibid., 3–4.
11. Ibid., 2.
12. Ibid.
13. Ibid.

CHAPTER

Pressure Hazards

Major Topics

- Pressure Hazards Defined
- Sources of Pressure Hazards
- Boilers and Pressure Hazards
- High-Temperature Water Hazards
- Hazards of Unfired Pressure Vessels
- Hazards of High-Pressure Systems
- Cracking Hazards in Pressure Vessels
- Nondestructive Testing of Pressure Vessels
- Pressure Dangers to Humans
- Decompression Procedures
- Measurement of Pressure Hazards
- Reduction of Pressure Hazards

Pressure Hazards Defined

Pressure is defined in physics as the force exerted against an opposing fluid or thrust distributed over a surface. This may be expressed in force or weight per unit of area, such as psi (pounds per square inch). A **hazard** is a condition with the potential of causing injury to personnel, damage to equipment or structures, loss of material, or lessening of the ability to perform a prescribed function. Thus, a **pressure hazard** is a hazard caused by a dangerous condition involving pressure. Critical injury and damage can occur with relatively little pressure. OSHA defines high-pressure cylinders as those designated with a service pressure of 900 pounds psi or greater.

We perceive pressure in relation to the earth's atmosphere. Approximately 21 percent of the atmosphere is oxygen, with most of the other 79 percent being nitrogen. In addition to oxygen and nitrogen, the atmosphere contains trace amounts of several inert gases: argon, neon, krypton, xenon, and helium.

At sea level, the earth's atmosphere averages 1,013 H (hydrogen) 10 N/m^2, or 1.013 millibars, or 760 mm Hg (29.92 inches), or 14.7 psi, depending on the measuring scale used.[1] The international system of measurement utilizes newtons per square meter (N/m^2). However, in human physiology studies, the typical unit is millimeters of mercury (mm Hg).

Atmospheric pressure is usually measured using a **barometer**. As the altitude above sea level increases, atmospheric pressure decreases in a nonlinear fashion. For example, at 5,486 meters (18,000 feet) above sea level, the barometric pressure is equal to 390 mm Hg. Half of this pressure, around 195 mm Hg, can be found at 2,010 meters (23,000 feet) above sea level.

Boyle's law states that the product of a given pressure and volume is constant with a constant temperature:

$$P_1V_1 = P_2V_2, \quad \text{when } T \text{ is constant}$$

Air moves in and out of the lungs because of a pressure gradient or difference in pressure. When atmospheric pressure is greater than pressure within the lungs, air flows down this pressure gradient from the outside into the lungs. This is called **inspiration**, inhalation, or breathing in, and occurs with greater lung volume than at rest. When pressure in the lungs is greater than atmospheric pressure, air moves down a pressure gradient outward from the lungs to the outside. **Expiration** occurs when air leaves the lungs and the lung volume is less than the relaxed volume, increasing pressure within the lungs.

Gas exchange occurs between air in the lung alveoli and gas in solution in blood. The pressure gradients causing this gas exchange are called partial pressures. **Dalton's law of partial pressures** states that, in a mixture of theoretically ideal gases, the pressure exerted by the mixture is the sum of the pressures exerted by each component gas of the mixture:

$$P_A = P_O + P_N + P_{\text{else}}$$

Air entering the lungs immediately becomes saturated with water vapor. Water vapor, although it is a gas, does not conform to Dalton's law. The partial pressure of water vapor in a mixture of gases is not dependent on its fractional concentration in that mixture. Water vapor partial pressure, instead, is dependent on its temperature. From this exception to Dalton's law comes the fact that at the normal body temperature of 37°C (98.6°F), water vapor maintains a partial pressure of 47 mm Hg as long as that temperature is maintained. With this brief explanation of how pressure is involved in human breathing, we now focus on the various sources of pressure hazards.

Sources of Pressure Hazards

There are many sources of pressure hazard—some natural, most created by humans. Because the human body is comprised of approximately 85 percent liquid, which is virtually incompressible, increasing pressure does not create problems by itself. Problems can result from air being trapped or expanded within body cavities.

When sinus passages are blocked so that air cannot pass easily from the sinuses to the nose, expansion of the air in these sinuses can lead to problems. The same complications can occur with air trapped in the middle ear's eustachian tube. As Boyle's law states, gas volume increases as pressure decreases. Expansion of the air in blocked sinus passages or the middle ear occurs with a rapid increase in altitude or rapid ascent underwater. This can cause pain and, if not eventually relieved, disease. Under extreme circumstances of rapid ascent from underwater diving or high-altitude decompression, lungs can rupture.

Nitrogen absorption into the body tissues can become excessive during underwater diving and breathing of nitrogen-enriched air. Nitrogen permeation of tissues occurs in proportion to the partial ppressure of nitrogen taken in. If the nitrogen is permeating tissues faster than the person can breathe it out, bubbles of gas may form in the tissues.

Decompression sickness can result from the decompression that accompanies a rapid rise from sea level to at least 5,486 meters (18,000 feet) or a rapid ascent from around 40 to 20 meters (132 to 66 feet) underwater. Several factors influence the onset of decompression sickness:

- *A history* of previous decompression sickness increases the probability of another attack.
- *Age* is a component. Being over 30 increases the chances of an attack.
- *Physical fitness* plays a role. People in better condition have a reduced chance of the sickness. Previously broken bones and joint injuries are often the sites of pain.
- *Exercise* during the exposure to decompression increases the likelihood and brings on an earlier onset of symptoms.
- *Low temperature* increases the probability of the sickness.
- *Speed of decompression* also influences the sickness. A rapid rate of decompression increases the possibility and severity of symptoms.
- *Length of exposure* of the person to the pressure is proportionately related to the intensity of symptoms. The longer the exposure, the greater the chances of decompression sickness.

A reduction in partial pressure can result from reduced available oxygen and cause a problem in breathing known as *hypoxia.* Too much oxygen or oxygen breathed under pressure that is too high is called **hyperoxia**. Another partial pressure hazard, *nitrogen narcosis,* results from a higher-than-normal level of nitrogen pressure.

When breathed under pressure, nitrogen causes a reduction of cerebral and neural activity. Breathing nitrogen at great depths underwater can cause a feeling of euphoria and loss of reality. At depths greater than 30 meters (100 feet), nitrogen narcosis can occur even when breathing normal air. The effects may become pathogenic at depths greater than 60 meters (200 feet), with motor skills threatened at depths greater than 91 meters (300 feet). Cognitive processes deteriorate quickly after reaching a depth of 99 meters (325 feet). Decompression procedures are covered later in this chapter.

Boilers and Pressure Hazards

A boiler is a closed vessel in which water is heated to form steam, hot water, or high-temperature water under pressure.[2] Potential safety hazards associated with boilers and other pressurized vessels include the following:

- Design, construction, or installation errors
- Poor or insufficient training of operators
- Human error
- Mechanical breakdown or failure
- Failure or blockage of control or safety devices
- Insufficient or improper inspections
- Improper application of equipment
- Insufficient preventive maintenance[3]

Through years of experience, a great deal has been learned about how to prevent accidents associated with boilers. OSHA recommends the following daily, weekly, monthly, and yearly accident prevention measures:

1. **Daily check.** Check the water to make sure that it is at the proper level. Vent the furnace thoroughly before starting the fire. Warm up the boiler using a small fire. When the boiler is operating, check it frequently.
2. **Weekly check.** At least once every week, test the low-water automatic shutdown control and record the results of the test on a tag that is clearly visible.
3. **Monthly check.** At least once every month, test the safety valve and record the results of the test on a tag that is clearly visible.
4. **Yearly check.** The low-level automatic shutdown control mechanism should be either replaced or completely overhauled and rebuilt. Arrange to have the vendor or a third-party expert test all combustion safeguards, including fuel pressure switches, limit switches, motor starter interlocks, and shutoff valves.[4]

High-Temperature Water Hazards

High-temperature water (HTW) is exactly what its name implies—water that has been heated to a very high temperature, but not high enough to produce steam.[5] In some cases, HTW can be used as an economical substitute for steam (e.g., in industrial heating systems). It has the added advantage of releasing less energy (pressure) than steam does.

In spite of this, there are hazards associated with HTW. Human contact with HTW can result in extremely serious burns and even death. The two most prominent sources of hazards associated with HTW are operator error and improper design. Proper training and careful supervision are the best guards against operator error.

Design of HTW systems is a highly specialized process that should be undertaken only by experienced engineers. Mechanical forces such as water hammer, thermal expansion, thermal shock, or faulty materials cause system failures more often than do thermodynamic forces. Therefore, it is important to allow for such causes when designing an HTW system.

The best designs are simple and operator-friendly. Designing too many automatic controls into an HTW system can create more problems than it solves by turning operators into mere attendants who are unable to respond properly to emergencies.

Hazards of Unfired Pressure Vessels

Not all pressure vessels are fired. Unfired pressure vessels include compressed air tanks, steam-jacketed kettles, digesters, and vulcanizers, as well as others that can create heat internally by various means rather than by external fire.[6] The various means of creating internal heat include (1) chemical action within the vessel, and (2) application of some heating medium (electricity, steam, hot oil, and so on) to the contents of the vessel. The potential hazards associated with unfired pressure vessels include hazardous interaction between the material of the vessel and the materials that will be processed in it; inability of the filled vessel to carry the weight of its contents and the corresponding internal pressure; inability of the vessel to withstand the pressure introduced into it plus pressure caused by chemical reactions that occur during processing; and inability of the vessel to withstand any vacuum that may be created accidentally or intentionally.

The most effective preventive measure for overcoming these potential hazards is proper design. Specifications for the design and construction of unfired pressure vessels include requirements in the following areas: working pressure range, working temperature range, type of materials to be processed, stress relief, welding or joining measures, and radiography. Designs that meet the specifications set forth for unfired pressure vessels in such codes as the ASME (American Society of Mechanical Engineers) Code (Section VIII) will overcome most predictable hazards.

Beyond proper design, the same types of precautions taken when operating fired pressure vessels can be used when operating unfired vessels. These include continual inspection, proper housekeeping, periodic testing, visual observation (for detecting cracks), and the use of appropriate safety devices.

Hazards of High-Pressure Systems

The hazards most commonly associated with high-pressure systems are leaks, pulsation, vibration, release of high-pressure gases, and whiplash from broken high-pressure pipe, tubing, or hose.[7] Strategies for reducing these hazards include limiting vibration through the use of vibration dampening (use of anchored pipe supports); decreasing the potential for leaks by limiting the number of joints in the system; using pressure gauges; placing shields or barricades around the system; using remote control and monitoring; and restricting access.

Cracking Hazards in Pressure Vessels

One of the most serious hazards in pressure vessels is the potential for cracking.[8] Cracking can lead to either a complete rupture or to leaks. The consequences of a complete rupture include (1) blast effects due to the sudden expansion of the contents of the vessel, and (2) possible injuries and damage from fragmentation. The consequences of a leak include (1) suffocation or poisoning of employees depending on the contents of the vessel, (2) explosion and fire, and (3) chemical and thermal burns from contact with the contents of the vessel.

Pressure vessels are used in many different applications to contain many different types of substances ranging from water to extremely toxic chemicals. Leakage or rupture may occur in welded seams, bolted joints, or at nozzles. Figure 13-1 shows a diagram of a typical pressure vessel showing the potential points of leakage and rupture. The types of vessels that are most susceptible to leakage and rupture, primarily because of the processes they are part of or their contents, are as follows:

Deaerator Vessels

Deaeration is the process of removing noncondensible gases, primarily oxygen, from the water used in steam generation. Deaerator vessels are used in such applications as power generation, pulp and paper processing, chemical processing, and petroleum refining. The most common failures associated with deaerator vessels are (1) cracks caused by water hammer at welded joints that were not postweld heat treated, and (2) cracks caused by corrosion fatigue.

FIGURE 13-1 Diagram of a typical pressure vessel showing potential points for leakage or rupture.

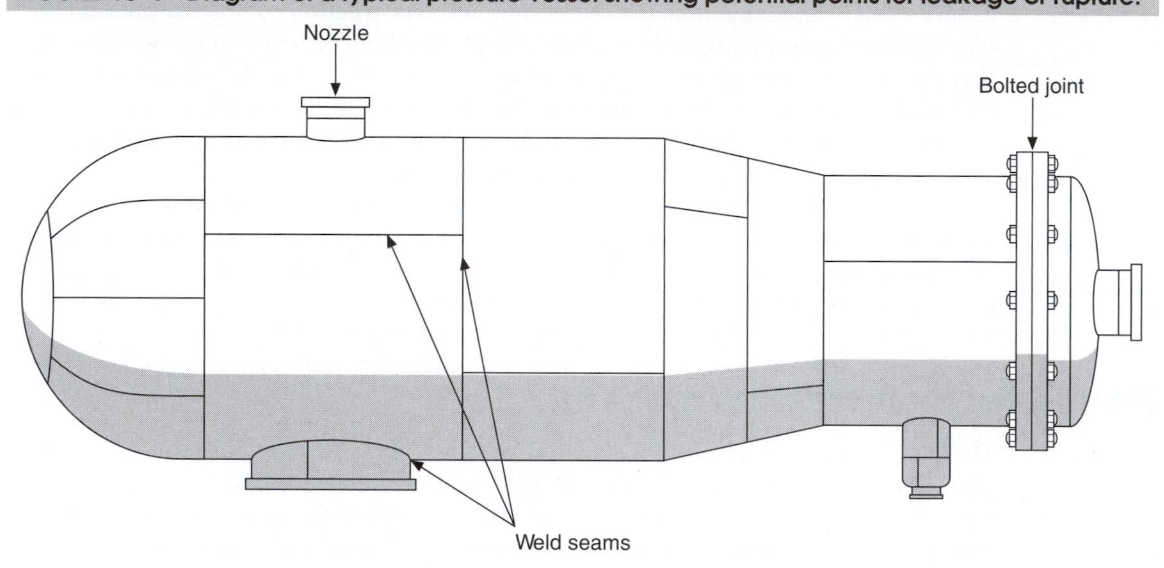

Amine Vessels

The *amine process* removes hydrogen sulfide from petroleum gases such as propane and butane. It can also be used for removing carbon dioxide in some processes. Amine vessels are used in petroleum refineries, gas treatment facilities, and chemical plants. The most common failures associated with amine vessels are cracks in stressed or unrelieved welds.

Wet Hydrogen Sulfide Vessels

Any fluid that contains water and hydrogen sulfide is considered wet hydrogen sulfide. Many of the vessels used to contain wet hydrogen sulfide are made of steel. Hydrogen is generated when steel is exposed to such a mixture. Dissolved hydrogen can cause cracking, blistering, and embrittlement, particularly in high-strength steels. Consequently, low-strength steels are recommended for wet hydrogen sulfide vessels.

Ammonia Vessels

Vessels for the containment of ammonia are widely used in commercial refrigeration systems and chemical processes. Such ammonia vessels are typically constructed as spheres of carbon steel. The water and oxygen content in ammonia can cause carbon steel to crack, particularly near welds.

Pulp Digester Vessels

The process used to digest pulp in the manufacture of paper involves the use of a weak water solution of sodium hydroxide and sodium sulfide in a temperature range of 110°C to 140°C (230°F to 285°F). The most common failure in pulp digester vessels is cracking along welded seams due primarily to caustic stress corrosion.

Nondestructive Testing of Pressure Vessels

To prevent leakage or rupture, it is necessary to examine pressure vessels periodically. There are five widely used nondestructive methods for testing: (1) visual examination, (2) liquid penetration test, (3) magnetic particle test, (4) gamma and X-ray radiography, and (5) ultrasonic test. Visual, liquid penetration, and magnetic particle tests can detect only those defects that are either on the surface or near it. Radiographic and ultrasonic tests can detect problems within the material. Consequently, the visual, liquid penetration, and magnetic particle tests are referred to as *surface tests.* Radiography and ultrasonic are called *volumetric tests.*

Visual Examination

A visual examination consists of taking a thorough look at the vessel to detect signs of corrosion, erosion, or hydrogen blistering. In order to conduct a dependable visual examination of a pressure vessel, it is necessary to have a clean surface and good lighting.

Liquid Penetration Test

This test involves placing a specially formulated liquid penetrant over an area and letting it seep in. When the penetrant is removed from the surface, some of it remains entrapped in the area of discontinuity. A developing agent is then applied, which draws out the entrapped penetrant and magnifies the discontinuity. The process can be enhanced by adding fluorescent chemicals to the penetrant to aid in the detection of problems.

Magnetic Particle Test

This test is based on the fact that discontinuities in or near the surface of a pressure vessel disturb magnetic flux lines that are induced in a ferromagnetic material. Disturbances are detected by applying fine particles of ferromagnetic material to the surface of the vessel. The necessary magnetic field is produced most frequently using the "prod" technique in which electric current is run through an area by applying opposing "prods" (contact probes). A drawback of this test is that corners and surface irregularities in the vessel material can produce the same disturbances as defects. Consequently, special care is needed when using this test in a region with corners or welded joints. Because this test works only with ferromagnetic material, its use is limited to vessels made of carbon and low-alloy steels.

Radiography Test

This test amounts to making an X-ray negative of a given portion of the vessel. The process works in the same way as those used by physicians and dentists. Irregularities such as holes, voids, or discontinuities produce a greater exposure (darker area) on the X-ray negative.

Ultrasonic Test

This test is similar to radar and other uses of electromagnetic and acoustic waves for detecting foreign objects. Short signals are induced into the material. Waves that are reflected back from discontinuities are detected by one or more transducers. Ultrasonic testing requires an electronic system for generating a signal, a transducer system for converting the electrical into mechanical vibrations and vice versa, and an electronic system for amplifying, processing, and displaying the return signal.

Pressure Dangers to Humans

The term *anoxia* refers to the rare case of a total lack of oxygen. Hypoxia, a condition that occurs when the available oxygen is reduced, can occur while ascending to a high altitude or when oxygen in air has been replaced with another gas, which may happen in some industrial situations.

Altitude sickness is a form of hypoxia associated with high altitudes. Ascent to an altitude of 10,000 feet above sea level can result in a feeling of malaise, shortness of breath, and fatigue. A person ascending to 14,000 to 15,000 feet may experience euphoria, along with a reduction in powers of reason, judgment, and memory. Altitude sickness includes a loss of **useful consciousness** at 20,000 to 25,000 feet. After approximately five minutes at this altitude, a person may lose consciousness. The loss of consciousness comes at approximately one minute or less at 30,000 feet. Over 38,000 feet, most people lose consciousness within 30 seconds and may fall into a coma and possibly die.

Hyperoxia, or an increased concentration of oxygen in air, is not a common situation. Hyperbaric chambers or improperly calibrated scuba equipment can create conditions that may lead to convulsions if pure oxygen is breathed for greater than three hours. Breathing air at a depth of around 300 feet can be toxic and is equivalent to breathing pure oxygen at a depth of 66 feet.

At high pressures of oxygen, around 2,000 to 5,000 mm Hg, dangerous cerebral problems such as dizziness, twitching, vision deterioration, and nausea may occur. Continued exposure to these high pressures will result in confusion, convulsion, and eventual death.

Changes in total pressure can induce *trapped gas effects.* With a decrease in pressure, trapped gases will increase in volume (according to Boyle's law). Trapped gases in the body include air pockets in the ears, sinuses, and chest. Divers refer to the trapped gas phenomenon as the *squeeze.* Jet travel causes the most commonly occurring instance of trapped gas effects. Takeoff and landing may cause relatively sudden shifts in pressure, which may lead to discomfort and pain. With very rapid ascent or descent, injury can develop.

Lung rupture can be caused by a swift return to the surface from diving or decompression during high-altitude flight. This event is rare and happens only if the person is holding his or her breath during the decompression.

Evolved gas effects are associated with the absorption of nitrogen into body tissues. When breathed, nitrogen can be absorbed into all body tissues in concentrations proportional to the partial pressure of nitrogen in air. When a person is ascending in altitude, on the ground, in flight, or under water, nitrogen must be exhaled at a rate equal to or exceeding the absorption rate to avoid evolved gas effects.

If the nitrogen in body tissues such as blood is being absorbed faster than it is being exhaled, bubbles of gas may form in the blood and other tissues. Gas bubbles in the tissues may cause decompression sickness, which can be painful and occasionally fatal. Early symptoms of this disorder occur in body bends or joints such as elbows, knees, and shoulders. The common name for decompression sickness is the **bends**.

When the formation of gas bubbles is due to rapid ambient pressure reduction, it is called **dysbarism**.[9] The major causes of dysbarism are (1) the release of gas from the blood, and (2) the attempted expansion of trapped gas in body tissues. The sickness may occur with the decompression associated with rapidly moving from sea level (considered zero) to approximately 20,000 feet above sea level. Dysbarism is most often associated with underwater diving or working in pressurized containers (such as airplanes). Obese and older people seem to be more susceptible to dysbarism and decompression sickness.

Dysbarism manifests itself in a variety of symptoms. The **creeps** are caused by bubble formation in the skin, which causes an itchy, crawling, rashy feeling in the skin. Coughing and choking, resulting from bubbles in the respiratory system, are called the **chokes**. Bubbles occurring in the brain, although rare, may cause tingling and numbing, severe headaches, spasticity of muscles, and in some cases, blindness and paralysis. Dysbarism of the brain is rare. Rapid pressure change may also cause pain in the teeth and sinuses.[10]

Aseptic necrosis of bone is a delayed effect of decompression sickness. Blood in the capillaries supplying the bone marrow may become blocked with gas bubbles, which can cause a collection of platelets and blood cells to build up in a bone cavity. The marrow generation of blood cells can be damaged as well as the maintenance of healthy bone cells. Some bone areas may become calcified with severe complications when the bone is involved in a joint.

Decompression Procedures

Employees who work in an environment that is under pressure must undergo decompression procedures before returning to a normal atmosphere.[11] Such procedures are planned based on the amount of pressure to which the employee is subjected and for how long. In 29 CFR 1926 (Subpart S, Appendix A), OSHA provides tables that can be used for planning appropriate decompression procedures for employees. Figure 13-2 is an example of a portion of such a table.

FIGURE 13-2 Portion of a table for planning a two-stage decompression.

Partial Table (Two-Stage Decompression)

Working Chamber Pressure (psig)	Working Period (hours)	Decompression Data						
		Stage No.	Pressure Reduction (psig) From	Pressure Reduction (psig) To	Time in Stage (minutes)	Pressure Reduction Rate (min/pound)	Total Time Decompress (minutes)
20	3	1	20	4	3	0.20	—
		2	4	0	12	3.00	15
	4	1	20	4	3	0.20	—
		2	4	0	40	10.00	43
	5	1	20	4	3	0.20	—
		2	4	0	60	15.00	63

(*Note:* Do not interpolate. Always use the next higher value for conditions that fall between numbers in the table.)

In most cases, decompression will need to occur in two stages. Figure 13-2 shows a portion of a table to be used for planning two-stage decompressions. The following example demonstrates how to use such a table:

> An employee will be working for four hours in an environment with a working chamber pressure of 20 pounds per square inch gauge (psig). Locate 20 psig and 4 working period hours in the table in Figure 17-2. Stage 1 of the decompression will require a reduction in pressure from 20 psig to 4 psig over a period of 3 minutes at the uniform rate of 0.20. Stage 2 of the decompression will require a reduction in pressure from 4 psig to 0 psig over a period of 40 minutes at the uniform rate of 10 minutes per pound. The total time for the decompression procedure is 43 minutes.

Decompression procedures are designed to prevent the various effects of decompression sickness that were explained in the previous section. For a complete set of decompression tables refer to the following web address: http://www.osha.gov

Measurement of Pressure Hazards

Confirming the point of pressurized gas leakage can be difficult. After a gas has leaked out to a level of equilibrium with its surrounding air, the symptoms of the leak may disappear. There are several methods of detecting pressure hazards:

- *Sounds* can be used to signal a pressurized gas leak. Gas discharge may be indicated by a whistling noise, particularly with highly pressurized gases escaping through small openings. Workers should not use their fingers to probe for gas leaks as highly pressurized gases may cut through tissue, including bone.
- *Cloth streamers* may be tied to the gas vessel to help indicate leaks. Soap solutions may be smeared over the vessel surface so that bubbles are formed when gas escapes. A stream of bubbles indicates gas release.
- *Scents* may be added to gases that do not naturally have an odor. The odor sometimes smelled in homes that cook or heat with natural gas is not the gas but a scent added to it.
- *Leak detectors* that measure pressure, current flow, or radioactivity may be useful for some types of gases.
- *Corrosion* may be the long-term effect of escaping gases. Metal cracking, surface roughening, and general weakening of materials may result from corrosion.

There are many potential causes of gas leaks. The most common of these are as follows:

- *Contamination* by dirt can prevent the proper closing of gas valves, threads, gaskets, and other closures used to control gas flow.
- *Overpressurization* can overstress the gas vessel, permitting gas release. The container closure may distort and separate from gaskets, leading to cracking.
- *Excessive temperatures* applied to dissimilar metals that are joined may cause unequal thermal expansion, loosening the metal-to-metal joint and allowing gas to escape. Materials may crack because of excessive cold, which may also result in gas escape. Thermometers are often used to indicate the possibility of gas release.

- *Operator errors* may lead to hazardous gas release from improper closure of valves, inappropriate opening of valves, or overfilling of vessels. Proper training and supervision can reduce operator errors.

Destructive as well as nondestructive methods may be used to detect pressure leaks and incorrect pressure levels. **Nondestructive testing** methods do not harm the material being tested. Nondestructive methods may include mixing dye penetrants and magnetic or radioactive particles with the gas and then measuring the flow of the gas. Ultrasonic and X-ray waves are another form of nondestructive testing and are often used to characterize materials and detect cracks or other leakage points.

Destructive testing methods destroy the material being checked. Proof pressures generate stresses to the gas container, typically 1.5 to 1.667 times the maximum expected operating pressure for that container. Strain measurements may also be collected to indicate permanent weakening changes to the container material that remain after the pressure is released. **Proof pressure tests** often call for the pressure to be applied for a specified time and released. Stress and strain tests are then applied to the material. Proof pressure tests may or may not result in the destruction of the container being tested.

Reduction of Pressure Hazards

The reduction of pressure hazards often requires better maintenance and inspection of equipment that measures or uses high-pressure gases. Proper storage of pressurized containers reduces many pressure hazards. Pressurized vessels should be stored in locations away from cold or heat sources, including the sun. Cryogenic compounds (those that have been cooled to unusually low temperatures) may boil and burst the container when not kept at the proper temperatures. The whipping action of pressurized flexible hoses can also be dangerous. Hoses should be firmly clamped at the ends when pressurized.

Gas compression can occur in sealed containers exposed to heat. For this reason, aerosol cans must never be thrown into or exposed to a fire. Aerosol cans may explode violently when exposed to heat, although most commercially available aerosols are contained in low-melting point metals that melt before pressure can build up.

Pressure should be released before working on equipment. Gauges can be checked before any work on the pressurized system is begun. When steam equipment is shut down, liquid may condense within the system. This liquid or dirt in the system may become a propellant, which may strike bends in the system, causing loud noises and possible damage.

Water hammer is a shock effect caused by liquid flow suddenly stopping.[12] The shock effect can produce loud noises. The momentum of the liquid is conducted back upstream in a shock wave. Pipe fittings and valves may be damaged by the shock wave. Reduction of this hazard involves using air chambers in the system and avoiding the use of quick-closing valves.

Negative pressures or **vacuums** are caused by pressures below atmospheric level. Negative pressures may result from hurricanes and tornadoes. Vacuums may cause collapse of closed containers. Building code specifications usually allow for a pressure differential. Vessel wall thickness must be designed to sustain the load imposed by the differential in pressure caused by negative pressure. Figure 13-3 describes several methods to reduce the hazards associated with pressurized containers.

FIGURE 13-3 Reduction of pressure hazards.

- Install valves so that failure of a valve does not result in a hazard.
- Do not store pressurized containers near heat or sources of ignition.
- Train and test personnel dealing with pressurized vessels. Only tested personnel should be permitted to install, operate, maintain, calibrate, or repair pressurized systems. Personnel working on pressure systems should wear safety face shields or goggles.
- Examine valves periodically to ensure that they are capable of withstanding working pressures.
- Operate pressure systems only under the conditions for which they were designed.
- Relieve all pressure from the system before performing any work.
- Label pressure system components to indicate inspection status as well as acceptable pressures and flow direction.
- Connect pressure relief devices to pressure lines.
- Do not use pressure systems and hoses at pressure exceeding the manufacturer's recommendations.
- Keep pressure systems clean.
- Keep pressurized hoses as short as possible.
- Avoid banging, dropping, or striking pressurized containers.
- Secure pressurized cylinders by a chain to prevent toppling.
- Store acetylene containers upright.
- Examine labels before using pressurized systems to ensure correct matching of gases and uses.
- Use dead man's switches on high-pressure hose wands.

Key Terms and Concepts

- Altitude sickness
- Amine vessels
- Ammonia vessels
- Aseptic necrosis
- Barometer
- Bends
- Boyle's law
- Chokes
- Creeps
- Dalton's law of partial pressures
- Deaerator vessels
- Decompression sickness
- Destructive testing
- Dysbarism
- Evolved gas effects
- Expiration
- Hazard
- Hyperoxia
- Hypoxia
- Inspiration
- Liquid penetration test
- Magnetic particle test
- Negative pressures
- Nitrogen narcosis
- Nondestructive testing
- Pressure
- Pressure hazard
- Proof pressure
- Pulp digester vessels
- Radiography test
- Trapped gas effects
- Ultrasonic testing
- Useful consciousness
- Vacuums
- Visual examination
- Water hammer
- Wet hydrogen sulfide vessels

Review Questions

1. Against which references is pressure measured? How are these references measured?
2. Define *inspiration* and *expiration.*
3. Explain Dalton's law of partial pressures.
4. Does water vapor conform to Dalton's law?
5. Briefly discuss decompression sickness.
6. What do length of exposure, the bends, the chokes, and aseptic necrosis of bone have in common?
7. Define *hypoxia* and *hyperoxia.*
8. Explain nitrogen narcosis.
9. Discuss altitude sickness.
10. What is the relationship between trapped gas effects and dysbarism?
11. What is the difference between destructive and nondestructive testing?
12. Briefly explain proof pressures.
13. What causes vacuums?
14. Explain three ways to conduct nondestructive testing of pressure vessels.
15. What is the total decompression time for an employee who works for four hours under pressure of 20 psig?

Endnotes

1. Occupational Safety and Health Administration, "Pressure Vessel Guidelines," *OSHA Technical Manual* (Washington, DC: Occupational Safety and Health Administration, 2000), sec. IV, chap. 3, 30.
2. Ibid., 31.
3. Ibid.
4. Ibid.
5. Ibid.
6. Ibid.
7. Ibid., 38.
8. Ibid., 39.
9. Ibid., 39.
10. Ibid.
11. OSHA Regulations 29 CFR 1926 (Subpart S, Appendix A), Decompression Tables.
12. Ibid.

CHAPTER 14

Electrical Hazards

Major Topics

- Electrical Hazards Defined
- Sources of Electrical Hazards
- Electrical Hazards to Humans
- Detection of Electrical Hazards
- Reduction of Electrical Hazards
- OSHA's Electrical Standards
- Electrical Safety Program
- Electrical Hazards Self-Assessment
- Prevention of Arc Flash Injuries (NFPA 70E)

Consider the following scenario: A textile mill in Massachusetts was fined $66,375 when an employee contacts OSHA and complains about unsafe conditions at the mill. The Region 1 Office of OSHA conducted an investigation in response to the complaint that uncovered the following willful violation: allowing employees to perform live electrical work without safe work procedures or appropriate personal protective equipment.[1] In addition, the investigation uncovered several serious violations including (1) storage of flammable materials near emergency exits, (2) improper storage of oxygen and acetylene cylinders, (3) failure to post load ratings, and (4) an exposed live electrical source and unsuitable electrical outlets for wet or damp locations.[2]

Electrical Hazards Defined

Electricity is the flow of negatively charged particles called *electrons* through an electrically conductive material. Electrons orbit the nucleus of an atom, which is located approximately in the atom's center. The negative charge of the electrons is neutralized by particles called **neutrons**, which act as temporary energy repositories for the interactions between positively charged particles called **protons** and **electrons**.

Figure 14-1 shows the basic structure of an atom, with the positively charged nucleus in the center. The electrons are shown as energy bands of orbiting negatively charged particles. Each ring of electrons contains a particular quantity of negative

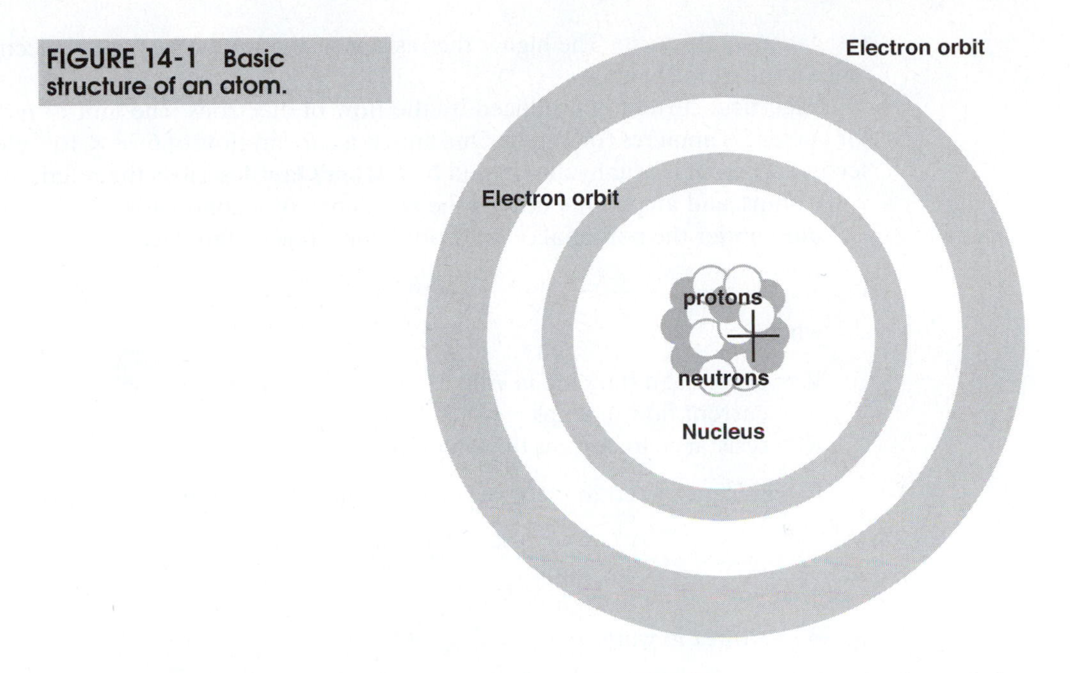

FIGURE 14-1 Basic structure of an atom.

charges. The basic characteristics of a material are determined by the number of electron rings and the number of electrons in the outer rings of its atoms. A *positive charge* is present when an atom (or group of atoms) in a material has too many electrons in its outer shell. In all other cases, the atom or material carries a *negative charge*.

Electrons that are freed from an atom and are directed by external forces to travel in a specific direction produce *electrical current,* also called *electricity*. **Conductors** are substances that have many free electrons at room temperature and can pass electricity. *Insulators* do not have a large number of free electrons at room temperature and do not conduct electricity. Substances that are neither conductors nor insulators can be called **semiconductors**.

Electrical current passing through the human body causes a shock. The quantity and path of this current determines the level of damage to the body. The path of this flow of electrons is from a negative source to a positive point, because opposite charges attract one another.

When a surplus or deficiency of electrons on the surface of a material exists, **static electricity** is produced. This type of electricity is called "static" because there is no positive material nearby to attract the electrons and cause them to move. Friction is not required to produce static electricity, although it can increase the charge of existing static electricity. When two surfaces of opposite static electricity charges are brought into close range, a discharge, or spark, will occur. The spark from static electricity is often the first clue that such static exists. A common example is the sparks that come from rustling woolen blankets in dry heated indoor air.

The *potential difference* between two points in a circuit is measured by **voltage**. The higher the voltage, the more likely it is that electricity will flow between the negative and positive points.

Pure conductors offer little **resistance** to the flow of electrons. Insulators, on the other hand, have very high resistance to electricity. Semiconductors have a medium-range resistance to electricity. The higher the resistance, the lower the flow of electrons. Resistance is measured in **ohms**.

Electrical current is produced by the flow of electrons. The unit of measurement for current is **amperes** (or amps). One amp is a current flow of 6.28×10^{18} electrons per second. Current is usually designated by *I*. **Ohm's law** describes the relationship among volts, ohms, and amps. One ohm is the resistance of a conductor that has a current of one amp under the potential of one volt. Ohm's law is stated as

$$V = IR$$

where

V = potential difference in volts
I = current flow in amps
R = resistance to current flow in ohms

Power is measured in wattage (or *watts*) and can be determined from Ohm's law:

$$W = VI \quad \text{or} \quad W = I^2R$$

where

W = power in watts

Most industrial and domestic use of electricity is supplied by **alternating current** (or **AC current**). In the United States, standard AC circuits cycle 60 times per second. The number of cycles per second is known as **frequency** and is measured in **hertz**. Because voltage cycles in AC current, an **effective current** for AC circuits is computed, which is slightly less than the peak current during a cycle.

A **direct current** (or **DC current**) has been found to generate as much heat as an AC current that has a peak current 41.4 percent higher than the DC. The ratio of effective current to peak current can be determined by

$$(\text{Effective current})/(\text{Peak current}) = (100\%)/(100\% + 41.4\%) = 0.707 \text{ or } 70.7\%$$

Effective voltages are computed using the same ratios as effective current. A domestic 110-volt circuit has an effective voltage of 110 volts, with peaks of voltage over 150 volts.

The path of electrical current must make a complete loop for the current to flow. This loop includes the source of electrical power, a conductor to act as the path, a device to use the current (called a **load**), and a path to the ground. The earth maintains a relatively stable electrical charge and is a good conductor. The earth is considered to have **zero potential** because of its massive size. Any electrical conductor pushed into the earth is said to have zero potential. The earth is used as a giant common conductor back to the source of power.

Electrocution occurs when a person makes contact with a conductor carrying a current and simultaneously contacts the ground or another object that includes a conductive path to the ground. This person completes the circuit loop by providing a load for the circuit and thereby enables the current to pass through his or her body. People can be protected from this danger by insulating the conductors, insulating the people, or isolating the danger from the people.

The National Electrical Code (NEC) is published by the National Fire Protection Association (NFPA). This code specifies industrial and domestic electrical safety

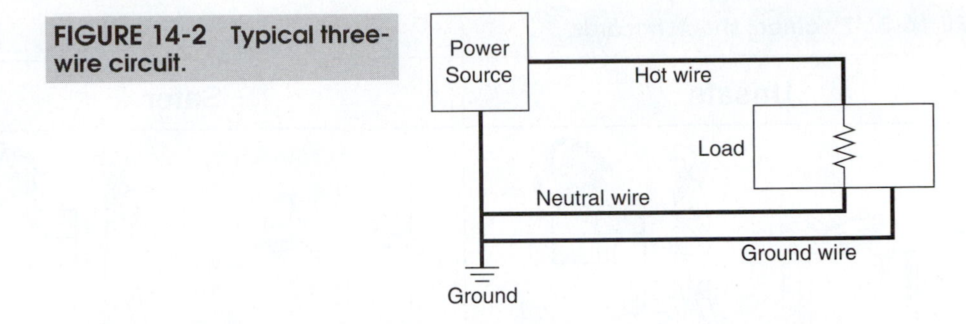

FIGURE 14-2 Typical three-wire circuit.

precautions. The NEC categorizes industrial locations and gases relative to their degree of fire hazard and describes in detail the safety requirements for industrial and home wiring. The NEC has been adopted by many jurisdictions as the local electrical code. The National Board of Fire Underwriters sponsors Underwriters Laboratories (UL). The UL determines whether equipment and materials for electrical systems are safe in the various NEC location categories. The UL provides labels for equipment that it approves as safe within the tested constraints.

Typical 110-volt circuit wiring has a **hot wire** carrying current, a **neutral wire**, and a **ground wire**. The neutral wire may be called a **grounded conductor**, with the ground wire being called a **grounding conductor**. Neutral wires usually have white insulation, hot wires have red or black insulation, and ground wires have green insulation or are bare. Figure 14-2 shows a typical three-wire circuit.

The hot wire carries an effective voltage of 110 volts with respect to the ground, whereas the neutral wire carries nearly zero voltage. If the hot wire makes contact with an unintended conductor, such as a metal equipment case, the current can bypass the load and go directly to the ground. With the load skipped, the ground wire is a low-resistance path to the earth and carries the highest current possible for that circuit.

A **short circuit** is a circuit in which the load has been removed or bypassed. The ground wire in a standard three-wire circuit provides a direct path to the ground, bypassing the load. Short circuits can be another source of electrical hazard if a human is the conductor to the ground, thereby bypassing the load.

Sources of Electrical Hazards

Short circuits are one of many potential electrical hazards that can cause electrical shock. Another hazard is water, which considerably decreases the resistance of materials, including humans. The resistance of wet skin can be as low as 450 ohms, whereas dry skin may have an average resistance of 600,000 ohms. According to Ohm's law, the higher the resistance, the lower the current flow. When the current flow is reduced, the probability of electrical shock is also reduced.

The major causes of electrical shock are

- Contact with a bare wire carrying current. The bare wire may have deteriorated insulation or be normally bare.
- Working with electrical equipment that lacks the UL label for safety inspection.
- Electrical equipment that has not been properly grounded. Failure of the equipment can lead to short circuits.

FIGURE 14-3 Electrical shock hazards.

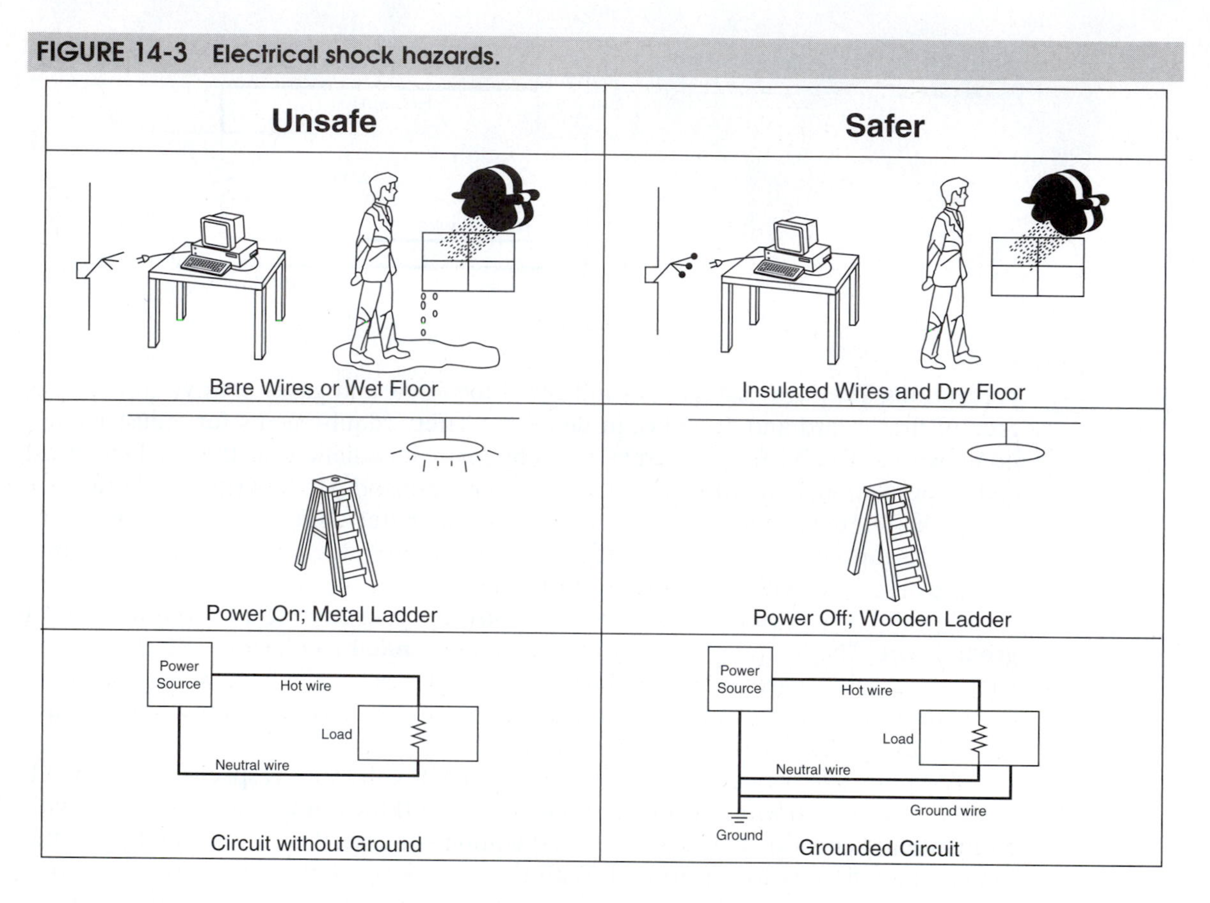

- Working with electrical equipment on damp floors or other sources of wetness.
- Static electricity discharge.
- Using metal ladders to work on electrical equipment. These ladders can provide a direct line from the power source to the ground, again causing a shock.
- Working on electrical equipment without ensuring that the power has been shut off.
- Lightning strikes.

Figure 14-3 depicts some of these electrical shock hazards.

Electrostatic Hazards

Electrostatic hazards may cause minor shocks. Shocks from static electricity may result from a single discharge or multiple discharges of static. Sources of electrostatic discharge include the following:

- Briskly rubbing a nonconductive material over a stationary surface. One common example of this is scuffing shoes across a wool or nylon carpet. Multilayered clothing may also cause static sparks.[3]
- Moving large sheets of plastic, which may discharge sparks.
- The explosion of organic and metallic dusts, which have occurred from static buildup in farm grain silos and mine shafts.

- Conveyor belts. Depending on their constituent material, they can rub the materials being transported and cause static sparks.
- Vehicle tires rolling across a road surface.
- Friction between a flowing liquid and a solid surface.[4]

The rate of discharge of electrical charges increases with lower humidity. Electrostatic sparks are often greater during cold, dry winter days. Adding humidity to the air is not commonly used to combat static discharge, however, because higher humidity may result in an uncomfortable working environment and adversely affect equipment.[5]

Arcs and Sparks Hazards

With close proximity of conductors or contact of conductors to complete a circuit, an electric **arc** can jump the air gap between the conductors and ignite combustible gases or dusts. When the electric arc is a discharge of static electricity, it may be called a **spark**. A spark or arc may involve relatively little or a great deal of power and is usually discharged into a small space.

Combustible and Explosive Materials

High currents through contaminated liquids may cause the contaminants to expand rapidly and explode. This situation is particularly dangerous with contaminated oil-filled circuit breakers or transformers. A poor match between current or polarity and capacitors can cause an explosion. In each of these cases, the conductor is not capable of carrying a current of such high magnitude. Overheating from high currents can also lead to short circuits, which in turn may generate fires or explosions.

Lightning Hazards

Lightning is static charges from clouds following the path of least resistance to the earth, involving very high voltage and current. If this path to the earth involves humans, serious disability may result, including electrocution. Lightning may also damage airplanes from intracloud and cloud-to-cloud flashes. Electrical equipment and building structures are commonly subject to lightning hazards. Lightning tends to strike the tallest object on the earth below the clouds. A tree is a common natural path for lightning.

Improper Wiring

Improper wiring permits equipment to operate normally but can result in hazardous conditions. The section of this chapter on detection of electrical hazards discusses tests to identify unsafe wiring practices. One common mistake is to "jump" the ground wire to the neutral wire. In this case, the ground wire is actually connected to the neutral wire. Equipment usually operates in a customary way, but the hazard occurs when low voltages are generated on exposed parts of the equipment, such as the housing. If the neutral circuit becomes corroded or loose, the voltage on the ground wire increases to a dangerous level.

Improper wiring (or miswiring) can cause other hazards. When the ground is connected improperly, the situation is referred to as **open ground**. Usually the equipment with this miswiring will operate normally. If a short occurs in the equipment circuitry without proper grounding, anyone touching that equipment may be severely shocked.

With **reversed polarity**, the hot and neutral wires have been reversed. A worker who is not aware that the black lead (hot) and white lead (neutral) have been reversed could be injured or cause further confusion by connecting the circuit to another apparatus. If a short between the on/off switch and the load occurred, the equipment may run indefinitely, regardless of the switch position. In a reversed polarity light bulb socket, the screw threads become conductors.[6]

Temporary wiring installations sometimes remain in place for years until an accident occurs. Flexible wiring should rarely be substituted for fixed wiring in permanent buildings. A loose knot should be tied in a flexible cord when the plug is installed or replaced. The knot can prevent a pull on the cord from being transmitted to electrical connections such as the plug.

Insulation Failure

The degradation of insulation can cause a bare wire and resulting shock to anyone coming in contact with that wire. Most insulation failure is caused by environments toxic to insulation. These environments include:

- Direct sunlight or other sources of ultraviolet light, which can induce gradual breakdown of plastic insulation material.
- Sparks or arcs from discharging static electricity, which can result in burned-through holes in insulation.
- Repeated exposure to elevated temperatures, which can produce slow but progressive degradation of insulation material.
- Abrasive surfaces, which can result in erosion of the material strength of the insulation.
- Substance incompatibility with the atmosphere around the insulation and the insulation material, which can induce chemical reactions. Such reactions may include oxidation or dehydration of the insulation and eventual breakdown.
- Animals such as rodents or insects chewing or eating the insulation material, leading to exposure of the circuit. Insects can also pack an enclosed area with their bodies so tightly that a short circuit occurs. This is a common occurrence with electrical systems near water, such as pump housings and television satellite dishes.
- Moisture and humidity being absorbed by the insulation material, which may result in the moisture on the insulation carrying a current.

Equipment Failure

There are several ways in which equipment failure can cause electrical shocks. Electrical equipment designers attempt to create devices that are explosion-proof, dust-ignition-proof, and spark-proof. Following are some of the more common types of equipment failure:

- Wet insulation can become a conductor and cause an electrical shock.
- Portable tool defects can result in the device's housing carrying an electric current. Workers do not expect tool housings to be charged and may be shocked when they touch a charged tool housing.
- Broken power lines carry great amperage and voltage and can cause severe disability.
- When equipment is not properly grounded or insulated, an unshielded worker may receive a substantial electrical shock.

FIGURE 14-4 Hazardous electrical equipment location categories.

Class	Group	Division I	Division II
I. Flammable vapors and gases	A. Acetylene B. Hydrogen C. Ether D. Hydrocarbon fuels and solvents	Normally explosive; flammable paint spray areas	Not normally in explosive concentration; adjacent to paint spray area
II. Combustible dusts	E. Metal dusts F. Carbon dusts G. Flour, starch, grain, plastic, or chemical dusts	Conductive or ignitable dusts may be present; grain mills or processors	Not normally in ignitable concentration; grain storage areas
III. Ignitable fibers	Textiles, woodworking	Handled or used in manufacturing; cotton gins	Stored or handled in storage, not in manufacturing; excelsior storage

Hazardous Locations for Electrical Equipment

The NEC classifies hazardous locations for electrical equipment. There are three basic classes: Class I for flammable vapors and gases, Class II for combustible dusts, and Class III for ignitable fibers. There are also two divisions of hazard categories. Division I has more stringent requirements for electrical installation than Division II does. Figure 14-4 gives examples for each location category.[7]

Electrical Hazards to Humans

The greatest danger to humans suffering electrical shock results from current flow. The voltage determines whether a particular person's natural resistance to current flow will be overcome. Skin resistance can vary between 450 ohms and 600,000 ohms, depending on skin moisture.[8] Some levels of current "**freeze**" a person to the conductor; the person cannot voluntarily release his or her grasp. **Let-go current** is the highest current level at which a person in contact with the conductor can release the grasp of the conductor. Figure 14-5 shows the relationship between amperage dosage and danger with a typical domestic 60-cycle AC current.

The severity of injury with electrical shock depends on the dosage of current, as shown in Figure 18-5, but also on the path taken through the body by the current. The path is influenced by the resistance of various parts of the body at the time of contact with the conductor. The skin is the major form of resistance to current flow. Current paths through the heart, brain, or trunk are generally much more injurious than paths through extremities.

FIGURE 14-5 Current effects on the human body (60-cycle AC current).

Dose in Current in Milliamps	Effect on Human Body
Less than 1	No sensation, no perceptible effect.
1	Shock perceptible, reflex action to jump away. No direct danger from shock but sudden motion may cause accident.
More than 3	Painful shock.
6	Let-go current for women.*
9	Let-go current for men.*
10–15	Local muscle contractions. Freezing to the conductor for 2.5% of the population.
30–50	Local muscle contractions. Freezing to the conductor for 50% of the population.
50–100	Prolonged contact may cause collapse and unconsciousness. Death may occur after three minutes of contact due to paralysis of the respiratory muscles.
100–200	Contact of more than a quarter of a second may cause ventricular fibrillation of the heart and death. AC currents continuing for more than one heart cycle may cause fibrillation.
Over 200	Clamps and stops the heart as long as the current flows. Heart beating and circulation may resume when current ceases. High current can produce respiratory paralysis, which can be reversed with immediate resuscitation. Severe burns to the skin and internal organs. May result in irreparable body damage.

*Difference between men and women is based on the relative body mass of the "average"-sized man and woman (60-cycle AC current).

Detection of Electrical Hazards

Several items of test equipment can be used to verify electrical equipment safety. A **circuit tester** is an inexpensive piece of test equipment with two wire leads capped by probes and connected to a small bulb. Most circuit testers test at least a 110- to 220-volt range. This simple tester can ensure that power has been turned off before electrical maintenance begins. The tester may also be used to determine whether housings and other equipment parts are carrying a current. When one of the leads makes contact with a hot wire and the other lead connects to a grounded conductor, the bulb lights.

A **receptacle wiring tester** is a device with two standard plug probes for insertion into an ordinary 110-volt outlet and a probe for the ground. Indicator lights show an improperly wired receptacle (outlet). However, there are several types of miswiring that are not disclosed by using this tester, including the ground wire to neutral wire mistake. Figure 14-6 shows the meaning of lit indicator lights on the receptacle wiring tester.

FIGURE 14-6 Receptacle wiring tester indicator lights.

Situation	Lights 1	2	3
Correct wiring	On	On	Off
Ground jumped to neutral	On	On	Off
Neutral and ground reversed	On	On	Off
Reversed polarity	On	Off	On
Open ground	On	Off	Off

A continuity tester may be used to determine whether a conductor is properly grounded or has a break in the circuit. Continuity is checked on circuits that are disconnected from a power source. Continuity testers often have an alligator clip on one end of a wire and a bulb and probe on the other end of the same wire. One terminal of the tester can be connected to the equipment housing; the other terminal is connected to a known ground. If the bulb does not light, the equipment is shown to be improperly grounded. With a circuit, the bulb lights when a current is capable of passing through the complete circuit. The unlit bulb of a continuity tester indicates a break in the circuit.

Infrared thermal imaging is another technique that can be used for detecting electrical hazards.

Reduction of Electrical Hazards

Grounding of electrical equipment is the primary method of reducing electrical hazards. The purpose of grounding is to safeguard people from electrical shocks, reduce the probability of a fire, and protect equipment from damage. Grounding ensures a path to the earth for the flow of excess current. Grounding also eliminates the possibility of a person being shocked by contact with a charged capacitor. The actual mechanism of grounding was discussed at the beginning of this chapter.

Electrical system grounding is achieved when one conductor of the circuit is connected to the earth. Power surges and voltage changes are attenuated and usually eliminated with proper system grounding. **Bonding** is used to connect two pieces of equipment by a conductor. Bonding can reduce potential differences between the equipment and thus reduce the possibility of sparking. Grounding, in contrast, provides a conducting path between the equipment and the earth. Bonding and grounding together are used for entire electrical systems.

Separate equipment grounding involves connecting all metal frames of the equipment in a permanent and continuous manner. If an insulation failure occurs, the current should return to the system ground at the power supply for the circuit. The equipment ground wiring will be the path for the circuit current, enabling circuit breakers and fuses to operate properly. The exposed metal parts of the equipment shown in Figure 14-7 must be grounded or provided with double insulation.[9]

FIGURE 14-7 Equipment requiring grounding or double insulation.

Portable electric tools such as drills and saws.
Communication receivers and transmitters.
Electrical equipment in damp locations.
Television antenna towers.
Electrical equipment in flammable liquid storage areas.
Electrical equipment operated with over 150 volts.

A **ground fault circuit interrupter (GFCI)**, also called a *ground fault interrupter (GFI)*, can detect the flow of current to the ground and open the circuit, thereby interrupting the flow of current. When the current flow in the hot wire is greater than the current in the neutral wire, a **ground fault** has occurred. The GFI provides a safety measure for a person who becomes part of the ground fault circuit. The GFI cannot interrupt current passing between two circuits or between the hot and neutral wires of a three-wire circuit. To ensure safety, equipment must be grounded as well as protected by a GFI. A GFI should be replaced periodically based on the manufacturer's recommendations.

There are several options for reducing the hazards associated with static electricity. The primary hazard of static electricity is the transfer of charges to surfaces with lower potential. Bonding and grounding are two means of controlling static discharge. **Humidification** is another mechanism for reducing electrical static; it was discussed in the section on sources of electrical hazards. Raising the humidity above 65 percent reduces charge accumulation.[10] However, when the relative humidity exceeds 65 percent, biological agents can begin to grow in heating, ventilation, and air conditioning (HVAC) ducts and unventilated areas.

Antistatic materials have also been used effectively to reduce electrical static hazards. Such materials either increase the surface conductivity of the charged material or absorb moisture, which reduces resistance and the tendency to accumulate charges.

Ionizers and electrostatic neutralizers ionize the air surrounding a charged surface to provide a conductive path for the flow of charges. Radioactive neutralizers include a radioactive element that emits positive particles to neutralize collected negative electrical charges. Workers need to be safely isolated from the radioactive particle emitter.

Fuses consist of a metal strip or wire that melts if a current above a specific value is conducted through the metal. Melting the metal causes the circuit to open at the fuse, thereby stopping the flow of current. Some fuses are designed to include a time lag before melting to allow higher currents during startup of the system or as an occasional event.

Magnetic circuit breakers use a solenoid (a type of coil) to surround a metal strip that connects to a tripping device. When the allowable current is exceeded, the magnetic force of the solenoid retracts the metal strip, opening the circuit. Thermal circuit breakers rely on excess current to produce heat and bending in a sensitive metal strip. Once bent, the metal strip opens the circuit. Circuit breakers differ from fuses in that they are usually easier to reset after tripping and often provide a lower time lag or none at all before being activated.

FIGURE 14-8 Lightning hazard control.

- Place lightning rods so that the upper end is higher than nearby structures.
- Avoid standing in high places or near tall objects. Be aware that trees in an open field may be the tallest object nearby.
- Do not work with flammable liquids or gases during electrical storms.
- Ensure proper grounding of all electrical equipment.
- If inside an automobile, remain inside the automobile.
- If in a small boat, lie down in the bottom of the boat.
- If in a metal building, stay in the building and do not touch the walls of the building.
- Wear rubber clothing if outdoors.
- Do not work touching or near conducting materials, especially those in contact with the earth such as fences.
- Avoid using the telephone during an electrical storm.
- Do not use electrical equipment during the storm.
- Avoid standing near open doors or windows where lightning may enter the building directly.

Double insulation is another means of increasing electrical equipment safety. Most double-insulated tools have plastic nonconductive housings in addition to standard insulation around conductive materials.

There are numerous methods of reducing the risk of electrocution by lightning. Figure 14-8 lists the major precautions to take.[11]

Another means of protecting workers is isolating the hazard from the workers or vice versa. **Interlocks** automatically break the circuit when an unsafe situation is detected. Interlocks may be used around high-voltage areas to keep personnel from entering the area. Elevator doors typically have interlocks to ensure that the elevator does not move when the doors are open. Warning devices to alert personnel about detected hazards may include lights, colored indicators, on/off blinkers, audible signals, or labels.

It is better to design safety into the equipment and system than to rely on human behavior such as reading and following labels. Figure 14-9 summarizes the many methods of reducing electrical hazards.

OSHA'S Electrical Standards

OSHA's standards relating to electricity are found in 29 CFR 1910 (Subpart S). They are extracted from the National Electrical Code. This code should be referred to when more detail is needed than appears in OSHA's excerpts. Subpart S is divided into the following two categories of standards: (1) Design of Electrical Systems, and (2) Safety-Related Work Practices. The standards in each of these categories are as follows:

DESIGN OF ELECTRICAL SYSTEMS

1910.302 Electric utilization systems
1910.303 General requirements

FIGURE 14-9 Summary of safety precautions for electrical hazards.

- Ensure that power has been disconnected from the system before working with it. Test the system for deenergization. Capacitors can store current after power has been shut off.
- Allow only fully authorized and trained people to work on electrical systems.
- Do not wear conductive material such as metal jewelry when working with electricity.
- Screw bulbs securely into their sockets. Ensure that bulbs are matched to the circuit by the correct voltage rating.
- Periodically inspect insulation.
- If working on a hot circuit, use the buddy system and wear protective clothing.
- Do not use a fuse with a greater capacity than was prescribed for the circuit.
- Verify circuit voltages before performing work.
- Do not use water to put out an electrical fire.
- Check the entire length of electrical cord before using it.
- Use only explosion-proof devices and nonsparking switches in flammable liquid storage areas.
- Enclose uninsulated conductors in protective areas.
- Discharge capacitors before working on the equipment.
- Use fuses and circuit breakers for protection against excessive current.
- Provide lightning protection on all structures.
- Train people working with electrical equipment on a routine basis in first aid and cardiopulmonary resuscitation (CPR).

1910.304 Wiring design and protection
1910.305 Wiring methods, components, and equipment for general use
1910.306 Specific-purpose equipment and installations
1910.307 Hazardous (classified) locations
1910.308 Special systems

SAFETY-RELATED WORK PRACTICES

1910.331 Scope
1910.332 Training
1910.333 Selection and use of work practices
1910.334 Use of equipment
1910.335 Safeguards for personal protection

Electrical Safety Program

With electrocution accounting for approximately 6 percent of all workplace deaths in the United States every year, it is important that employers have instituted an effective electrical safety program. The National Institute for Occupational Safety and Health (NIOSH) recommends the following strategies for establishing such a program:

- Develop and implement a comprehensive safety program and, when necessary, revise existing programs to address thoroughly the area of electrical safety in the workplace.
- Ensure compliance with existing OSHA regulations, Subpart S of 29 CFR 1910.302 through 1910.399 of the General Industry Safety and Health Standards, and Subpart K of 29 CFR 1926.402 through 1926.408 of the OSHA Construction Safety and Health Standards.
- Provide all workers with adequate training in the identification and control of the hazards associated with electrical energy in their workplace.
- Provide additional specialized electrical safety training to those working with or around exposed components of electric circuits. This training should include, but not be limited to, training in basic electrical theory, proper safe work procedures, hazard awareness and identification, proper use of personal protective equipment (PPE), proper lockout/tagout procedures, first aid including CPR, and proper rescue procedures. Provide periodic retraining as necessary.
- Develop and implement procedures to control hazardous electrical energy that include lockout and tagout procedures. Ensure that workers follow these procedures.
- Provide testing or detection equipment for those who work directly with electrical energy that ensure their safety during performance of their assigned tasks.
- Ensure compliance with the National Electrical Code and the National Electrical Safety Code.
- Conduct safety meetings regularly.
- Conduct scheduled and unscheduled safety inspections at work sites.
- Actively encourage all workers to participate in workplace safety.
- In a construction setting, conduct a job site survey before starting any work to identify all electrical hazards, implement appropriate control measures, and provide training to employees specific to all identified hazards.
- Ensure that proper personal protective equipment is available and worn by workers where required (including fall protection equipment).
- Conduct job hazard analyses of all tasks that may expose workers to the hazards associated with electrical energy and implement control measures that will adequately insulate and isolate workers from electrical energy.
- Identify potential electrical hazards and appropriate safety interventions during the planning phase of construction or maintenance projects. This planning should address the project from start to finish to ensure that workers have the safest possible work environment.[12]

Electrical Hazards Self-Assessment

Even the best safety professional cannot be everywhere at once. Consequently, one of the best strategies for safety personnel is to enlist the assistance of supervisors. After all, helping ensure a safe and healthy work environment is part of every supervisor's job description, or at least it should be. To help prevent accidents and injuries from electrical hazards, safety personnel should consider developing checklists supervisors can use to undertake periodic self-assessments in their areas of responsibility. What follows are the types of questions that should be contained in such checklists:

1. Are all electricians in your company up-to-date with the latest requirements of the National Electrical Code (NEC)?
2. Does your company specify compliance with the NEC as part of its contracts for electrical work with outside personnel?
3. Do all electrical installations located in the presence of hazardous dust or vapors meet the NEC requirements for hazardous locations?
4. Are all electrical cords properly strung (i.e., so that they do not hang on pipes, nails, hooks, etc.)?
5. Is all conduit, BX cable, and so on properly attached to supports and tightly connected to junction boxes and outlet boxes?
6. Are all electrical cords free of fraying?
7. Are rubber cords free of grease, oil, chemicals, and other potentially damaging materials?
8. Are all metallic cables and conduit systems properly grounded?
9. Are all portable electric tools and appliances grounded or double insulated?
10. Are all ground connections clean and tightly made?
11. Are all fuses and circuit breakers the proper size and type for the load on each circuit?
12. Are all fuses free of "jumping" (i.e., with pennies or metal strips)?
13. Are all electrical switches free of evidence of overheating?
14. Are all switches properly mounted in clean, tightly closed metal boxes?
15. Are all electrical switches properly marked to show their purpose?
16. Are all electric motors kept clean and free of excessive grease, oil, or potentially damaging materials?
17. Are all electric motors properly maintained and provided with the necessary level of overcurrent protection?
18. Are bearings in all electrical motors in good condition?
19. Are all portable lights equipped with the proper guards?
20. Are all lamps kept free of any and all potentially combustible materials?
21. Is the organization's overall electrical system periodically checked by a person competent in the application of the NEC?[13]

Prevention of Arc Flash Injuries

Arc flash injuries occur in the workplace every day in this country. Many of these injuries lead to severe burns and even death, which is doubly tragic because the accidents could have been prevented. An **arc flash** is an electrical short circuit that travels through the air rather than flowing through conductors, bus bars, and other types of equipment. The uncontrolled energy released by an arc flash can produce high levels of heat and pressure. It can also cause equipment to explode, sending dangerous shrapnel flying through the air.[14]

Arc flashes are sometimes produced by electrical equipment malfunctions, but a more common cause is accidental human contact with an electrical circuit or conductor.

For example, a person working near a piece of energized electrical equipment might accidentally drop a tool that then makes contact with an electrical circuit or conductor. The result is an arc flash that can injure or even kill the worker, not to mention the equipment damage.

Arc flashes become even more hazardous when workers are wearing flammable clothing instead of appropriate personal protective equipment (PPE). Arc flashes can produce sufficient heat to easily ignite clothing, cause severe burns, and even damage hearing (hearing damage is caused by the high level of pressure that can be released by an arc flash). The best and most obvious way to prevent arc flash injuries is to deenergize the electrical equipment in question and lock or tag it out before beginning maintenance or service work on it.

However, this is not always possible. Some maintenance and service functions such as troubleshooting require that the equipment being worked on be energized. When this is the case, it is important to consult the National Fire Protection Association's *Handbook for Electrical Safety in the Workplace* (NFPA 70E) and proceed as follows:

- Perform a flash hazard analysis in accordance with NFPA 70E, Article 130.3, or use Table 130.7(C)(9)(a) (Hazard/Risk Category Classifications) to identify the hazard/risk category of the job tasks that must be performed on the energized equipment.
- Establish a flash protection boundary around the equipment in question in accordance with NFPA 70E, Article 130.3(A).
- Select the PPE that will be worn by the worker(s) who will perform the tasks in question on the energized equipment from Table 130.7(C)(10) (Protective Clothing and Personal Protective Equipment Matrix) based on the level of risk identified for these tasks in Table 130.7(C)(9)(a) in the first step above.

For example, if you determine that a worker must perform tasks on a piece of energized electrical equipment and these tasks are rated in "Hazard/Risk Category 3," Table 130.7(C)(10) would require the following PPE:

1. Cotton underwear.
2. Fire-resistant pants and shirt.
3. Fire-resistant coverall.[15]

Key Terms and Concepts

- Alternating current (AC)
- Amperes (amps)
- Antistatic materials
- Arc
- Arc flash
- Bonding
- Circuit tester
- Conductors
- Continuity tester
- Direct current (DC)
- Double insulation
- Effective current
- Electrical current
- Electrical hazards
- Electrical safety program
- Electrical system grounding
- Electricity
- Electrons
- Electrostatic hazards
- Electrostatic neutralizers
- Equipment failure
- Freeze
- Frequency
- Fuses
- Ground fault
- Ground fault circuit interrupter (GFCI)
- Ground fault interrupter (GFI)
- Ground wire
- Grounded conductor
- Grounding conductor
- Hazardous locations
- Hertz
- Hot wire
- Humidification
- Improper wiring

- Insulation failure
- Insulators
- Interlocks
- Ionizers
- Isolating the hazard
- Jump the ground wire
- Let-go current
- Lightning
- Load
- Magnetic circuit breaker
- National Board of Fire Underwriters
- National Electrical Code (NEC)
- Negative charge
- Neutral wire
- Neutrons
- NFPA 70E
- Ohms
- Ohm's law
- Open ground
- Positive charge
- Potential difference
- Power
- Protons
- Radioactive neutralizers
- Receptacle wiring tester
- Resistance
- Reversed polarity
- Semiconductor
- Separate equipment grounding
- Shock
- Short circuit
- Spark
- Static electricity
- Thermal circuit breaker
- Underwriters Laboratories (UL)
- Voltage
- Warning devices
- Watts
- Zero potential

Review Questions

1. Define *zero potential*. Explain the relationship between zero potential and grounding.
2. Explain what each of the following terms measure: *volt, amp, ohm, hertz*, and *watt*.
3. Briefly discuss the difference between DC and AC using the concept of effective current.
4. What is the relationship between the NEC and UL?
5. Briefly state the relationship among potential difference, lightning, and grounding.
6. Define *open ground*.
7. List at least five lightning hazard control measures.
8. Explain the relationship between circuit load and short circuits.
9. How do ionizers, radioactive neutralizers, and antistatic materials work? Why does humidification work?
10. Discuss how bonding and grounding work together to increase electrical safety.
11. How do continuity testers, circuit testers, and receptacle wiring testers operate?
12. Explain freeze and let-go current.
13. Describe the structure of an atom.
14. Discuss the proper wiring of a three-wire circuit.
15. Why is jumping the ground wire a hazard?
16. Explain reversed polarity.
17. Why are warning devices less effective than designed-in safety precautions?
18. Explain five strategies for establishing an effective electrical safety program.
19. Explain why it is important for safety personnel to help employees and supervisors conduct self-assessments.
20. Explain how to reduce arc flash hazards.

Endnotes

1. "Massachusetts Mill Cited for Safety and Health Violations," Occupational Health & Safety News 20, no. 9: 10.
2. Ibid.
3. National Fire Protection Association, Handbook for Electrical Safety in the Workplace (Quincy, MA: National Fire Protection Association, 2004), 57.
4. Ibid., 59.
5. Ibid., 60.
6. Ibid., 61.
7. National Fire Protection Association, Handbook for Electrical Safety, 103.
8. Ibid., 104
9. Ibid., 105
10. National Fire Protection Association, *Handbook for Electrical Safety*, 109.
11. Ibid., 110
12. National Institute for Occupational Safety and Health, Worker Deaths by Electrocution: A Summary of Surveillance Findings and Investigative Case Reports, 2006, 13–16. Retrieved from http://www.cdc.gov/niosh/homepage.html.
13. Retrieved from http://online.migu.nodak.edu/19577/BADM309checklist.htm
14. W. Wallace, "NFPA 70E: Performing the Electrical Flash Hazard Analysis," Occupational Safety & Health 74, no. 8: 38–44.
15. National Fire Protection Association, Handbook for Electrical Safety, Table 130.7(C)(9)(a).

Fire Hazards and Life Safety

Major Topics

- Fire Hazards Defined
- Sources of Fire Hazards
- Fire Dangers to Humans
- Detection of Fire Hazards
- Reduction of Fire Hazards
- Development of Fire Safety Standards
- OSHA Fire Standards
- Life Safety
- Flame-Resistant Clothing
- Fire Safety Programs
- Explosive Hazards
- OSHA's Firefighting Options
- Self-Assessment in Fire Protection

The assistant manager of a fuel storage plant was killed while trying to repair a broken-down piece of equipment. According to the *NFPA Journal,*

> The victim had been called to investigate a strong odor of gasoline that had been detected by a gasoline tank driver. . . . It was suspected that the facility's vapor-recovery system, which captures gasoline vapors displaced from tank trucks being filled with product, had malfunctioned. Soon after the manager entered the area, a violent explosion occurred. The victim's severely burned body and repair tools were found near the damaged equipment.[1]

In another recent example, facility damage was held to a minimum and human injury was avoided when a sprinkler system suppressed a fire that broke out on the second floor of a polyurethane foam manufacturing plant in North Carolina. The *NFPA Journal* stated,

> The fire broke out just after 4:00 P.M., an hour after all employees but one had left for the day. The ballast of a ceiling-mounted fluorescent light fixture short-circuited and cracked open, allowing burning ballast material to drop to the floor and the conveyor below. The burning material ignited scrap urethane.[2]

Fortunately, in this case, the company had a properly functioning automatic fire suppression system, and tragedy was avoided.

Two employees were killed and another received second- and third-degree burns over 45 percent of his body when an explosion and fire occurred at a chemical plant in Ohio. According to the *NFPA Journal,*

> The employees inserted a long metal rod into the hopper and attempted to dislodge the blockage. As the men moved the rod around, an explosion occurred inside the hopper.[3]

What all these tragedies have in common is that they did not have to happen. These incidents could have been prevented. The damage and injuries that can result from fire can be both physically and psychologically damaging. The resulting trauma can affect even those employees who are not physically injured. Therefore, modern safety and health professionals should be familiar with fire hazards and their prevention.

Fire Hazards Defined

Fire hazards are conditions that favor fire development or growth. Three elements are required to start and sustain fire: (1) oxygen, (2) fuel, and (3) heat. Because oxygen is naturally present in most earth environments, fire hazards usually involve the mishandling of fuel or heat.

Fire, or **combustion**, is a chemical reaction between oxygen and a combustible fuel. Combustion is the process by which fire converts fuel and oxygen into energy, usually in the form of heat. By-products of combustion include light and smoke. For the reaction to start, a source of ignition, such as a spark or open flame, or a sufficiently high temperature is needed. Given a sufficiently high temperature, almost every substance will burn. The **ignition temperature** or **combustion point** is the temperature at which a given fuel can burst into flame.

Fire is a chain reaction. For combustion to continue, there must be a constant source of fuel, oxygen, and heat (see Figure 15-1). The flaming mode is represented by the tetrahedron on the left (heat, oxidizing agent, and reducing agent) that results from a chemical chain reaction. The smoldering mode is represented by the triangle on the right. **Exothermic** chemical reactions create heat. Combustion and fire are exothermic reactions and can often generate large quantities of heat. **Endothermic** reactions consume more heat than they generate. An ongoing fire usually provides its own sources of heat. It is important to remember that cooling is one of the principal ways to control a fire or put it out.

All chemical reactions involve forming and breaking chemical bonds between atoms. In the process of combustion, materials are broken down into basic elements. Loose atoms form bonds with each other to create molecules of substances that were not originally present.

FIGURE 15-1 Fire tetrahedron (L) and fire triangle (R).

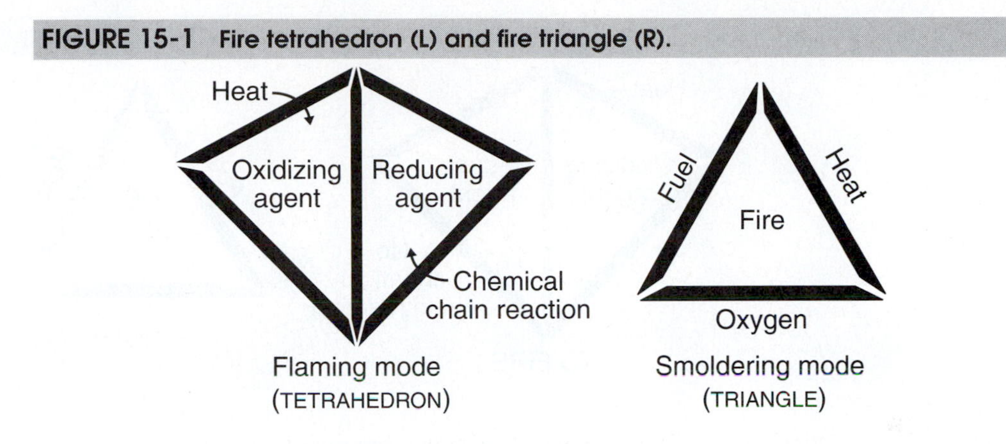

Carbon is found in almost every flammable substance. When a substance burns, the carbon is released and then combines with the oxygen that *must* be present to form either **carbon dioxide** or **carbon monoxide**.

Carbon dioxide is produced when there is more oxygen than the fire needs. It is not toxic, but it can be produced in such volumes that it seriously reduces the concentration of oxygen in the air surrounding the fire site. Carbon monoxide—a colorless, odorless, deadly gas—is the result of incomplete combustion of a fuel. It is produced when there is insufficient oxygen to burn the fuel present efficiently. In general, most fires have insufficient oxygen and therefore produce large quantities of carbon monoxide. It is important in any intentional industrial fire that the fuel be consumed as completely as possible. This will reduce ash and minimize smoke and gases, including carbon monoxide.

Hydrogen, found in most fuels, combines with oxygen to form water. Synthetic polymers, found in plastics and vinyls, often form deadly fumes when they are consumed by fire, or when they melt or disintegrate from being near fire or high heat. Burning, melting, or disintegrating plastic at a fire site should be presumed to be releasing toxic fumes.

Liquids and solids, such as oil and wood, do not burn directly but must first be converted into a flammable **vapor** by heat. Hold a match to a sheet of paper, and the paper will burst into flames. Look closely at the paper, and you will see that the paper is not burning. The flames reside in a vapor area just above the surface of the sheet.

Vapors will burn only at a specific range of mixtures of oxygen and fuel, determined by the composition of the fuel. At the optimum mixture, a fire burns, generates heat and some light, and produces no other by-products. In an unintentional fire, the mixture is constantly changing as more or less oxygen is brought into the flames and more or less heat is generated, producing more or fewer vapors and flammable gases.

Remove the fire's access to fuel or remove the oxygen, and the fire dies. Although a spark, flame, or heat may start a fire, the heat that a fire produces is necessary to sustain it. Therefore, a fire may be extinguished by removing the fuel source, starving it of oxygen, or cooling it below the combustion point. Even in an oxygen-rich, combustible environment, such as a hospital oxygen tent, fire can be avoided by controlling heat and eliminating sparks and open flames (see Figure 15-2). The broken lines in the tetrahedron and the triangle indicate that the necessary elements are removed.

An **explosion** is a very rapid, contained fire. When the gases produced exceed the pressure capacity of the vessel, a rupture or explosion must result. The simplest example

FIGURE 15-2 The broken tetrahedron and broken triangle.

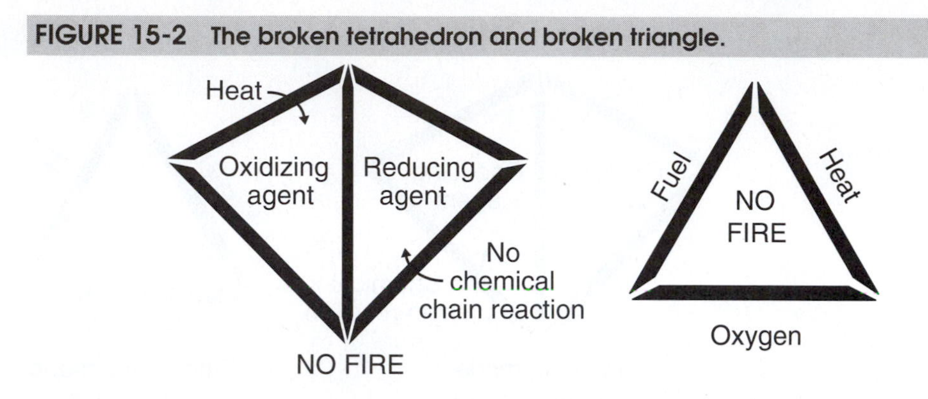

is a firecracker. The fuse, which usually contains its own source of oxygen, burns into the center of a firecracker. The surrounding powder ignites, and the heat produced vaporizes the balance of the explosive material and ignites it. The tightly wrapped paper of the firecracker cannot contain the expanding gases. The firecracker explodes, in much less time than was required to read about it.

Heat always flows from a higher temperature to a lower temperature, never from a lower temperature to a higher temperature without an outside force being applied. Fires generate heat, which is necessary to sustain the fire. Excess heat is then transferred to surrounding objects, which may ignite, explode, or decompose. **Heat transfer** is accomplished by three means, usually simultaneously: (1) conduction, (2) radiation, and (3) convection.

Conduction is direct thermal energy transfer. On a molecular level, materials near a source of heat absorb the heat, raising their kinetic energy. **Kinetic energy** is the energy resulting from a moving object. Energy in the form of heat is transferred from one molecule to the next. Materials conduct heat at varying rates. Metals are very good conductors of heat. Concrete and plastics are poor conductors, hence good insulators. Nevertheless, a heat buildup on one side of a wall will transfer to the other side of the wall by conduction.

Radiation is electromagnetic wave transfer of heat to a solid. Waves travel in all directions from the fire and may be reflected off a surface, as well as absorbed by it. Absorbed heat may raise the temperature beyond a material's combustion point, and then a fire erupts. Heat may also be conducted through a vessel to its contents, which will expand and may explode. An example is the spread of fire through an oil tank field. A fire in one tank can spread to nearby tanks through radiated heat, raising the temperature and pressure of the other tank contents.

Convection is heat transfer through the movement of hot gases. The gases may be the direct products of fire, the results of chemical reaction, or additional gases brought to the fire by the movement of air and heated at the fire surfaces by conduction. Convection determines the general direction of the spread of a fire. Convection causes fires to rise as heat rises and move in the direction of the prevailing air currents.

All three forms of heat transfer are present at a campfire. A metal poker left in a fire gets red hot at the flame end. Heat is conducted up the handle, which gets progressively hotter until the opposite end of the poker is too hot to touch. People around the fire are warmed principally by radiation, but only on the side facing the fire. People farther away from the fire will be warmer on the side facing the fire than the backs of people closer to the fire. Marshmallows toasted above the flames are heated by convection (see Figure 15-3).

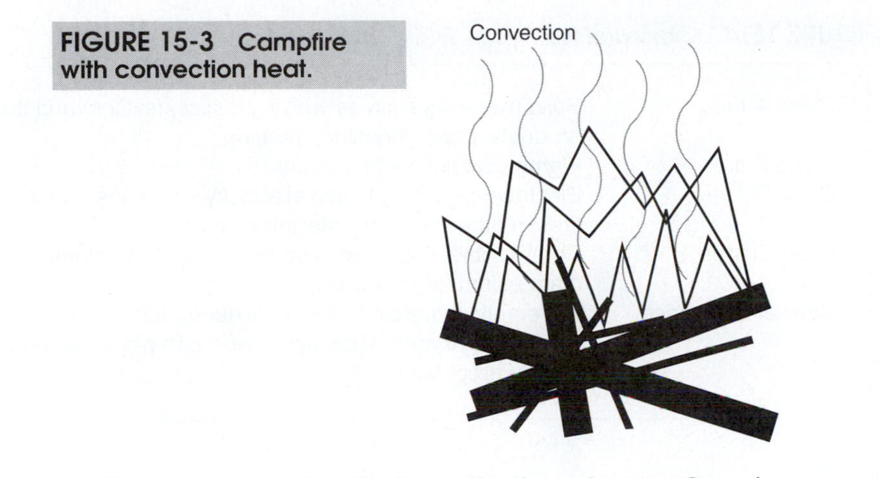

FIGURE 15-3 Campfire with convection heat.

Spontaneous combustion is rare, but it can happen. Organic compounds decompose through natural chemical processes. As they degrade, they release methane gas (natural gas), an excellent fuel. The degradation process—a chemical reaction—produces heat. In a forest, the concentrations of decomposing matter are relatively minimal, and both the gas and the heat vent naturally.

A classic example of spontaneous combustion is a pile of oil-soaked rags. A container of oil seldom ignites spontaneously. A collection of clean fabrics seldom bursts into flames. Rags soaking completely within oil are usually safe. One oil-soaked rag is unlikely to cause a problem. However, in a pile of oil-soaked rags—especially in a closed container—the chemistry is quite different.

The fibers of the rags expose a large surface area of oil to oxidation. The porous nature of rags allows additional oxygen to be absorbed, replacing the oxygen already consumed. When the temperature rises sufficiently, the surfaces of the oil on the rags vaporize.

Hypergolic reactions occur when mixing fuels. Oxidizers produce just such a rapid heat buildup, causing immediate combustion at room temperature with no apparent source of ignition. Although the term *hypergolic* originated with rocket propellants, the phenomenon has been around for a long time. **Pyrophor hypergolic fuels** are those that self-ignite in the presence of oxygen found at normal atmospheric concentrations. One example is white phosphorus, which is kept underwater. If it starts to dry out, the phosphorus erupts in flames.

Sources of Fire Hazards

Almost everything in an industrial environment can burn. Metal furniture, machines, plaster, and concrete block walls are usually painted. Most paints and lacquers will easily catch fire. Oxygen is almost always present. Therefore, the principal method of fire suppression is passive—the absence of sufficient heat. Within our environment, various conditions elevate the risk of fire and so are termed *fire hazards*.

For identification, fires are classified according to their properties, which relate to the nature of the fuel. The properties of the fuel directly correspond to the best means of combating a fire (see Figure 15-4).

Without a source of fuel, there is no fire hazard. However, almost everything in our environment can be a fuel. Fuels occur as solids, liquids, vapors, and gases.

FIGURE 15-4 Classes of fire.

Class A fires	Solid materials such as wood, plastics, textiles, and their products: paper, housing, clothing.
Class B fires	Flammable liquids and gases.
Class C fires	Electrical (referring to live electricity situations, not including fires in other materials started by electricity).
Class D fires	Combustible, easily oxidized metals such as aluminum, magnesium, titanium, and zirconium.
Special categories	Extremely active oxidizers or mixtures, flammables containing oxygen, nitric acid, hydrogen peroxide, and solid missile propellants.

Solid fuels include wood, building decorations and furnishings such as fabric curtains and wall coverings, and synthetics used in furniture. What would an office be without paper? What would most factories be without cardboard and packing materials such as Styrofoam molds and panels, shredded or crumpled papers, bubble wrap, and shrink wrap? All these materials easily burn.

Few solid fuels are, or can be made, fireproof. Even fire walls do not stop fires, although they are defined by their ability to slow the spread of fire. Wood and textiles can be treated with fire- or flame-retardant chemicals to reduce their flammability.

Solid fuels are involved in most industrial fires, but mishandling flammable liquids and flammable gases is a major cause of industrial fires. Two often-confused terms applied to flammable liquids are *flash point* and *fire point*. The **flash point** is the lowest temperature for a given fuel at which vapors are produced in sufficient concentrations to flash in the presence of a source of ignition. The **fire point** is the minimum temperature at which the vapors continue to burn, given a source of ignition. The **auto-ignition temperature** is the lowest point at which the vapors of a liquid or solid self-ignite *without* a source of ignition.

Flammable liquids have a flash point below 37.7°C (100°F). **Combustible liquids** have a flash point at or higher than that. Both flammable and combustible liquids are further divided into the three classifications shown in Figure 15-5.

As the temperature of any flammable liquid increases, the amount of vapor generated on the surface also increases. Safe handling, therefore, requires both a knowledge of the properties of the liquid and an awareness of ambient temperatures in the work or storage place. The **explosive range**, or **flammable range**, defines the concentrations of a vapor or gas in air that can ignite from a source. The auto-ignition temperature is the lowest temperature at which liquids spontaneously ignite.

Most flammable liquids are lighter than water. If the flammable liquid is lighter than water, water cannot be used to put out the fire.[4] The application of water floats the fuel and spreads a gasoline fire. Crude oil fires burn even while floating on fresh or sea water.

Unlike solids (which have a definite shape and location) and unlike liquids (which have a definite volume and are heavier than air), gases have no shape. Gases expand to fill the volume of the container in which they are enclosed, and they are frequently lighter than air. Released into air, gas concentrations are difficult to monitor due to the

FIGURE 15-5 Classes of flammable and combustible liquids.

Flammable Liquids

Class I–A	Flash point below 73°F, boiling point below 100°F.
Class I–B	Flash point below 73°F, boiling point at or above 100°F.
Class I–C	Flash point at or above 73°F, but below 100°F.

Combustible Liquids

Class II	Flash point at or above 100°F, but below 140°F.
Class III–A	Flash point at or above 140°F, but below 200°F.
Class III–B	Flash point at or above 200°F.

changing factors of air, current direction, and temperature. Gases may stratify in layers of differing concentrations but often collect near the top of whatever container in which they are enclosed. Concentrations found to be safe when sampled at workbench level may be close to, or exceed, flammability limits if sampled just above head height.

The products of combustion are gases, flame (light), heat, and smoke. Smoke is a combination of gases, air, and suspended particles, which are the products of incomplete combustion. Many of the gases present in smoke and at a fire site are toxic to humans. Other, usually nontoxic, gases may replace the oxygen normally present in air. Most fatalities associated with fire are from breathing toxic gases and smoke and from being suffocated because of oxygen deprivation. Gases that may be produced by a fire include acrolein, ammonia, carbon monoxide, carbon dioxide, hydrogen bromide, hydrogen cyanide, hydrogen chloride, hydrogen sulfide, sulfur dioxide, and nitrogen dioxide. Released gases are capable of traveling across a room and randomly finding a spark, flame, or adequate heat source, **flashing back** to the source of the gas.

The National Fire Protection Association (NFPA) has devised the NFPA 704 system for quick identification of hazards presented when substances burn (see Figure 15-6). The NFPA's red, blue, yellow, and white diamond is used on product labels, shipping cartons, and buildings. Ratings within each category are 0 to 4, where zero represents

FIGURE 15-6 Identification of fire hazards.

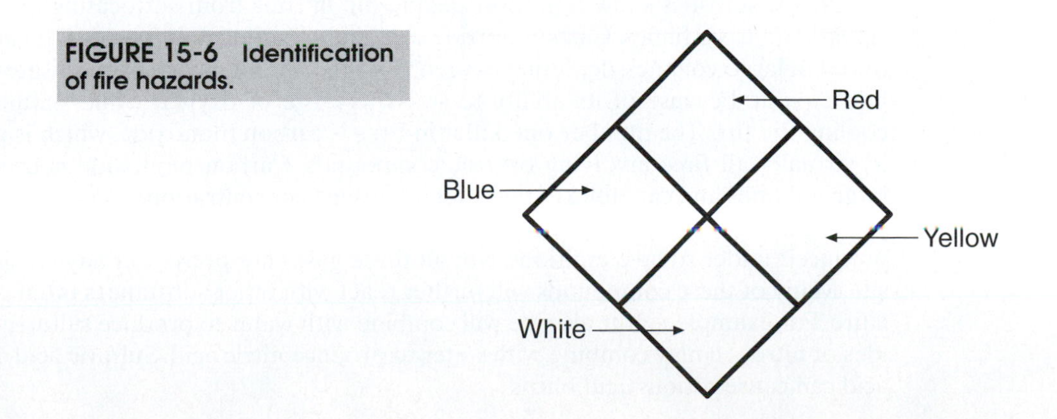

no hazard; 4, the most severe hazard level. The colors refer to a specific category of hazard:

Red = Flammability
Blue = Health
Yellow = Reactivity
White = Special information

Although we do not think of electricity as burning, natural and generated electricity play a large role in causing fires. Lightning strikes cause many fires every year. In the presence of a flammable gas or liquid mixture, one spark can produce a fire.

Electrical lines and equipment can cause fires either by a short circuit that provides an ignition spark, by arcs, or by resistances generating a heat buildup. Electrical switches and relays commonly arc as contact is made or broken.

Another source of ignition is heat in the form of hot surfaces. It is easy to see the flame hazard present when cooking oil is poured on a very hot grill. The wooden broom handle leaning up against the side of a hot oven may not be as obvious a hazard. Irons used in textile manufacturing and dry-cleaning plants also pose a heat hazard.

Space heaters frequently have hot sides, tops, backs, and bottoms, in addition to the heat-generating face. Hotplates, coffee pots, and coffee makers often create heated surfaces. Many types of electric lighting generate heat, which is transferred to the lamp housing.

Engines produce heat, especially in their exhaust systems. Compressors produce heat through friction, which is transferred to their housings. Boilers produce hot surfaces, as do steam lines and equipment using steam as power. Radiators, pipes, flues, and chimneys all have hot surfaces. Metal stock that has been cut by a blade heats up as the blade does. Surfaces exposed to direct sunlight become hot surfaces and transmit their heat by conduction to their other side. Heated surfaces are a potential source of fire.

Fire Dangers to Humans

Direct contact with flame is obviously dangerous to humans. Flesh burns, as do muscles and internal organs. The fact that we are 80 percent water, by some estimations, does not mitigate the fact that virtually all the other 20 percent burns. Nevertheless, burns are not the major cause of death in a fire.

NFPA statistics show that most people die in fires from suffocating or breathing smoke and toxic fumes. Carbon dioxide can lead to suffocation because it can be produced in large volumes, depleting oxygen from the air. Many fire extinguishers use carbon dioxide because of its ability to starve the fire of oxygen while simultaneously cooling the fire. The number one killer in fires is carbon monoxide, which is produced in virtually all fires involving organic compounds. Carbon monoxide is produced in large volumes and can quickly reach lethal dosage concentrations.

Figure 15-7 shows the major chemical products of combustion. Other gases may be produced under some conditions. Not all these gases are present at any particular fire site. Many of these compounds will further react with other substances often present at a fire. For example, sulfur dioxide will combine with water to produce sulfuric acid. Oxides of nitrogen may combine with water to produce nitric acid. Sulfuric acid and nitric acid can cause serious acid burns.

FIGURE 15-7 Major chemical products of combustion.

Product	Fuels	Pathology
Acrolein	Cellulose, fatty substances, woods, and paints	Highly toxic irritant to eyes and respiratory system
Ammonia (NH_3)	Wool, silk, nylon, melamine, refrigerants, hydrogen–nitrogen compounds	Somewhat toxic irritant to eyes and respiratory system
Carbon dioxide (CO_2)	All carbon and organic compounds	Not toxic, but depletes available oxygen
Carbon monoxide (CO)	All carbon and organic compounds	Can be deadly
Hydrogen chloride (HCl)	Wool, silk, nylon, paper, polyurethane, rubber, leather, plastic, wood	Quickly lethal asphyxiant
Hydrogen sulfide (H_2S)	Sulfur-containing compounds, rubber, crude oil	Highly toxic gas; strong odor of rotten eggs, but quickly destroys sense of smell
Nitrogen dioxide (NO_2)	Cellulose nitrate, celluloid, textiles, other nitrogen oxides	Lung irritant, causing death or damage
Sulfur dioxide (SO_2)	Sulfur and sulfur-containing compounds	Toxic irritant

Detection of Fire Hazards

Many automatic fire detection systems are used in industry today. Many systems can warn of the presence of smoke, radiation, elevated temperature, or increased light intensity. **Thermal expansion detectors** use a heat-sensitive metal link that melts at a predetermined temperature to make contact and ultimately sound an alarm. Heat-sensitive insulation can be used, which melts at a predetermined temperature, thereby initiating a short circuit and activating the alarm.

Photoelectric fire sensors detect changes in infrared energy that is radiated by smoke, often by the smoke particles obscuring the photoelectric beam. A relay is open under acceptable conditions and closed to complete the alarm circuit when smoke interferes.

Ionization or **radiation sensors** use the tendency of a radioactive substance to ionize when exposed to smoke. The substance becomes electrically conductive with the smoke exposure and permits the alarm circuit to be completed.

Ultraviolet or **infrared detectors** sound an alarm when the radiation from fire flames is detected. When rapid changes in radiation intensities are detected, a fire alarm signal is given.

OSHA has mandated the monthly and annual inspection and recording of the condition of fire extinguishers in industrial settings. A hydrostatic test to determine the integrity of the fire extinguisher metal shell is recommended according to the type of fire extinguisher. The hydrostatic test measures the capability of the shell to contain internal pressures and the pressure shifts expected to be encountered during a fire.

Reduction of Fire Hazards

The best way to reduce fires is to prevent them. A major cause of industrial fires is hot, poorly insulated machinery and processes. One means of reducing a fire hazard is the isolation of the three triangle elements: fuel, oxygen, and heat. In the case of fluids, closing a valve may stop the fuel element.

Fires may also be prevented by the proper storage of flammable liquids. Liquids should be stored as follows:

- In flame-resistant buildings that are isolated from places where people work. Proper drainage and venting should be provided for such buildings.
- In tanks below ground level.
- On the first floor of multistory buildings.

Substituting less-flammable materials is another effective technique for fire reduction. A catalyst or fire inhibitor can be employed to create an endothermic energy state that eventually smothers the fire. Several ignition sources can be eliminated or isolated from fuels:

- Prohibit smoking near any possible fuels.
- Store fuels away from areas where electrical sparks from equipment, wiring, or lightning may occur.
- Keep fuels separate from areas where there are open flames. These may include welding torches, heating elements, or furnaces.
- Isolate fuels from tools or equipment that may produce mechanical or static sparks.

Other strategies for reducing the risk of fires are as follows:

- Clean up spills of flammable liquids as soon as they occur. Properly dispose of the materials used in the cleanup.
- Keep work areas free from extra supplies of flammable materials (e.g., paper, rags, boxes, and so on). Have only what is needed on hand with the remaining inventory properly stored.
- Run electrical cords along walls rather than across aisles or in other trafficked areas. Cords that are walked on can become frayed and dangerous.
- Turn off the power and completely deenergize equipment before conducting maintenance procedures.
- Don't use spark- or friction-prone tools near combustible materials.
- Routinely test fire extinguishers.

Fire Extinguishing Systems

In larger or isolated industrial facilities, an employee fire brigade may be created. (See OSHA requirements in next section.) **Standpipe and hose systems** provide the hose and pressurized water for firefighting. Hoses for these systems usually vary from 1 inch to 2.5 inches in diameter.[5]

Automatic sprinkler systems are an example of a fixed extinguishing system because the sprinklers are fixed in position. Water is the most common fluid released from the sprinklers. Sprinkler supply pipes may be kept filled with water in heated buildings; in warmer climates, valves are used to fill the pipes with water when the sprinklers are activated. When a predetermined heat threshold is breached, water flows to the heads and is released from the sprinklers.

FIGURE 15-8 Fire extinguisher characteristics.

Fire Class	Extinguisher Contents	Mechanism	Disadvantages
A	Foam, water, dry chemical	Chain-breaking cooling, smothering, and diluting	Freezes if not kept heated.
B	Dry chemical, bromotrifluoromethane, and other halogenated compounds, foam, CO_2, dry chemical	Chain-breaking smothering, cooling, and shielding	Halogenated compounds are toxic.
C	Bromotrifluoromethane, CO_2, dry chemical	Chain-breaking smothering, cooling, and shielding	Halogenated compounds are toxic; fires may ignite after CO_2 dissipates.
D	Specialized powders such as graphite, sand, limestone, soda ash, sodium chloride	Cooling, smothering	Expensive cover of powder may be broken with resultant reignition.

Portable fire extinguishers are classified by the types of fire that they can most effectively reduce. Figure 15-8 describes the four major fire extinguisher classifications. Blocking or shielding the spread of fire can include covering the fire with an inert foam, inert powder, nonflammable gas, or water with a thickening agent added. The fire may suffocate under such a covering. Flooding a liquid fuel with nonflammable liquid can dilute this fire element. Figures 15-9, 15-10, and 15-11 are photographs of effective fire prevention equipment.

FIGURE 15-9 Fireproof storage cabinet.

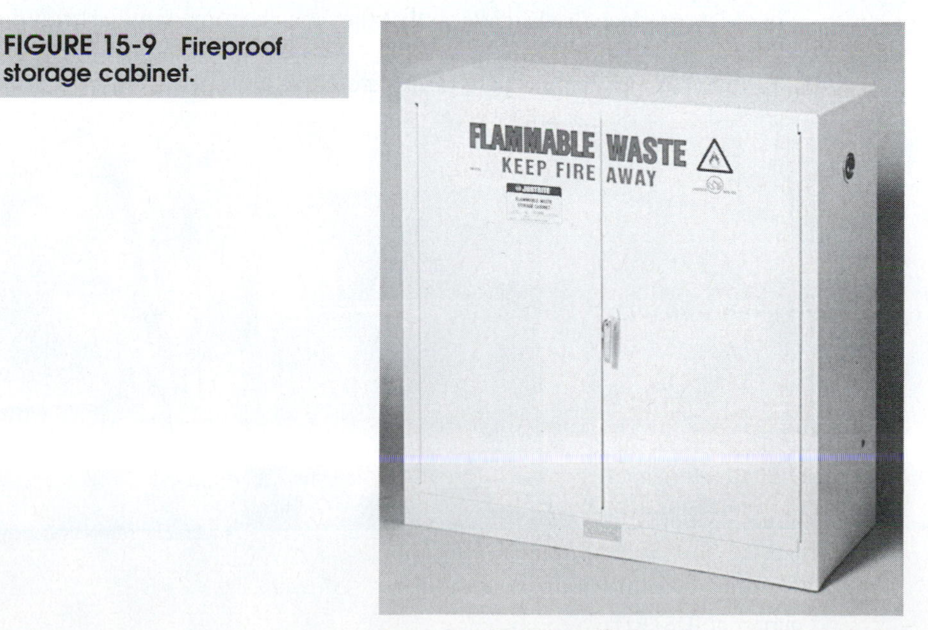

Courtesy of JUSTRITE®.

FIGURE 15-10 Fire-protective drums on a poly spill pallet.

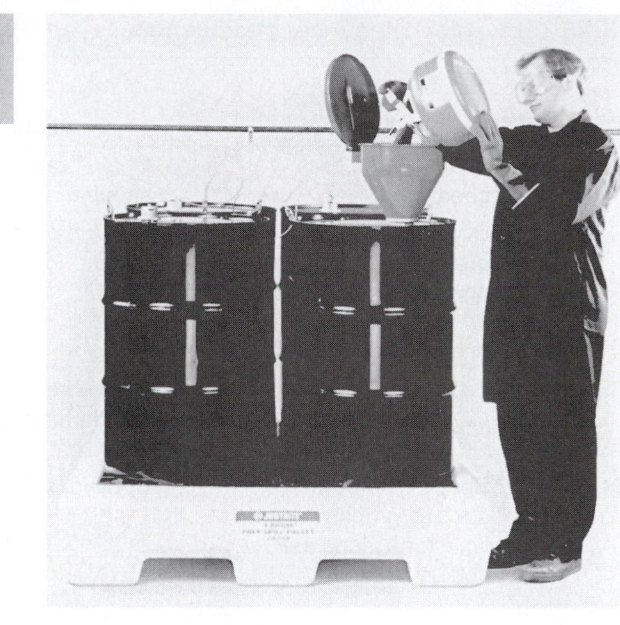

Courtesy of JUSTRITE®.

OSHA Regulations for Fire Brigades

Fire brigade regulations are covered in 29 CFR 1910.156 (Subpart L, Appendix A). Relevant requirements from the regulations are as follows:

1. **Scope.** Employers are not required to form a fire brigade. However, if an employer does decide to organize a fire brigade, the requirements of this section apply.
2. **Prefire planning.** Have prefire planning conducted by the local fire department or the workplace fire brigade so that they may become familiar with the workplace

FIGURE 15-11 Fire-protective drums stored in an outside storage cabinet.

Courtesy of JUSTRITE®.

and process hazards. Involvement with the local fire department or fire prevention bureau is encouraged to facilitate coordination and cooperation between members of the fire brigade and those who may be called upon for assistance during a fire emergency.

3. **Organizational statement.** The organizational statement should contain the following information: a description of the duties that the fire brigade members are expected to perform; the line authority of each fire brigade officer; the number of the fire brigade officers and number of training instructors; and a list and description of the types of awards or recognition that brigade members may be eligible to receive.
4. **Physical capability.** The physical capability requirement applies only to those fire brigade members who perform interior structural firefighting. Employees who cannot meet the physical capability requirement may still be members of the fire brigade as long as such employees do not perform interior structural firefighting. It is suggested that fire brigade members who are unable to perform interior structural firefighting be assigned less stressful and physically demanding fire brigade duties (e.g., certain types of training, record keeping, fire prevention inspection and maintenance, and fire pump operations).

   "Physically capable" can be defined as being able to perform those duties specified in the training requirements of Section 1910.156(c). Physical capability can also be determined by physical performance tests or by a physical examination when the examining physician is aware of the duties that the fire brigade member is expected to perform.
5. **Training and education.** Training and education must be commensurate with those functions that the fire brigade is expected to perform (i.e., those functions specified in the organizational statement). Such a performance requirement provides the necessary flexibility to design a training program that meets the needs of individual fire brigades.

   At a minimum, hands-on training is required to be conducted annually for all fire brigade members. However, for those fire brigade members who are expected to perform interior structural firefighting, some type of training or education session must be provided at least quarterly.
6. **Firefighting equipment.** It is important to remove from service and replace any firefighting equipment that is damaged or unserviceable. This prevents fire brigade members from using unsafe equipment by mistake.

   Firefighting equipment, except portable fire extinguishers and respirators, must be inspected at least annually. Portable fire extinguishers and respirators are required to be inspected at least monthly.
7. **Protective clothing.** Paragraph (e) of 1910.156 does not require all fire brigade members to wear protective clothing. It is not the intention of the standards to require employers to provide a full ensemble of protective clothing for every fire brigade member without consideration given to the types of hazardous environments to which the fire brigade member may be exposed. It is the intention of the standards to require adequate protection for those fire brigade members who may be exposed to fires in an advanced stage, smoke, toxic gases, and high temperatures. Therefore, the protective clothing requirements apply only to those fire brigade members who perform interior structural firefighting operations.

Additionally, the protective clothing requirements do not apply to the protective clothing worn during outside firefighting operations (brush and forest fires, crash crew operations) or other special firefighting activities. It is important that the protective clothing to be worn during these types of firefighting operations reflect the hazards that may be encountered by fire brigade members.

8. **Respiratory protective devices.** Respiratory protection is required to be worn by fire brigade members while working inside buildings or confined spaces where toxic products of combustion or an oxygen deficiency are likely to be present; respirators are also to be worn during emergency situations involving toxic substances. When fire brigade members respond to emergency situations, they may be exposed to unknown contaminants in unknown concentrations. Therefore, it is imperative that fire brigade members wear proper respiratory protective devices during these situations. Additionally, there are many instances where toxic products of combustion are still present during mop-up and overhaul operations. Therefore, fire brigade members should continue to wear respirators during these types of operations.[6]

Disaster Preparations

Training employees may be the most successful lifesaving preparation for a fire disaster. Company fire brigade members should be trained and tested at least quarterly. Disaster preparation initially requires management commitment and planning and continued response and recovery practice by the fire brigade on a regular basis. Also necessary are regular, but less frequent, fire drills for all personnel.

Disaster preparations also include the integration of company planning with community plans. Community disaster relief agencies such as the police, fire department, Red Cross, and hospitals should be consulted and informed of company disaster preparation plans.

Preventing Office Fires

The shop floor is not the only part of the plant where fire hazards exist. Offices are also susceptible to fires. According to Vogel, "Every year about 7,000 fires occur in office buildings, which cause injuries, deaths, and millions of dollars in fire damages."[7] The following strategies are helpful in preventing office fires:

- Confine smoking to designated areas that are equipped with nontip ashtrays and fire-resistant furnishings.
- Periodically check electrical circuits and connections. Replace frayed or worn cords immediately.
- Make sure that extension cords and other accessories are UL-approved and used only as recommended.
- Make sure there is plenty of air space left around copying machines and other office machines that can overheat.
- Locate heat-producing appliances away from the wall or anything else that can ignite.
- Frequently inspect personal appliances such as hotplates, coffee pots, and cup warmers. Assign responsibility for turning off such appliances every day to a specific person.
- Keep aisles, stairwells, and exits clear of paper, boxes, and other combustible materials.[8]

Development of Fire Safety Standards

The purpose of modern fire safety standards is the protection of life and the prevention of property damage. However, the impetus for developing standards has always been and continues to be the occurrence of major disasters. Typically, standards are developed after a major tragedy occurs in which property is damaged on a large scale and lives are lost. Public shock turns into an outcry for action. A flurry of political activity follows, and agencies and organizations that develop standards are called on to develop new standards.

According to Arthur E. Cote,

> There are approximately 89,000 standards in the United States. Of these, 50,000 of them actually are procurement specifications developed and used by the government. The remaining 39,000 are voluntary standards developed in the U.S. Most codes and standards in the fire protection field are developed by three organizations: the National Fire Protection Association (NFPA), the American Society for Testing Materials (ASTM), and Underwriter's Laboratories (UL). In addition, two model code organizations—the International Conference of Building Officials and Code Administration (BOCA), and the Southern Building Code Congress International (SBCC)—develop model building and fire prevention codes.[9]

The trend in fire safety standards is toward performance-based standards and away from the traditional specification-based approach. An example of each type of standard will help illustrate the difference. A specification-based standard may require that brick, concrete, or steel material be used in a given type of building. A performance-based standard may specify that materials used have a one-, two-, or four-hour fire resistance rating.[10] Advances in the testing of engineering materials will help overcome most of the barriers to full development and implementation of performance-based standards.

Cote summarizes his views on the future of fire safety standards as follows:

> Codes and standards will survive in the 21st century. They may, however, be considerably different from the codes and standards we now have. They certainly will be based more on standardized fire tests, models, data, and related science and engineering than on consensus judgment. How much more will depend on the extent to which there is widespread acceptance of the anticipated breakthroughs in fire science and in related modeling and calculation methods.[11]

OSHA Fire Standards

OSHA standards for fire protection appear in 29 CFR 1910.156 (Subpart L). This subpart contains the standards for fire brigades, fixed fire suppression equipment, and other fire protection systems. Employers are not required to form fire brigades, but those who choose to do so must meet a number of specific requirements. There are other fire-related requirements in other subparts. For example, fire exits, emergency action plans, and means of egress are covered in Subpart E. The standards in Subpart L are as follows:

FIRE PROTECTION

1910.155 Scope, application, and definitions
1910.156 Fire brigades

FIGURE 15-12 Fire prevention and suppression summary.

- Use the least-flammable materials whenever possible.
- Analyze the company to determine types of potential fires and provide appropriate sprinklers and/or extinguishers.
- Develop a database of the flammability of materials available in the company.
- Store containers or flammable materials away from sources of heat or sparks and away from humans.
- Do not permit smoking near flammable materials.
- Include a venting mechanism in storage containers and locate them near a drain.
- Minimize fuel storage container size to reduce the size of the fire that may involve those fuels.
- Isolate fuels from sources of heat.
- Include a smoke detection system and portable fire extinguishers in the facility. Extinguishers should be easily available to every workstation.
- Make sure sources of heat have controlling mechanisms and are near fire detection equipment.
- Check fire extinguishing equipment regularly.
- Perform periodic inspections for fire hazards and reappraisal of fire hazards.
- Train plant personnel in basic fire prevention, which should include periodic fire drills.
- Make sure fire brigade personnel are well trained, tested, and regularly practice fire control.
- Stress cleanliness and an organized method of disposal of flammable materials.

PORTABLE FIRE SUPPRESSION EQUIPMENT

1910.157 Portable fire extinguishers
1910.158 Standpipe and hose systems

FIXED FIRE SUPPRESSION EQUIPMENT

1910.159 Automatic sprinkler systems
1910.160 Fixed extinguishing systems, general
1910.161 Fixed extinguishing systems, dry chemical
1910.162 Fixed extinguishing systems, gaseous agent
1910.163 Fixed extinguishing systems, water spray and foam

OTHER FIRE PROTECTION SYSTEMS

1910.164 Fire detection systems
1910.165 Employee alarm systems

Figure 15-12 summarizes fire prevention and suppression strategies.

Life Safety

Life safety involves protecting the vehicles, vessels, and lives of people in buildings and structures from fire. The primary reference source for life safety is the *Life Safety Code,* published by the National Fire Protection Association. The code applies to new and existing buildings. It addresses the construction, protection, and occupancy features necessary to minimize the hazards of fire, smoke, fumes, and panic. A major part of the code is devoted to the minimum requirements for design of egress necessary to ensure that occupants can quickly evacuate a building or structure.

Basic Requirements

In this section, the term *structure* refers to a structure or building.

- Every structure, new and existing, that is to be occupied by people must have a **means of egress** and other fire protection safeguards that (1) ensure that occupants can promptly evacuate or be adequately protected without evacuating; and (2) provide sufficient backup safeguards to ensure that human life is not endangered if one system fails.
- Every structure must be constructed or renovated, maintained, and operated in such a way that occupants are (1) protected from fire, smoke, or fumes; (2) protected from fire-related panic; (3) protected long enough to allow a reasonable amount of time for evacuation; and (4) protected long enough to defend themselves without evacuating.
- In providing structures with means of egress and other fire protection safeguards, the following factors must be considered: (1) character of the occupancy, (2) capabilities of occupants, (3) number of occupants, (4) available fire protection, (5) height of the structure, (6) type of construction, and (7) any other applicable concerns.
- No lock or other device may be allowed to obstruct egress in any part of a structure at any time that it is occupied. The only exceptions to this requirement are mental health detention and correctional facilities. In these, the following criteria are required: (1) responsible personnel must be available to act in the case of fire or a similar emergency; and (2) procedures must be in place to ensure that occupants are evacuated in the event of an emergency.
- All exits in structures must satisfy the following criteria: (1) be clearly visible or marked in such a way that an unimpaired individual can readily discern the route of escape; (2) all routes to a place of safety must be arranged or clearly marked; (3) any doorway and passageway that may be mistaken as a route to safety must be arranged or clearly marked in such a way as to prevent confusion in an emergency; and (4) all appropriate steps must be taken to ensure that occupants do not mistakenly enter a dead-end passageway.
- Egress routes and facilities must be included in the lighting design wherever artificial illumination is required in a structure.
- Fire alarm systems must be provided in any facility that is large enough or so arranged that a fire itself may not adequately warn occupants of the danger. Fire alarms should alert occupants to initiate appropriate emergency procedures.
- In any structure or portion of a structure in which a single means of egress may be blocked or overcrowded in an emergency situation, at least two means of egress must be provided. The two means of egress must be arranged in such a way as to minimize the possibility of both becoming impassable in the same emergency situation.
- All stairs, ramps, and other means of moving from floor to floor must be enclosed (or otherwise protected) to afford occupants protection when used as a means of

egress in an emergency situation. These means of vertical movement should also serve to inhibit the spread of fire, fumes, and smoke from floor to floor.

- Compliance with the requirements summarized herein does not eliminate or reduce the need to take other precautions to protect occupants from fire hazards, nor does it permit the acceptance of any condition that could be hazardous under normal occupancy conditions.[12]

The information in this section is a summary of the broad fundamental requirements of the *Life Safety Code* of the NFPA. More specific requirements relating to means of egress and features of fire protection are explained in the following sections.

Means of Egress

This section explains some of the more important issues in the *Life Safety Code* relating to means of egress. Students and practitioners who need more detailed information are encouraged to refer to the *Life Safety Code*.

1. **Doors.** Doors that serve as exits must be designed, constructed, and maintained in such a way that the means of egress is direct and obvious. Windows that could be mistaken for doors in an emergency situation must be made inaccessible to occupants.
2. **Capacity of means of egress.** The means of egress must have a capacity sufficient to accommodate the occupant load of the structure calculated in accordance with the requirements of the *Life Safety Code*.
3. **Number of means of egress.** Any component of a structure must have a minimum of two means of egress (with exceptions as set forth in the code). The minimum number of means of egress from any story or any part of a story is three for occupancy loads of 500 to 1,000 and four for occupancy loads of more than 1,000.
4. **Arrangement of means of egress.** All exits must be easily accessible at all times in terms of both location and arrangement.
5. **Measurement of travel distance to exits.** The travel distance to at least one exit must be measured on the walking surface along a natural path of travel beginning at the most remote occupied space and ending at the center of the exit. Distances must comply with the code.
6. **Discharge from exits.** All exits from a structure must terminate at a public way or at yards, courts, or open spaces that lead to the exterior of the structure.
7. **Illumination of means of egress.** All means of egress shall be illuminated continuously during times when the structure is occupied. Artificial lighting must be used as required to maintain the necessary level of illumination. Illumination must be arranged in such a way that no area is left in darkness by a single lighting failure.
8. **Emergency lighting.** Emergency lighting for all means of egress must be provided in accordance with the code. In cases where maintaining the required illumination depends on changing from one source of power to another, there shall be no appreciable interruption of lighting.
9. **Marking of means of egress.** Exits must be marked by readily visible, approved signs in all cases where the means of egress is not obviously apparent to occupants. No point in the exit access corridor shall be more than 100 feet from the nearest sign.

10. **Special provisions for high hazard areas.** If an area contains contents that are classified as highly hazardous, occupants must be able to exit by traveling no more than 75 feet. At least two means of egress must be provided, and there shall be no dead-end corridors.[13]

The requirements summarized in this section relate to the fundamental specifications of the *Life Safety Code* relating to means of egress. For more detailed information concerning general requirements, means of egress, and other factors such as fire protection and fire protection equipment, refer to the actual code.

Flame-Resistant Clothing

For employees who work in jobs in which flames or electric arcs may occur, wearing flame-resistant clothing can be a lifesaver.[14] Electric arcs are the result of electricity passing through ionized air. Although electric arcs last for only a few seconds, during that time they can produce extremely high levels of heat and flash flame.

OSHA's standards relating to flame-resistant clothing are found in CFR 1910.269, paragraph 1. Key elements of paragraph 1 explain the employer's responsibilities regarding personal protective equipment and flame-resistant clothing.

Apparel

CFR 1910.269, paragraph 1(6) reads as follows:

i. When work is performed within reaching distance of exposed energized parts of equipment, the employer shall ensure that each employee removes or renders nonconductive all exposed conductive articles, such as key or watch chains, rings, or wrist watches or bands, unless such articles do not increase the hazards associated with contact with the energized parts.

ii. The employer shall train each employee who is exposed to the hazards or flames or electric arcs in the hazards involved.

iii. The employer shall ensure that each employee who is exposed to the hazards of flames or electric arcs does not wear clothing that, when exposed to flames or electric arcs, could increase the extent of injury that would be sustained by the employee.
Note: Clothing made from the following types of fabrics, either alone or in blends, is prohibited by this paragraph, unless the employer can demonstrate that the fabric has been treated to withstand the conditions that may be encountered or that the clothing is worn in such a manner as to eliminate the hazard involved: acetate, nylon, polyester, rayon.

iv. Fuse Handling. When fuses must be installed or removed with one or both terminals energized at more than 300 volts or with exposed parts energized at more than 50 volts, the employer shall ensure that tools or gloves rated for the voltage are used. When expulsion-type fuses are installed with one or both terminals energized at more than 300 volts, the employer shall ensure that each employee wears eye protection meeting the requirements of Subpart I of this Part, uses a tool rated for the voltage, and is clear of the exhaust path of the fuse barrel.[15]

Fire Safety Programs

Organizations that are interested in protecting their employees from fire hazards should remember the Boy Scouts' motto: Be prepared. The best way to be prepared is to establish a comprehensive fire safety program that encompasses all the functional activities required for being prepared.[16] A comprehensive fire safety program should have at least the following components: assessment, planning, awareness/prevention, and response.

An effective way to develop, implement, and maintain a comprehensive fire safety program is to establish a cross-functional fire safety committee. "Cross-functional" means that it should have members from all the organization's various functional units. It should also have at least one executive-level manager to ensure and demonstrate that level of support. This approach has several advantages including the following: (1) it focuses the eyes and ears of a broad cross section of the workforce on fire safety, (2) it ensures a broad base of input, and (3) it ensures executive level commitment. This committee should be staffed and chaired by the organization's highest ranking safety and health professional.

Assessment

Assessment of the workplace for fire hazards should be continuous and ongoing. Although the organization's safety and health professional will have primary responsibility for this, committee members also need to be involved and involve the departments that they represent. Members of the safety committee should be trained in the fundamentals of fire hazard assessment by the safety and health professional. They should then pass on this knowledge to employees in their departments, units, and teams. In this way, all employees are involved in continually looking for fire hazards and communicating their concerns to the safety committee.

Planning

OSHA requires that an organization's emergency fire safety plan have at least the following components:

- Emergency escape procedures and routes
- Critical "shutdown" procedures
- Employee headcount procedures
- Rescue and medical procedures
- Procedures for reporting fires and emergencies
- Important contact personnel for additional information

Once the plan is in place, it should be reviewed at least annually and updated as necessary.

Awareness and Prevention

After the fire safety committee has completed the emergency plan and upper management has approved it, employees must become acquainted with it. All employees should receive awareness training so that they understand their role in carrying out the emergency plan. The fire safety committee should evaluate the training program periodically, using guidelines such as the following:

- Do all employees know the role they play in implementing the emergency plan?
- How are employees with disabilities provided for?

- Do all employees understand the escape plans? Evacuation procedures?
- Do all new and temporary employees receive training?
- Are all employees informed when the plan is revised?
- Is a comprehensive drill undertaken at least once each year?
- Are all employees familiar with the sound of the alarm system?
- Is the alarm system checked periodically?
- Are sufficient fire detection devices in place? Are they tested periodically?
- Do all employees know the most likely causes of fires?

Response

Accidents can happen in even the safest organizations. Therefore, it is very important that employees understand the emergency plan and periodically practice responding. Just knowing what the plan says is not sufficient. People do not always think clearly in an emergency situation. They will, however, do what they have learned to do through practice. Consequently, one of the fire safety committee's most important responsibilities is to arrange periodic drills so that employees automatically respond properly.

Explosive Hazards

Many chemical and toxic substances used in modern organizations are flammable or combustible. Consequently, under certain conditions, they can explode. Working in these conditions involves hazards that require special precautions for handling, storing, transporting, and using such substances.

Explosives-Related Concepts

Safety relating to explosive materials is a highly specialized field. This section discusses terms and concepts used in this field with which modern safety and health professionals should be familiar.

- A **flammable substance** is any substance with a flash point below 37.8°C (100°F) and a vapor pressure of less than 40 psi at that temperature. Such liquids are also known as *Class I liquids*. They tend to be compositions of hydrogen and carbon such as crude oil and its numerous by-products.
- A **combustible substance** is any substance with a flash point of 37.8°C (100°F) or higher. Such liquids are known as *Class II liquids*. They also tend to be compositions of hydrogen and carbon such as crude oil and its numerous by-products.
- The *flash point* is the lowest temperature at which a substance gives off sufficient vapors to combine with air to form an ignitable mixture. Ignition can be precipitated by a spark.
- The *auto-ignition temperature* is the lowest temperature at which a vapor-producing substance or a flammable gas ignites even without the presence of a spark or a flame. This is sometimes known as *spontaneous ignition*.
- In most cases, a certain amount of oxygen must be present in a vapor–air mixture for an explosion to occur. The amount that must be present for a given substance is the **oxygen limit** for that substance.
- **Volatility** is the evaporation (vaporization) capability of a given substance. The greater the tendency of a substance to vaporize, the more volatile it is.[17]

Common Uses of Flammable and Combustible Substances

Flammable and combustible substances are widely used in modern organizations. Therefore, the hazards associated with them are not limited to the industries producing such materials and substances. The National Safety Council lists the following as common uses of flammable and combustible substances in modern industry and specific related precautions that should be taken with each.

- **Dip tanks.** Dipping operations involving flammable or combustible substances should take place in a stand-alone, one-story building constructed of noncombustible materials. The building should be (1) well ventilated, (2) clearly marked as a hazardous area, (3) free of ignition sources, and (4) large. The dip tank itself should be covered and should contain an automatic fire extinguishing system.
- **Japanning and drying ovens.** Ovens used to evaporate varnish, japan enamel, and any other combustible substance should be (1) well ventilated, (2) equipped with an automatic fire protection system, and (3) have a shutdown system that activates automatically in the event of a fire or explosion.
- **Oil burners.** Selecting the proper type of fuel for use in an oil burner is the best precaution to prevent the accumulation and potential ignition of soot. The safest fuel to use in an oil burner is one that meets the following criteria: (1) flash point higher than 37.8°C (100°F), (2) hydrocarbon based, and (3) acid- and grit-free. In addition, the supply tank should be located outside the building housing the oil burner and should be underground. The top of the storage tank should be lower than all pipes entering it. Finally, the oil burner should have an automatic system for preventing the discharge of unburned oil into a hot firebox.
- **Cleaning solvents.** Many of the solvents used to clean metal parts are combustible. The primary precaution when cleaning metal parts is selecting substances that are not easily ignited. Additional precautions include ventilation and the selection of a cleaning area that is free of ignition sources.
- **Internal combustion engines.** Internal combustion engines are widely used in modern industry for powering such equipment as forklifts and lift trucks. Because they are typically fueled with gasoline or diesel fuel, there are fire and explosive hazards associated with this operation. Precautions include (1) proper maintenance, (2) good housekeeping, (3) shutdown of engines and cooling of exhaust pipes before filling fuel tanks, and (4) a well-ventilated area for filling fuel tanks.
- **Spray-painting booths.** The hazard associated with spray-painting booths is that an explosive mixture of paint vapor and air can occur. To prevent such occurrences, proper ventilation is critical. Regular cleaning of the booth to remove accumulated spray deposits is also important. Paint booths should be equipped with automatic fire protection systems.[18]

The preceding lists only some of the many ways that potentially explosive materials are used in modern organizations.

Other Health Hazards of Explosive Materials

The health hazards associated with explosions and fires are well known. The potential for serious injury or death from the force of a blast or from burns is very high. However, there are other hazards associated with explosive and combustible materials. These include skin irritation, intoxication, and suffocation.

Irritation can occur when the skin comes in contact with hazardous substances. The degree of irritation can range from minor to severe, depending on the type of substance, its concentration, and the duration of contact. Intoxication can occur when an employee breathes the vapors of combustible substances. This can cause impaired judgment, performance, and reaction time, which can, in turn, result in an accident. Finally, the vapors from combustible materials can accumulate in confined spaces. When this happens, the air becomes contaminated and is both toxic and explosive. In such cases, the hazard of suffocation must be added to those associated with explosives.[19]

OSHA's Firefighting Options

Even with the best fire prevention program, and even with the best engineering controls in place, it may still be necessary to manually fight fires at some point. Some companies prefer to have their employees evacuate the premises in the event of a fire. However, for some companies the potential for fires is so much a part of daily operations that they prefer to equip their employees to fight fires. Companies that either allow or expect employees to help fight fires should follow the guidelines set forth by OSHA for manual firefighting.[20]

There are three options available to companies that wish to have their employees participate in firefighting:

1. All employees are involved.
2. Only designated employees are involved.
3. Only employees who are part of an established fire brigade are involved.

Each of these options has its own set of requirements.

Option 1: All Employees Fight Fires

With this option, all employees are allowed to fight fires. However, first they are required to

1. Have and understand an emergency action plan provided by the company.
2. Have and understand a fire prevention plan provided by the company.
3. Complete annual training and refresher training concerning their duties in fighting fires and in the proper use of fire extinguishers.

Option 2: Designated Employees Fight Fires

With this option, only selected employees are allowed to fight fires. First they are required to

1. Have and understand an emergency action plan provided by the company.
2. Have and understand a fire prevention plan provided by the company.
3. Complete annual training and refresher training concerning their duties fighting fires and in how to properly use fire extinguishers.

Option 3: Fire Brigades Fight Fires

With this option, only those employees who are part of an established fire brigade are allowed to fight fires. Fire brigades are divided into two types—**incipient** and **interior**

structural. An incipient fire brigade is used to control only small fires. It requires no special protective clothing or equipment. An interior structural fire brigade may fight any type of fire provided it has been issued the appropriate protective clothing and equipment.

The requirements for each type of fire brigade are different. Employees who are part of an incipient fire brigade are required to

1. Have and understand an emergency action plan provided by the company.
2. Have and understand a fire protection plan provided by the company.
3. Have and understand an organizational statement that establishes the scope, organizational structure, training, equipment, and functions of the fire brigade.
4. Have and understand standard operating procedures for the fire brigade to follow during emergencies.
5. Complete annual training and refresher training that is hands-on in nature.

The requirements of an interior structural fire brigade are the same as those for an incipient fire brigade through the standard operating procedures. However, there are also additional requirements including

- Satisfactory completion of medical examinations that verify their fitness to participate.
- Special protective clothing and equipment of the type used by local fire departments including self-contained breathing equipment.
- Quarterly, as opposed to annual, training and retraining that is hands-on in nature.

Self-Assessment in Fire Protection

Safety and health personnel cannot be everywhere at the same time. Consequently, it is wise to enlist the assistance of supervisors and employees in fire protection. An excellent way to do this is to provide them with a self-assessment checklist that will guide them in scanning their areas of responsibility for fire hazards. Such checklists should contain at least the following questions:

1. Are portable fire extinguishers properly mounted, readily accessible, and available in adequate number and type?
2. Are fire extinguishers inspected monthly for both operability and general condition with appropriate notation made on their respective tags?
3. Are fire extinguishers recharged regularly and are the dates noted on their tags?
4. Are interior standpipes and valves inspected regularly?
5. Is the fire alarm system tested regularly?
6. Are employees trained in the proper use of fire extinguishers?
7. Are employees trained concerning under what conditions they should help fight fires and under what conditions they should evacuate?
8. Are the nearest fire hydrants flushed annually?
9. Are the nearest fire hydrants maintained regularly?
10. Are avenues and ingress and egress clearly marked?
11. Are all avenues of ingress and egress kept free of clutter and other types of obstructions?
12. Are fire doors and shutters in good working condition?

13. Are fusible links in place and readily accessible?
14. Is the local fire department familiar with the facility and any specific hazards?
15. Is the automatic sprinkler system in good working order, maintained on a regular basis, given the proper overhead clearance, and protected from inadvertent contact damage?[21]

Key Terms and Concepts

- Auto-ignition temperature
- Automatic sprinkler systems
- Carbon
- Carbon dioxide
- Carbon monoxide
- Combustible liquids
- Combustible substance
- Combustion
- Combustion point
- Conduction
- Convection
- Endothermic
- Exothermic
- Explosion
- Explosive range
- Fire
- Fire brigade
- Fire point
- Fixed extinguishing system
- Flame-resistant clothing
- Flammable gases
- Flammable liquids
- Flammable range
- Flammable substance
- Flash point
- Flashing back
- Fuel
- Heat
- Heat transfer
- Hydrogen
- Hypergolic
- Ignition temperature
- Incipient fire brigade
- Infrared detectors
- Interior structural fire brigade
- Ionization sensors
- Isolation
- Kinetic energy
- *Life Safety Code*
- Means of egress
- Oxygen
- Oxygen limit
- Photoelectric fire sensors
- Polymers
- Products of combustion
- Pyrophor hypergolic fuels
- Radiation
- Radiation sensors
- Smoke
- Source of ignition
- Spontaneous combustion
- Standpipe and hose systems
- Thermal energy
- Thermal expansion detectors
- Ultraviolet detectors
- Vapor
- Volatility

Review Questions

1. What are the three elements of the fire triangle?
2. Fire is a chemical reaction. What is going on?
3. Where is carbon found?
4. Compare and contrast carbon monoxide and carbon dioxide.
5. How is combustion of liquids and solids different from gases?
6. What are the three methods of heat transfer? Describe each.
7. What can happen to a pile of oil-soaked rags in a closed container? Describe the process.
8. In which direction does a fire normally travel?
9. Name something in this room that will not burn.
10. What are the classes of fires?
11. What property do almost all packing materials share?
12. What are the differences among flash point, fire point, and auto-ignition temperature?
13. Which are more stable: combustible liquids or flammable liquids?
14. Which way do gases usually travel?
15. Describe the NFPA hazards identification system.
16. In what four ways can electricity cause a fire?
17. What are the leading causes of fire-related deaths?
18. What are some of the toxic chemicals often produced by fires?
19. What are some of the systems utilized by smoke detectors?
20. What is the most successful life-saving preparation for a fire disaster?
21. Explain the best ways to prevent an office fire.
22. What is the trend with regard to future fire safety standards?
23. Define the term *life safety*.
24. What types of fabrics are prohibited in environments that are flame or arc prone?

25. Explain briefly OSHA's regulations for fire brigades.
26. Summarize the key components of a fire safety program.
27. Describe the precautions that should be taken for dip tanks, oil burners, and spray-painting booths.
28. Explain all three of OSHA's manual firefighting options.

Endnotes

1. "Worker Killed in Vapor Cloud Ignition at Tank Storage Facility," *NFPA Journal* 85, no. 3: 29.
2. Ibid., 29–30.
3. Ibid., 31.
4. National Fire Protection Association, *Fire Protection Handbook,* 19th ed. (Quincy, MA: 2003), 112.
5. Ibid., 132.
6. 29 CFR 1910.156 (Subpart L, Appendix A).
7. C. Vogel, "Fires Can Raze Office Buildings," *Safety & Health* 144, no. 3: 27.
8. Ibid., 26–27.
9. A. E. Cote, "Will Fire Safety Standards Survive in the 21st Century?" *NFPA Journal* 85, no. 5: 37.
10. Ibid., 42.
11. Ibid.
12. National Fire Protection Association, *Life Safety Code* (NFPA 101) (Quincy, MA: 2006), 101–119.
13. Ibid., 101–26 through 101–50.
14. National Fire Protection Association, *Life Safety Code,* 151.
15. CFR 1910.269, paragraph 1(6).
16. A. Burke, "Before the Fire," *Occupational Health & Safety* 69, no. 2: 50–52.
17. National Fire Protection Association, *Life Safety Code,* 162.
18. Ibid.
19. Ibid.
20. C. Schroll, "Manual Fire Control," *Occupational Health & Safety* 71, no. 2: 27–28.
21. Retrieved from http://online.misu.nodak.edu/19577/BADM309checklist.htm.

CHAPTER 16

Industrial Hygiene and Confined Spaces

Major Topics

- Overview of Industrial Hygiene
- Industrial Hygiene Standards
- OSH Act and Industrial Hygiene
- Hazards in the Workplace
- Toxic Substances Defined
- Entry Points for Toxic Agents
- Effects of Toxic Substances
- Relationship of Doses and Responses
- Airborne Contaminants
- Effects of Airborne Toxics
- Effects of Carcinogens
- Asbestos Hazards
- Indoor Air Quality and "Sick-Building" Syndrome
- Toxic Mold and Indoor Air Quality
- Threshold Limit Values
- Hazard Recognition and Evaluation
- Prevention and Control
- NIOSH and Industrial Hygiene
- NIOSH Guidelines for Respirators
- Standards and Regulations
- General Safety Precautions
- Confined Space Hazards
- OSHA Confined Space Standard
- OSHA Standards for Toxic and Hazardous Materials
- OSHA's Hazard Communication Standard

Industrial hygiene is an area of specialization within the broader field of industrial safety and health. This chapter provides prospective and practicing safety and health professionals with the information they need to know about this area of specialization.

Overview of Industrial Hygiene

Industrial hygiene is a safety and health profession that is concerned with predicting, recognizing, assessing, controlling, and preventing environmental stressors in the workplace that can cause sickness or serious discomfort to workers. An environmental stressor is any factor in the workplace that can cause enough discomfort to result in lost time or illness. Common stressors include gases, fumes, vapors, dusts, mists, noise, and radiation.

The Code of Ethics of the American Academy of Industrial Hygiene describes the responsibilities of industrial hygienists:

- To ensure the health of employees
- To maintain an objective approach in recognizing, assessing, controlling, and preventing health hazards regardless of outside pressure and influence
- To help employees understand the precautions that they should take to avoid health problems
- To respect employers' honesty in matters relating to industrial hygiene
- To make the health of employees a higher priority than obligations to the employer[1]

Role of the Safety and Health Professional

The role of modern safety and health professionals vis-à-vis industrial hygiene often depends on the size of the company employing them. Large companies often employ professionals who specialize in industrial hygiene. These specialists have titles such as occupational physician, industrial hygienist, industrial toxicologist, and health physicist. In smaller companies, safety and health professionals often have responsibility for all safety and health matters, including industrial hygiene.

In companies that employ specialists, their recommendations are used by safety and health professionals to develop, implement, monitor, and evaluate the overall safety and health program. If specialists are not employed, safety and health professionals are responsible for seeking the advice and assistance necessary to predict, recognize, assess, control, and overcome environmental stressors that may cause sickness or serious discomfort to employees.

Industrial Hygiene Standards

Industrial hygiene is a broad field encompassing many different areas of specialization. As such, there are a variety of different organizations and agencies that develop standards of practice relating to some aspect of the field. Safety and health professionals need to know how to contact these organizations in order to stay current with their latest standards. Contact information for the most prominent standards-developing organizations in the broad field of industrial hygiene are as follows:

Air-Conditioning and Refrigeration Institute: http://www.ari.org

Air Movement and Control Association International, Inc.: http://www.amca.org

American Conference of Governmental Industrial Hygienists: http://www.amca.org

American Industrial Hygiene Association: http://www.aiha.org

American National Standards Institute: http://www.ansi.org

American Society of Heating, Refrigerating and Air-Conditioning Engineers: http://www.ashrae.org

American Society of Safety Engineers: http://www.asse.org

American Society for Testing Materials: http://www.astm.org

International Organization for Standardization: http://www.iso.ch

National Air Duct Cleaners Association: http://www.nadca.com

National Environmental Balancing Bureau: http://www.nebb.org

National Fire Protection Association: http://www.nfpa.org

National Institute for Occupational Safety and Health: http://www.cdc.gov/niosh

Occupational Safety and Health Administration: http://www.osha.gov

Underwriters Laboratories: http://www.ul.com

U.S. Department of Energy: http://tis.eh.doe.gov

U.S. Environmental Protection Agency: http://www.epa.gov

OSH Act and Industrial Hygiene

The principal federal legislation relating to industrial hygiene is the Occupational Safety and Health Act of 1970 (OSH Act) as amended. The OSH Act sets forth the following requirements relating to industrial hygiene:

- Use of warning labels and other means to make employees aware of potential hazards, symptoms of exposure, precautions, and emergency treatment
- Prescription of appropriate personal protective equipment and other technological preventive measures (29 CFR 1910.133 and 1910.134 [Subpart I])
- Provision of medical tests to determine the effect on employees of exposure to environmental stressors
- Maintenance of accurate records of employee exposures to environmental stressors that are required to be measured or monitored
- Accessibility of monitoring tests and measurement activities to employees
- Availability of monitoring tests and measurement activities records to employees on request
- Notification of employees who have been exposed to environmental stressors at a level beyond the recommended threshold and corrective action being taken

OSHA Process Safety Standard

The OSHA Process Safety Standard is found in 29 CFR 1910.119. Its purpose is to prevent *catastrophic* accidents caused by major releases of highly hazardous chemicals. To comply with this standard, companies must have written operating procedures, mechanical integrity programs, and formal incident investigation procedures. Other key elements are as follows:

1. **Coverage.** Although the process safety standard is typically associated with large chemical and petrochemical processing plants, its coverage is actually

much broader than this. Any company is covered that uses the threshold amount of a chemical listed in the standard—or 10,000 pounds or more of a flammable material at one site in one location.

2. **Employee participation.** Section (c) of the standard requires that employees be involved in all aspects of the process safety management program. In addition, employees must be given access to information developed as part of the program.
3. **Process safety information (PSI).** Section (d) of the standard requires organizations to establish and maintain process safety information files. Information included in the files includes chemical, process, and equipment data.
4. **Process hazard analyses (PHAs).** Section (e) of the standard requires that companies conduct process hazard analyses for all processes covered by the standard. Like any other hazard analysis, the PHAs are supposed to identify potential problems so that prompt corrective action or preventive measures can be taken.
5. **Standard operating procedures (SOPs).** Section (f) of the standard requires employers to establish and maintain written standard operating procedures for using chemicals safely. The requirement applies to handling, processing, transporting, and storing chemicals.
6. **Requirements for contractors.** Section (h) of the standard describes the special requirements imposed on companies that contract portions of their work to other companies. Complying with the standard is a matter of making sure that contractors comply. The following requirements are imposed by Section (h):
    - Screen contractors before issuing a contract to ensure that they have a comprehensive safety and health program.
    - Orient contractors concerning the chemicals with which they may be required to work or be around, the emergency action plan, and other pertinent information.
    - Evaluate contractors periodically to ensure that their safety performance is acceptable.
    - Maintain an OSHA injury and illness log for the contractor that is separate from, and in addition to, that of the host company.

OSHA Regulation for Chemical Spills

OSHA issues a special regulation dealing with chemical spills. The standard (29 CFR 1910.120) is called the Hazardous Waste Operations and Emergency Response Standard, or *HAZWOPER.* **HAZWOPER** gives organizations two options for responding to a chemical spill. The first is to evacuate all employees in the event of a spill and to call in professional emergency response personnel. Employers who use this option must have an emergency action plan (EAP) in place in accordance with 29 CFR 1010.38(a). The second option is to respond internally. Employers using this must have an emergency response plan that is in accordance with 29 CFR 1010.120.

1. **Emergency action plans (EAPs).** An **emergency action plan** should have at least the following elements: alarm systems, evacuation plan, a mechanism or procedure for emergency shutdown of the equipment, and a procedure for notifying emergency response personnel.

2. **Emergency response plan.** Companies that opt to respond internally to chemical spills must have an **emergency response plan** that includes the provision of comprehensive training for employees. OSHA Standard 29 CFR 1910.120 specifies the type and amount of training required, ranging from awareness to in-depth technical training for employees who will actually deal with the spill. It is important to note that OSHA forbids the involvement of untrained employees in responding to a spill. The following topics are typical of those covered in up-to-date HAZWOPER courses.

   - Summary of key federal laws
   - Overview of impacting regulations
   - Classification and categorization of hazardous waste

     Definition of hazardous waste

     Characteristics

     Toxic characteristic leaching procedure (TCLP)

     Lists of hazardous wastes
   - Hazardous waste operations

     Definitions

     Levels of response
   - Penalties for noncompliance

     Civil penalty policy
   - Responses to spills

     Groundwater contamination

     Sudden releases

     Clean-up levels

     Risk assessment

     Remedial action
   - Emergency response

     Workplan

     Site evaluation and control

     Site specific safety and health plan

     Information and training program

     Personal protective equipment

     Monitoring

     Medical surveillance

     Decontamination procedures

     Emergency response

     Other provisions
   - Contingency plans

     Alarm systems

     Action plan

- Personal protective equipment
  Developing a PPE program
  Respiratory equipment
  Protective clothing
  Donning PPE
  Doffing PPE
- Material safety data sheets
  Introduction
  Preparing MSDSs
  MSDS information
  Hazardous ingredients
  Physical/chemical characteristics
  Fire and explosion hazard data
  Reactivity data
  Health hazard data
  Precautions for safe handling and use
  Control measures
- Site control
  Site maps
  Site preparation
  Work zones
  Buddy system
  Site security
  Communications
  Safe work practices
- Hazardous waste containers
  Emergency control
  Equipment
  Tools
  Safety
- Decontamination
  Types
  Decontamination plan
  Prevention of contamination
  Planning
  Emergencies
  Physical injury
  Heat stress

Chemical exposure
Medical treatment area
Decontamination of equipment
Decontamination procedures
Sanitation of PPE
Disposal of contaminated materials

In a given year, industrial facilities in the United States release almost 6 billion pounds of toxic substances into the environment. The Environmental Protection Agency (EPA) began measuring such emissions in 1987. Since that time, the trend in toxic emissions has been downward. Even with the downward trend, however, in a typical year, toxic substances will be released into the environment in the following amounts (according to the EPA):

- 2.3 billion pounds released into the air
- billion pounds injected into underground wells
- 910 million pounds placed in treatment and disposal facilities
- 550 million pounds placed in municipal wastewater treatment plants
- 440 million pounds placed in landfills
- 100 million pounds released into the water[2]

People are exposed to a variety of substances every day in the home and at work—paints, paint remover, detergent, cleaning solvents, antifreeze, and motor oil, to name just a few. Most substances with which we interact are not dangerous in small amounts or limited exposure. However, high levels of exposure to certain substances in high concentrations can be dangerous. Levels of exposure and concentration, as well as how we interact with substances, help determine how hazardous substances are. In addition, some substances used frequently in certain industrial settings can explode. Following is some information that modern safety and health professionals need to know about toxic and explosive substances.

Hazards in the Workplace

The environmental stressors on which industrial hygiene focuses can be divided into the following broad categories: chemical, physical, biological, and ergonomic hazards. Typical **chemical hazards** include mists, vapors, gases, dusts, and fumes. Chemical hazards are either inhaled or absorbed through the skin or both. **Physical hazards** include noise, vibration, extremes of temperature, and excessive radiation (electromagnetic or ionizing). **Biological hazards** come from molds, fungi, bacteria, and insects. Bacteria may be introduced into the workplace through sewage, food waste, water, or insect droppings. **Ergonomic hazards** are related to the design and condition of the workplace. Poorly designed workstations and tools are ergonomic hazards. Conditions that put workers in awkward positions and impair their visibility are also hazards.

Material Safety Data Sheets

Employees should be warned of chemical hazards by labels on containers or material safety data sheets (MSDSs). MSDSs are special sheets that summarize all pertinent information about a specific chemical. The hazard communication standard of the

OSH Act requires that chemical suppliers provide users with an MSDS for each chemical covered by the standard.

An MSDS should contain the following information as appropriate: manufacturer's name, address, and telephone number; a list of hazardous ingredients; physical and chemical characteristics; fire and explosion hazard information; reactivity information; health hazard information; safety precautions for handling; and recommended control procedures.

An MSDS must contain specified information in eight categories:

- **Section I: General information.** This section contains directory information about the manufacturer of the substance, including the manufacturer's name and address, telephone number of an emergency contact person, a nonemergency telephone number for information, and a dated signature of the person who developed or revised the MSDS.
- **Section II: Hazardous ingredients.** This section should contain the common name, chemical name, and Chemical Abstracts Service (CAS) number for the substance. Chemical names are the scientific designations given in accordance with the nomenclature system of the International Union of Pure and Applied Chemistry. The CAS number is the unique number for a given chemical that is assigned by the Chemical Abstracts Service.
- **Section III: Physical and chemical characteristics.** Data relating to the vaporization characteristics of the substance are contained in this section.
- **Section IV: Fire and explosive hazard data.** Data relating to the fire and explosion hazards of the substance are contained in this section. Special firefighting procedures are also included in this section.
- **Section V: Reactivity data.** Information concerning the stability of the substance as well as the potential for hazardous decomposition or polymerization of the substance is contained in this section.
- **Section VI: Health hazards.** This section contains a list of the symptoms that may be suffered as a result of overexposure to the substance. Emergency first-aid procedures are also explained in this section.
- **Section VII: Safe handling and use.** This section explains special handling, storage, spill, and disposal methods and precautions relating to the substance.
- **Section VIII: Control measures.** The types of ventilation, personal protective equipment, and special hygienic practices recommended for the substance are explained in this section.[3]

Environmental Stressors

Noise is sound that is unwanted or that exceeds safe limits. It can cause problems ranging from annoyance to hearing loss. Acceptable levels of noise have been established by OSHA, NIOSH, and the EPA. OSHA mandates that an employee's exposure level be limited to 90 decibels (dB) calculated as an eight-hour, time-weighted average. Applying the OSHA recommendation, 14 percent of workers are employed in situations that expose them to excess noise levels.[4] Chapter 22 contains the latest data from NIOSH and OSHA on noise-level standards.

Temperature control is the most basic way to eliminate environmental hazards:

> People function efficiently only in a very narrow body temperature range, a "core" temperature measured deep inside the body, not on the skin or at body extremities. Fluctuations in core temperatures exceeding about 2°F below, or

3°F above, the normal core temperature of 37.6°C (99.6°F), which is 37°C mouth temperature (98.6°F mouth temperature), impair performance markedly. If this five-degree range is exceeded, a health hazard exists.[5]

Radiation hazards are increasingly prevalent in the age of high technology. In the category of **ionizing radiation**, safety and health professionals are concerned with five kinds of **radiation** (alpha, beta, X-ray, gamma, and neutron). Of these, alpha radiation is the least penetrating, making shielding simple, whereas the others are more difficult to shield against. Meters and other instruments are available to measure radiation levels in the workplace. The greatest risk for nonionizing radiation in the modern workplace comes from lasers. Shielding requirements for lasers are described in the construction regulations of the OSH Act.[6]

Extremes of pressure also represent a potential hazard in the workplace. According to Olishefski,

> One of the most common troubles encountered by workers under compressed air is pain and congestion in the ears from inability to ventilate the middle ear properly during compression and decompression. As a result, many workers subjected to increased air pressures suffer from temporary hearing loss; some have permanent hearing loss. This damage is believed to be caused by obstruction of the eustachian tubes, which prevents proper equalization of pressure from the throat to the middle ear. The effects of reduced pressure on the worker are much the same as the effects of decompression from a high pressure. If pressure is reduced too rapidly, decompression sickness and ear disturbances similar to the diver's conditions can result.[7]

Biological hazards from various biological organisms can lead to disease in workers. The now-famous outbreak of what has come to be known as Legionnaire's disease is an example of what can result from biological hazards. This disease first surfaced at a convention where numerous participants became sick and soon died. The cause was eventually traced back to bacteria that grew in the cooling/air-moving systems serving the convention center. That bacteria has since been named *Legionnella.*

Ergonomic hazards are conditions that require unnatural postures and unnatural movement. The human body can endure limited amounts of unnatural postures or motions. However, repeated exposure to such conditions can lead to physical stress and injury. Design of tools, workstations, and jobs can lead to or prevent ergonomic hazards.

Toxic Substances Defined

A **toxic substance** is one that has a negative effect on the health of a person or animal. Toxic effects are a function of several factors, including the following: (1) properties of the substance, (2) amount of the dose, (3) level of exposure, (4) route of entry, and (5) resistance of the individual to the substance. The issue of toxic substances can be summarized as follows:

> When a toxic chemical acts on the human body, the nature and extent of the injurious response depends upon the dose received—that is, the amount of the chemical that actually enters the body or system and the time interval during which this dose was administered. Response can vary widely and might be as little as a cough or mild respiratory irritation or as serious as unconsciousness and death.[8]

Entry Points for Toxic Agents

The development of preventive measures to protect against the hazards associated with industrial hygiene requires first knowing how toxic agents enter the body. A toxic substance must first enter the bloodstream to cause health problems. The most common routes of entry for toxic agents are inhalation, absorption, injection, and ingestion (see Figure 16-1). These routes are explained in the following paragraphs.

Inhalation

The route of entry about which safety and health professionals should be most concerned is **inhalation**. Airborne toxic substances such as gases, vapors, dust, smoke, fumes, aerosols, and mists can be inhaled and pass through the nose, throat, bronchial tubes, and lungs to enter the bloodstream. The amount of a toxic substance that can be inhaled depends on the following factors: (1) concentration of the substance, (2) duration of exposure, and (3) breathing volume.

According to Olishefski,

> Inhalation, as a route of entry, is particularly important because of the rapidity with which a toxic material can be absorbed in the lungs, pass into the bloodstream, and reach the brain. Inhalation is the major route of entry for hazardous chemicals in the work environment.[9]

Absorption

The second most common route of entry in an industrial setting is **absorption**, or passage through the skin and into the bloodstream. The human skin is a protective barrier against many hazards. However, certain toxic agents can penetrate the barrier through absorption. Of course, unprotected cuts, sores, and abrasions facilitate the process, but even healthy skin will absorb certain chemicals. Humans are especially susceptible to absorbing such chemicals as organic lead compounds, nitro compounds, organic phosphate pesticides, TNT, cyanides, aromatic amines, amides, and phenols.[10]

With many substances, the rate of absorption and, in turn, the hazard levels increase in a warm environment. The extent to which a substance can be absorbed through the skin depends on the factors shown in Figure 16-2. Another factor is body

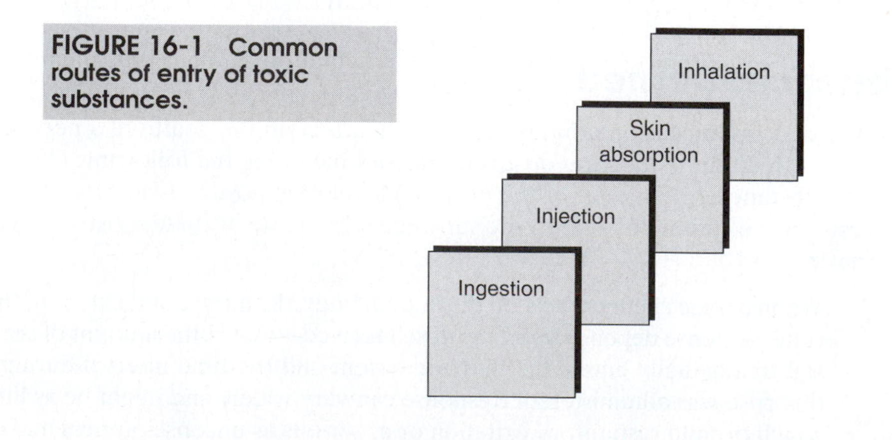

FIGURE 16-1 Common routes of entry of toxic substances.

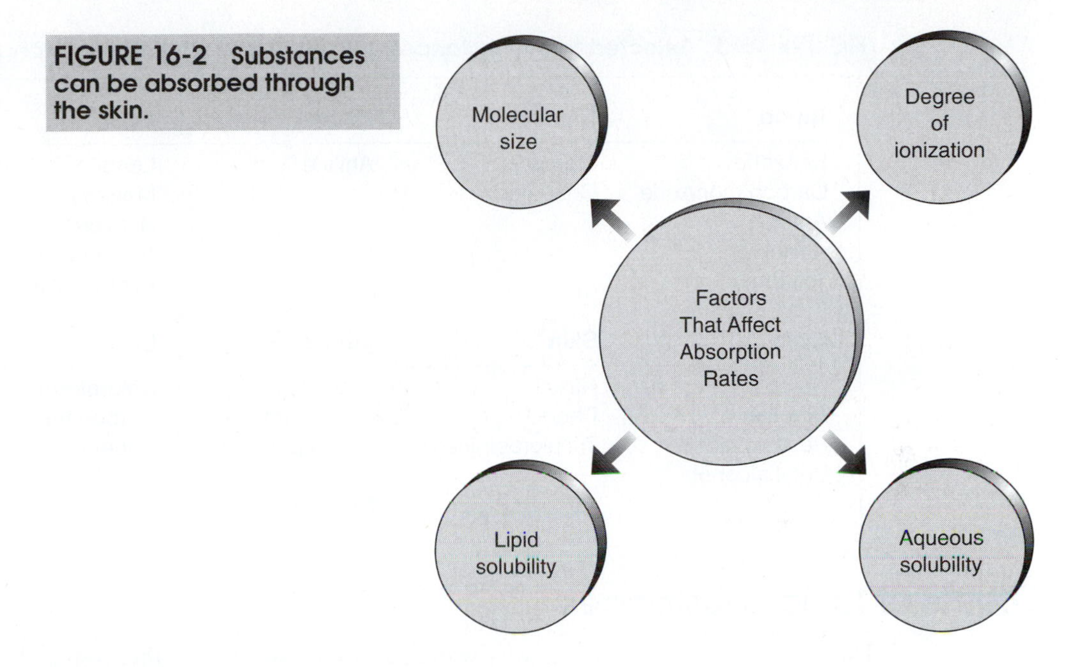

FIGURE 16-2 Substances can be absorbed through the skin.

site. Different parts of the body have different absorption capabilities. For example, the forearms have a lower absorption potential than do the scalp and forehead.

Ingestion

Ingestion, not a major concern in an industrial setting, is entry through the mouth.[11] An ingested substance is swallowed. It moves through the stomach into the intestines and from there into the bloodstream. Toxic agents sometimes enter the body by ingestion when they are accidentally consumed by workers eating lunch or a snack. Airborne contaminants can also rest on food or the hands and, as a result, be ingested during a meal or snack. The possibility of ingesting toxic agents makes it critical to confine eating and drinking to sanitary areas away from the work site and to make sure that workers practice good personal hygiene such as washing their hands thoroughly before eating or drinking.

As it moves through the gastrointestinal tract, the toxic substance's strength may be diluted. In addition, depending on the amount and toxicity of the substance, the liver may be able to convert it to a nontoxic substance. The liver can, at least, decrease the level of toxicity and pass along the substance to the kidneys, where some of the substance is eliminated in the urine.

Injection

Injection involves the introduction of a substance into the body by way of a needle and syringe, compressed air, high-pressure hydraulic leaks, or any other capable medium. Consequently, this is not often a route of entry for a toxic substance in the workplace. Injection is sometimes used for introducing toxic substances in experiments involving animals. However, this approach can produce misleading research results because the needle bypasses some of the body's natural protective mechanisms.

FIGURE 16-3 Selected toxic substances and the organs that they endanger most.

Blood	Kidneys	Heart	Brain
Benzene Carbon monoxide Arsenic Aniline Toluene	Mercury Chloroform	Aniline	Lead Mercury Benzene Manganese Acetaldehyde
Eyes	**Skin**	**Lungs**	**Liver**
Cresol Acrolein Benzyl chloride Butyl alcohol	Nickel Phenol Trichloroethylene	Asbestos Chromium Hydrogen sulfide Mica Nitrogen dioxide	Chloroform Carbon tetrachloride Toluene

Effects of Toxic Substances

The effects of toxic substances vary widely, as do the substances themselves. However, all the various effects and exposure times can be categorized as being either acute or chronic.

Acute effects and exposures involve a sudden dose of a highly concentrated substance. They are usually the result of an accident (a spill or damage to a pipe) that results in an immediate health problem ranging from irritation to death. Acute effects and exposures are (1) sudden, (2) severe, (3) typically involve just one incident, and (4) cause immediate health problems. Acute effects and exposures are not the result of an accumulation over time.

Chronic effects and exposures involve limited continual exposure over time. Consequently, the associated health problems develop slowly. The characteristics of chronic effects and exposures are (1) continual exposure over time, (2) limited concentrations of toxic substances, (3) progressive accumulation of toxic substances in the body and progressive worsening of associated health problems, and (4) little or no awareness of exposures on the part of affected workers.

When a toxic substance enters the body, it eventually affects one or more body organs. Part of the liver's function is to collect such substances, convert them to nontoxics, and send them to the kidneys for elimination in the urine. However, when the dose is more than the liver can handle, toxics move on to other organs, producing a variety of different effects. The organs that are affected by toxic substances are the blood, kidneys, heart, brain, central nervous system, skin, liver, lungs, and eyes. Figure 16-3 lists some of the more widely used toxic substances and the organs that they endanger most.

Relationship of Doses and Responses

Safety and health professionals are interested in predictability when it comes to toxic substances. How much of a given substance is too much? What effect will a given dose of a given substance produce? These types of questions concern dose–response relationships. A **dose** of a toxic substance can be expressed in a number of different ways depending on the characteristics of the substance; for example, amount per unit of body weight, amount per body surface area, or amount per unit of volume of air breathed. The dose–response relationship may be expressed mathematically as follows:[12]

$$(C) \times (T) = K$$

where

C = concentration
T = duration (time) of exposure
K = constant

Note that in this relationship, C times T is *approximately* equal to K. The relationship is not exact.

Three important concepts to understand relating to doses are dose threshold, lethal dose, and lethal concentration. These concepts are explained in the following paragraphs.

Dose Threshold

The **dose threshold** is the minimum dose required to produce a measurable effect. Of course, the threshold is different for different substances. In animal tests, thresholds are established using such methods as (1) observing pathological changes in body tissues, (2) observing growth rates (are they normal or retarded?), (3) measuring the level of food intake (has there been a loss of appetite?), and (4) weighing organs to establish body weight to organ weight ratios.

Lethal Dose

A **lethal dose** of a given substance is a dose that is highly likely to cause death. Such doses are established through experiments on animals. When lethal doses of a given substance are established, they are typically accompanied by information that is of value to medical professionals and industrial hygienists. Such information includes the type of animal used in establishing the lethal dose, how the dose was administered to the animal, and the duration of the administered dose. Lethal doses do not apply to inhaled substances. With these substances, the concept of lethal concentration is applied.

Lethal Concentration

A **lethal concentration** of an inhaled substance is the concentration that is highly likely to result in death. With inhaled substances, the duration of exposure is critical because the amount inhaled increases with every unprotected breath.

Airborne Contaminants

It is important to understand the different types of airborne contaminants that may be present in the workplace.[13] Each type of contaminant has a specific definition that must be understood in order to develop effective safety and health measures to protect against it. The most common types of airborne contaminants are dusts, fumes, smoke, aerosols, mists, gases, and vapors (Figure 16-4).

- **Dusts.** **Dusts** are various types of solid particles that are produced when a given type of organic or inorganic material is scraped, sawed, ground, drilled, handled,

FIGURE 16-4 Common airborne contaminants.

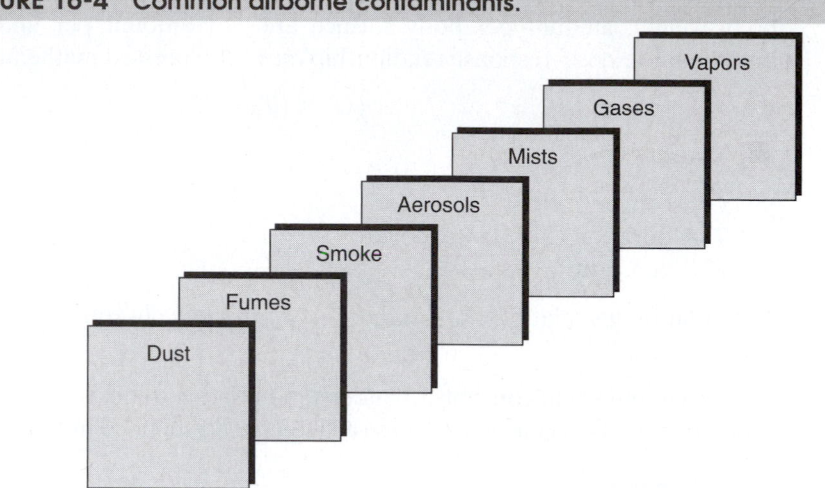

heated, crushed, or otherwise deformed. The degree of hazard represented by dust depends on the toxicity of the parent material and the size and level of concentration of the particles.

- **Fumes.** The most common causes of fumes in the workplace are such manufacturing processes as welding, heat treating, and metalizing, all of which involve the interaction of intense heat with a parent material. The heat volatilizes portions of the parent material, which then condenses as it comes in contact with cool air. The result of this reaction is the formation of tiny particles that can be inhaled.
- **Smoke.** Smoke is the result of the incomplete combustion of carbonaceous materials. Because combustion is incomplete, tiny soot or carbon particles remain and can be inhaled.
- **Aerosols.** **Aerosols** are liquid or solid particles that are so small they can remain suspended in air long enough to be transported over a distance. They can be inhaled.
- **Mists.** **Mists** are tiny liquid droplets suspended in air. Mists are formed in two ways: (1) when vapors return to a liquid state through condensation; and (2) when the application of sudden force or pressure turns a liquid into particles.
- **Gases.** Unlike other airborne contaminants that take the form of either tiny particles or droplets, gases are actually formless fluids. Gases become particularly hazardous when they fill a confined, unventilated space. The most common sources of gases in an industrial setting are from welding and the exhaust from internal combustion engines.
- **Vapors.** Certain materials that are solid or liquid at room temperature and at normal levels of pressure turn to **vapors** when heated or exposed to abnormal pressure. Evaporation is the most common process by which a liquid is transformed into a vapor.

In protecting workers from the hazards of airborne contaminants, it is important to know the permissible levels of exposure for a given contaminant and to monitor continually the level of contaminants using accepted measurement practices and technologies. The topic of exposure thresholds is covered later in this chapter.

Effects of Airborne Toxics

Airborne toxic substances are also classified according to the type of effect they have on the body. The primary classifications are shown in Figure 16-5 and explained in the paragraphs that follow. With all airborne contaminants, concentration and duration of exposure are critical concerns.

Irritants

Irritants are substances that cause irritation to the skin, eyes, and the inner lining of the nose, mouth, throat, and upper respiratory tract. However, they produce no irreversible damage:

> Irritants can be subdivided into primary and secondary irritants. A primary irritant is a material that exerts little systemic toxic action, either because the products formed on the tissues of the respiratory tract are nontoxic or because the irritant action is far in excess of any systemic toxic action. A secondary irritant produces irritant action on mucous membranes, but this effect is overshadowed by systemic effects resulting from absorption. Normally, irritation is a completely reversible phenomenon.[14]

Asphyxiants

Asphyxiants are substances that can disrupt breathing so severely that suffocation results. Asphyxiants may be simple or chemical in nature. A simple asphyxiant is an inert gas that dilutes oxygen in the air to the point that the body cannot take in enough air to satisfy its needs for oxygen. Common simple asphyxiants include carbon dioxide, ethane, helium, hydrogen, methane, and nitrogen. Chemical asphyxiants, by chemical action, interfere with the passage of oxygen into the blood or the movement of oxygen from the lungs to body tissues. Either way, the end result is suffocation due to insufficient or no oxygenation. Common chemical asphyxiants include carbon monoxide, hydrogen cyanide, and hydrogen sulfide.

Narcotics and Anesthetics

Narcotics and **anesthetics** are similar in that carefully controlled dosages can inhibit the normal operation of the central nervous system without causing serious or irreversible effects. This makes them particularly valuable in a medical setting. Dentists

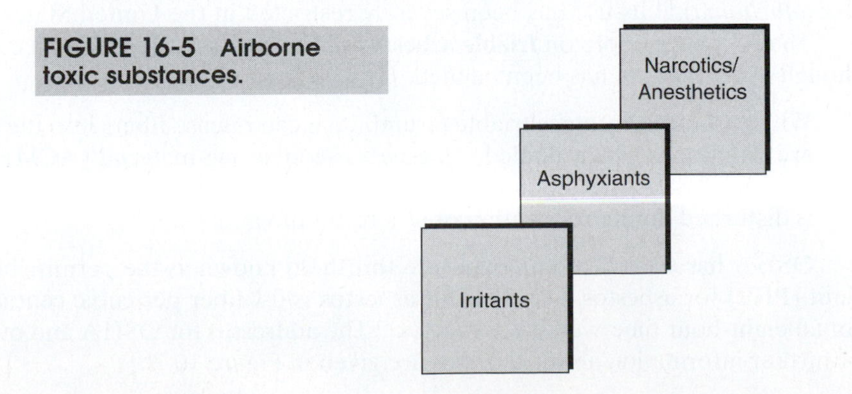

FIGURE 16-5 Airborne toxic substances.

and physicians use narcotics and anesthetics to control pain before and after surgery. However, if the concentration of the dose is too high, narcotics and anesthetics can cause unconsciousness and even death. When this happens, death is the result of asphyxiation. Widely used narcotics and anesthetics include acetone, methyl-ethyl-ketone, acetylene hydrocarbons, ether, and chloroform.

Effects of Carcinogens

A **carcinogen** is any substance that can cause a malignant tumor or a **neoplastic growth**. A *neoplasm* is cancerous tissue or tissue that may become cancerous. Other terms used synonymously for carcinogen are *tumorigen, oncogen,* and *blastomogen*:

> It is well established that exposure to some chemicals can produce cancer in laboratory animals and man. There are a number of factors that have been related to the incidence of cancer—the genetic pattern of the host, viruses, radiation including sunshine, and hormone imbalance, along with exposure to certain chemicals. Other factors such as cocarcinogens and tumor accelerators are involved. It is also possible that some combination of factors must be present to induce cancers. There is pretty good clinical evidence that some cancers are virus-related. It may be that a given chemical in some way inactivates a virus, activates one, or acts as a cofactor.[15]

Medical researchers are not sure exactly how certain chemicals cause cancer. However, there are a number of toxic substances that are either known, or are strongly suspected, to be carcinogens. These include coal tar, pitch, creosote oil, anthracene oil, soot, lamp black, lignite, asphalt, bitumen waxes, paraffin oils, arsenic, chromium, nickel compounds, beryllium, cobalt, benzene, and various paints, dyes, tints, pesticides, and enamels.[16]

Asbestos Hazards

The EPA estimates that approximately 75 percent of the commercial buildings in use today contain asbestos in some form.[17] Asbestos was once thought to be a miracle material because of its many useful characteristics, including fire resistance, heat resistance, mechanical strength, and flexibility. As a result, asbestos was widely used in commercial and industrial construction between 1900 and the mid-1970s (see Figure 16-6).[18]

In the mid-1970s, medical research clearly tied asbestos to respiratory cancer, scarring of the lungs (now known as *asbestosis*), and cancer of the chest or abdominal lining (*mesothelioma*).[19] Its use has been severely restricted in the United States since 1989.

The following quote on **friable asbestos** shows why asbestos is still a concern even though its further use has been banned:

> When asbestos becomes friable (crumbly), it can release fibers into the air that are dangerous when inhaled. As asbestos-containing material (ACM) ages, it becomes less viable and more friable. Asbestos can be released into the air if it is disturbed during renovation or as a result of vandalism.[20]

OSHA has established an exposure threshold known as the permissible exposure limit (PEL) for asbestos. The PEL for asbestos is 0.1 fiber per cubic centimeter of air for an eight-hour time-weighted average.[21] The addresses for OSHA and other sources of further information about asbestos are given in Figure 16-7.

FIGURE 16-6 Asbestos use, 1900 to present.

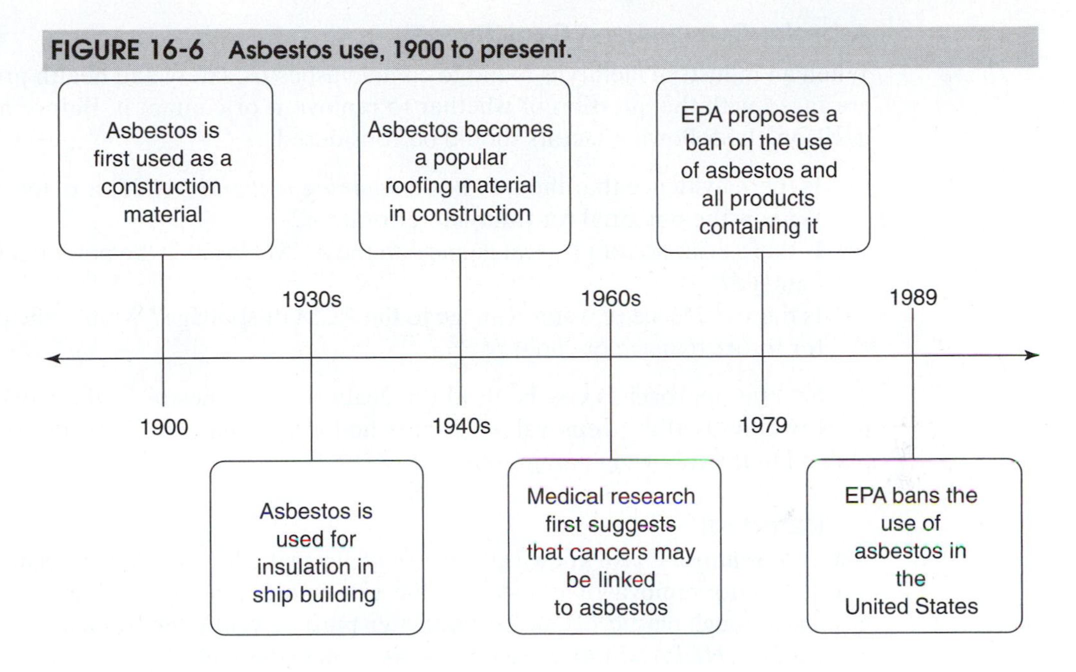

FIGURE 16-7 Sources of information about asbestos in the workplace.

Asbestos Action Program
EPA
Mail Code TS-799
401 M St. SW
Washington, DC 20460
202-382-3949

Cancer Information Service
National Cancer Institute
Bldg. 31, Room 10A24
9000 Rockville Pike
Bethesda, MD 20892
800-4CANCER

NIOSH Publications Office
4676 Columbia Pkwy.
Cincinnati, OH 45226
513-533-8287

OSHA Publications Office
Room N–3101
200 Constitution Ave. NW
Washington, DC 20210
202-523-9649

Asbestos Abatement Council of AWCI
1600 Cameron St.
Alexandria, VA 22314
703-684-2924

Asbestos Information Association of North America
1745 Jefferson Davis Hwy., Suite 509
Arlington, VA 22202
703-979-1150

The National Asbestos Council
1777 Northeast Expressway, Suite 150
Atlanta, GA 30329
404-633-2622

Asbestos Removal and Containment

When an industrial facility is found to contain asbestos, safety and health professionals are faced with the question of whether to remove it or contain it. Before making this decision, the following factors should be considered:

- Is there evidence that the *asbestos-containing material (ACM)* is deteriorating? What is the potential for future deterioration?
- Is there evidence of physical damage to the ACM? What is the potential for future damage?
- Is there evidence of water damage to the ACM or spoilage? What is the potential for future damage or spoilage?[22]

Several approaches can be used for dealing with asbestos in the workplace. The most widely used are removal, enclosure, and encapsulation. These methods are explained in the following paragraphs.

Removal

Asbestos removal is also known as *asbestos abatement.* The following procedures are recommended for removal of asbestos: (1) the area in question must be completely enclosed in walls of tough plastic; (2) the enclosed area must be ventilated by *high-efficiency particle absolute (HEPA)* filtered negative air machines (these machines work somewhat like a vacuum cleaner in eliminating asbestos particles from the enclosed area); (3) the ACM must be covered with a special liquid solution to cut down on the release of asbestos fibers; and (4) the ACM must be placed in leakproof containers for disposal.[23]

Enclosure

Enclosure of an area containing ACMs involves completely encapsulating the area in airtight walls.[24] The following procedures are recommended for enclosing asbestos: (1) use HEPA-filtered negative air machines in conjunction with drills or any other tools that may penetrate or otherwise disturb ACMs; (2) construct the enclosing walls of impact-resistant and airtight materials; (3) post signs indicating the presence of ACMs within the enclosed area; and (4) note the enclosed area on the plans of the building.

Encapsulation

Encapsulation of asbestos involves spraying the ACMs with a special sealant that binds them together, thereby preventing the release of fibers.[25] The sealant should harden into a tough, impact-resistant skin. This approach is generally used only on acoustical plaster and similar materials.

Personal Protective Equipment for Asbestos Removal

It is important to use the proper types of personal protective clothing and respiratory devices. Clothing should be disposable and should cover all parts of the body.[26] Respirators used when handling asbestos should be "high-efficiency cartridge filter type (half-and-full-face types); any powered-air purifying respirator; any type C continuous-flow supplied-air, pressure-demand respirator, equipped with an auxiliary positive pressure self-contained breathing apparatus."[27]

Medical Records and Examinations

It is important that employees who handle ACMs undergo periodic medical monitoring. Medical records on such employees should be kept current and maintained for at least 20 years.[28] They should contain a complete medical history on the employee. These records must be made available on request to employees, past employees, health care professionals, employee representatives, and OSHA personnel.

Medical examinations, conducted at least annually, should also be required for employees who handle ACMs. These examinations should include front and back chest X-rays that are at least 7 inches by 14 inches. The examination should also test pulmonary function, including forced vital capacity and forced expiratory volume at one second.

Indoor Air Quality and "Sick-Building" Syndrome

A key concept relating to indoor air quality is what has come to be called "sick-building syndrome." In reality, a "sick" building is one that makes people sick because it has become infested with mold, mildew, spores, and other airborne microorganisms. Although much is still unknown about sick-building syndrome, the EPA estimates that as many as 30 percent of the buildings in the United States have air quality problems.

Poor **indoor air quality (IAQ)** can cause a variety of health problems ranging from the temporary to the long term. Health problems commonly associated with poor indoor air quality include allergic reactions, respiratory problems, eye irritation, sinusitis, bronchitis, and pneumonia. Often, the cause of poor indoor air quality can be slipshod maintenance such as failure to keep fans, ductwork, and filters clean. Other contributors are the particles and gases that can be released by office equipment, carpets, paints, cleaning solvents, and office supplies.

One of the keys to preventing sick-building syndrome is air exchange. Important factors in a building's ability to eliminate contaminated air and bring in fresh air are

- Ventilation
- Air infiltration rates
- Airflow rates in ducts
- Airflow patterns
- Fume exhaust

The most accurate methods available for measuring these factors fall under the broad heading of **tracer gas techniques**. A tracer gas is any gas or vapor not normally found in a building. The best have the following characteristics:

- Nontoxic
- Nonallergic at the levels used
- Chemically inert
- Odorless and tasteless
- Nonflammable and nonexplosive
- Easily transported
- Easily dispersed as an atmospheric gas
- Easily and economically measured with a high degree of reliability

The most widely used tracer gases are sulfur hexafluoride, halogenated refrigerants, and perfluorocarbons. To perform a tracer gas test, the following materials and equipment are needed:

- A suitable tracer gas
- A device for measuring tracer gas concentrations
- An air-sampling system
- A tracer gas injection system
- A data acquisition and control system

There are several different types of tracer gas tests including tracer decay, constant concentration, buildup/decay, CO_2 measurement, and reentrainment/recirculation. Regardless of the type of test used, the testing process involves the following steps:

1. Inject the tracer gas into the building.
2. Measure the concentration of the tracer gas in different parts of the building at different times over a certain period.

The data collected during a tracer gas test can give safety and health professionals the following types of information:

- Total air exchange rate for the building
- Air change rate due to the operation of the building's HVAC system
- Air change rate due to air infiltration and leakage
- Percentage of outside air supplied by the building's HVAC system
- Effectiveness of the ventilation system in removing contaminants
- The distribution of the ventilation air throughout the building

With this type of information available, safety and health personnel can determine if there are pockets where contaminated air is trapped; if the ventilation and air infiltration rates are sufficient; if airflow rates through ducts are sufficient; if airflow patterns are what they should be; and if fume hoods are performing as they should. This type of information is needed to detect and prevent indoor air quality problems.

ANSI's Indoor Air Quality/HVAC Standard

The American National Standards Institute (ANSI) developed a standard specifically addressing indoor air quality (IAQ). The standard, ANSI Z9.8, carries the following title: "Fundamentals Governing the Management, Operation, Testing, and Maintenance of HVAC Systems for Maintaining Acceptable Indoor Air Quality In Employee Occupancies Through Dilution Ventilation." Key components of the standard are summarized in the following paragraphs:

General coverage. ANSI Z9.8 is very specifically written to apply primarily to office space. It applies specifically to employee occupancies in nonindustrial spaces including general office spaces, commercial operations, and office spaces within industrial facilities.

Application flexibility. The requirements set forth in the standard are minimums. Employers may use demonstrably equal or better approaches. Where the provisions of ANSI Z9.8 conflict with other standards, the more stringent standard is to take precedence. If employers deviate from the standard, they must justify the deviation(s) in writing.

Acceptable air quality. The standard does not prescribe or define what is or is not "acceptable" air quality. Employers are required to determine and define "acceptable." Employers may use threshold limit values (TLVs) or PELs as guidelines or they may establish ceiling limits (e.g., if more than 2 percent of employees complain about air quality problems the ceiling limit has been reached).

Tobacco smoke. The standard requires employers to evaluate smoking in the workplace and apply whatever management controls are appropriate. This gives employers a great deal of latitude. The attitude of the committee that developed the standard (ANSI Z9.8) seems to be that most employers already have smoking controls established.

Toxic Mold and Indoor Air Quality

Toxic mold has surfaced as an issue relating to IAQ. The issue is complicated by at least two factors. First, there are thousands of types of molds, but only a few are toxic. Second, different people have different levels of sensitivity to mold. On OSHA's list of the 13 most critical indoor air quality hazards, mold is the last entry. On the other hand, in those limited instances in which molds are toxic, they can cause coughing, atypical asthma, nasal congestion, sinusitis, rhinitis, skin rashes, and fatigue. In severe cases, toxic molds can be deadly.

OSHA recommends a three-step process for investigating the possibility of the presence of toxic mold in the workplace. Step one involves interviewing employees about symptoms. Step two involves conducting an on-site review of the workplace in question. Step three, which is undertaken only if the symptoms and physical evidence warrant it, involves an on-site environmental evaluation in which a specialist takes samples.

The principal causal factor in most cases of toxic mold is inadequate ventilation. Consequently, an effective approach for preventing the accumulation of toxic mold in the workplace is to apply the following steps: (1) check outdoor intakes and make sure they are not near trash storage areas, standing water, exhausts, or anything else that might contribute to the growth of mold; (2) make sure the drip pans are sloped sufficiently to prevent the accumulation of standing water; and (3) check ductwork regularly to ensure that the lining is dry and clean.

Toxic Mold Assessment and Remediation

Modern safety and health professionals must be prepared to deal with moisture and mold issues. It is important to investigate periodically to identify sources of moisture and mold. The following procedures may be used to guide investigations:

- *Look* for mold in likely locations such as around pipes, drains, windows, and dark, poorly ventilated areas.
- *Listen* to the feedback and comments of employees who might complain about allergies that could be affected by mold or about any aspect of indoor air quality.
- *Smell* the air in the work environment. You can tell if the air is damp, stale, or musty. If it is, there is a moisture problem, even if it is hidden under floors or behind walls.

- *Train* employees how to be "mold investigators" by showing them how to look, listen, and smell in their work environment.
- *Inventory* the principal areas of moisture and mold risk in your facility and monitor these high-risk areas continually.[29]

When mold is found, it is important to act. Mold remediation, in general terms, proceeds as follows: (1) stop the moisture intrusion, (2) contain and isolate the moisture that is already present, (3) dry and filter the affected area, (4) remove anything in the affected area that cannot be dried, (5) kill existing bacteria with disinfectants and sanitizing agents, (6) clean and then reclean the area, and (7) take whatever steps are necessary to prevent any further moisture intrusion.

According to Alfred Draper III, an industrial hygienist who specializes in mold-related restoration and remediation, mold remediation projects can be divided into four classes ranging from low impact to high. Class I remediation is just good housekeeping (e.g., minimizing dust, using drop cloths, and cleaning up with HEPA-filtered vacuums). Class II requires the use of EPA-registered disinfectants, containing construction waste, and limiting access to work areas.

Class III projects are where remediation begins to be more challenging. For Class III projects, Draper recommends the following procedures:

- Remove or isolate the affected area's HVAC system to prevent ductwork contamination.
- Seal off the affected area using hard critical barriers (e.g., plywood, sheetrock, etc.).
- Require all personnel who must enter the affected area to wear the proper personal protective equipment (PPE) including disposable full-body coverings, gloves, and half-mask HEPA filter respirators.
- Contain all remediated waste in tightly covered containers before it is shipped.
- Clean and decontaminate all equipment prior to making the final wipedown and cleanup of the affected area.

Class IV projects are the most challenging and demanding of moisture or mold remediations. For Class IV, Draper recommends all the Class III procedures plus the following:

- Cover all equipment, structures, and surfaces not being cleaned.
- Require all remediation personnel to wear the PPE which should include disposable full-body coverings, gloves, rubber boots, and full-face respirators.
- Construct a decontamination facility and require all personnel to pass through it when leaving the affected area.
- Monitor exterior air throughout the duration of the remediation project.
- Evaluate high-risk personnel who work near the affected area, at least during the more hazardous aspects of the remediation.
- Thoroughly "air wash" any materials in the affected room that cannot be disposed of.
- Use negative pressure throughout the cleaning process for local exhaust at in the affected area.
- Use air scrubbing throughout the remediation project to remove any mold spores that might be introduced into the air.
- Have all air ducts professionally cleaned.
- Collect cultured and noncultured samples that can be tested to ensure that the remediation process was effective.

Threshold Limit Values

How much exposure to a toxic substance is too much? How much is acceptable? Guidelines that answer these questions for safety and health professionals are developed and issued annually by the American Conference of Governmental Industrial Hygienists (ACGIH). The guidelines are known as **threshold limit values (TLVs)**. The ACGIH describes threshold limit values as follows:

> Threshold limit values refer to airborne concentrations of substances and represent conditions under which it is believed that nearly all workers may be repeatedly exposed day after day without adverse effect. Because of wide variation in individual susceptibility, however, a small percentage of workers may experience discomfort from some substances at concentrations at or below the threshold limit; a smaller percentage may be affected more seriously by aggravation of a preexisting condition or by development of an occupational illness.
>
> Threshold limits are based on the best available information from industrial experience, from experimental human and animal studies, and, when possible, from a combination of the three. The basis on which the values are established may differ from substance to substance; protection against impairment of health may be a guiding factor for some, whereas reasonable freedom from irritation, narcosis, nuisance, or other forms of stress may form the basis for others.[30]

ACGIH's Classifications of TLVs and BEIs

The ACGIH develops threshold limit values (TLVs) and **biological exposure indices (BEIs)** to help safety and health professionals control certain chemical, biological, and physical health hazards in the workplace.[31] TLVs and BEIs are not legal standards and are not intended to be; rather, they are guidelines. However, their impact is increasingly felt, and in a positive way. As government organizations and agencies continue to find that political considerations make it difficult to promulgate legally authorized standards in a timely manner, the ability of the ACGIH to produce TLV guidelines that are updated annually makes its guidelines more and more valuable. Key concepts about TLVs and BEIs that should be understood by safety and health professionals are as follows:

Threshold limit value–time-weighted average (TLV-TWA). The time-weighted average for a conventional eight-hour workday and a 40-hour workweek for a given substance to which it is believed that nearly all workers may be repeatedly exposed on a daily basis without suffering ill effects. For example, the TLV-TWA for liquefied petroleum gas is 1,000 parts per million (ppm).

Threshold limit value–short-term exposure limit (TLV-STEL). The concentration of a given substance to which it is believed that workers may be exposed continuously for short periods without suffering ill effects. A STEL is defined as a 15-minute TWA exposure that should not be exceeded at any time during the workday period. Also, exposures above the TLV-TWA up to the STEL should not exceed 15 minutes and should not occur more than four times in a day (with at least 60 minutes between exposures). For example, the TLV-STEL for isopropyl ether is 310 ppm.

Threshold limit value–ceiling (TLV-C). The concentration of a given substance that should not be exceeded at any point during an exposure period.

Biological exposure indices. The levels of determinants that are expected to be present in specimens taken from healthy workers who have been exposed to selected substances to the same extent as other workers with inhalation exposure to the substance at the TLV. For example, the BEI for acetone in the urine of a worker is 50 milligrams per liter (mg/L).

Physical agents. Substances or factors that can introduce added stress on the human body so that the effects of a given substance at the TLV might be magnified. Physical factors include acoustics (noise), ergonomic conditions, ionizing radiation, lasers, nonionizing radiation, subfrequency and static electric fields, and thermal stress (cold and heat). For example, the TLV for noise of 94 dB (average) is one hour per day.

Calculating a TWA

Time-weighted averages (TWAs) can be calculated for exposures to given substances.[32] Olishefski gives the following formula for calculating the TWA for an eight-hour day:

$$\frac{TWA = CaTa + CbTb + \ldots CnTn}{8}$$

where

Ta = time of the first exposure period during the eight-hour shift

Ca = concentration of the substance in question in period a

Tb = another time period during the same shift

Cb = concentration of the substance in question in period b

Tn = nth or final time period in the eight-hour shift

Cn = concentration during period n

Hazard Recognition and Evaluation

The degree and nature of the hazard must be understood before effective hazard control procedures can be developed. This involves recognizing that a hazard exists and then making judgments about its magnitude with regard to chemical, physical, biological, and ergonomic stresses.

Questions that can be used for recognizing hazards in the workplace are as follows:

- What is produced?
- What raw materials are used in the process?
- What additional materials are used in the process?
- What equipment is used?
- What operational procedures are involved?
- What dust control procedures are involved?
- How are accidental spills cleaned up?
- How are waste by-products disposed?
- Is there adequate ventilation?
- Are processes equipped with exhaust devices?
- How does the facility layout contribute to employee exposure?
- Are properly working personal protective devices available?
- Are safe operating procedures recorded, made available, monitored, and enforced?[33]

Olishefski recommends that all processes be subjected to the following hazard recognition procedures:

- Determine the exposure threshold for each hazardous substance identified when applying the questions just listed, including airborne contaminants.
- Determine the level of exposure to each hazardous substance.
- Determine which employees are exposed to each hazardous material, how frequently, and for how long.
- Calculate the TWAs to the exposure thresholds identified earlier.[34]

For hazard evaluation, the following considerations are important: the nature of the material or substance involved, the intensity of the exposure, and the duration of the exposure. Key factors to consider are how much exposure is required to produce injury or illness; the likelihood that enough exposure to produce injury or illness will take place; the rate of generation of airborne contaminants; the total duration of exposure; and the prevention and control measures used.[35]

For example, the textile industry was once ranked fifth on a list of 43 industries in terms of preventing workplace injuries and fatalities. The industry improved this ranking to first, a feat attributed to hazard evaluation and recognition efforts within the industry adopted to comply with the OSHA cotton-dust standard (1910.1043).[36]

The textile industry has become a leader in the area of industrial safety and health. Even so, there are still problems. According to *Safety & Health,*

Although the textile industry has stopped the use of cancer-associated dyes, as many as 50,000 current and former industry workers have had significant major exposure during their work lives and are at high risk for developing bladder cancer.[37]

One of the ways in which textile companies are dealing with their industrial hygiene problems is by adding more safety and health professionals to their staffs. According to Bone, there will be "a growth of safety positions at the plant level. Each plant is going to need one safety-and-health professional to keep up with regulations."[38]

Honeywell (GE) includes industrial hygiene as a major component in its overall safety and health program. A corporatewide environmental auditing program for recognizing and evaluating hazards is the mainstay of Honeywell's industrial hygiene effort.

As with all effective programs of this nature, Honeywell's program has the involvement and support of the board of directors and higher management. The program came into being in response to what came to be known as the "Kepone tragedy," an incident in which employees at a company in Hopewell, Virginia, suffered Kepone poisoning. The tragic results of this incident motivated Honeywell to establish a rigorous environmental auditing component to its overall industrial hygiene program. Honeywell now conducts approximately 50 environmental audits annually in such areas as occupational health, medical programs, solid waste disposal, hazardous waste disposal, safety/loss prevention, and product safety. The results of the audits are used to develop prevention and control strategies.

Prevention and Control

Most prevention and control strategies can be placed in one of the following four categories: (1) engineering controls, (2) ventilation, (3) personal protective equipment, and (4) administrative controls.[39] Examples of strategies in each category are given in the following paragraphs.

Engineering Controls

The category of **engineering controls** includes such strategies as replacing a toxic material with one that is less hazardous or redesigning a process to make it less stressful or to reduce exposure to hazardous materials or conditions. Other engineering controls are isolating a hazardous process to reduce the number of people exposed to it and introducing moisture to reduce dust.[40]

For example, exhaust ventilation, which involves trapping and removing contaminated air, is an engineering control. This type of ventilation is typically used with such processes as abrasive blasting, grinding, polishing, buffing, and spray painting or finishing. It is also used in conjunction with open-surface tanks. Dilution ventilation involves simultaneously removing and adding air to dilute a contaminant to acceptable levels.[41]

Personal Protection from Hazards

When the work environment cannot be made safe by any other method, **personal protective equipment (PPE)** is used as a last resort. PPE imposes a barrier between the worker and the hazard but does nothing to reduce or eliminate the hazard. Typical equipment includes safety goggles, face shields, gloves, boots, earmuffs, earplugs, full-body clothing, barrier creams, and respirators.[42]

Occasionally, in spite of an employee's best efforts in wearing PPE, his or her eyes or skin will be accidentally exposed to a contaminant. When this happens, it is critical to wash away or dilute the contaminant as quickly as possible. Specially designed eyewash and emergency wash stations such as those shown in Figures 16-8, 16-9, and 16-10 should be readily available and accessible in any work setting where contaminants may be present.

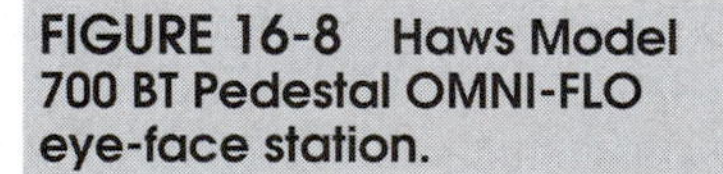

FIGURE 16-8 Haws Model 700 BT Pedestal OMNI-FLO eye-face station.

Courtesy of Haws Drinking Faucet Company.

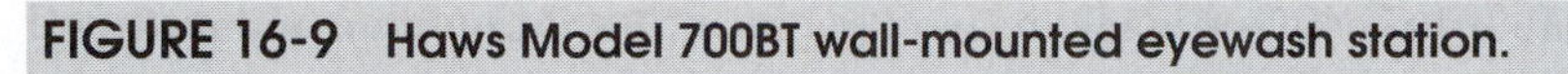

FIGURE 16-9 Haws Model 700BT wall-mounted eyewash station.

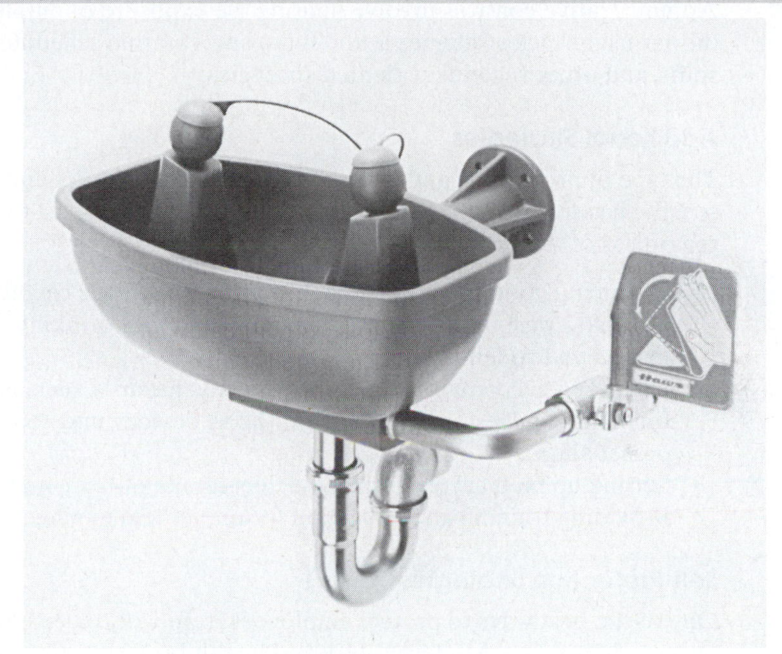

Courtesy of Haws Drinking Faucet Company.

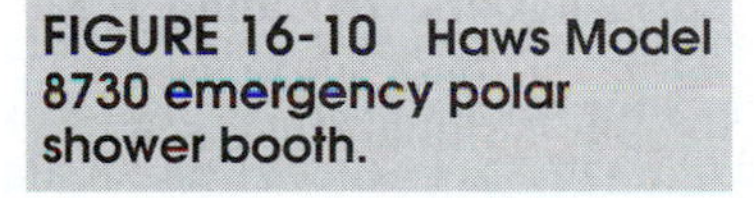

FIGURE 16-10 Haws Model 8730 emergency polar shower booth.

Courtesy of Haws Drinking Faucet Company.

Administrative Controls

Administrative controls involve limiting the exposure of employees to hazardous conditions using such strategies as the following: rotating schedules, required breaks, work shifts, and other schedule-oriented strategies.[43]

Additional Strategies

The type of prevention and control strategies used will depend on the evaluation of the specific hazards present in the workplace. The following list of generic strategies apply regardless of the setting:

- Practicing **good housekeeping**, including workplace cleanliness, waste disposal, adequate washing and eating facilities, healthful drinking water, and control of insects and rodents
- Using special control methods for specific hazards, such as reduction of exposure time, film badges and similar monitoring devices, and continuous sampling with preset alarms
- Setting up medical programs to detect intake of toxic materials
- Providing training and education to supplement engineering controls[44]

Self-Protection Strategies

One of the best ways to protect employees from workplace hazards is to teach them to protect themselves. Modern safety and health professionals should ensure that all employees are familiar with the following rules of self-protection:

1. **Know the hazards in your workplace.** Take the time to identify all hazardous materials and conditions in your workplace and know the safe exposure levels for each.
2. **Know the possible effects of hazards in your workplace.** Typical effects of workplace hazards include respiratory damage, skin disease and irritation, injury to the reproductive system, and damage to the blood, lungs, central nervous system, eyesight, and hearing.
3. **Use personal protective equipment properly.** Proper use of personal protective equipment means choosing the right equipment, getting a proper fit, correctly cleaning and storing equipment, and inspecting equipment regularly for wear and damage.
4. **Understand and obey safety rules.** Read warning labels before using any contained substance, handle materials properly, read and obey signs, and do only authorized work.
5. **Practice good personal hygiene.** Wash thoroughly after exposure to a hazardous substance, shower after work, wash before eating, and separate potentially contaminated work clothes from others before washing them.[45]

NIOSH and Industrial Hygiene

The National Institute for Occupational Safety and Health (NIOSH) is part of the Department of Health and Human Services (DHHS). This agency is important to industrial hygiene professionals. The main focus of the agency's research is on toxicity levels and human tolerance levels of hazardous substances. NIOSH prepares recommendations for OSHA standards dealing with hazardous substances, and NIOSH studies are made available to employers.

The areas of research of NIOSH's four major divisions—Biomedical and Behavioral Science; Respiratory Disease Studies; Surveillance, Hazard Evaluations, and Field Studies; and Training and Manpower Development—were discussed in Chapter 5. The results of these divisions' studies and their continually updated lists of toxic materials and recommended tolerance levels are extremely helpful to industrial hygienists concerned with keeping the workplace safe.

NIOSH Guidelines for Respirators

The respirator is one of the most important types of personal protective equipment available to individuals who work in hazardous environments (see Figures 16-11 and 16-12). Because the performance of a respirator can mean the difference between life

FIGURE 16-11 Breathing protection devices.

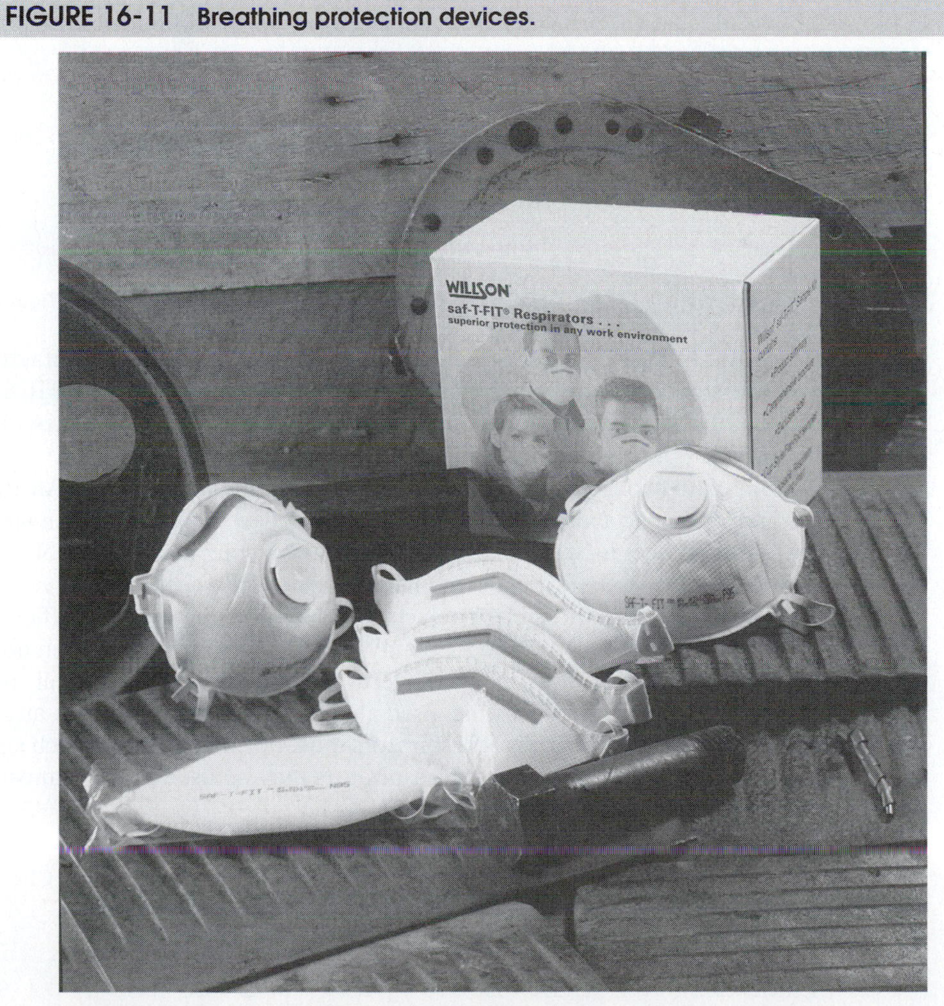

Courtesy of Dalloz Safety (Willson Safety Equipment).

FIGURE 16-12 Respirator.

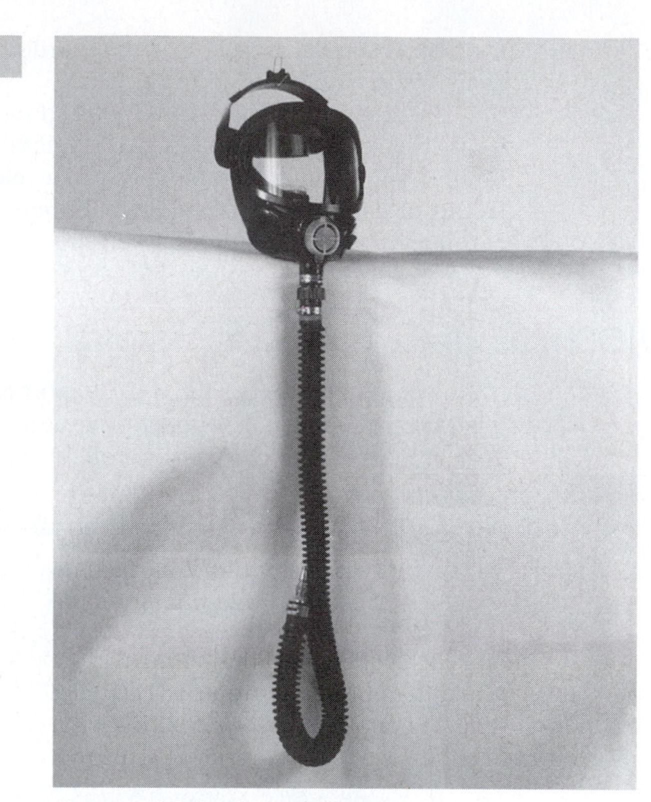

Courtesy of Dalloz Safety (Willson Safety Equipment).

and death, NIOSH publishes strict guidelines regulating the manufacture of respirators. The standard with which manufacturers must comply is 42 CFR 84. In addition, safety and health professionals must ensure that employees are provided respirators that meet all of the specifications set forth in 42 CFR 84.

There are two types of respirators: air filtering and air supplying. Air-filtering respirators filter toxic particulates out of the air. To comply with 42 CFR 84, an air-filtering respirator must protect its wearer from the most penetrating aerosol size of particle, which is 0.3 micron aerodynamic mass in median diameter. The particulate filters used in respirators are divided into three classes, each class having three levels of efficiency as follows:

Class N respirators may be used only in environments that contain no oil-based particulates. They may be used in atmospheres that contain solid or nonoil contaminants.

Class R respirators may be used in atmospheres containing any contaminant. However, the filters in Class R respirators must be changed after each shift if oil-based contaminants are present. Class P respirators may be used in any atmosphere containing any particulate contaminant.

	Class N(Not Oil Resistant)	Class R(Oil Resistant)	Class P(Oil Proof)
Efficiency	95%	95%	95%
Efficiency	99%	99%	99%
Efficiency	99.97%	99.97%	99.97%

If there is any question about the viability of an air-filtering respirator in a given setting, employees should use air-supplying respirators. This type of respirator works in much the same way as an air tank for a scuba diver. Air from the atmosphere is completely blocked out, and fresh air is provided via a self-contained breathing apparatus.

In addition to the NIOSH regulations regarding respirators (42 CFR 84), there are regulations published by OSHA in 29 CFR 1010.134. Key provisions of the OSHA regulations are as follows:

- Respirators, when they are required, must be provided by the employer.
- Medical evaluations must be provided for respirator users.
- Fit testing according to standards must be conducted.
- Respirators must be used in reasonably foreseeable emergency situations.
- Respirators must be properly cleaned and maintained.
- Adequate air quality, quantity, and flow for atmosphere-supplied respirators must be ensured.
- Training and evaluation programs must be provided to ensure effectiveness.

Air Safety Program Elements

Companies with facilities in which fumes, dust, gases, vapors, or other potentially harmful particulates are present should have an air safety program as part of their overall safety and health program. The program should have at least the following elements:

- Accurate hazard identification and analysis procedures to determine what types of particulates are present and in what concentration.
- Standard operating procedures (in writing) for all elements of the air safety program.
- Respirators that are appropriate in terms of the types of hazards present and that are included in 42 CFR 84.
- Training including fit testing, limitations, use, and maintenance of respirators.
- Standard procedures for routine maintenance and storage of respirators.

Standards and Regulations

Standards and regulations relating to toxic substances come from a number of different sources. Prominent among these are OSHA and the EPA. Important standards and regulations in this area include the OSHA Chemical Process Standard, the EPA Clean Air Act, Title III of the Superfund Amendments and Reauthorization Act of 1986 (SARA), the Hazardous Materials Transportation Uniform Safety Act of 1990, the Toxic Substances Control Act, and the HAZMAT Act of 1975. These and other standards and regulations are explained in this section.

OSHA Chemical Process Guidelines

OSHA developed its Chemical Process Guidelines (Process Safety Management of Highly Hazardous Chemicals—Compliance Guidelines and Enforcement Procedures) in response to incidents in which fires and explosions resulted from accidental chemical releases. The guidelines require chemical procedures to analyze their processes to identify potentially hazardous situations and to assess the extent of the hazard. Having completed their analysis, the results must be used in their emergency response plans and to take action to minimize the hazards identified. Specific additional requirements include the following:

- Compiling process safety information
- Maintaining safe operating procedures
- Training and educating employees
- Maintaining equipment
- Conducting accident investigations
- Developing emergency response plans
- Conducting safety compliance audits[46]

EPA Clean Air Act

President George H. W. Bush signed the Clean Air Act amendments into law in November 1990, thereby renewing and extending the original Clean Air Act of 1970.

A key element of the Clean Air Act is its requirement that companies use the **maximum achievable control technology (MACT)**. Such technologies represent the current state of the art in pollution control on an ever-changing, ever-improving basis. This means that as technological improvements occur, companies will have to upgrade their pollution control systems to stay in compliance.

The standard also focuses on preventing accidental releases of toxic substances. In addition to identifying potential hazards and taking steps to control them, companies are required to develop plans for minimizing the damage that results when, in spite of controls, releases occur.

The Clean Air Act should enhance a downward trend in air pollution that was already occurring when it was renewed in 1990. That trend saw the amount of toxics released into the air being reduced at a rate of approximately 650 million pounds per year. Of course, this is far short of the 56 billion pounds per year reduction required by the Clean Air Act Amendments of 1990, but it does represent movement in the right direction.

Superfund Amendments and Reauthorization Act

Title III of the **Superfund Amendments and Reauthorization Act (SARA)** is also known as the Emergency Planning and Community Right-to-Know Act. This law is designed to allow individuals to obtain information about hazardous chemicals in their communities so that they can protect themselves in case of an emergency. It applies to all companies that use, make, transport, or store chemicals.

Safety and health professionals involved in developing emergency response plans for their companies should be familiar with SARA and its requirements relating to emergency planning. The major components of the Emergency Planning and Community Right-to-Know Act are discussed in the following paragraphs.

Emergency Planning (Sections 301–303)

Communities are required to form local **emergency planning** committees (LEPCs), and states are required to form state emergency response commissions (SERCs). LEPCs must develop emergency response plans for their local communities, host public forums, select a planning coordinator for the community, and work with the coordinator in developing local plans. SERCs must oversee LEPCs and review their emergency response plans. Plans for individual companies in a given community should be part of that community's larger plan. Local emergency response professionals should use their community's plan as the basis for simulating emergencies and practicing their response.

Emergency Notification (Section 304)

Chemical spills or releases of toxic substances that exceed established allowable limits must be reported to appropriate LEPCs and SERCs. Immediate **emergency notification** may be verbal, provided a written notification is filed promptly thereafter. The report must contain at least the following information: (1) names of the substances released, (2) where the release occurred, (3) when the release occurred, (4) the estimated amount of the release, (5) known hazards to people and property, (6) recommended precautions, and (7) name of a contact person in the company.

Information Requirements (Section 311)

Local companies are required to keep their LEPCs and SERCs—and, through them, the public—informed about the hazardous substances that they store, handle, transport, and use. These information requirements include keeping comprehensive, up-to-date records of the substances on file and readily available; providing copies of MSDSs for all hazardous substances; recording general storage locations for all hazardous substances; estimating the amount of each hazardous substance on hand on a given day; and estimating the average annual amount of hazardous substances kept on hand.

Toxic Chemical Release Reporting (Section 313)

Local companies must report the total amount of toxic substances released into the environment as either emissions or hazardous waste. Toxic chemical release reports go to the EPA and the state-level environmental agency. Section 313 applies to companies that meet the following criteria:

- Employ 10 or more full-time personnel.
- Produce or process more than 25,000 pounds of a given toxic substance or use more than 10,000 pounds of a given toxic substance in any capacity.
- Conduct 50 percent or more of their business in areas defined by Standard Industrial Classification (SIC) Codes 20–39.

Section 313 of SARA requires companies that produce, store, use, or transport chemicals to estimate and report their releases of toxic substances.

Hazardous Materials Transportation and Uniform Safety Act

Transportation of toxic substances always involves a certain amount of hazard. According to the NSC, in a given year,

> hazardous materials incidents caused 165 injuries, 17 deaths and more than $21 million in damages. Most of these happened on the nation's highways where more than 60 percent of the materials are transported. . . . Human error caused most of the incidents. Package failure caused another 30 percent of incidents and vehicle crashes caused 8 percent.[47]

The Hazardous Materials Transportation and Uniform Safety Act of 1975 was passed in response to statistics such as these. The basic provisions of the act are shown

FIGURE 16-13 Basic provisions of the Hazardous Materials Transportation and Uniform Safety Act.

- Training relating directly to specific functions for private manufacturers, shippers, and carriers.
- Training for employees of public agencies.
- National registration with registration fees used to help pay for the costs of emergency response training.
- A national permit system for motor carriers.

in Figure 16-13. In addition to these provisions, the act makes companies that transport hazardous materials partially liable for damages when an accident occurs and the carrier does not have a satisfactory rating from the Department of Transportation.

The following steps can be used for minimizing the risks associated with transporting hazardous materials:

- Hire personnel who are certified and experienced in dealing with hazardous materials. Then make sure to brief them properly on the hazardous materials that they will be handling and keep them up-to-date on the latest federal regulations.
- Establish a training program that covers at least the following topics: safety equipment, identification of hazardous shipments, routine handling procedures, and emergency procedures.
- At each location where hazardous materials will be loaded or unloaded, name one person who is responsible for making periodic inspections, instructing new employees, meeting with representatives of companies that ship major hazardous materials, maintaining comprehensive records, filing all necessary reports, and contacting the shipper and cosigner when there is a problem.
- Identify all hazardous materials and post a diamond-shaped placard on the outside of all vehicles used to transport hazardous materials.[48]

General Safety Precautions

Following are a number of general safety precautions that apply in any settings where explosive and combustible materials are present.

1. **Prohibit smoking.** Smoking should be prohibited in any areas of a plant where explosive and combustible materials are present. Eliminating potential sources of ignition is a standard safety precaution in settings where explosions and fire are possible. In the past, such areas were marked off as restricted, and No Smoking signs were posted. It is becoming common practice to prohibit smoking on the premises altogether or to restrict smoking to designated areas that are well removed from hazard areas.
2. **Eliminate static electricity.** Static electricity occurs when dissimilar materials come into contact and then separate. If these materials are combustible or are near other materials that are combustible, an explosion can occur. Therefore, it is important to eliminate static electricity. The potential for the occurrence of

FIGURE 16-14 How static electricity occurs.

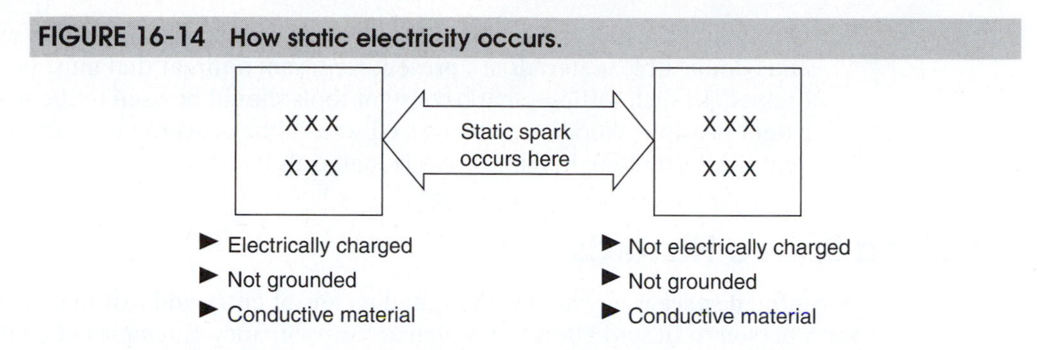

static electricity can be reduced substantially by the processes of grounding and bonding. **Bonding** involves eliminating the difference in static charge potential between materials. *Grounding* involves eliminating the difference in static charge potential between a material and the ground. Figures 16-14, 16-15, and 16-16 illustrate these concepts.

FIGURE 16-15 Bonding prevents static spark.

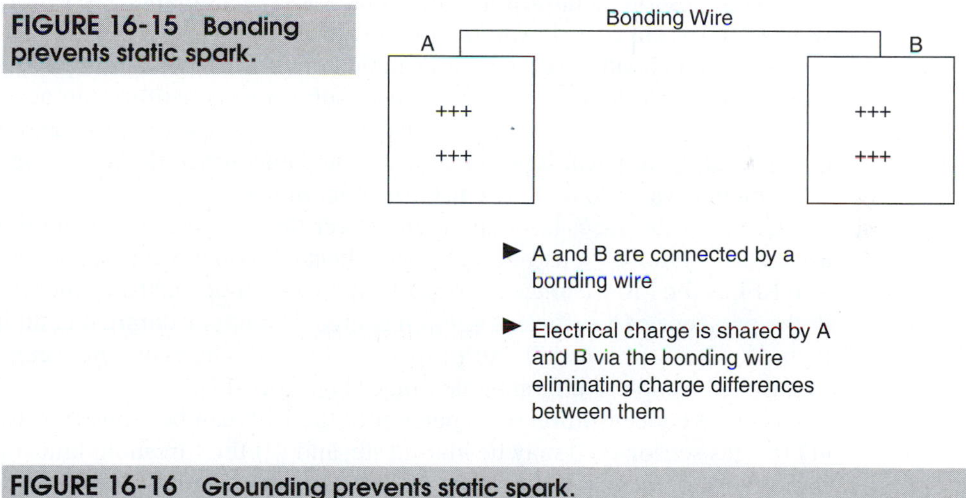

FIGURE 16-16 Grounding prevents static spark.

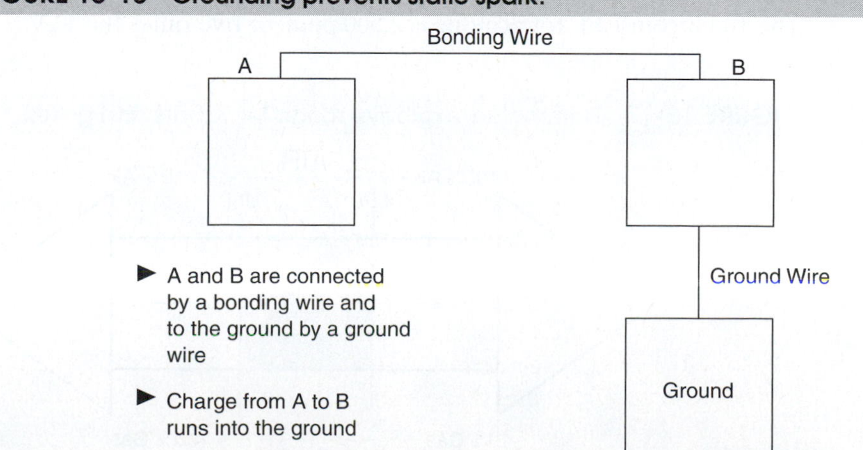

3. **Use spark-resistant tools.** Sparks from tools used in settings where explosive and combustible materials are present represent a threat that must be addressed. In such settings, spark-resistant tools should be used to the maximum extent possible. Wooden, leather-faced, and rubber-covered tools can help prevent sparks that might ignite volatile materials.[49]

Confined Space Hazards

A **confined space** is any area with limited means of entry and exit that is large enough for a person to fit into but is not designed for occupancy. Examples of confined spaces include vaults, vats, silos, ship compartments, train compartments, sewers, and tunnels. What makes confined spaces hazardous, beyond those factors that define the concept, is their potential to trap toxic and explosive vapors and gases.

OSHA's Confined Space Standard (29 CFR 1910.146, paragraph (b)) defines a hazardous atmosphere as one "that may expose employees to the risk of death, incapacitation, impairment of ability to self-rescue, injury, or acute illness."[50] The definition goes on to describe a hazardous atmosphere as a "flammable gas, vapor, or mist in excess of 10 percent of its lower flammable limit."[51]

Safety and health professionals should be cautious in assuming that 10 percent of a lower flammable limit (LFL) for a given substance constitutes an acceptable atmosphere. Even 10 percent of the LFL may exceed the TLV for gases and vapors. Consequently, safety and health professionals should encourage the use of the most sensitive instruments available to detect airborne contaminants.

Certain substances have upper and lower flammable limits (sometimes referred to as "explosive limits," a less-accurate label because some will flame without exploding). The LFL is the lowest concentration of a gas or a vapor that can generate a flame when in the presence of a sufficient ignition source. The upper flammable limit (UFL) is the highest concentration that can propagate a flame. The range between these two extremes is the explosive/flammable range (Figure 16-17).

Using a concentration of 10 percent of the LFL can be dangerous for two reasons: (1) the gas sensor used may be inaccurate; and (2) the threshold limit value (TLV) for some substances is exceeded at the LFL. For example, the TLV for acetone is 500 ppm. The 10 percent LFL for acetone is 2,500 ppm, or five times the TLV.

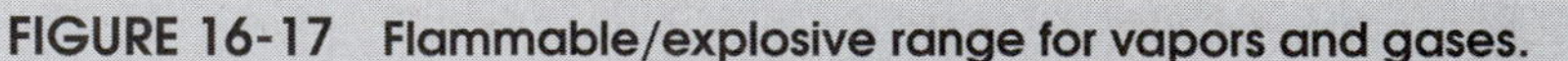

FIGURE 16-17 Flammable/explosive range for vapors and gases.

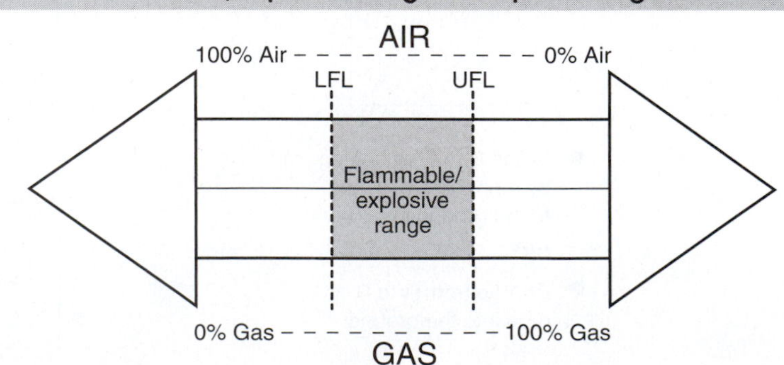

Safety and health professionals are well advised to take the following precautions when dealing with confined spaces that may have a toxic environment. First, use the most sensitive detection instrument available. Detector tubes, portable gas chromatographs, and infrared analyzers are all more accurate than combustible gas sensors are. Second, compare the 10 percent LFL for any substance in question with its TLV, and let the TLV take precedence.

To ensure that a confined space is safe, the following questions should be asked and answered in the affirmative before allowing entry. A negative response to even one of these questions means that entry into the confined space in question is not yet safe.

- Are access and exit equipment such as ladders and steps in good working condition?
- Has the confined space been properly purged of the toxic vapors and other toxic substances?
- Are all lines that transport potentially hazardous substances into or through the confined space turned off and properly capped?
- Are all moving equipment and moving parts of equipment in the confined space shut down and locked out?
- Has proper ventilation (natural or mechanical) been provided?
- Has the atmosphere inside the confined space been checked by appropriately sensitive detection devices?
- Have provisions been made to monitor continually the atmosphere inside the confined space during work?

In addition to the toxic and explosive hazards associated with confined spaces, there are often physical hazards. For example, tunnels often contain pipes that can trip an employee or that can leak and cause a fall. Empty liquid or gas storage vessels may contain mechanical equipment or pipes that must be carefully maneuvered around, often in the dark.

OSHA Confined Space Standard

The OSHA standard relating to confined spaces is found in 29 CFR 1910.146. This standard mandates that entry permits be required before employees are allowed to enter a potentially hazardous confined space. This means that an employee must have a written permit to enter a confined space. Before the permit is issued, a supervisor, safety or health professional, or some other designated individual should do the following:

1. **Shut down equipment/power.** Any equipment, steam, gas, power, or water in the confined space should be shut off and locked or tagged to prevent its accidental activation.
2. **Test the atmosphere.** Test for the presence of airborne contaminants and to determine the oxygen level in the confined space. Fresh, normal air contains 20.8 percent oxygen. OSHA specifies the minimum and maximum safe levels of oxygen as 19.5 and 23.5 percent, respectively. Atmospheric tests indicate whether a respirator is required and, if so, what type, classification, and level. Figures 16-18, 16-19, 16-20, and 16-21 are examples of devices used for checking the atmosphere.
3. **Ventilate the space.** Spaces containing airborne contaminants should be purged to remove them. Such areas should also be ventilated to keep contaminants from building up again while an employee is working in the space.

FIGURE 16-18 SAFE T NET 2000 detection device.

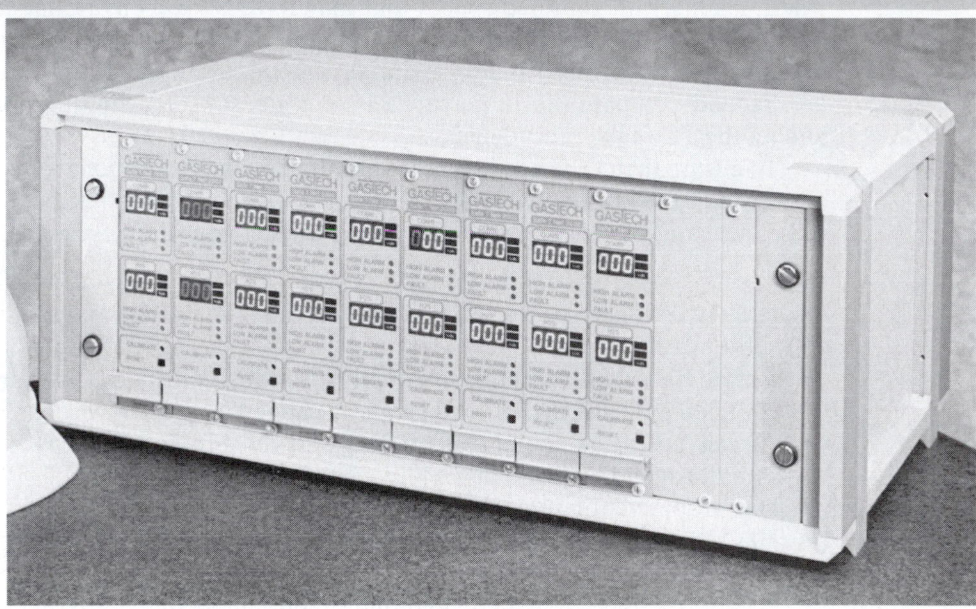

Courtesy of Gas Tech, Inc.

FIGURE 16-19 GT Land Surveyor detection device.

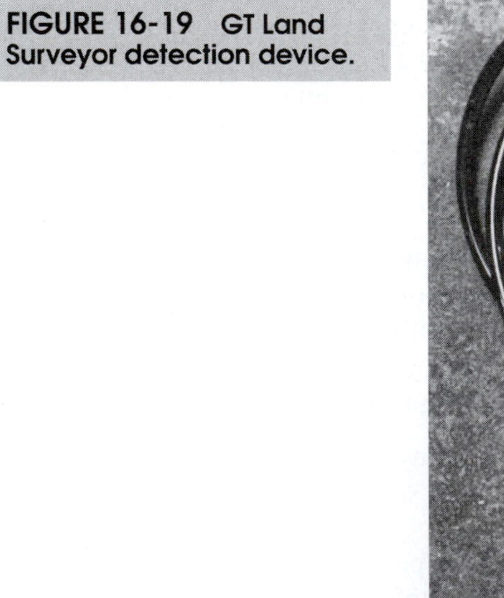

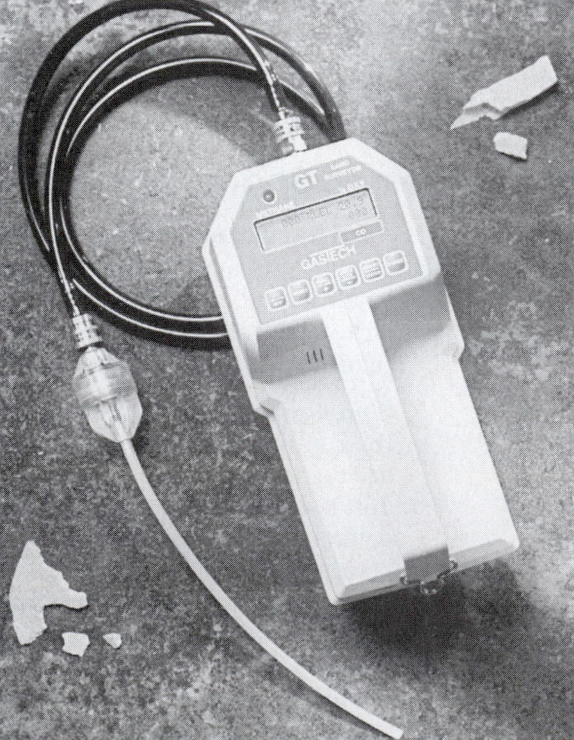

Courtesy of Gas Tech, Inc.

FIGURE 16-20 STM 2100 detection device.

Courtesy of Gas Tech, Inc.

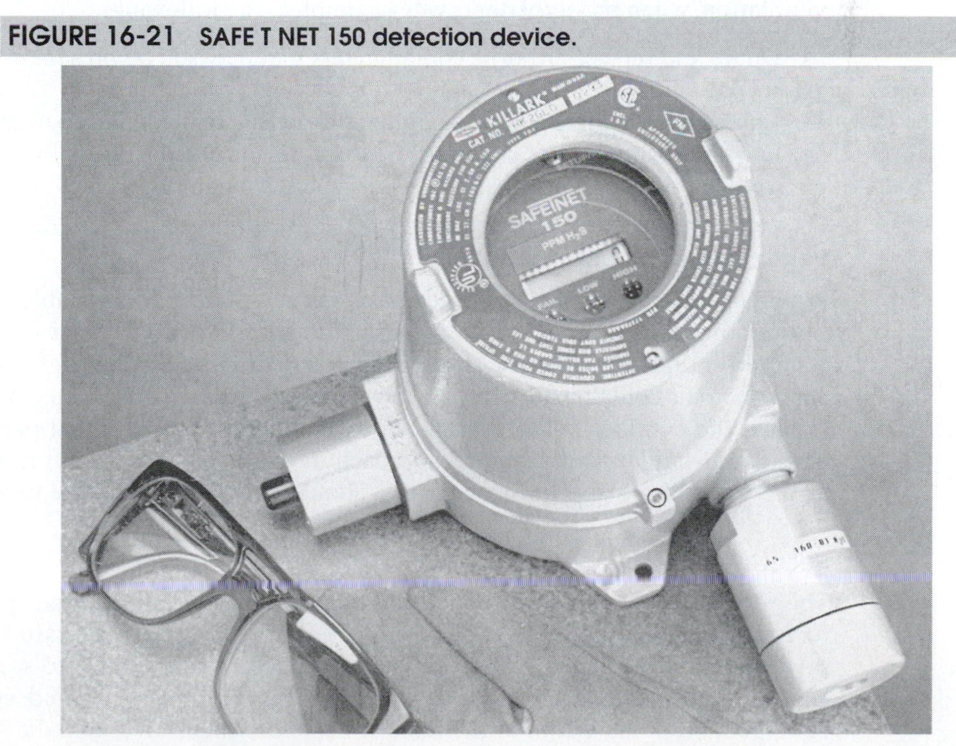

FIGURE 16-21 SAFE T NET 150 detection device.

Courtesy of Gas Tech, Inc.

4. **Have rescue personnel stand by.** Never allow an employee to enter a confined space without having rescue personnel standing by in the immediate vicinity. These personnel should be fully trained and properly equipped. It is not uncommon for an untrained, improperly equipped employee to be injured or killed trying to rescue a colleague who gets into trouble in a confined space.
5. **Maintain communication.** An employee outside the confined space should stay in constant communication with the employee inside. Communication can be visual, verbal, or electronic (radio, telephone) depending on the distance between the employee inside and the entry point.
6. **Use a lifeline.** A lifeline attached to a full-body harness and a block and tackle will ensure that the employee who is inside can be pulled out should he or she lose consciousness. The apparatus should be rigged so that one employee working alone can pull an unconscious employee out of the confined space.

Ventilation of Confined Spaces

Before allowing employees to enter a confined space, it is important to make the space as safe as possible. One of the most effective strategies for doing so is ventilation. Because confined spaces vary in size, shape, function, and hazard potential, there must be a number of different methods for ventilating them.

Before ventilating a confined space, it should be *purged.* Purging is the process of initially clearing the space of contaminants. Once the area has been purged, ventilation can begin. Ventilation is the process of continually moving fresh air through a space. Ventilation, when properly done, will accomplish the following:

- Dilute and replace airborne contaminants that may still be present in the confined space.
- Ensure an adequate supply of oxygen (between 19.5 and 23.5 percent).
- Exhaust contaminants produced by work performed in the confined space (e.g., welding, painting, and so on).

Ventilation and Local Exhaust

Providing ventilation in a confined space can maintain a comfortable temperature, it can remove odors, and it can dilute contaminants. However, never depend solely on general ventilation to remove toxic contaminants from the air. To eliminate the hazards posed by toxic contaminants such as solvent vapors and welding fumes, it is necessary to exhaust the confined space aggressively. The combination of initial purging, local exhaust, and ventilation is the ideal approach. If contaminant concentrations remain too high even with this approach, employees should wear an appropriate respirator.

Rescue Preparation

The time to think about getting injured employees out of a confined space is well before they enter the space in the first place. Every year, employees are killed trying to save an injured colleague inside a confined space. In an attempt to save injured colleagues, well-meaning employees, who are neither properly trained nor adequately equipped, often fall victim to the toxic atmosphere themselves and die. This is a tragic circumstance made even more so because it is unnecessary and avoidable.

With the right amount of planning and training, employees can be quickly and effectively rescued from confined spaces. Planning should answer the following questions:

- What types of injuries or incidents may occur in a given space?
- What types of hazards may be present in the space?
- What precautions should be taken by rescue personnel entering the space (e.g., lifelines, hoist, respirator, and so on)?
- How much maneuvering room is in the confined space?
- What if the victim needs first aid before he or she can be moved?

All these questions should be answered in the organization's emergency action plan. In addition, all members of the rescue team should have received the training necessary to respond quickly, safely, and effectively. An effective response is one that is appropriate to the magnitude of the incident and is carried out safely.

OSHA's Confined Space Rescue Requirements

OSHA's Confined Space Standard (29 CFR 1910.146) sets forth the following procedures for effecting the rescue of a worker from a confined space:

1. Preplan a confined space incident.
2. Assess the incident.
3. Conduct monitoring of the confined space environment.
4. Control any hazards present.
5. Prepare for entry—then enter the confined space.
6. Package and remove the victim from the confined space.
7. Remove entrants (rescuers) from the confined space.
8. Secure the confined space.[52]

Confined Space Management Policy

Organizations that expose workers to confined spaces in the course of doing their jobs should adopt a comprehensive confined space management policy and enforce it carefully and consistently. The policy should cover at least the following areas of concern: (1) administrative controls, (2) training for all applicable personnel, (3) permitting procedures, and (4) work-team requirements.[53]

Administrative Controls

Organizations that expose workers to confined spaces should include the following administrative controls in their confined space management policy:

1. Identification of all confined spaces and related hazards with signs and placards posted to warn employees appropriately.
2. Establishment of an entry-permitting process for controlling and restricting entry into confined spaces.
3. Selection of confined space work teams including posting of the names of team members at all sites in question.
4. Training for all members of confined space work teams.
5. Provision of the proper personal protective equipment to all members of confined space work teams as well as training in its proper use.

6. Communication of all requirement and cautions to all members of confined space work teams before their work begins.
7. Provision of stand-alone rescue equipment as well as periodic rescue drills.
8. Establishment of safe exposure levels inside the confined spaces in question including posting of these levels at the sites.
9. Monitoring and testing of atmospheric conditions inside confined spaces.
10. Issuance of evacuation orders when safe exposure levels are exceeded in confined spaces.
11. Maintenance of all applicable records for the confined space management program.

Training for Applicable Personnel

All personnel who will be assigned to a confined space work team as well as their supervisors should receive periodic training that covers at least the following areas of concern:

1. Entry permitting
2. Hazard awareness
3. Selection and proper use of monitoring equipment
4. Selection and proper use of PPE
5. Selection and proper use of stand-alone rescue equipment as well as all applicable rescue procedures
6. Communication procedures
7. Performance auditing
8. Documentation and recordkeeping

Permitting Procedures

Organizations that expose workers to confined spaces should have a comprehensive permitting procedure that is adhered to without exception. What follows is a checklist that should be completed before issuing a permit allowing any member of a confined space work team to begin work:

- Has an entry supervisor been designated to ensure compliance with all applicable procedures?
- Has confined space monitoring been completed and documented?
- Have all applicable PPE requirements been established for the confined space in question?
- Are all entrants wearing the proper PPE and have they been trained in its proper use?
- Have all concerns of work-team members have satisfactorily resolved?
- Has an attendant who will monitor from outside the confined space been designated and properly trained?
- Is the attendant properly situated to carry out his responsibilities?
- Does the attendant know how and when to order an evacuation of the confined space?
- Has access to the confined space in question been properly controlled?
- Have communication and evacuation procedures been discussed and practiced?
- Has an emergency rescue team been established, properly trained, and made ready to act?

- Is stand-alone rescue equipment available outside the entrance to the confined space?
- Has the emergency rescue team established its procedures and practiced them?
- Has an expiration date and time for work in the confined space been established and made known to all members of the work team?
- Have those who are authorized to enter the confined space been designated and are there names known by all members of the work team?
- Has the work that is authorized to be done in the confined space been outlined and communicated to all members of the work team?
- Have procedures for monitoring the confined space during work been established and practiced?

When the answer is "Yes" to all the questions on the checklist, the entry supervisor is allowed to issue the necessary permits for entry into the confined space in question. The checklist should be completed every time work must be done in a confined space. Attempting to save time by skipping steps can be a fatal mistake.

Work-Team Requirements

Organizations that expose workers to confined spaces should establish work teams that consist of the following: (1) entry supervisor who issues the work permit, (2) attendant who monitors from outside the confined space and enforces the work permit, (3) entrants who actually enter the confined space to perform the work, (4) monitor who regularly monitors the atmosphere in the confined space and records the results, (5) rescue team, and (6) safety and health professional to coordinate the team and all its activities.

OSHA Standards for Toxic and Hazardous Materials

The OSHA standards for hazardous materials are contained in 29 CFR (Subpart H). Nine of the standards apply to specific materials. Four of the standards have broader applications. The standards applying to specific materials are as follows:

HAZARDOUS MATERIALS (SPECIFIC STANDARDS)

1910.101	Compressed gases
1910.102	Acetylene
1910.103	Hydrogen
1910.104	Oxygen
1910.105	Nitrous oxide
1910.108	Dip tanks
1910.109	Explosives and blasting agents (An amendment to the OSH Act passed in 1992 now requires that manufacturers of explosives and pyrotechnics must observe the requirements of the Process Safety Management Standards in 1910.119 in addition to this standard.)
1910.110	Liquefied petroleum gases
1910.111	Anhydrous ammonia

In addition to these specific standards, Subpart H contains four standards that have broad applications. Standard 1910.106 parallels the National Fire Protection Association's NFPA 30: *Flammable and Combustible Liquids Code.* Standard 1910.107

regulates processes in which paint is applied by compressed air, electrostatic steam, or other continuous or intermittent processes. Standard 1910.119 regulates process safety management relating to 125 specific chemicals. Standard 1910.120 regulates both hazardous waste operations and spills or accidental releases.

Toxic and hazardous substances are covered in 29 CFR (Subpart Z). The standards in this subpart establish permissible exposure limits (PELs) for over 450 toxic and hazardous substances. Each standard deals with a specific substance or substances. The standards contained in Subpart Z begin with 1910.1000 and run through 1910.1500.

OSHA'S Hazard Communication Standard

Any organization that uses hazardous materials in the workplace is required to fully inform employees and on-site contractors of the hazards and to provide training concerning the safe handling, storage, and use of the materials. OSHA's Hazard Communication Standard is 29 CFR 1910.1200.[54] Organizations that use any of the following types of substances should be especially attentive to complying with this standard: acids, cleaning solvents and compounds, flammable gases and liquids, paints, lacquers, and enamels, insecticides, lubricants and oils, fumigants, fungicides and herbicides, and adhesives. This list is not exhaustive, but it does provide a good start concerning what substances to watch for.

Written Hazard Communication Program

Organizations that use hazardous materials are required by 29 CFR 1910.1200 to develop a written program and to communicate applicable elements of it to employees and on-site contractors. An organization's written hazard communication program should contain at least the following elements:

1. Purpose and objectives
2. Inventory (list) of hazardous chemicals and substances
3. Labels and other forms for warning (prevention procedures)
4. Material safety data sheets (procedures and sheets)
5. Hazards of nonroutine tasks (methods of informing)
6. On-site contractors (methods of informing)
7. Training (procedures and elements)

Employee Right-to-Know

The purpose of 29 CFR 1910.1200 is to prevent accidents and injuries by making sure that employees and on-site contractors know about any hazardous substances they might come in contact with in the course of their work, and how to use those substances safely. Those portions of an organization's written hazard communication program that must be shared with employees and on-site contractors are as follows:

1. **Hazardous product inventory.** A comprehensive list of all hazardous substances found in the workplace in question.
2. **Material safety data sheets file.** A master file of data sheets for all hazardous substances in the inventory must be maintained both at the main offices of the organization and any branch sites.

FIGURE 16-22 Employees must be notified of this right to know.

ATTENTION ALL EMPLOYEES!
As an employee, or on-site contractor, you have the right to be informed of all hazardous chemical substances to which you may be exposed in the workplace.
29 CFR 1910.1200

3. **Proper labeling.** All containers—original, immediate-use, and storage— must be properly labeled. The label must contain the name of the hazardous substance, the manufacturer, emergency procedures, and instructions for the safe use of the substance in question. "Immediate-use" containers (those filled by an employee for use during a shift) must have a label that contains the name of the substance and the nature of the hazard (e.g., caustic, corrosive, flammable, etc.).
4. **Emergency plan.** Such plans must contain all the necessary actions to be taken in case of an incident (e.g., fire, spill, accidental release, reaction, etc.). The plan must also list all equipment and materials required to properly respond to an incident, an indication of their locations, and a description of the procedures for using the equipment and materials.
5. **Employee training program.** Training provided to employees and onsite contractors must cover at least the following: (a) discussion of material safety data sheets for all substances in the inventory; (b) right-to-know requirements so that employees and on-site contractors understand what they have a right to know (see Figure 16-22); (c) explanation of labeling requirements; (d) proper handling and use procedures; (e) emergency procedures; and (f) completion of a standardized form verifying completion of the training.

Key Terms and Concepts

- Absorption
- Acceptable air quality
- Acute effects and exposures
- Aerosols
- Airborne contaminants
- Air-filtering respirators
- Air safety program
- Air-supplying respirators
- Anesthetics
- Application flexibility
- Asbestos-containing material (ACM)
- Asbestos removal
- Asphyxiants
- Biological exposure index (BEI)
- Biological hazards
- Bonding
- Carcinogen
- Ceiling
- Chemical hazards
- Chemical Process Standard
- Chronic effects and exposures
- Clean Air Act
- Confined space
- Dose
- Dose threshold
- Dusts
- Emergency action plan
- Emergency notification

- Emergency planning
- Emergency response plan
- Encapsulation
- Enclosure
- Engineering controls
- Ergonomic hazards
- Exposure ceiling
- Exposure thresholds
- Friable asbestos
- Good housekeeping
- Grounding
- Hazard recognition
- HAZWOPER
- High-efficiency particle absolute (HEPA)
- Indoor air quality (IAQ)
- Industrial hygiene
- Ingestion
- Inhalation
- Injection
- Ionizing radiation
- Irritants
- Isolating or enclosing
- Lethal concentration
- Lethal dose
- Local exhaust
- Material safety data sheet (MSDS)
- Maximum achievable control technology (MACT)
- Mists
- Narcotics
- Neoplastic growth
- Noise
- Nonionizing radiation
- OSHA Process Safety Standard
- OSHA's Hazard Communication Standard
- Personal hygiene
- Personal protective equipment (PPE)
- Physical hazards
- Radiation
- Respirator
- Route of entry
- Self-protection strategies
- Short-term exposure limit
- Sick-building syndrome
- Static electricity
- Substituting
- Superfund Amendments and Reauthorization Act (SARA)
- Temperature control
- Threshold limit value (TLV)
- Time-weighted average (TWA)
- Toxic substance
- Tracer gas techniques
- Vapors
- Ventilation
- Volatility
- Wet methods

Review Questions

1. Define the term *industrial hygiene.*
2. Briefly explain the responsibilities of the modern industrial hygienist.
3. What is the role of the safety and health professional regarding industrial hygiene?
4. List five OSHA requirements relating to industrial hygiene.
5. Briefly explain the typical categories of hazards in the workplace.
6. What are the most common routes of entry for toxic agents?
7. Describe the following types of airborne contaminants: dusts, fumes, smoke, mists, and gases.
8. What factors should be considered in deciding whether to remove or contain asbestos?
9. Summarize the various elements of ANSI Z9.8 (indoor air quality/HVAC).
10. Explain the following ways of dealing with asbestos in the workplace: removal, enclosure, and encapsulation.
11. What types of medical examinations should be required of employees who handle ACMs?
12. Briefly explain the following concepts relating to exposure thresholds: time-weighted average, short-term exposure limit, and exposure ceiling.
13. List the most important considerations when evaluating hazards in the workplace.
14. List five generic prevention and control strategies that can be used in any workplace.
15. Give an example of a prevention/control strategy in each of the following categories: engineering controls, ventilation, and personal protective equipment.
16. Explain five self-protection strategies that employees can use in the workplace.
17. How does NIOSH relate to industrial hygiene?
18. What is a toxic substance?
19. List the factors that determine the effect that a toxic substance will have.
20. Describe the most common routes of entry for toxic substances.
21. Explain the mathematical expression of the dose–response relationship.
22. Define the following terms: *dose threshold, lethal dose,* and *lethal concentration.*
23. Differentiate between acute and chronic effects and exposures.
24. List and describe the various classifications of airborne toxics.
25. What is a carcinogen?
26. Describe the basic provisions of the following standards: OSHA Chemical Process Standard, EPA Clean Air Act, and SARA.
27. What is a threshold limit value?

28. Define the following terms: *time-weighted average* and *ceiling.*
29. Explain the three NIOSH categories of respirators.
30. What is "sick-building" syndrome?
31. Explain the major tenets of the OSHA Confined Space Standard.
32. Explain the various elements that should be included in an organization's confined space management policy.
33. Summarize the requirements of OSHA's Hazard Communication Standard.

ENDNOTES

1. J. B. Olishefski, "Overview of Industrial Hygiene," in *Fundamentals of Industrial Hygiene,* 3rd ed. (Chicago: National Safety Council), 5.
2. Retrieved from http://www.epa.gov/air/oagps/peg-caa/pegcaain.html.
3. Olishefski, "Overview of Industrial Hygiene," 160–161.
4. Ibid., 164–165.
5. Ibid., 10.
6. Ibid., 12–14.
7. Ibid., 14.
8. Ibid., 152.
9. Ibid., 17.
10. Ibid.
11. Ibid., 14.
12. Ibid., 148.
13. Ibid., 18–19.
14. Ibid., 22–23.
15. Ibid., 368.
16. Ibid., 369.
17. Centers for Disease Control and Prevention, National Institute for Occupational Safety and Health, *Selected Topics: Asbestos, May 25,* 2006, 4–7. Retrieved from http://www.cdc.gov/niosh/asbestos.html.
18. Ibid.
19. Ibid.
20. Ibid.
21. Ibid.
22. Ibid., 3–4.
23. Ibid., 48.
24. Ibid.
25. Ibid.
26. Ibid., 49.
27. Ibid.
28. Ibid.
29. A. Draper, "I Think It's Mold (Now What?)," *Occupational Health & Safety* 74, no. 5: 67–69.
30. American Conference of Governmental Industrial Hygienists, *Threshold Limit Values* (Cincinnati, OH: American Conference of Governmental Industrial Hygienists, 2006), 3.
31. Ibid., v–vi, 45, 44, 96, and 112.
32. Olishefski, "Overview of Industrial Hygiene," 21.
33. Ibid.
34. Ibid.
35. Ibid., 22.
36. J. Bone, "Textile Industry Weaves a Safety Future," *Safety & Health* 144, no. 3: 17.
37. Ibid., 52.
38. Ibid., 53.
39. Olishefski, "Overview of Industrial Hygiene," 25.
40. Ibid.
41. Ibid., 26.
42. Ibid.
43. Ibid., 27.
44. Olishefski, "Overview of Industrial Hygiene," 24.
45. National Institute for Occupational Safety and Health, Publication No. 90–109. Retrieved June 2006 from http://www.cdc.gov/NIOSH.
46. Retrieved from http://www.osha.gov/pls/oshaweb/owadisp.show.
47. National Safety Council, Accident Facts, 2006, 19.
48. Hazardous Materials Transportation and Uniform Safety Act. Retrieved June 2006 from http://www.eh.doe.gov/oepa/laws/hmta.htm.
49. National Institute for Occupational Safety and Health, Publication 90–109.
50. 29 CFR 1910.146, paragraph (b).
51. Ibid.
52. 29 CFR 1910.146.
53. S. V. Magyar, Jr., "Confined Space Entry," *Occupational Health & Safety* 75, no. 2: 38–50.
54. Retrieved from http://online.misu.nokak.edu/19577/hazMater.htm.

Violence in the Workplace

Major Topics

- Occupational Safety and Workplace Violence: The Relationship
- Workplace Violence: Definitions
- Workplace Violence: Cases
- Size of the Problem
- Legal Considerations
- Risk-Reduction Strategies
- Contributing Social and Cultural Factors
- OSHA's Voluntary Guidelines
- Conflict Resolution and Workplace Violence
- Do's and Don'ts for Supervisors
- Emergency Preparedness Plan

> America has been hard at work in the past 10 days and here is what happened: A Federal Express pilot took a claw hammer and attacked three others in the cockpit, forcing one of them to put the fully loaded DC-10 cargo plane through a series of violent rolls and nose dives in a melee that brought the whole crew back bleeding. A purchasing manager in suburban Chicago stabbed his boss to death because, police say, they couldn't agree on how to handle some paperwork. And a technician who quit because he had trouble working for a woman sneaked back inside the fiber optics laboratory, pulled out a 9-mil semiautomatic pistol and started firing at workers, who ducked or fled and curled up in closets and file cabinets. By the time he finished the job, two were dead, two were injured; he walked upstairs to an office and shot himself in the head.[1]

Workplace violence has emerged as a critical safety and health issue. According to the Bureau of Labor Statistics, homicide is the second leading cause of death to American workers, "accounting for 16 percent of the 6,588 fatal work injuries in the United

States."[2] Although more than 80 percent of workplace homicide victims are men, workplace violence is not just a male problem. In fact, workplace homicide is the leading cause of death on the job for women in the United States.

Almost one million people are injured or killed in workplace-violence incidents every year in the United States, and the number of incidents is on the rise. In fact, according to the U.S. Department of Justice, the workplace is the most dangerous place to be in the United States.[3] Clearly, workplace violence is an issue of concern to safety and health professionals.

Occupational Safety and Workplace Violence: The Relationship

The prevention of workplace violence is a natural extension of the responsibilities of safety and health professionals. Hazard analysis, records analysis and tracking, trend monitoring, incident analysis, and prevention strategies based on administrative and engineering controls are all fundamental to both concepts. In addition, emergency response and employee training are key elements of both. Consequently, occupational safety and health professionals are well suited to add the prevention of workplace violence to their normal duties.

Workplace Violence: Definitions

Safety and health professionals should be familiar with the language that has developed around the issue of workplace violence. This section contains the definitions of several concepts as they relate specifically to workplace violence.

- **Workplace violence.** Violent acts, behavior, or threats that occur in the workplace or are related to it. Such acts are harmful or potentially harmful to people, property, or organizational capabilities.
- **Occupational violent crime (OVC).** Intentional battery, rape, or homicide during the course of employment.[4]
- **Employee.** An individual with an employment-related relationship (present or past) with the victim of a workplace-violence incident.
- **Outsider.** An individual with no relationship of any kind with the victim of a workplace-violence incident or with the victim's employer.
- **Employee-related outsider.** An individual with some type of personal relationship (past or present) with an employee, but who has no work-related relationship with the employee.

SAFETY FACT

High-Risk Occupations and People

High-risk occupations in terms of workplace violence are taxicab drivers; retail workers; police and security officers; and finance, insurance, real estate, health care, and community service employees. Employees 65 years and older are more likely to be victims than are younger employees.

- **Customer.** An individual who receives products or services from the victim of a workplace-violence incident or from the victim's employer.

Each of these terms has other definitions. Those presented here reflect how the terms are used in the language that has evolved around workplace violence.

Workplace Violence: Cases

This section contains numerous cases of workplace violence that occurred in the United States during a one-year period. These cases are provided to give students of occupational safety and health a better understanding of the types of violent incidents that occur frequently in today's workplace. The names of individuals involved have been changed, but the incidents are real.

1. Jackson, Mississippi. A 32-year-old firefighter shot his wife through the head and then proceeded to a firehouse. Using an assault rifle, the man shot six coworkers—all supervisors—killing four of them and seriously wounding another two. He then fled the scene and exchanged fire with a police officer, wounding the officer, before being shot in the head himself and critically injured. The president of the union representing the shooter described him as "a time bomb waiting to go off."

2. Fort Lauderdale, Florida. Shouting "Everyone is going to die," a maintenance employee who had been fired walked into a meeting of his former coworkers and began shooting, killing five people and injuring another. The employees were inside a temporary trailer office when the disgruntled former employee showed up with two handguns. Police say that he chased workers around the office and shot them methodically, pausing only to reload.

3. Honolulu, Hawaii. An employee returned to his former workplace after being fired from his job. He held five coworkers hostage—including his former boss—for up to six hours. During the incident, the disgruntled former employee shot and seriously wounded his former supervisor. He was eventually killed by police after he held a shotgun to the head of one coworker for several hours while negotiating with officers. Prior to being shot by police, the perpetrator threatened to kill the coworker and started a countdown to pulling the trigger of the gun. The countdown prompted the hostage to grab the barrel of the gun and gave police the opportunity to shoot. The perpetrator died of his wounds.

4. Waterville, Maine. A former patient of a mental institution was accused of beating and stabbing four nuns, killing two of them. Police said that they caught the perpetrator in the convent's chapel Saturday evening, standing over one of the nuns and beating her with a religious figurine. Officers said that they had to pull him off the woman and that he had also beaten and stabbed three others in an adjacent part of the convent. Nuns at the convent of the Servants of the Blessed Sacrament said that the man had applied for a job but had been turned down. Nuns had just finished a prayer service when the perpetrator smashed the glass on a locked door, opened it, and walked inside. The perpetrator was described as an accomplished musician who had played trumpet in a local jazz combo and studied at the University of Maine at Augusta.

5. Evensdale, Ohio. A male in his early fifties returned to the offices from which he had recently been fired. Brandishing two pistols, he shot and killed three employees and wounded a fourth. A witness quoted the perpetrator as saying he was "going after someone who had screwed him over." After the murders, he surrendered quietly to authorities.

6. Harlem, New York. A man who was apparently involved in protesting the closure of an electronics store entered a clothing store with a gun and arson materials. He immediately set the store on fire and began shooting at employees. Eight individuals were killed (including the perpetrator), and another four were wounded—three seriously—by gunfire.

7. Columbus, Ohio. A man was fired from his job at a bank credit card center following charges of sexual harassment against coworkers. A year later, he forced his way into at least two homes of former coworkers, fatally shooting four individuals (including a four-month-old child) and wounding two others. He was captured by police attempting to flee the city in his automobile.

8. San Jose, California. A young accountant, on the job for just six weeks, shot and killed his female supervisor, then committed suicide with the same gun. This happened one day after he received his first performance counseling session. The killer was a 28-year-old Asian male. The victim was a 32-year-old white female.

9. Palatine, Illinois. A postal worker reported to work with a handgun and shot two coworkers who he claimed were his friends. The shooter was a 53-year-old white male who had worked at the same location for almost 20 years and had an exemplary service record.

10. Los Angeles, California. A city electrician with a 12-year work history shot and killed four of his supervisors when he learned he was facing possible dismissal for poor performance. After shooting his supervisors, the killer, 42 years old, quietly waited for police to arrive and arrest him. He was heard saying that he had specifically targeted the four murdered individuals because he "felt he was being picked on and singled out" by them.

11. Industry, California. A "quiet, unassuming" postal worker who had been on the job for 22 years shot and killed his supervisor. The perpetrator was 58 years old and was easily disarmed by coworkers after the shooting.

12. Asheville, North Carolina. A "classic loner" just fired from his job at a machine tool company returned the next day with a rifle and a pistol. The perpetrator, 47 years old, killed three workers and wounded another four before quietly surrendering to police.

13. Littleton, Colorado. A man distraught over marital problems opened fire in a crowded grocery store, killing three people before he was subdued by a bystander. The perpetrator was a 35-year-old white man. Among the victims was his wife, age 37, and the store manager, age 39. This type of incident is becoming more common. It is often referred to as *spillover* violence.

14. Richmond, California. After a Richmond Housing Authority employee was fired from his job, he went to his automobile, retrieved a handgun, and returned to shoot and kill a supervisor and a coworker. The perpetrator was 38 years old. Both victims were women, aged 47 and 24 years.

Size of the Problem

Violence in the workplace no longer amounts to just isolated incidents that are simply aberrations. In fact, workplace violence should be considered a common hazard worthy of the attention of safety and health professionals. In a report on the subject, the U.S. Department of Justice revealed the following information:

- About one million individuals are the direct victims of some form of violent crime in the workplace every year. This represents approximately 15 percent of all violent crimes committed annually in America. Approximately 60 percent of these violent crimes were categorized as *simple assaults* by the Department of Justice.
- Of all workplace violent crimes reported, over 80 percent were committed by males; 40 percent were committed by complete strangers to the victims; 35 percent by casual acquaintances, 19 percent by individuals well known to the victims, and 1 percent by relatives of the victims.
- More than half of the incidents (56 percent) were not reported to police, although 26 percent were reported to at least one official in the workplace.
- In 62 percent of violent crimes, the perpetrator was not armed; in 30 percent of the incidents, the perpetrator was armed with a handgun.
- In 84 percent of the incidents, there were no reported injuries; 10 percent required medical intervention.
- More than 60 percent of violent incidents occurred in private companies, 30 percent in government agencies, and 8 percent to self-employed individuals.
- It is estimated that violent crime in the workplace caused 500,000 employees to miss 1,751,000 days of work annually, or an average of 3.5 days per incident. This missed work equates to approximately $55 million in lost wages.[5]

The Society for Human Resource Management (SHRM) periodically surveys its members on the issue of workplace violence. One such survey produced the following results:[6]

REGARDING VIOLENT INCIDENTS IN THE WORKPLACE

- 33 percent of all managers surveyed experienced at least one violent incident in the workplace.
- 54 percent of these managers reported between two and five acts of violence in the five years prior to the survey.

REGARDING THE TYPE OF VIOLENCE EXPERIENCED

- 75 percent of the reported incidents were fistfights.
- 17 percent of the incidents were shootings.
- 8 percent of the incidents were stabbings.
- 6 percent of the incidents were sexual assaults.

REGARDING THE VICTIMS OF THE INCIDENTS

- 54 percent of the incidents were employee against employee.
- 13 percent of the incidents were employee against a supervisor.
- 7 percent of the incidents were customer against worker(s).

REGARDING THE GENDER OF THE PERPETRATOR

- 80 percent of all violent acts were committed by males.

REGARDING THE INJURIES SUSTAINED BY THE VICTIMS

- 22 percent of the incidents involved serious harm.
- 42 percent of the incidents required medical intervention.

REGARDING THE REASONS FOR THE VIOLENT INCIDENTS

- 38 percent were attributed to personality conflicts.
- 15 percent were attributed to marital or family problems.
- 10 percent were attributed to drug or alcohol abuse.
- 7 percent were nonspecific as to attribution.
- 7 percent were attributed to firings or layoffs.

REGARDING CRISIS MANAGEMENT PROGRAMS

- 28 percent of the organizations had a crisis management program in place prior to the violent incident.
- 12 percent of the organizations implemented a crisis management program after the violent incident occurred.

REGARDING THE EFFECT OF A VIOLENT INCIDENT ON THE WORKPLACE

- 41 percent of the organizations reported increased stress levels in the workplace after a violent incident.
- 20 percent reported higher levels of paranoia.
- 18 percent reported increased distrust among employees.

Legal Considerations

Most issues relating to safety and health have legal ramifications, and workplace violence is no exception. The legal aspects of the issue revolve around the competing rights of violent employees and their coworkers (Figure 17-1). These conflicting rights create potential liabilities for employers.

Rights of Violent Employees

It may seem odd to be concerned about the rights of employees who commit violent acts on the job. After all, logic suggests that in such situations the only concern would be the protection of other employees. However, even violent employees have rights. Remember, the first thing that law enforcement officers must do after taking criminals into custody is to read them their rights. According to Stephen C. Yohay,

> Employee rights are granted by a number of sources, including individual employment contracts, collective bargaining agreements, statutes such as the

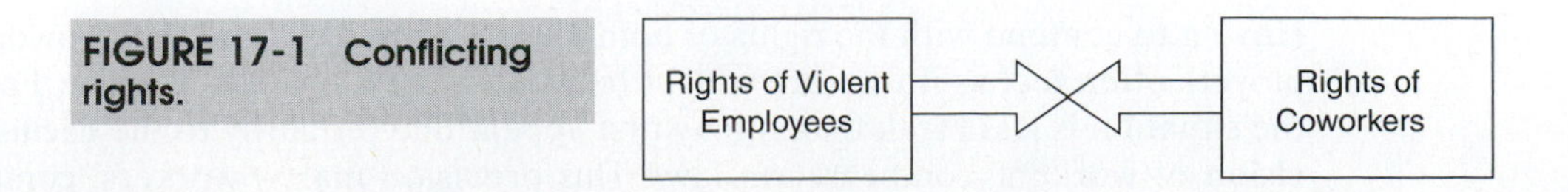

FIGURE 17-1 Conflicting rights.

SAFETY MYTH

Workplace Violence—An Isolated Problem?

Most working Americans are unconcerned about workplace violence. They see it as an isolated problem that happens in other organizations, but not theirs. Right? Not any more. A Time/CNN poll of the general population in the United States found that 37 percent of respondents viewed workplace violence as a growing problem. Of those responding to the poll, 18 percent had personally witnessed some form of violence at work. The same percentage of respondents (19 percent) said they feared for their own safety while at work.

various state, local, and federal civil rights statutes, and sometimes from policies and procedures utilized by the employer. To the extent possible under the circumstances, the employer should consider following the procedures outlined in applicable contracts or policies before taking any adverse employment action against a threatening or violent employee.[7]

This does not mean that an employer cannot take the immediate action necessary to prevent a violent act or the recurrence of such an act. In fact, failure to act prudently in this regard can subject an employer to charges of negligence. However, before taking long-term action that will adversely affect the violent individual's employment, employers should follow applicable laws, contracts, policies, and procedures. Failure to do so can serve to exacerbate an already difficult situation.

For example, in *International Union v. Micro Manufacturing, Inc.,* the employer was required to reinstate an employee who had been fired for assaulting and battering the owner of the company.[8] The reason? The violent employee had been terminated on the spot without a union representative present. This approach violated the terms of the union's collective bargaining agreement.

In addition to complying with all applicable laws, policies, and procedures, it is also important to apply these consistently when dealing with violent employees. Dealing with one violent employee one way while dealing with another in a different way puts the employer at a disadvantage legally. Consequently, it is important for employers to be prepared to deal both promptly and properly with violent employees. According to Yohay,

> The courts and the EEOC as a general matter do not require an employer to retain an employee who is violent or threatening to the other employees. However, . . . the violent and threatening employee does have some competing rights. Employers must check the law in their jurisdictions to ascertain exactly where the line is drawn regarding the rights of the threatening or violent employee.[9]

Employer Liability for Workplace Violence

Having to contend with the rights of both violent employees and their coworkers, employers often feel as if they are caught between a rock and a hard place. Fortunately, the situation is less bleak than it may first appear due primarily to the **exclusivity provision** of workers' compensation laws. This provision makes workers' compensation the employee's exclusive remedy for injuries that are work related. This means that even in cases of workplace violence, as long as the violence occurs within the scope of the victim's employment, the employer is protected from civil lawsuits and the excessive jury verdicts that have become so common.

The key to enjoying the protection of the exclusivity provision of workers' compensation laws lies in determining that violence-related injuries are within the scope of the victim's employment—a more difficult undertaking than one may expect. For example, if the violent act occurred at work but resulted from a non-work-related dispute, does the exclusivity provision apply? What if the dispute was work related, but the violent act occurred away from the workplace?

Making Work-Related Determinations

The National Institute for Occupational Safety and Health (NIOSH) developed the following guidelines for categorizing an injury as being work related:

- If the violent act occurred on the employer's premises, it is considered an on-the-job event if one of the following criteria apply:

  The victim was engaged in work activity, apprenticeship, or training.

  The victim was on break, in hallway, restrooms, cafeteria, or storage areas.

  The victim was in the employer's parking lots while working, arriving at, or leaving work.

- If the violent act occurred off the employer's premises, it is still considered an on-the-job event, if one of the following criteria apply:

  The victim was working for pay or compensation at the time, including working at home.

  The victim was working as a volunteer, emergency services worker, law enforcement officer, or firefighter.

  The victim was working in a profit-oriented family business, including farming.

  The victim was traveling on business, including to and from customer–business contacts.

  The victim was engaged in work activity in which the vehicle is part of the work environment (e.g., taxi driver, truck driver, and so on).[10]

Risk-Reduction Strategies

Figure 17-2 is a checklist that can be used by employers to reduce the risk of workplace violence in their facilities. Most of these risk-reduction strategies grow out of the philosophy of *crime reduction through environmental design (CRTED)*.[11] CRTED has the following four major elements, to which the author has added a fifth (administrative controls):

- Natural surveillance
- Control of access
- Establishment of territoriality
- Activity support
- Administrative controls

FIGURE 17-2 Checklist for workplace-violence risk reduction.

- ✓ Identify high-risk areas and make them visible. Secluded areas invite violence.
- ✓ Install good lighting in parking lots and inside all buildings.
- ✓ Minimize the handling of cash by employees and the amount of cash available on the premises.
- ✓ Install silent alarms and surveillance cameras where appropriate.
- ✓ Control access to all buildings (employee badges, visitor check-in and check-out procedure, visitor passes, and so on).
- ✓ Discourage working alone, particularly late at night.
- ✓ Provide training in conflict resolution as part of a mandatory employee orientation.
- ✓ Conduct background checks before hiring new employees.
- ✓ Train employees how to handle themselves and respond when a violent act occurs on the job.
- ✓ Develop policies that establish ground rules for employee behavior and responses in threatening or violent situations.
- ✓ Nurture a positive, harmonious work environment.
- ✓ Encourage employees to report suspicious individuals and activities or potentially threatening situations.
- ✓ Deal with allegations of harassment or threatened violence promptly before the situation escalates.
- ✓ Take threats seriously and act appropriately.
- ✓ Adopt a *zero-tolerance* policy toward threatening or violent behavior.
- ✓ Establish a *violence hot line* so that employees can report potential problems anonymously.
- ✓ Establish a *threat-management team* with responsibility for preventing and responding to violence.
- ✓ Establish an *emergency response team* to deal with the immediate trauma of workplace violence.

The following explains how these elements can help avoid workplace violence:

1. Natural surveillance. This strategy involves designing, arranging, and operating the workplace in a way that minimizes secluded areas. Making all areas inside and outside the facility easily observable allows for natural surveillance.

2. Control of access. One of the most common occurrences of workplace violence involves an outsider entering the workplace and harming employees. The most effective way of stopping this type of incident is to control access to the workplace. Channeling the flow of outsiders to an access-control station, requiring visitor's passes, issuing access badges to employees, and isolating pickup and delivery points can minimize the risk of violence perpetrated by outsiders.

3. Establishment of territoriality. This strategy involves giving employees control over the workplace. With this approach, employees move freely within their established

DISCUSSION CASE

What Is Your Opinion?

A man walks into an office building and asks to see his wife. The man is well known to the other employees, one of whom escorts him to his wife's workstation. Suddenly, the man pulls a gun and shoots his wife and another employee who tries to intervene. Is this an on-the-job event? Is the employer at fault? What is your opinion about this incident involving an employee-related outsider?

territory but are restricted in other areas. Employees come to know everyone who works in their territory and can, as a result, immediately recognize anyone who shouldn't be there.

4. Activity support. Activity support involves organizing work flow and natural traffic patterns in ways that maximize the number of employees conducting natural surveillance. The more employees observing the activity in the workplace, the better.

5. Administrative controls. Administrative controls consist of management practices that can reduce the risk of workplace violence. These practices include establishing policies, conducting background checks, and providing training for employees.

Contributing Social And Cultural Factors

Another way to reduce the risk of workplace violence is to ensure that managers understand the social and cultural factors that can lead to it. These factors fall into two broad categories: individual and environmental factors.

Individual Factors Associated with Violence

The factors explained in this section can be predictors of the potential for violence. Employees and individuals with one or more of the following factors may respond to anger, stress, or anxiety in a violent way.

1. Record of violence. Past violent behavior is typically an accurate predictor of future violent behavior. Consequently, thorough background checks should be a normal part of the employment process.

2. Membership in a hate group. Hate groups often promote violence against the subjects of their prejudice. Hate-group membership on the part of an employee should raise a red flag in the eyes of management.

3. Psychotic behavior. Individuals who incessantly talk to themselves, express fears concerning conspiracies against them, say that they hear voices, or become increasingly disheveled over time may be violence prone.

4. Romantic obsessions. Workplace violence is often the result of romantic entanglements or love interests gone awry. Employees who persist in making unwelcome advances may eventually respond to rejection with violence.

SAFETY FACT

Eight Steps for Preventing Workplace Violence

- Complete a risk-assessment survey of the entire workplace.
- Review existing security procedures.
- Develop and publish a policy statement that explains expectations, rules for behavior, roles, duties, and responsibilities.
- Develop work-site-specific prevention procedures.
- Train all managers, supervisors, and employees.
- Establish incident reporting and investigation procedures.
- Establish incident follow-up procedures (trauma plan, counseling services, and disciplinary guidelines).
- Monitor, evaluate, and adjust.

5. Depression. People who suffer from depression are prone to hurt either themselves or someone else. An employee who becomes increasingly withdrawn or overly stressed may be suffering from depression.

6. Finger pointers. Refusal to accept responsibility is a factor often exhibited by perpetrators of workplace violence. An employee's tendency to blame others for his or her own shortcomings should raise the caution flag.

7. Unusual frustration levels. The workplace has become a competitive, stressful, and sometimes frustrating place. When frustration reaches the boiling point, the emotional explosion that results can manifest itself in violence.

8. Obsession with weapons. Violence in the workplace often involves a weapon (gun, knife, or explosive device). A normal interest in guns used for hunting or target practice need not raise concerns. However, an employee whose interest in weapons is unusually intense and focused is cause for concern.

9. Drug dependence. It is common for perpetrators of workplace violence to be drug abusers. Consequently, drug dependence should cause concern not only for all the usual reasons but also for its association with violence on the job.

Environmental Factors Associated with Violence

The environment in which employees work can contribute to workplace violence. An environment that produces stress, anger, frustration, feelings of powerlessness, resentment, and feelings of inadequacy can increase the potential for violent behavior. The following factors can result in such an environment:

1. **Dictatorial management.** Dictatorial, overly authoritative management that shuts employees out of the decision-making process can cause them to feel powerless, as if they have little or no control over their jobs. Some people respond to powerlessness by striking out violently—a response that gives them power, if only momentarily.
2. **Role ambiguity.** One of the principal causes of stress and frustration on the job is role ambiguity. Employees need to know for what they are responsible, how they will be held accountable, and how much authority they have. When these questions are not clear, employees become stressed and frustrated, factors often associated with workplace violence.

3. **Partial, inconsistent supervision.** Supervisors who play favorites engender resentment in employees who aren't the favorite. Supervisors who treat one employee differently than another or one group of employees differently from another group also cause resentment. Employees who feel that they are being treated unfairly or unequally may show their resentment in violent ways.
4. **Unattended hostility.** Supervisors who ignore hostile situations or threatening behavior are unwittingly giving them their tacit approval. An environment that accepts hostile behavior will have hostile behavior.
5. **No respect for privacy.** Supervisors and managers who go through the desks, files, tool boxes, and work areas of employees without first getting their permission can make them feel invaded or even violated. Violent behavior is a possible response to these feelings.
6. **Insufficient training.** Holding employees accountable for performance on the job without providing the training that they need to perform well can cause them to feel inadequate. People who feel inadequate can turn their frustration inward and become depressed or turn it outward and become violent.

The overriding message in this section is twofold. First, managers should establish and maintain a positive work environment that builds up employees rather than tearing them down. Second, managers should be aware of the individual factors that can contribute to violent behavior and respond promptly if employees show evidence of responding negatively to these factors.

OSHA's Voluntary Guidelines

The U.S. Department of Labor, working through the Occupational Safety and Health Administration (OSHA), has established *advisory guidelines* relating to workplace violence.[12] Two key points to understand about these guidelines are:

- The guidelines are *advisory* in nature and *informational* in content. The guidelines *do not* add to or enhance in any way the requirements of the general duty clause of the OSH Act.
- The guidelines were developed with night retail establishments in mind. Consequently, they have a service-oriented emphasis. However, much of the advice contained in the guidelines can be adapted for use in manufacturing, processing, and other settings.

Figure 17-3 is a checklist of those elements of the specifications that have broader applications. Any management program relating to safety and health in the workplace should have at least these four elements.

Management Commitment and Employee Involvement

Management commitment and employee involvement are fundamental to developing and implementing any safety program, but they are especially important when trying to prevent workplace violence. The effectiveness of a workplace-violence prevention program may be a life-or-death proposition.

Figure 17-4 is a checklist that explains what management commitment means in practical terms. Figure 17-5 describes the practical application of employee involvement.

FIGURE 17-3 Broadly applicable elements of OSHA's advisory guidelines on workplace violence.

- ✓ Management commitment and employee involvement
- ✓ Workplace analysis
- ✓ Hazard prevention and control
- ✓ Safety and health training
- ✓ Record keeping and evaluation

FIGURE 17-4 Elements of management commitment to workplace-violence prevention.

- ✓ Hands-on involvement of executive management in developing and implementing prevention strategies.
- ✓ Sincere, demonstrated concern for the protection of employees.
- ✓ Balanced commitment to both employees and customers.
- ✓ Inclusion of safety, health, and workplace-violence prevention in the job descriptions of all executives, managers, and supervisors.
- ✓ Inclusion of safety, health, and workplace-violence prevention criteria in the performance evaluations of all executives, managers, and supervisors.
- ✓ Assignment of responsibility for providing coordination and leadership for safety, health, and workplace-violence prevention to a management-level employee.
- ✓ Provision of the resources needed to prevent workplace violence effectively.
- ✓ Provision of or guaranteed access to appropriate medical counseling and trauma-related care for employees affected physically or emotionally by workplace violence.
- ✓ Implementation, as appropriate, of the violence-prevention recommendations of committees, task forces, and safety professionals.

FIGURE 17-5 Elements of employee involvement in workplace-violence prevention.

- ✓ Staying informed concerning all aspects of the organization's safety, health, and workplace-violence program.
- ✓ Voluntarily complying—in both letter and spirit—with all applicable workplace-violence prevention strategies adopted by the organization.
- ✓ Making recommendations—through proper channels—concerning ways to prevent workplace violence and other hazardous conditions.
- ✓ Prompt reporting of all threatening or potentially threatening situations.
- ✓ Accurate and immediate reporting of all violent or threatening incidents.
- ✓ Voluntary participation on committees, task forces, and focus groups concerned with preventing workplace violence.
- ✓ Voluntary participation in seminars, workshops, and other educational programs relating to the prevention of workplace violence.

FIGURE 17-6 Checklist for incorporating workplace-violence prevention in strategic plans and operational practices.

- ✓ Include the prevention of workplace violence in the safety and health component of the organization's strategic plan.
- ✓ Adopt, disseminate, and implement a *no-tolerance* policy concerning workplace violence.
- ✓ Adopt, disseminate, and implement a policy that protects employees from reprisals when they report violent, threatening, or potentially threatening situations.
- ✓ Establish procedures for reporting violent and threatening incidents.
- ✓ Establish procedures for making recommendations for preventing workplace violence.
- ✓ Establish procedures for monitoring reports of workplace violence so that trends can be identified and incidents predicted and prevented.
- ✓ Develop a comprehensive workplace-violence prevention program that contains operational procedures and standard practices.
- ✓ Develop a workplace-violence component to the organization's emergency response plan.
- ✓ Train all employees in the application of standard procedures relating to workplace violence.
- ✓ Conduct periodic emergency-response drills for employees.

Figure 17-6 is a checklist that can be used to ensure that violence prevention becomes a standard component of organizational plans and operational practices. These three checklists can be used by any type of organization to operationalize the concepts of management commitment and employee involvement.

Workplace Analysis

Workplace analysis is the same process used by safety and health professionals to identify potentially hazardous conditions unrelated to workplace violence. Work-site analysis should be ongoing and have at least four components (Figure 17-7). An effective way to conduct an ongoing program of workplace analysis is to establish a threat-assessment team with representatives from all departments and led by the organization's chief safety and health professional.

Records Monitoring and Tracking

The purpose of **records monitoring and tracking** is to identify and chart all incidents of violence and threatening behavior that have occurred within a given time frame. Records to analyze include the following: incident reports, police reports, employee evaluations, and letters of reprimand. Of course, individual employees' records should be analyzed in confidence by the human resources member of the team. The type of information that is pertinent includes the following:

- Where specifically did the incident occur?
- What time of day or night did the incident occur?

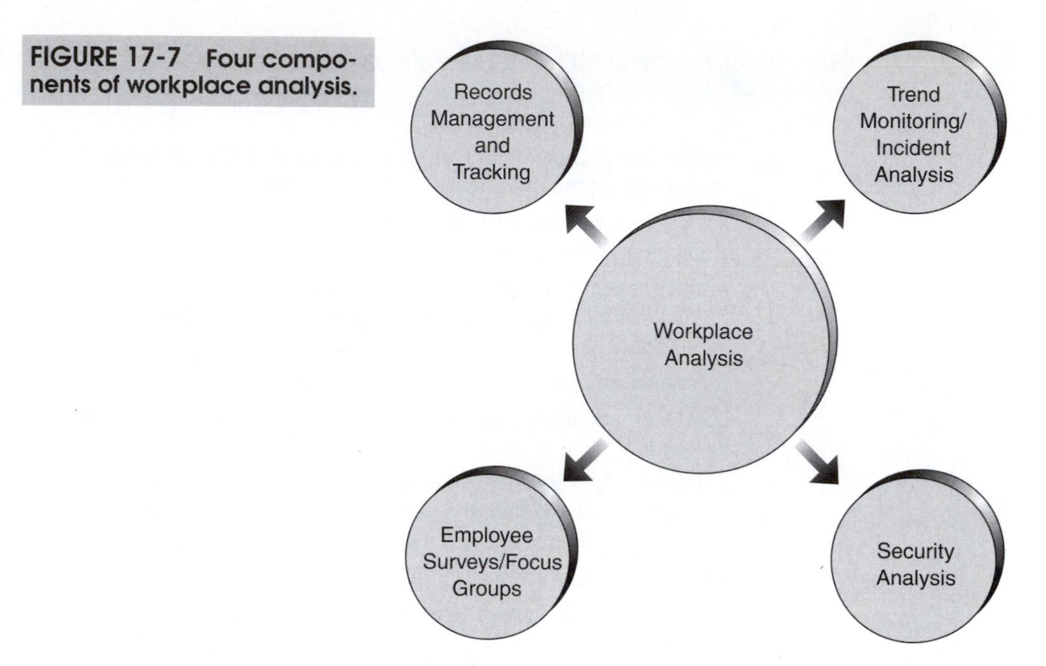

FIGURE 17-7 Four components of workplace analysis.

- Was the victim an employee? Customer? Outsider?
- Was the incident the result of a work-related grievance? Personal?

Trend Monitoring and Incident Analysis

Trend monitoring and incident analysis may prove helpful in determining patterns of violence. If there have been enough incidents to create one or more graphs, the team will want to determine if the graphs suggest a trend or trends. If the organization has experienced only isolated incidents, the team may want to monitor national trends. By analyzing both local and national incidents, the team can generate information that will be helpful in predicting and, thereby, preventing workplace violence. The team should look for trends in severity, frequency, and type of incidents.

Employee Surveys and Focus Groups

Employees are one of the best sources of information concerning workplace hazards. This is also true when it comes to identifying vulnerabilities to workplace violence. Employee input should be solicited periodically through written **employee surveys and focus groups** or one of these methods. Where are we vulnerable? What practices put our employees at risk? These are the types of questions that should be asked of employees. An effective strategy for use with focus groups is to give participants case studies of incidents that occurred in other organizations. Then ask such questions as, "Could this happen here? Why? Or why not? How can we prevent such incidents from occurring here?"

Security Analysis

Is the workplace secure, or could a disgruntled individual simply walk in and harm employees? It is important to ask this question. The team should periodically perform a **security analysis** of the workplace to identify conditions, situations, procedures, and practices that make employees vulnerable. The types of questions to ask include the following:

- Are there physical factors about the facility that make employees vulnerable (e.g., isolated, poorly lighted, infrequently trafficked, or unobservable)?
- Is there a process for handling disgruntled customers? Does it put employees at risk?
- Are the prevention strategies already implemented working?
- Is the training provided to employees having a positive effect? Is more training needed? Who needs the training? What kind of training is needed?
- Are there situations in which employees have substantial amounts of money in their possession, on- or off-site?
- Are there situations in which employees are responsible for highly valuable equipment or materials late at night or at isolated locations?

Hazard Prevention and Control

Once hazardous conditions have been identified, the strategies and procedures necessary to eliminate them must be put in place. The two broad categories of prevention strategies are engineering controls and administrative controls, just as they are with other safety and health hazards. In addition to these, organizations should adopt postincident response strategies as a way to prevent future incidents.

Engineering Controls

Engineering controls relating to the prevention of workplace violence serve the same purpose as engineering controls relating to other hazards. They either remove the hazard, or they create a barrier between it and employees. Engineering controls typically involve changes to the workplace. Examples of engineering controls include:

- Installing devices and mechanisms that give employees a complete view of their surroundings (e.g., mirrors, glass or clear plastic partitions, interior windows, and so on)
- Installing surveillance cameras and television screens that allow for monitoring of the workplace
- Installing adequate lighting, particularly in parking lots
- Pruning shrubbery and undergrowth outside and around the facility
- Installing fencing so that routes of egress and ingress to company property can be channeled and, as a result, better controlled
- Arranging outdoor sheds, storage facilities, recycling bins, and other outside facilities for maximum visibility

Administrative Controls

Whereas engineering controls involve making changes to the workplace, **administrative controls** involve making changes to how work is done. This amounts to changing work procedures and practices. Administrative controls fall into four categories (Figure 17-8).

- **Proper work practices** are those that minimize the vulnerability of employees. For example, if a driver has to make deliveries in a high-crime area, the company may employ a security guard to go along, change delivery schedules to daylight hours only, or both.

FIGURE 17-8 Categories of administrative controls.

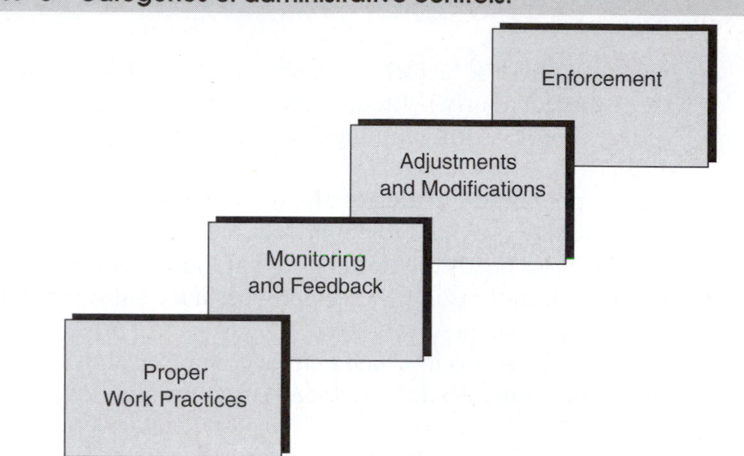

- **Monitoring and feedback** ensures that proper work practices are being used and that they are having the desired effect. For example, say, a company established a controlled access system in which visitors must check in at a central location and receive a visitor's pass. Is the system being used? Are all employees sticking to specified procedures? Has unauthorized access to the workplace been eliminated?
- **Adjustments and modifications** are made to violence prevention practices if it becomes clear from monitoring and feedback that they are not working or that improvements are needed.
- **Enforcement** involves applying meaningful sanctions when employees fail to follow the established and proper work practices. An employee who has been fully informed concerning a given administrative control, has received the training needed to practice it properly, but consciously decides not to follow the procedure should be disciplined appropriately.

Postincident Response

Postincident response relating to workplace violence is the same as postincident response relating to traumatic accidents. The first step is to provide immediate medical treatment for injured employees. The second step involves providing psychological treatment for traumatized employees. This step is even more important in cases of workplace violence than with accidents. Employees who are present when a violent incident occurs in the workplace, even if they don't witness it, can experience the symptoms of psychological trauma shown in Figure 17-9. Employees experiencing such symptoms or any others growing out of psychological trauma should be treated by professionals such as psychologists, psychiatrists, clinical nurse specialists, or certified social workers. In addition to one-on-one counseling, employees may also be enrolled in support groups. The final aspect of postincident response is the investigation, analysis, and report. In this step, safety professionals determine how the violent incident occurred and how future incidents may be prevented, just as postaccident investigations are handled.

FIGURE 17-9 Symptoms of psychological trauma in cases of workplace violence.

- ✓ Fear of returning to work
- ✓ Problems in relationships with fellow employees and/or family members
- ✓ Feelings of incompetence
- ✓ Guilt feelings
- ✓ Feelings of powerlessness
- ✓ Fear of criticism by fellow employees, supervisors, and managers

Training and Education

Training and education are as fundamental to the prevention of workplace violence as they are to the prevention of workplace accidents and health-threatening incidents. A complete safety and health training program should include a comprehensive component covering all aspects of workplace violence (e.g., workplace analysis, hazard prevention, proper work practices, and emergency response). Such training should be provided on a mandatory basis for supervisors, managers, and employees.

Record Keeping and Evaluation

Maintaining accurate, comprehensive, up-to-date records is just as important when dealing with violent incidents as it is when dealing with accidents and nonviolent incidents. By evaluating records, safety personnel can determine how effective their violence prevention strategies are, where deficiencies exist, and what changes need to be made. Figure 17-10 shows the types of records that should be kept.

- **OSHA log of injury and illness.** OSHA regulations require inclusion in the Injury and Illness Log of any injury that requires more than first aid, is a lost-time injury, requires modified duty, or causes loss of consciousness. Of course, this applies only to establishments required to keep OSHA logs. Injuries caused by assaults which are otherwise recordable also must be included in the log. A fatality or catastrophe that results in the hospitalization of three or more employees must be reported to OSHA within eight hours. This includes those resulting from workplace violence and applies to all establishments.
- **Medical reports.** Medical reports of all work injuries should be maintained. These records should describe the type of assault (e.g., unprovoked sudden attack), who was assaulted, and all other circumstances surrounding the incident. The records should include a description of the environment or location, potential or actual cost, lost time, and the nature of injuries sustained.
- **Incidents of abuse.** Incidents of abuse, verbal attacks, aggressive behavior—which may be threatening to the employee but do not result in injury, such as pushing, shouting, or acts of aggression—should be evaluated routinely by the affected department.
- **Minutes of safety meetings.** Minutes of safety meetings, records of hazard analyses, and corrective actions recommended and taken should be documented.
- **Records of all training programs.** Records of all training programs, attendees, and qualifications of trainers should be maintained.

FIGURE 17-10 Types of records that should be kept.

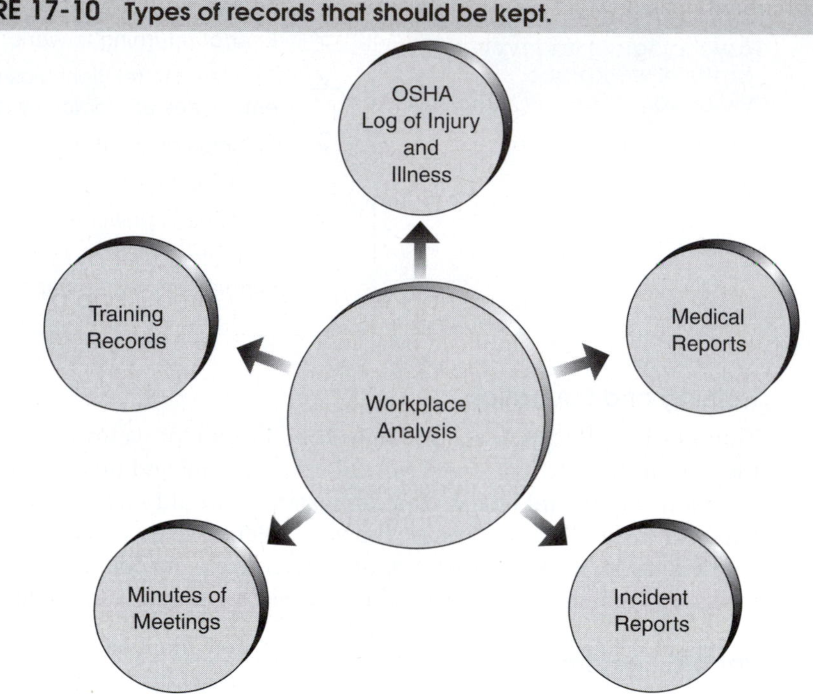

As part of its overall program, an employer should regularly evaluate its safety and security measures. Top management should review the program regularly, as well as each incident, to determine the program's effectiveness. Responsible parties (managers, supervisors, and employees) should collectively evaluate policies and procedures on a regular basis. Deficiencies should be identified, and corrective action taken. An evaluation program should involve the following activities:

- Establishing a uniform violence reporting system and regular review of reports
- Reviewing reports and minutes from staff meetings on safety and security issues
- Analyzing trends and rates of illness or injury or fatalities caused by violence relative to initial or *baseline* rates
- Measuring improvements based on lowering the frequency and severity of workplace violence
- Keeping up-to-date records of administrative and work practice changes to prevent workplace violence to evaluate their effectiveness
- Surveying employees before and after making job or workplace changes or installing security measures or new systems to determine their effectiveness
- Keeping abreast of new strategies available to deal with violence as they develop
- Surveying employees who experience hostile situations about the medical treatment they received initially and, again, several weeks afterward, and then several months later
- Complying with OSHA and state requirements for recording and reporting deaths, injuries, and illnesses

- Requesting periodic law enforcement or outside consultant review of the workplace for recommendations on improving employee safety

Management should share violence prevention evaluation reports with all employees. Any changes in the program should be discussed at regular meetings of the safety committee, union representatives, or other employee groups.

Conflict Resolution and Workplace Violence

When developing a violence prevention program for an organization, the natural tendency is to focus on protecting employees from outsiders. This is important. However, increasingly with workplace violence, the problem is internal. All too often in the modern workplace, conflict between employees is turning violent. Consequently, a violence prevention program is not complete without the following elements: **conflict management** and **anger management**.[13]

Conflict Management Component

Disagreements on the job can generate counterproductive conflict. This is one of the reasons why managers in organizations should do what is necessary to manage conflict properly. However, it is important to distinguish between just conflict and counterproductive conflict. Not all conflict is bad. In fact, properly managed conflict that has the improvement of products, processes, people, and the work environment as its source is positive conflict.

Counterproductive conflict—the type associated with workplace violence—occurs when employees behave in ways that work against the interests of the overall organization and its employees. This type of conflict is often characterized by deceitfulness, vindictiveness, personal rancor, and anger. *Productive conflict* occurs when right-minded, well-meaning people disagree, without being disagreeable, concerning the best way to support the organization's mission. Conflict management has the following components:

- Establishing conflict guidelines
- Helping all employees develop conflict prevention and resolution skills
- Helping all employees develop anger management skills

Establishing Conflict Guidelines

Conflict guidelines establish ground rules for discussing and debating differing points of view, differing ideas, and differing opinions concerning how best to accomplish the organization's vision, mission, and broad objectives. Figure 17-11 is an example of an organization's conflict guidelines. Guidelines such as these should be developed with a broad base of employee involvement from all levels in the organization.

Develop Conflict Prevention and Resolution Skills

If managers are going to expect employees to disagree without being disagreeable, they are going to have to ensure that all employees are skilled in the art and science of conflict resolution. The first guideline in Figure 17-11 is an acknowledgment of human nature. It takes advanced human relation skills and constant effort to disagree without being disagreeable. Few people are born with this ability. Fortunately, it can be learned.

FIGURE 17-11 Sample conflict guidelines.

Conflict Guidelines
Micro Electronics Manufacturing (MEM)

Micro Electronics Manufacturing encourages discussion and debate among employees at all levels concerning better ways to improve continually the quality of our products, processes, people, and work environment. This type of interaction, if properly handled, will result in better ideas, policies, procedures, practices, and decisions. However, human nature is such that conflict can easily get out of hand, take on personal connotations, and become counterproductive. Consequently, in order to promote productive conflict, MEM has adopted the following guidelines. These guidelines are to be followed by all employees at all levels:

- The criteria to be applied when discussing/debating any point of contention is as follows: Which recommendation is most likely to move our company closer to accomplishing its mission?
- Disagree, but don't be disagreeable. If the debate becomes too hot, stop and give all parties an opportunity to cool down before continuing. Apply your conflict resolution skills and anger management skills. Remember, even when we disagree about how to get there, we are all trying to reach the same destination.
- Justify your point of view by tying it to our mission and require others to do the same.
- In any discussion of differing points of view, ask yourself the following question: "Am I just trying to win the debate for the sake of winning (ego), or is my point of view really the most valid?"

The following strategies are based on a three-phase model developed by Tom Rusk and described in his book *The Power of Ethical Persuasion*.[14]

Explore the Other Person's Viewpoint

Allow the other person to present his or her point of view. The following strategies will help make this phase of the discussion more positive and productive.

1. Establish that your goal at this point is mutual understanding.
2. Elicit the other person's complete point of view.
3. Listen nonjudgmentally and do not interrupt.

SAFETY FACT

Aggressive Employees Threaten Productivity

No one wants to work with or around an aggressive person. As a result, aggressive employees cause tardiness, absenteeism, and turnover. All three of these factors are known to harm productivity. Consequently, it is important to deal with aggressive employees through counseling, aggression management training, or even termination.

4. Ask for clarification if necessary.
5. Paraphrase the other person's point of view and restate it to show that you understand.
6. Ask the other person to correct your understanding if it appears to be incomplete.

Explain Your Viewpoint

After you accurately and fully understand the other person's point of view, present your own. The following strategies will help make this phase of the discussion more positive and productive:

1. Ask for the same type of fair hearing for your point of view that you gave the other party.
2. Describe how the person's point of view affects you. Don't point the finger of blame or be defensive. Explain your reactions objectively, keeping the discussion on a professional level.
3. Explain your point of view accurately and completely.
4. Ask the other party to paraphrase and restate what you have said.
5. Correct the other party's understanding if necessary.
6. Review and compare the two positions (yours and that of the other party). Describe the fundamental differences between the two points of view and ask the other party to do the same.

Agree on a Resolution

Once both viewpoints have been explained and are understood, it is time to move to the resolution phase. This is the phase in which both parties attempt to come to an agreement. Agreeing to disagree—in an agreeable manner—is an acceptable solution. The following strategies will help make this phase of the discussion more positive and productive:

1. Reaffirm the mutual understanding of the situation.
2. Confirm that both parties are ready and willing to consider options for coming to an acceptable solution.
3. If it appears that differences cannot be resolved to the satisfaction of both parties, try one or more of the following strategies:

    Take time out to reflect and try again.

    Agree to third-party arbitration or neutral mediation.

    Agree to a compromise solution.

    Take turns suggesting alternative solutions.

    Yield (this time), once your position has been thoroughly stated and is understood. The eventual result may vindicate your position.

    Agree to disagree while still respecting each other.

Develop Anger Management Skills

It is difficult, if not impossible, to keep conflict positive when anger enters the picture. If individuals in an organization are going to be encouraged to question, discuss, debate, and even disagree, they must know how to manage their anger. Anger is an intense, emotional reaction to conflict in which self-control may be lost. Anger is a major cause of workplace violence. Anger occurs when people feel that one or more of their fundamental needs are being threatened. These needs include the following:

- Need for approval
- Need to be valued
- Need to be appreciated
- Need to be in control
- Need for self-esteem

When one or more of these needs is threatened, a normal human response is to become angry. An angry person can respond in one of four ways:

1. Attacking. With this response, the source of the threat is attacked. This response often leads to violence, or at least verbal abuse.

2. Retaliating. With this response, you fight fire with fire, so to speak. Whatever is given, you give back. For example, if someone calls your suggestion ridiculous (threatens your need to be valued), you may retaliate by calling his or her suggestion dumb. Retaliation can escalate into violence.

3. Isolating. This response is the opposite of venting. With the isolation response, you internalize your anger, find a place where you can be alone, and simmer. The childhood version of this response was to go to your room and pout. For example, when someone fails to even acknowledge your suggestion (threatens your need to be appreciated), you may swallow your anger, return to your office, and boil over in private.

4. Coping. This is the only positive response to anger. Coping does not mean that you don't become angry. Rather, it means that, even when you do, you control your emotions instead of letting them control you. A person who copes well with anger is a person who, in spite of his or her anger, stays in control. The following strategies will help employees manage their anger by becoming better at coping:

Avoid the use of anger-inducing words and phrases including the following: *but, you should, you made me, always, never, I can't, you can't,* and so on.

Admit that others don't make you angry; you allow yourself to become angry. You are responsible for your emotions and your responses to them.

Don't let pride get in the way of progress. You don't have to be right every time.

Drop your defenses when dealing with people. Be open and honest.

Relate to other people as equals. Regardless of position or rank, you are no better than they, and they are no better than you.

Avoid the human tendency to rationalize your angry responses. You are responsible and accountable for your behavior.

If employees in an organization can learn to manage conflict properly and to deal with anger positively, the potential for workplace violence will be diminished substantially. Conflict and anger management will not prevent violent acts from outsiders. There are other methods for dealing with outsiders. However, properly managing conflict and anger can protect employees from each other.

Do's and Don'ts for Supervisors

Supervisors can play a pivotal role in the prevention of workplace violence. Following are some rules-of-thumb that will enhance the effectiveness of supervisors in this regard:

- *Don't* try to diagnose the personal, emotional, or psychological problems of employees.
- *Don't* discuss an employee's drinking unless it occurs on the job. Restrict comments to performance.
- *Don't* preach to employees. Counsel employees about attendance, tardiness, and job performance, not about how they should live their lives.
- *Don't* cover up for employees or make excuses for inappropriate behavior. Misguided kindness may allow problems to escalate and get out of hand.
- *Don't* create jobs to get problem employees out of the way. Stockpiling an employee simply gives him or her more time to brood and to allow resentment to build.
- *Don't* ignore the warning signs explained earlier in this chapter. The problems that they represent will not simply go away. Sooner or later, they will have to be handled. Sooner is better.
- *Do* remember that chemical dependence and emotional problems tend to be progressive. Left untreated, they get worse, not better.
- *Do* refer problem employees to the employee assistance program or to other mental health service providers.
- *Do* make it clear to employees that job performance is the key issue. They are expected to do what is necessary to maintain and improve their performance.
- *Do* make it clear that inappropriate behavior will not be tolerated.[15]

Emergency Preparedness Plan

To be prepared for properly handling a violent incident in the workplace, employers should form a crisis management team.[16] The team should have only one mission—immediate response to violent acts on the job—and be chaired by a safety and health professional. Team members should receive special training and be updated regularly. The team's responsibilities should be as follows:

- Undergo trauma response training
- Handle media interaction
- Operate telephone and communication teams
- Develop and implement, as necessary, an emergency evacuation plan
- Establish a backup communication system
- Calm personnel after an incident
- Debrief witnesses after an incident
- Ensure that proper security procedures are established, kept up-to-date, and enforced
- Help employees deal with posttraumatic stress
- Keep employees informed about workplace violence as an issue, how to respond when it occurs, and how to help prevent it

SUMMARY

1. Preventing workplace violence is a natural extension of the responsibilities of safety and health professionals. Like the traditional responsibilities of such professionals, dealing with workplace violence involves such activities as hazard analysis, records analysis and tracking, trend monitoring, and incident analysis.
2. Key concepts relating to workplace violence include the following: occupational violent crime (OVC), employee, outsider, employee-related outsider, and customer. These terms have definitions relating specifically to workplace violence.
3. Approximately one million people are victims of workplace violence every year. These incidents result in more than 1.75 million lost days of work annually.
4. Almost 40 percent of all violent acts in the workplace are committed by males. The majority of violent incidents reported each year (75 percent) are fistfights.
5. When dealing with violent incidents on the job, it is important to remember that even the perpetrator has rights. Employee rights are protected by employment contracts, collective bargaining agreements, and various local, state, and federal civil rights statutes. To the extent possible, when dealing with a violent employee, follow the procedures stipulated in contracts, agreements, and statutes.
6. Although it is important that employers consider perpetrators' rights when dealing with workplace violence, it is equally important that they act prudently to prevent harm to other employees and customers.
7. The exclusivity provision of workers' compensation laws provide employers with some protection from liability in cases of workplace violence, provided that the incident is work related. When this is the case, workers' compensation is the injured employee's exclusive remedy.
8. A violent act can be considered an on-the-job incident, even if it is committed away from the workplace. Specific guidelines have been established by NIOSH for determining whether a violent act can be classified as an on-the-job incident.
9. The concept of crime reduction through environmental design (CRTED) has four major elements as follows: natural surveillance, control of access, establishment of territoriality, and activity support. The author has added another: administrative controls.
10. OSHA has produced voluntary advisory guidelines relating to workplace violence. Although the guidelines are aimed specifically at the night retail industry, they provide an excellent framework that can be used in other industries including manufacturing, transportation, and processing. The framework has the following broad elements: management commitment and employee involvement, workplace analysis and hazard prevention control, safety and health training, and record keeping and evaluation.

KEY TERMS AND CONCEPTS

- Activity support
- Adjustments and modifications
- Administrative controls
- Anger management
- Conflict guidelines
- Conflict management
- Control of access
- Customer
- Employee
- Employee-related outsider
- Employee surveys and focus groups
- Enforcement
- Engineering controls
- Establishment of territoriality
- Exclusivity provision
- Management commitment and employee involvement
- Monitoring and feedback
- Natural surveillance
- Occupational violent crime (OVC)
- Outsider
- Postincident response
- Proper work practices
- Record keeping and evaluation
- Records monitoring and tracking
- Security analysis
- Trend monitoring and incident analysis
- Workplace analysis
- Workplace violence

Review Questions

1. Define the following terms as they relate to violence in the workplace: *occupational violent crime, employee,* and *outsider.*
2. Approximately how many people are direct victims of workplace violence annually?
3. Defend or refute the following statement: Employees who commit violent acts forfeit their rights and can be dealt with accordingly.
4. What is the exclusivity provision of workers' compensation laws? Why is this provision significant?
5. Defend or refute the following statement: A violent act that occurs away from the employer's premises cannot be considered work related.
6. Explain the concept of crime reduction through environmental design (CRTED).
7. Defend or refute the following statement: A manufacturer must comply with OSHA's guidelines on workplace violence.
8. What elements of OSHA's guidelines on workplace violence can be adapted for future use in most types of business and industrial firms?
9. What are the primary causes of conflict on the job?
10. Explain the four ways in which an angry person may respond in a work setting.

Endnotes

1. U.S. Department of Labor, Occupational Safety and Health Administration, *OSHA Workplace Violence Guidelines*. Retrieved June 2003 from http://www.osha.gov/workplaceviolence/viol.html.
2. Anastasia Toufexis, "Workers Who Fight Firing with Fire," *Time* 143, no. 17: 34.
3. U.S. Department of Justice, *Violence and Theft in the Workplace* (NCJ-148199) (Annapolis Junction, MD: Bureau of Justice Statistics Clearinghouse).
4. Lloyd G. Nigro, "Violence in the American Workplace: Challenges to the Public Employees," *Public Administration Review* 56, no. 4: 326.
5. U.S. Department of Justice, *Violence and Theft in the Workplace*.
6. Society for Human Resource Management, "Workplace Violence Survey," results published in *USA Today Magazine,* January 2006.
7. Stephen C. Yohay and Melissa L. Peppe, "Workplace Violence: Employer Responsibilities," *Occupational Hazards*, July 1996, 22.
8. *International Union v. Micro Manufacturing, Inc.,* 895 F. Supp. 170, 171 (E.D. Mich. 1995).
9. Yohay and Peppe, "Workplace Violence," 24.
10. National Institute for Occupational Safety and Health and U.S. Department of Health and Human Services, *Homicide in U.S. Workplaces: A Strategy for Prevention and Research* (Washington, DC: Centers for Disease Control and Prevention, NIOSH).
11. Janice L. Thomas, "A Response to Occupational Violent Crime," *Professional Safety*, 27–31, July 2006.
12. U.S. Department of Labor, Occupational Safety and Health Administration, *Guidelines for Workplace Violence Prevention Programs for Night Retail Establishments*. Retrieved July 2003 from http://www.osha.gov/workplace evidence.
13. Ibid.
14. Tom Rusk, *The Power of Ethical Persuasion* (New York: Penguin Books), xv–xvii.
15. Illinois State Police, *Do's and Don'ts for the Supervisor*. Retrieved from http://www.state.il.us/isp/viowkplc/vwpp6c.htm.
16. Illinois State Police, *Do's and Don'ts for the Supervisor*. Retrieved from http://www.state.il.us/isp/viowkplc/vwpp8.htm.

Noise and Vibration Hazards

Major Topics

- Hearing Loss Prevention Terms
- Characteristics of Sound
- Hazard Levels and Risks
- Standards and Regulations
- Workers' Compensation and Noise Hazards
- Identifying and Assessing Hazardous Noise Conditions
- Noise Control Strategies
- Vibration Hazards
- Other Effects of Noise Hazards
- Corporate Policy
- Evaluating Hearing Loss Prevention Programs

The modern workplace can be noisy. This poses two safety- and health-related problems. First, there is the problem of distraction. Noise can distract workers and disrupt their concentration, which can lead to accidents. Second, there is the problem of hearing loss. Exposure to noise that exceeds prescribed levels can result in permanent hearing loss.

Modern safety and health professionals need to understand the hazards associated with noise and vibration, how to identify and assess these hazards, and how to prevent injuries related to them. This chapter provides the necessary information for prospective and practicing safety and health professionals to do so.

Hearing Loss Prevention Terms

There are certain terms common to hearing loss prevention that must be understood by safety and health professionals. You may find the definitions in this section helpful when trying to understand the content of this chapter.

- **Attenuation**: Real-world **baseline audiogram**. Estimated sound protection provided by hearing protective devices as worn in "real-world" environments.

- **Baseline audiogram.** A valid audiogram against which subsequent audiograms are compared to determine if hearing thresholds have changed. The baseline audiogram is preceded by a quiet period to obtain the best estimate of the person's hearing at that time.
- **Continuous noise.** Noise of a constant level measured over at least one second using the "slow" setting on a sound level meter. Note that an intermittent noise (e.g., on for over a second and then off for a period) is both variable and continuous.
- **Decibel (dB).** The unit used to express the intensity of sound. The decibel was named after Alexander Graham Bell. The decibel scale is a logarithmic scale in which 0 dB approximates the threshold of hearing in the midfrequencies for young adults and in which the threshold of discomfort is between 85 and 95 dB and the threshold for pain is between 120 and 140 dB.
- **Dosimeter.** When applied to noise, the instrument that measures sound levels over a specified interval, stores the measures, and calculates the sound as a function of sound level and sound **duration**. It describes the results in terms of dose, time-weighted average, and other parameters such as peak level, equivalent sound level, sound exposure level, and so on.
- **Exchange rate.** The relationship between intensity and dose. OSHA uses a 5-dB exchange rate. Thus, if the intensity of an exposure increases by 5 dB, the dose doubles. This may also be referred to as the *doubling rate.* The U. S. Navy uses a 4-dB exchange rate; the U. S. Army and U. S. Air Force use a 3-dB exchange rate.
- **Hazardous noise.** Any sound for which any combination of **frequency**, intensity, or duration is capable of causing permanent hearing loss in a specified population.
- **Conductive and sensorineural loss.** Hearing loss is often characterized by the area of the auditory system responsible for the loss. For example, when injury or a medical condition affects the *outer ear* or *middle ear* (i.e., from the pinna, ear canal, and eardrum to the cavity behind the eardrum—which includes the ossicles), the resulting hearing loss is referred to as a conductive loss. When an injury or medical condition affects the inner ear or the auditory nerve that connects the *inner ear* to the brain (i.e., the cochlea and the VIIIth cranial nerve), the resulting hearing loss is referred to as a sensorineural loss. Thus, a welder's spark that damages the eardrum causes a conductive hearing loss. Because noise can damage the tiny hair cells located in the cochlea, it causes a sensorineural hearing loss.
- **Hearing threshold level (HTL).** The hearing level, above a reference value, at which a specified sound or tone is heard by an ear in a specified fraction of the trials. Hearing threshold levels have been established so that dB HTL reflects the best hearing of a group of persons.
- **Hertz (Hz).** The unit measurement for audio frequencies. The frequency range for human hearing lies between 20 Hz and approximately 20,000 Hz. The sensitivity of the human ear drops off sharply below about 500 Hz and above 4,000 Hz.
- **Impulsive noise.** Generally used to characterize impact or impulse noise typified by a sound that rapidly rises to a sharp peak and then quickly fades. The sound may or may not have a "ringing" quality (such as striking a hammer on a metal plate or a gunshot in a reverberant room). Impulsive noise may be repetitive, or may be a single event (as with a sonic boom). *Note:* If impulses occur in very rapid succession (e.g., some jackhammers), the noise is not described as impulsive.

- **Material hearing impairment.** As defined by OSHA, a material hearing impairment is an average hearing threshold level of 25 dB HTL at the frequencies of 1,000, 2,000, and 3,000 Hz.
- **Noise.** Any unwanted sound.
- **Noise dose.** The noise exposure expressed as a percentage of the allowable daily exposure. For OSHA, a 100 percent dose equals an eight-hour exposure to a continuous 90-dBA noise; a 50 percent dose equals an eight-hour exposure to an 85-dBA noise or a four-hour exposure to a 90-dBA noise. If 85 dBA is the maximum permissible level, an eight-hour exposure to a continuous 85-dBA noise equals a 100 percent dose. If a 3-dB exchange rate is used in conjunction with an 85-dBA maximum permissible level, a 50 percent dose equals a two-hour exposure to 88 dBA or an eight-hour exposure to 82 dBA.
- **Noise-induced hearing loss.** A sensorineural hearing loss that is attributed to noise and for which no other etiology can be determined.
- **Standard threshold shift.** OSHA uses this term to describe a change in hearing threshold relative to the baseline audiogram of an average of 10 dB or more at 2,000, 3,000, and 4,000 Hz in either ear. Used by OSHA to trigger additional audiometric testing and related **follow-up**.
- **Significant threshold shift.** NIOSH uses this term to describe a change of 15 dB or more at any frequency, 400 through 6,000 Hz, from baseline levels that is present on a retest in the same ear and at the same frequency. NIOSH recommends a confirmation audiogram within 30 days with the confirmation audiogram preceded by a quiet period of at least 14 hours.
- **Time-weighted average (TWA).** A value, expressed in dBA, computed so that the resulting average is equivalent to an exposure resulting from a constant noise level over an eight-hour period.

Characteristics of Sound

Sound is any change in pressure that can be detected by the ear. Typically, sound is a change in **air pressure**. However, it can also be a change in water pressure or any other pressure-sensitive medium. **Noise** is unwanted sound. Consequently, the difference between noise and sound is in the perception of the person hearing it (e.g., loud rock music may be considered "sound" by a rock fan but "noise" by a shift worker trying to sleep).

What we think of as sound, the eardrum senses as fluctuations in atmospheric pressure. The eardrum responds to these fluctuations in atmospheric pressure by vibrating. These vibrations are carried to the brain in the form of neural sensations and interpreted as sound.

Sound and vibration are very similar. Sound typically relates to a sensation that is perceived by the inner ear as hearing. **Vibration**, on the other hand, is inaudible and is perceived through the sense of touch. Sound can occur in any medium that has both mass and elasticity (air, water, and so on). It occurs as elastic waves that cross over (above and below) a line representing normal atmospheric pressure (Figure 18-1).

Normal atmospheric pressure is represented in Figure 18-1 by a straight horizontal line. Sound is represented by the wavy line that crosses above and below the line. The more frequently the sound waves cross the normal atmospheric pressure line (the shorter the cycle), the higher the pitch of the sound. The greater the vertical distance

FIGURE 18-1 Sound waves.

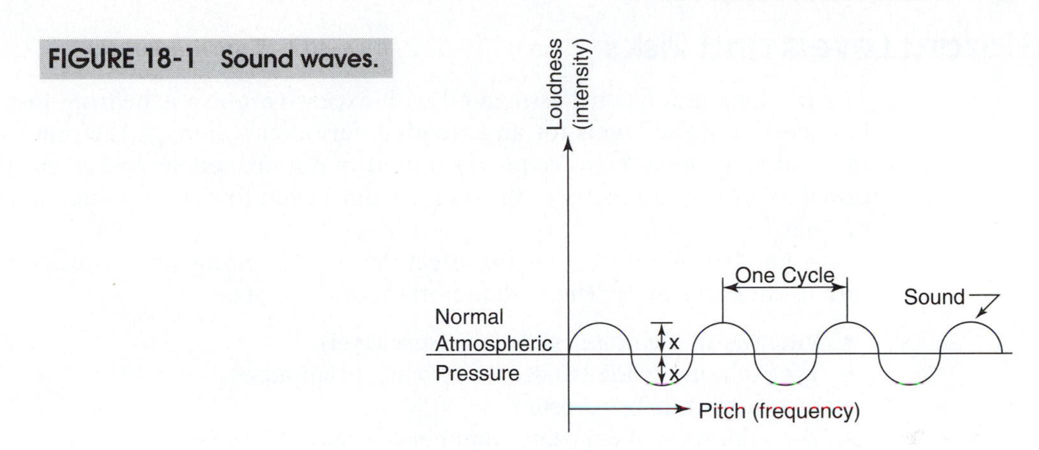

above and below the atmospheric pressure line (distance *X*), the louder or more intense the sound.

The unit of measurement used for discussing the level of sound and, correspondingly, what noise levels are hazardous is the **decibel**, or one-tenth of a bel. One decibel represents the smallest difference in the level of sound that can be perceived by the human ear. Figure 18-2 shows the decibel levels for various common sounds. The weakest sound that can be heard by a healthy human ear in a quiet setting is known as the **threshold of hearing** (1 dBA). The maximum level of sound that can be perceived without experiencing pain is known as the **threshold of pain** (140 dBA).

The three broad types of industrial noise are described as follows: **Wide band noise** is noise that is distributed over a wide range of **frequencies**. Most noise from manufacturing machines is wide band noise. **Narrow band noise** is noise that is confined to a narrow range of frequencies. The noise produced by power tools is narrow band noise. Finally, **impulse noise** consists of transient pulses that can occur repetitively or nonrepetitively. The noise produced by a jackhammer is repetitive impulse noise.[1]

FIGURE 18-2 Selected sound levels.

Source	Decibels (dBA)
Whisper	20
Quiet library	30
Quiet office	50
Normal conversation	60
Vacuum cleaner	70
Noisy office	80
Power saw, lawn mower	90
Chain saw	90
Grinding operations	100
Passing truck	100
Gunshot blast	140
Jet aircraft	150
Rocket launching	180

Hazard Levels and Risks

The fundamental hazard associated with excessive noise is hearing loss. Exposure to excessive noise levels for an extended period can damage the inner ear so that the ability to hear high-frequency sound is diminished or lost altogether. Additional exposure can increase the damage until even lower frequency sounds cannot be heard.[2]

A number of different factors affect the risk of hearing loss associated with exposure to excessive noise. The most important of these are

- Intensity of the noise (**sound pressure level**)
- Type of noise (wide band, narrow band, or impulse)
- Duration of daily exposure
- Total duration of exposure (number of years)
- Age of the individual
- Coexisting hearing disease
- Nature of environment in which exposure occurs
- Distance of the individual from the **source** of the noise
- Position of the ears relative to the sound waves[3]

Of these various factors, the most critical are the sound level, frequency, duration, and distribution of noise (Figure 18-3). The unprotected human ear is at risk when exposed to sound levels exceeding 115 dBA. Exposure to sound levels below 80 dBA is generally considered safe. Prolonged exposure to noise levels higher than 80 dBA should be minimized through the use of appropriate personal protective devices.

To decrease the risk of hearing loss, exposure to noise should be limited to a maximum eight-hour time-weighted average of 85 dBA. The following general rules should be applied for dealing with noise in the workplace:

- Exposures of less than 80 dBA may be considered safe for the purpose of risk assessment.
- A time-weighted average (threshold) of 85 dBA should be considered the maximum limit of continuous exposure over eight-hour days without protection.[4]

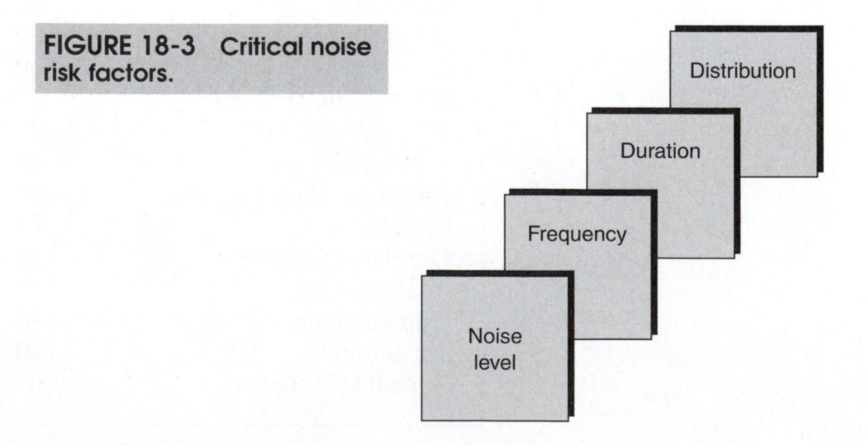

FIGURE 18-3 Critical noise risk factors.

Standards and Regulations

The primary sources of standards and regulations relating to noise hazards are OSHA, the American National Standards Institute (ANSI), and the National Institute for Occupational Safety and Health (NIOSH). OSHA regulations require the implementation of **hearing conservation** programs under certain conditions. OSHA's regulations should be considered as minimum standards. ANSI's standard provides a way to determine the effectiveness of hearing conservation programs such as those required by OSHA. The ANSI standard and OSHA regulations are discussed in the following sections. NIOSH bases most of its materials on OSHA regulations, although NIOSH does make recommendations that exceed OSHA regulations in some cases.

ANSI Standard

The American National Standards Institute (ANSI) published ANSI Standard S12.13, Evaluation of Hearing Conservation Programs. This standard is designed to help safety and health professionals determine if hearing conservation programs work as they are intended.[5]

Federal regulations require that employees be protected from excessive noise in the workplace. However, they provide no methodology for determining the effectiveness of hearing conservation programs. The primary reason for the development of **ANSI S12.13** was because hearing conservation programs were not actually protecting employees but were only recording their steadily declining hearing ability.

The working group that developed the standard used **audiometric database analysis (ADBA)** to identify procedures for measuring variability in hearing threshold levels. The two procedures identified were as follows:

- **Percent worse sequential.** This procedure identifies the percentage of subjects who show a deterioration of 15 dBA or more in their ability to hear at least one test frequency (500 through 6000 Hz) in either ear between two sequential audiograms.
- **Percent better or worse sequential.** This procedure identifies the percentage of subjects who show either a deterioration or an improvement of 15 dBA or more in thresholds for at least one test frequency (500 through 6000 Hz) in either ear between two sequential audiograms.[6]

There are important differences between the traditional approaches to measuring the effectiveness of hearing conservation programs and those required by ANSI S12.13. The ANSI approach can be summarized as follows:

- Results of tests are compared in sequence. For example, the results of year 4 are compared with those of year 3. The results of year 3 are compared with those of year 2, and so on. In this way, a current audiogram is compared against an earlier audiogram. The results of the earlier test are used as a baseline for comparison.
- Test results from several employees in a given work unit are examined individually and compared with past results sequentially. If enough employees show hearing loss, the conclusion may be that the work unit's hearing conservation program is ineffective.[7]

OSHA Regulations

OSHA's regulations relating to occupational noise exposure and hearing conservation are found in 29 CFR 1910.95. The basic requirements generated from this standard for hearing conservation programs are as follows:

- Hearing hazards monitoring
- Engineering and **administrative controls**
- Audiometric evaluation
- Personal hearing protection devices
- Education and motivation
- Record keeping
- Program evaluation

Hearing Hazards Monitoring

As with any health hazard, it is important to determine accurately the nature of the hearing hazard and to identify the affected employees. Those responsible for this aspect of the program must ensure that the exposures of all employees have been properly evaluated and that reevaluations are conducted when changes in equipment or operations significantly alter working conditions. Recent evidence has indicated that aromatic solvents, metals, and petrochemicals may be associated with occupational hearing loss. Although studies are exploring the relationship between hearing loss and chemical exposures, there is insufficient information about this relationship to speculate on potential risk factors. Therefore, this section focuses on monitoring noise exposure, the major factor associated with occupational hearing losses. Hearing hazard exposure monitoring is conducted for various purposes, including

- Determining whether hazards to hearing exist
- Determining whether noise presents a safety hazard by interfering with speech communication or the recognition of audible warning signals
- Identifying employees for inclusion in the hearing loss prevention program
- Classifying employees' noise exposures for prioritizing noise control efforts and defining and establishing hearing protection practices
- Evaluating specific noise sources for noise control purposes
- Evaluating the success of noise control efforts

Various kinds of incrementation and measurement methods may be used, depending on the type of measurements being conducted. The most common measurements are area surveys, dosimetry, and engineering surveys.

In an area survey, environmental noise levels are measured using a sound level meter to identify work areas where exposures are above or below hazardous levels and where more thorough exposure monitoring may be needed. The result is often plotted in the form of a "noise map," showing noise level measurements for the different areas of the workplace.

Dosimetry involves the use of body-worn instruments (dosimeters) to monitor an employee's noise exposure over the work shift. Monitoring results for one employee can also represent the exposures of other workers in the area with similar noise exposures. It may also be possible to use task-based exposure methods to represent the exposures of other workers in different areas whose exposures result from having performed the same tasks.

Engineering surveys typically employ more sophisticated acoustical equipment in addition to sound level meters. These may include octave-band analyzers and sound level recorders that furnish information on the frequency–intensity composition of the noise being emitted by machinery or other sound sources in various modes of operation. These measurements are used to assess options for applying **engineering controls**.

Engineering and Administrative Controls

Engineering and administrative controls are essential to achieve an effective hearing loss prevention program. Engineering and administrative controls represent the first two echelons in the hierarchy of controls: (1) remove the hazard, and (2) remove the worker. The use of these controls should reduce hazardous exposure to the point where the risk to hearing is eliminated or at least more manageable. Engineering controls are technologically feasible for most noise sources, but their economic feasibility must be determined on a case-by-case basis. In some instances, the application of a relatively simple noise control solution reduces the hazard to the extent that the other elements of the program, such as audiometric testing and the use of hearing protection devices, are no longer necessary. In other cases, the noise reduction process may be more complex and must be accomplished in stages over a period of time. Even so, with each reduction of a few decibels, the hazard to hearing is reduced, communication is improved, and noise-related annoyance is reduced as well.

It is especially important that organizations specify low noise levels when purchasing new equipment. Many types of previously noisy equipment are now available in noise-controlled versions. Consequently, a "buy quiet" purchasing policy should not require new engineering solutions in many cases.

For hearing loss prevention purposes, *engineering controls* are defined as any modification or replacement of equipment or related physical change at the noise source or along the transmission path (with the exception of hearing protectors) that reduces the noise level at the employee's ear. Typical engineering controls involve:

- Reducing noise at the source (e.g., installing a muffler)
- Interrupting the noise path (e.g., erecting acoustical enclosures and barriers)
- Reducing reverberation (e.g., installing sound-absorbing material)
- Reducing structure-borne vibration (e.g., installing vibration mounts and providing proper lubrication)

Assessing the applicability of engineering controls is a sophisticated process. First, the noise problem must be thoroughly defined. This necessitates measuring the noise levels and developing complete information on employee noise exposure and the need for noise reduction. Next, an assessment of the effect of these controls on overall noise levels should be made. Once identified and analyzed, the preceding controls can be considered. Choices are influenced, to some extent, by the cost of purchasing, operating, servicing, and maintaining the control. For this reason, engineering, safety, and industrial hygiene personnel, as well as employees who operate, service, and maintain equipment, must be involved in the noise control plan. Employees who work with the equipment on a daily basis should be asked to provide valuable guidance on such important matters as the positioning of monitoring indicators and panels, lubrication and servicing points, control switches, and the proper location of access doors for operation and maintenance. An acoustical consultant may be hired to assist in the design, implementation, installation, and evaluation of these controls.

Administrative controls, defined as changes in the work schedule or operations that reduce noise exposure, may also be used effectively. Examples include operating a noisy machine on the second or third shift when fewer people are exposed, or shifting an employee to a less noisy job once a hazardous daily noise dose has been reached. Generally, administrative controls have limited use in industry because employee contracts seldom permit shifting from one job to another. Moreover, the practice of rotating employees between quiet and noisy jobs, although it may reduce the risk of substantial hearing loss in a few workers, may actually increase the risk of small hearing losses in many workers. A more practical administrative control is to provide for quiet areas where employees can gain relief from workplace noise. Areas used for work breaks and lunch rooms should be located away from noise.

Audiometric Evaluation

Audiometric evaluation is crucial to the success of the hearing loss prevention program in that it is the only way to determine whether occupational hearing loss is being prevented. When the comparison of audiograms shows temporary threshold shift (a temporary hearing loss after noise exposure), early permanent threshold shift, or progressive occupational hearing loss, it is time to take swift action to halt the loss before additional deterioration occurs. Because occupational hearing loss occurs gradually and is not accompanied by pain, the affected employee may not notice the change until a large threshold shift has accumulated. However, the results of audiometric tests can trigger changes in the hearing loss prevention program more promptly, initiating protective measures, and motivating employees to prevent further hearing loss.

OSHA and NIOSH presently have differing definitions of the amount of change in hearing indicated by repeated audiometry that should trigger additional audiometric testing and related follow-up. OSHA uses the term *standard threshold shift* to describe an average change in hearing from the baseline levels of 10 dB or more for the frequencies of 2,000, 3,000, and 4,000 Hz. NIOSH uses the term *significant threshold shift* to describe a change of 15 dB or more at any frequency 500 through 6,000 Hz from baseline levels that is present on an immediate retest in the same ear and at the same frequency. NIOSH recommends a confirmation audiogram within 30 days with the confirmation audiogram preceded by a quiet period of at least 14 hours. The NIOSH STS, called "15 dB twice" (same ear, same frequency), can be tested only if the baseline audiogram is available at the time of the annual audiometric test.

For maximum protection of employees, audiograms should be performed on the following occasions:

- Preemployment
- Prior to initial assignment in a hearing hazardous work area
- Annually as long as the employee is assigned to a noisy job (a time-weighted average exposure level equal to or greater than 85 dBA)
- At the time of reassignment out of a hearing hazardous job
- At the termination of employment

In addition, it is suggested that employees who are not exposed be given periodic audiograms as part of the organization's health care program. The audiograms of these employees can be compared to those of the exposed employees whenever the overall effectiveness of the hearing loss prevention program is evaluated. In an optimally effective program, the two employee groups show essentially the same amount of audiometric change.

Personal Hearing Protection Devices

A personal hearing protection device (or "hearing protector") is anything that can be worn to reduce the level of sound entering the ear. Earmuffs, ear canal caps, and earplugs are the three principal types of devices. Each employee reacts individually to the use of these devices, and a successful hearing loss prevention program should be able to respond to the needs of each employee. Ensuring that these devices protect hearing effectively requires the coordinated effort of management, the hearing loss prevention program operators, and the affected employees.[8]

There are several different types of earmuffs with which the modern safety and health professional should be familiar:

- **Passive earmuffs.** This kind of device is what most people think of when they think of earmuffs. They consist of ear cups lined with foam and block noise using nothing but the foam-lined cups. The primary weakness of passive earmuffs is they tend to block out not just unwanted noise, but also certain advantageous sounds such as voices trying to warn of danger.
- **Uniform attenuation earmuffs.** These earmuffs not only block noise, but also attenuate the noise more uniformly within several key octave bands (250 Hz–4 KHz). This allows employees wearing them to hear certain important sounds such as spoken instructions or warnings, thus reducing one of the main safety risks associated with earmuffs.
- **Electronic earmuffs.** This type of earmuff uses electronic technology to both block and modulate sound. Some of the more popular brands of electronic earmuffs can receive AM/FM radio signals or have a wireless connection to a CD or MP3 player.

Regardless of the kind of ear protection device used, it is important to remember the four Cs: *comfort, convenience, communication* (the device should not interfere with the worker's ability to communicate), and *caring* (workers must care enough about protecting their hearing to wear the devices).

Education and Motivation

Training is a critical element of a good hearing loss prevention program. In order to obtain sincere and energetic support by management and active participation by employees, it is necessary to educate and motivate both groups. A hearing loss prevention program that overlooks the importance of education and motivation is likely to fail because employees will not understand why it is in their best interest to cooperate, and management will fail to make the necessary commitment. Employees and managers who appreciate the precious sense of hearing and understand the reasons for, and the mechanics of, the hearing loss prevention program will be more likely to participate for their mutual benefit, rather than viewing the program as an imposition.

Record Keeping

Records often get the least attention of any of the program's components. However, audiometric comparisons, reports of hearing protector use, and the analysis of hazardous exposure measurements all involve the keeping of records. Unfortunately, records are often kept poorly because there is no organized system in place, and in many cases, those responsible for maintaining the records do not understand their value. People tend to assume that if they merely place records in a file or enter them into a computer, adequate record-keeping procedures are being followed. OSHA's latest version of the Form 300 Log has a column for recording hearing loss.

Many companies have found that their record-keeping system is inadequate only when they discover that they need accurate information. This sometimes occurs during the processing of compensation claims. Problems can be avoided by implementing an effective record-keeping system, in which (1) management encourages that the system be kept active and accessible; (2) hearing loss prevention program implementers make sure that all the information entered is accurate and complete; and (3) employees validate the information.

Hearing loss prevention program records should include all items for each phase of the program: (1) hearing loss prevention audit, (2) monitoring hearing hazards, (3) engineering and administrative controls, (4) audiometric evaluation, (5) personal hearing protective devices, (6) education and motivation, (7) record keeping, and (8) program evaluation. Each phase must be considered in order to evaluate the effectiveness of the hearing loss prevention program.

Program Evaluation

The primary goal of any hearing loss prevention program must be to reduce, and eventually eliminate, hearing loss due to workplace exposures. Although management may have the best intentions of implementing this goal and a company's hearing loss prevention program may have the appearance of being complete and complying with OSHA's requirements, the program still may not achieve this goal. A thorough evaluation of the effectiveness of all the program's components is necessary to determine the extent to which the hearing loss prevention program is really working.

Management and program implementers should conduct periodic program evaluations to assess compliance with federal and state regulations and to ensure that hearing is being conserved. There are two basic approaches to following program evaluation: (1) assess the completeness and quality of the program's components, and (2) evaluate the audiometric data.

Workers' Compensation and Noise Hazards

Hearing loss claims are being covered by state workers' compensation laws. Some states have written hearing loss into their workers' compensation law. Others are covering claims whether hearing loss is in the law or not.

Medical professionals have established a procedure for determining if there is a causal relationship between workplace noise and hearing loss. In making determinations of such relationships, physicians consider the following factors:

- Onset and progress of the employee's history of hearing loss
- The employee's complete work history
- Results of the employee's otological examination
- Results of hearing studies that have been performed
- Determination of whether causes of hearing loss originated outside the workplace

Because approximately 15 percent of all working people are exposed to noise levels exceeding 90 dBA, hearing loss may be as significant in workers' compensation costs in the future as back injuries, carpal tunnel syndrome, and stress are now.

Identifying and Assessing Hazardous Noise Conditions

Identifying and assessing hazardous noise conditions in the workplace involve (1) conducting periodic noise surveys, (2) conducting periodic audiometric tests, (3) record keeping, and (4) follow-up action. Each of these components is covered in the following sections.

Noise Surveys

Conducting **noise surveys** involves measuring noise levels at different locations in the workplace. The devices that are most widely used to measure noise levels are sound level meters and dosimeters. A **sound level meter** produces an immediate reading that represents the noise level at a specific instant in time.[9] A **dosimeter** provides a time-weighted average over a period of time such as one complete work shift. A dosimeter also calculates the sound as a function of sound level and duration and describes the results in terms of dose, peak level, equivalent sound level, sound exposure level, and time-weighted average.[10] The dosimeter is the most widely used device because it measures total exposure, which is what OSHA and ANSI standards specify. Using a dosimeter in various work areas and attaching a personal dosimeter to one or more employees is the recommended approach to ensure dependable, accurate readings.

Audiometric Testing

Audiometric testing measures the hearing threshold of employees. Tests conducted according to ANSI S12.13 can detect changes in the hearing threshold of the employee. A negative change represents hearing loss within a given frequency range.

The initial **audiogram** establishes a baseline hearing threshold. After that, audiometric testing should occur at least annually. Testing should not be done on an employee who has a cold, an ear infection, or who has been exposed to noise levels exceeding 80 dBA within 14 to 16 hours prior to a test. Such conditions can produce invalid results.[11]

When even small changes in an employee's hearing threshold are identified, more frequent tests should be scheduled and conducted as specified in ANSI S12.13. "For those employees found to have standard threshold shift—a loss of 10 dBA or more averaged at 2,000, 3,000, and 4,000 **hertz (Hz)** in either ear—the employer is required to fill out the OSHA 300 Log in which the loss is recorded as a worktime illness."[12]

Record Keeping

Figure 18-4 is an example of an audiometric form that can be used to record test results for individual employees. Such forms should be completed and kept on file to allow for sequential comparisons. It is also important to retain records containing a worker's employment history, including all past positions and the working conditions in those positions.

Follow-Up

Follow-up is critical. Failure to take prompt corrective action at the first sign of hearing loss can lead to permanent debilitating damage.

FIGURE 18-4 Sample audiometric test form.

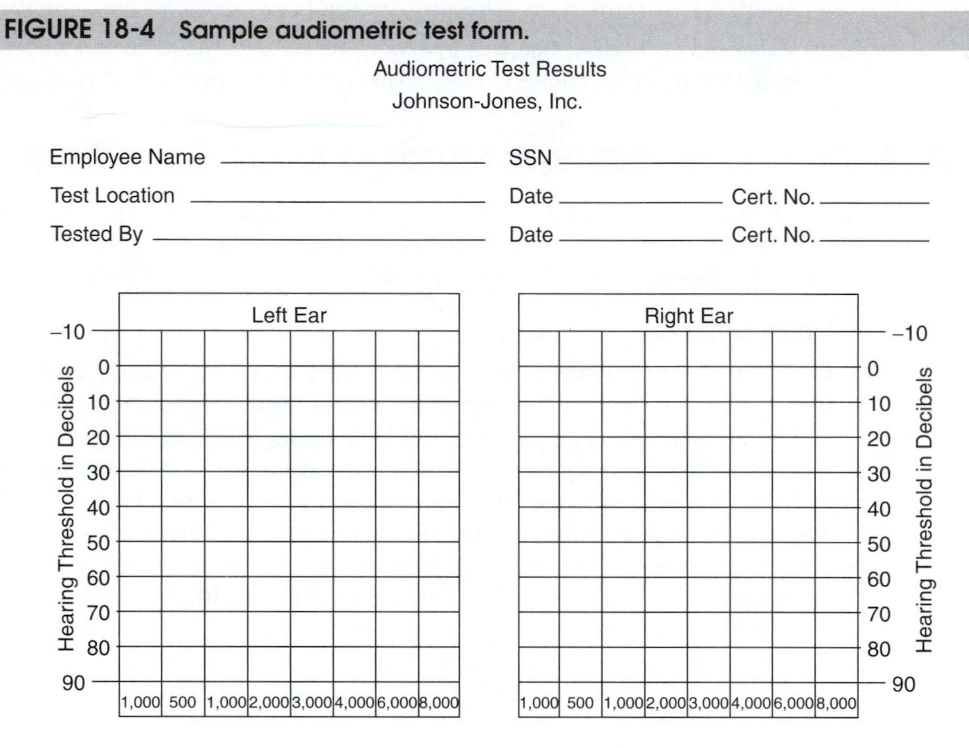

Audiometric Test Results
Johnson-Jones, Inc.

Employee Name ______________ SSN ______________
Test Location ______________ Date ______________ Cert. No. ______________
Tested By ______________ Date ______________ Cert. No. ______________

Hearing loss can occur without producing any evidence of physiological damage. Therefore, it is important to follow up on even the slightest evidence of a change in an employee's hearing threshold.

Follow-up can take a number of different forms. The following would all be appropriate follow-up responses:

- Administering a retest to verify the hearing loss
- Changing or improving the type of **personal protection** used
- Conducting a new noise survey in the employee's work area to determine if engineering controls are sufficient
- Testing other employees to determine if the hearing loss is isolated to the one employee in question or if other employees have been affected

Noise Control Strategies

Figure 18-5 illustrates the three components of a noise hazard. Noise can be reduced by engineering and administrative controls applied to one or more of these components. The most desirable **noise controls** are those that reduce noise at the source. The second priority is to reduce noise along its path. The last resort is noise reduction at the receiver using personal protective devices. The latter approach should never be substituted for the two former approaches.

FIGURE 18-5 Three parts of a noise hazard.

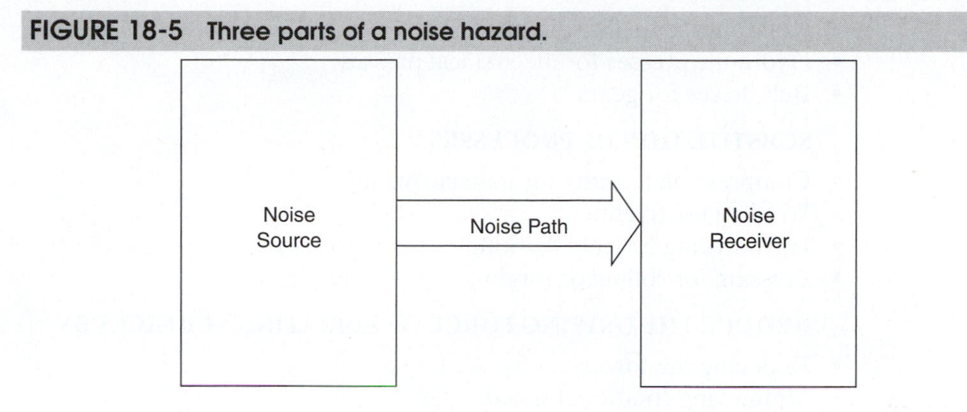

The following paragraphs explain widely used strategies for reducing workplace noise at the source, along its path, and at the receiver:

- Noise can be reduced at its source by enclosing the source, altering the acoustical design at the source, substituting equipment that produces less noise, making alterations to existing equipment, or changing the process so that less noisy equipment can be used.
- Noise can be reduced along its **path** by moving the source farther away from receivers and improving the acoustical design of the path so that more sound is absorbed as it travels toward receivers.
- Noise can be reduced at the **receiver** by enclosing the worker, using personal protective devices, and changing job schedules so that exposure time is reduced.

Some of the noise reduction strategies explained in the preceding paragraphs are engineering controls; others are administrative controls. For example, enclosing a noise source and substituting less noisy equipment are both examples of engineering controls. Changing job schedules is an example of an administrative control. Safety and health professionals should be familiar with both types of controls.

Engineering Controls

Engineering controls consist of facility and equipment adjustments other than administrative and personal protection strategies made to reduce the noise level either at the source or within the worker's hearing zone. Following are some commonly used engineering controls. All these controls are designed to reduce noise at the source, along its path, or at the receiver. They focus primarily on the noise rather than the employees who are exposed to it.

MAINTENANCE

- Replacement or adjustment of worn, loose, or unbalanced parts of machines
- Lubrication of machine parts and use of cutting oils
- Use of properly shaped and sharpened cutting tools

SUBSTITUTION OF MACHINES

- Larger, slower machines for smaller, faster ones
- Step dies for single-operation dies
- Presses for hammers

- Rotating shears for square shears
- Hydraulic presses for mechanical presses
- Belt drives for gears

SUBSTITUTION OF PROCESSES

- Compression riveting for impact riveting
- Welding for riveting
- Hot working for cold working
- Pressing for rolling or forging

REDUCE THE DRIVING FORCE OF VIBRATING SURFACES BY

- Reducing the forces
- Minimizing rotational speed
- Isolating

REDUCE THE RESPONSE OF VIBRATING SURFACES BY

- Damping
- Additional support
- Increasing the stiffness of the material
- Increasing the mass of vibrating members
- Changing the size to change resonance frequency

REDUCE THE SOUND RADIATION FROM THE VIBRATING SURFACES BY

- Reducing the radiating area
- Reducing overall size
- Perforating surfaces

REDUCE THE SOUND TRANSMISSION THROUGH SOLIDS BY USING

- Flexible mounting
- Flexible sections in pipe runs
- Flexible-shaft couplings
- Fabric sections in ducts
- Resilient flooring

REDUCE THE SOUND PRODUCED BY GAS FLOW BY

- Using intake and exhaust mufflers
- Using fan blades designed to reduce turbulence
- Using large, low-speed fans instead of smaller, high-speed fans
- Reducing the velocity of fluid flow (air)
- Increasing the cross-section of streams
- Reducing the pressure
- Reducing the air turbulence

REDUCE NOISE BY REDUCING ITS TRANSMISSION THROUGH AIR BY

- Using sound-absorptive material on walls and ceiling in work areas
- Using sound barriers and sound absorption along the transmission path
- Completely enclosing individual machines
- Using baffles
- Confining high-noise machines to insulated rooms

Administrative Controls

Administrative controls are controls that reduce the exposure of employees to noise rather than reducing the noise itself.

Administrative controls should be considered a second-level approach, with engineering controls given top priority. Smaller companies that cannot afford to reduce noise through engineering measures may use administrative controls instead. However, this approach should be avoided if at all possible.

Hearing Protection Devices

In addition to engineering and administrative controls, employees should be required to use appropriate **hearing protection devices (HPDs)**. It should be noted, however, that such devices are effective only if worn properly. Enforcement of the proper use of HPDs is difficult in some settings. The following four classifications of HPDs are widely used: enclosures, earplugs, superaural caps, and earmuffs.

Enclosures are devices that completely encompass the employee's head, much like the helmets worn by motorcycle riders. **Earplugs** (also known as *aurals*) are devices that fit into the ear canal. Custom-molded earplugs are designed and molded for the individual employee. Premolded earplugs are generic in nature, are usually made of a soft rubber or plastic substance, and can be reused. **Formable earplugs** can be used by anyone. They are designed to be formed individually to a person's ears, used once, and then discarded.

Superaural caps fit over the external edge of the ear canal and are held in place by a headband. Earmuffs, also known as *circumaurals,* cover the entire ear with a cushioned cup that is attached to a headband. Earplugs and earmuffs are able to reduce noise by 20 to 30 dB. By combining earplugs and earmuffs, an additional 3 to 5 dB of blockage can be gained.

Figures 18-6 through 18-9 are examples of various types of HPDs. Figure 18-6 illustrates ear, face, and head protection combined into one comprehensive device. Figure 18-7 contains two types of superaural caps. Figure 18-8 shows an earmuff-style HPD equipped with an FM radio capability. Figure 18-9 displays soft, moldable earplugs.

The effectiveness of HPDs can be enhanced through the use of technologies that reduce noise levels. These **active noise reduction (ANR)** technologies reduce noise by manipulating sound and signal waves. Such waves are manipulated by creating an electronic mirror image of sound waves that tends to cancel out the unwanted noise in the same way that negative numbers cancel out positive numbers in a mathematical equation. Using ANR in conjunction with enclosure devices or earmuffs can be an especially effective strategy. Safety and health professionals should know the noise reduction rating (NRR) of all devices and technologies they recommend for employee use.

Traditional, or passive, HPDs can distort or muffle sounds at certain frequencies, particularly high-pitched sounds. *Flat-attenuation HPDs* solve this problem by using electronic devices to block all sound frequencies equally. This eliminates, or at least reduces, the distortion and muffling problems. Flat-attenuation HPDs are especially helpful for employees in settings where high-pitched sound is present that they should be able to hear, as well as for employees who have already begun to lose their ability to hear such sounds. The ability to hear high-pitched sounds is significant because warning signals and human voices can be high-pitched.

FIGURE 18-6 Earmuff-style HPD.

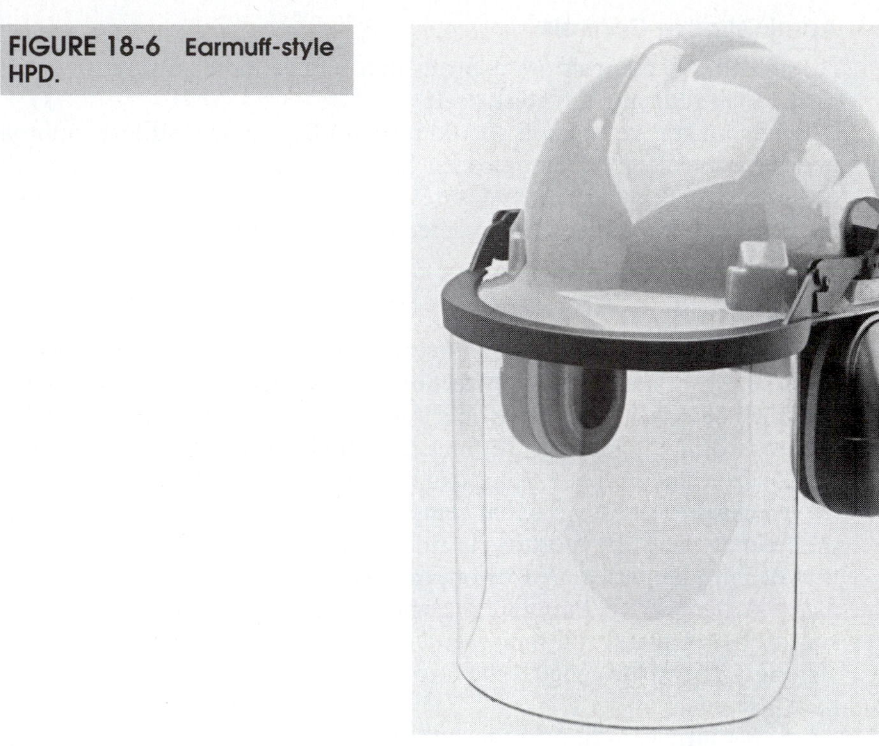

Courtesy of ELVEX Corporation.

FIGURE 18-7 Canal-cap-style HPD.

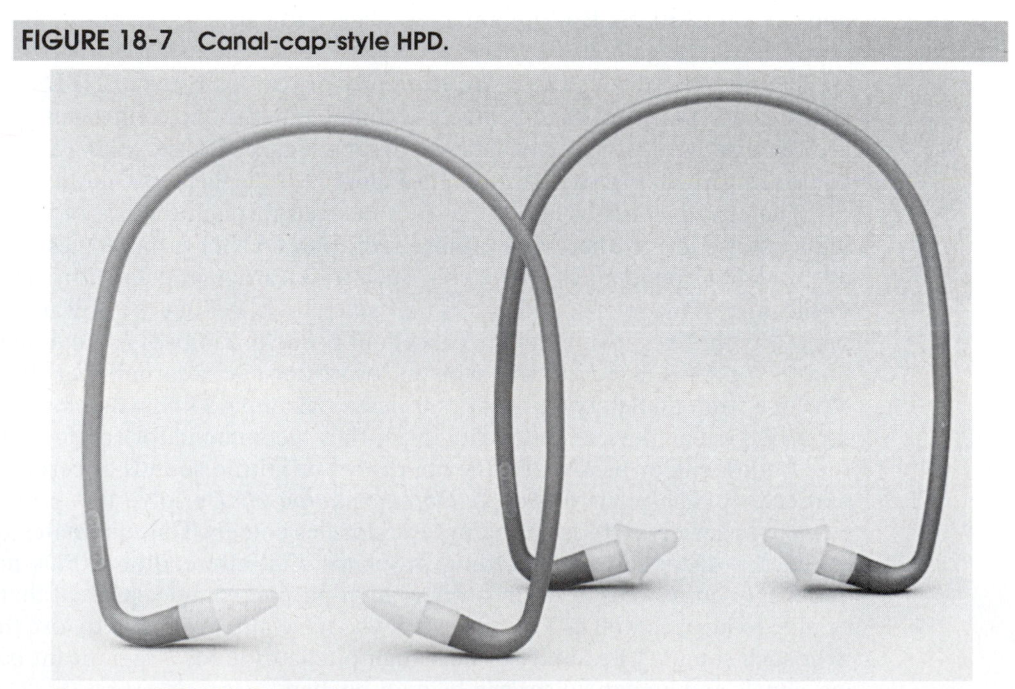

Courtesy of ELVEX Corporation.

FIGURE 18-8 Earmuff-style HPDs with FM radio capability.

Courtesy of ELVEX Corporation.

FIGURE 18-9 Earplug-style HPDs.

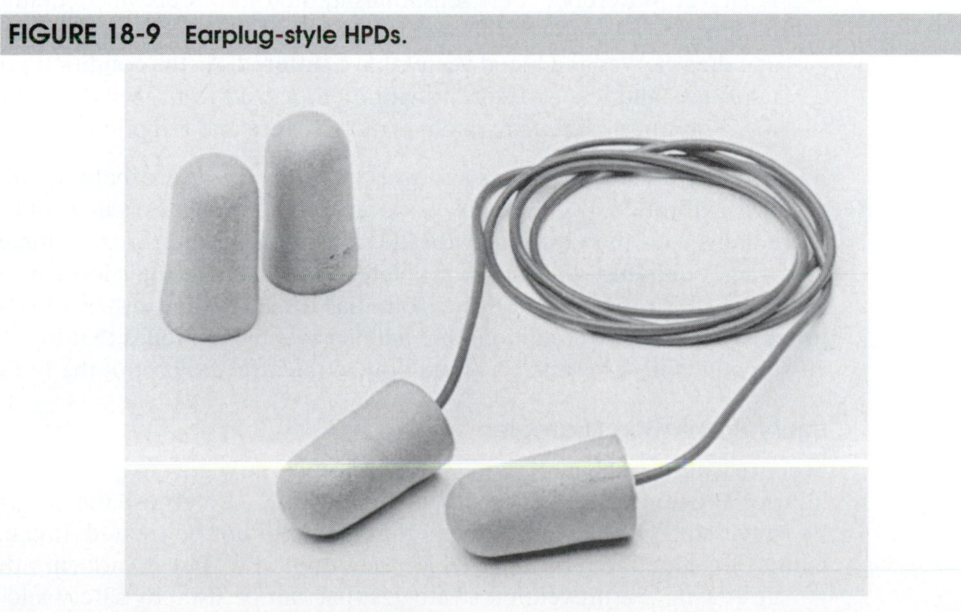

Courtesy of ELVEX Corporation.

A benefit of ANR technologies is **optimization**. The amount of noise protection can be adjusted so that employees can hear as much as they should, but not too much. Too much noise can cause employees to suffer hearing loss. Too little noise can mean that they may not hear warning signals.

Vibration Hazards

Vibration hazards are closely associated with noise hazards because tools that produce vibration typically also produce excessive levels of noise. The strategies for protecting employees against the noise associated with vibrating tools are the same as those presented so far in this chapter. This section focuses on the other safety and health hazards associated with vibration.

Eastman explains the problems associated with vibration:

> Vibration related problems are not only serious, they are widespread. Donald Wasserman, author of *Human Aspects of Occupational Vibration,* says that up to 8 million workers are exposed to some type of vibration hazard. Of these, it has been estimated that more than half will show some signs of injury.[13]

The types of injuries associated with vibration depend on its source. For example, workers who operate heavy equipment often experience vibration over the whole body. This can lead to problems ranging from motion sickness to spinal injury. However, the most common vibration-related problem is known as **hand-arm vibration syndrome (HAV)**. Eastman describes HAV as follows:

> The condition, a form of Reynaud's Syndrome, strikes an alarming number of workers who use vibrating power tools day in and day out as part of their jobs. For HAV sufferers, . . . the sensations in their hands are more than just minor, temporary discomforts. They are symptoms of the potentially irreversible damage their nerves and blood vessels have suffered. As the condition progresses, it takes less and less exposure to vibration or cold to trigger the symptoms, and the symptoms themselves become more severe and crippling.[14]

Environmental conditions and worker habits can exacerbate the problems associated with vibration. For example, working with vibrating tools in a cold environment is more dangerous than working with the same tools in a warm environment. Gripping a vibrating tool tightly will lead to problems sooner than using a loose grip. Smoking and excessive noise also increase the potential for HAV and other vibration-related injuries. What all these conditions and habits have in common is that they constrict blood vessels which in turn restricts blood flow to the affected part of the body.[15]

Injury Prevention Strategies

Modern safety and health professionals should know how to prevent vibration-related injuries. Prevention is especially important with HAV because the disease is thought to be irreversible. This does not mean that HAV cannot be treated. It can, but the treatments developed to date only reduce the symptoms. They do not cure the disease.

Following are prevention strategies that can be used by safety and health professionals in any company regardless of its size.

Purchase Low-Vibration Tools

Interest in producing low-vibration tools is relatively new but growing. Since the 1960 s, only a limited number of manufacturers produced low-vibration tools. However, a lawsuit filed against three prominent tool manufacturers generated a higher level of interest in producing low-vibration tools.

According to Eastman, the suit was filed on behalf of 300 employees of General Dynamics's Electric Boat Shipyard in Connecticut, most of whom now suffer from HAV. The case claimed that three predominant tool manufacturers failed to (1) warn users of their tools of the potential for vibration-related injury and (2) produce low-vibration tools even though the technology to do so has been available for many years.[16] As a result of this lawsuit and the potential for others like it, low-vibration tools are becoming more commonplace.

Limit Employee Exposure

Although a correlation between cumulative exposure to vibration and the onset of HAV has not been scientifically quantified, there is strong suspicion in the safety and health community that such a link exists. For example, NIOSH recommends that companies limit the exposure of their employees to no more than four hours per day, two days per week.[17] Until the correlation between cumulative exposure and HAV has been quantified, safety and health professionals are well advised to apply the NIOSH recommendation.

Change Employee Work Habits

Employees can play a key role in protecting themselves if they know how. Safety and health professionals should teach employees who use vibration-producing tools the work habits that will protect them from HAV and other injuries. These work habits include the following: (1) wearing properly fitting thick gloves that can partially absorb vibration; (2) taking periodic breaks (at least 10 minutes every hour); (3) using a loose grip on the tool and holding it away from the body; (4) keeping tools properly maintained (i.e., replacing vibration-absorbing pads regularly); (5) keeping warm; and (6) using vibration-absorbing floor mats and seat covers as appropriate.[18]

Modern safety and health professionals should also encourage higher management to require careful screening of applicants for jobs involving the use of vibration-producing tools and equipment. Applicants who smoke or have other conditions that constrict blood vessels should be guided away from jobs that involve excessive vibration.

Other Effects of Noise Hazards

Hearing loss is the principal concern of safety and health professionals relating to noise hazards. However, hearing loss is not the only detrimental effect of excess noise. Noise can also cause communication problems, isolation, and productivity problems.

Noise can also be detrimental to productivity by interfering with an employee's ability to think, reason, and solve problems. Not all employees respond to noise in this way, but many do. Can you concentrate on your studies in a noisy room? Some students can, whereas others cannot. If excessive noise makes it difficult for you to study, you will probably have the same problem on the job.

Isolation is another problem for some employees in a noisy environment. Employees can begin to feel left out and uninformed, the antithesis of the goal of a modern teamwork-oriented organization.

Corporate Policy

Corporate policies relating to hearing loss prevention should be carefully planned and executed to benefit the affected employee and the employer.[19] Organizations with the most successful programs address the following areas of concern:

- The organizational environment should promote a safety culture where employees are empowered to protect their own health and that of coworkers.
- Policies should be based on effective practices rather than on minimum compliance with government regulations.
- The hearing prevention program must be a functional part of the overall company safety and health program. It should not be a stand-alone, separate-budget operation.
- A key individual (or *program implementor*) should have ultimate responsibility for the program. This person may not necessarily perform all the functions of the hearing loss prevention program but is in charge of the overall program. Experience with successful hearing loss prevention programs shows that a single individual often makes the crucial difference between success and failure. This person may be a safety and health officer, a supervisor, or a designated employee. This program implementor acts as the *conscience* and *champion* of the hearing loss prevention program.
- The program implementor should work with management and employees to develop and implement hearing loss prevention plans and policies for an effective program. As a team leader, the program implementor should be given the authority to establish hearing loss prevention provisions that meet or exceed the letter and intent of OSHA's noise control and hearing conservation regulations.
- Employee and administrative compliance with the company's hearing loss prevention program.
- Hearing loss prevention program policies should clearly describe standard operating procedures for each phase of the program. Specific policy statements should be developed for the important elements of the program. For example, company policy should require the participation of all noise-exposed employees in the audiometric program and require the consistent and proper wearing of hearing protectors in posted areas, even if employees or supervisors are only passing through these areas. These requirements should be conditions of employment. Other important policy statements should be written to cover:

    Adoption of a prescribed schedule for monitoring employee noise exposure levels and other risks, including ensuring that equipment and personnel training are appropriate to the task.

    Counseling of employees immediately following each audiometric test, whether it is the initial, annual, retest, threshold-shift confirmation, or termination examination.

    Determining the adequacy and correct use of hearing protection devices by on-site equipment checks.

Educating, training, and motivating employees to support the company's hearing loss prevention program provisions; assessing employee attitudes and assessing knowledge gained from periodic training.

Establishing a program of quality assurance for the performance of audiometry and management of audiometric records.

Reviewing audiometric data to verify the effectiveness of the hearing loss prevention program.

Encouraging employees to use company-provided hearing protectors for off-the-job exposure.

Purchasing hearing protectors, audiometers, noise measuring equipment, and quieter machinery. This policy should address the reasons why the program implementor responsible for the hearing loss prevention program, not the purchasing department, should have final authority about anticipated purchases.

If services such as noise surveys, employee education, audiometric testing, medical counseling, or the fitting of hearing protection devices cannot be handled by in-house staff, outside vendors or contractors may be required. They should be selected carefully to ensure that their services complement the abilities of the company staff and functional conduct of the in-house program elements. Vendors must understand and agree to abide by the company's hearing loss prevention program policies and standards of operation. On-site personnel must supervise contractors to make sure that they carry out their obligations. Regardless of whether outside vendors or contractors are used, responsibility for the program stays with the program implementor.

Companies that issue clearly defined hearing loss prevention policies and then adhere to these policies consistently will have smoothly running hearing loss prevention programs. Employees will be fully informed, comprehend their functional role, and know what is expected of them. Equipment will be appropriate, hearing protection will be used by the right people in the right places, and the program elements will be implemented in a timely fashion.

Evaluating Hearing Loss Prevention Programs

Hearing loss prevention programs should be evaluated periodically to ensure their effectiveness.[20] Such evaluations should have at least the following components: (1) training and education; (2) supervisor involvement; (3) noise measurement; (4) engineering and administrative controls; (5) monitoring and record keeping; (6) referrals; (7) hearing protection devices; and (8) administration. Following are checklists for each of these components.

Training and Education

Failures or deficiencies in hearing conservation programs (hearing loss prevention programs) can often be traced to inadequacies in the training and education of noise-exposed employees and those who conduct elements of the program.

- Has training been conducted at least once a year?
- Was the training provided by a qualified instructor?
- Was the success of each training program evaluated?
- Is the content revised periodically?
- Are managers and supervisors directly involved?

- Are posters, regulations, handouts, and employee newsletters used as supplements?
- Are personal counseling sessions conducted for employees having problems with hearing protection devices or showing hearing threshold shifts?

Supervisor Involvement

Data indicate that employees who refuse to wear hearing protectors or who fail to show up for hearing tests frequently work for supervisors who are not totally committed to the hearing conservation programs.

- Have supervisors been provided with the knowledge required to supervise the use and care of hearing protectors by subordinates?
- Do supervisors wear hearing protectors in appropriate areas?
- Have supervisors been counseled when employees resist wearing protectors or fail to show up for hearing tests?
- Are disciplinary actions enforced when employees repeatedly refuse to wear hearing protectors?

Noise Measurement

Noise measurements, to be useful, should be related to noise exposure risks or the prioritization of noise control efforts, rather than merely filed away. In addition, the results should be communicated to the appropriate personnel, especially when follow-up actions are required.

- Were the essential or critical noise studies performed?
- Was the purpose of each noise study clearly stated? Have noise-exposed employees been notified of their exposures and apprised of auditory risks?
- Are the results routinely transmitted to supervisors and other key individuals?
- Are results entered into health or medical records of noise-exposed employees?
- Are results entered into shop folders?
- If noise maps exist, are they used by the proper staff?
- Are noise measurement results considered when contemplating procurement of new equipment? Modifying the facility? Relocating employees?
- Have there been changes in areas, equipment, or processes that have altered noise exposure? Have follow-up noise measurements been conducted?
- Are appropriate steps taken to include (or exclude) employees in the hearing loss prevention programs whose exposures have changed significantly?

Engineering and Administrative Controls

Controlling noise by engineering and administrative methods is often the most effective means of reducing or eliminating the hazard. In some cases, engineering controls will remove requirements for other components of the program, such as audiometric testing and the use of hearing protectors.

- Have noise control needs been prioritized?
- Has the cost effectiveness of various options been addressed?
- Are employees and supervisors apprised of plans for noise control measures?
- Are they consulted on various approaches?
- Will in-house resources or outside consultants perform the work?

- Have employees and supervisors been counseled on the operation and maintenance of noise control devices?
- Are noise control projects monitored to ensure timely completion?
- Has the full potential for administrative controls been evaluated? Are noisy processes conducted during shifts with fewer employees? Do employees have sound-treated lunch or break areas?

Monitoring Audiometry and Record Keeping

The skills of audiometric technicians, the status of the audiometer, and the quality of audiometric test records are crucial to hearing loss prevention program success. Useful information may be ascertained from the audiometric records as well as from those who actually administer the tests.

- Has the audiometric technician been adequately trained, certified, and recertified as necessary?
- Do on-the-job observations of the technicians indicate that they perform a thorough and valid audiometric test, instruct and consult the employee effectively, and keep appropriate records?
- Are records complete?
- Are follow-up actions documented?
- Are hearing threshold levels reasonably consistent from test to test? If not, are the reasons for inconsistencies investigated promptly?
- Are the annual test results compared to baseline to identify the presence of an OSHA standard threshold shift?
- Is the annual incidence of standard threshold shift greater than a few percent? If so, are problem areas pinpointed and remedial steps taken?
- Are audiometric trends (deteriorations) being identified, both in individuals and in groups of employees? (NIOSH recommends no more than 5 percent of workers showing 15-dB significant threshold shift, same ear, same frequency.)
- Do records show that appropriate audiometer calibration procedures have been followed?
- Is there documentation showing that the background sound levels in the audiometer room were low enough to permit valid testing?
- Are the results of audiometric tests being communicated to supervisors and managers as well as to employees?
- Has corrective action been taken if the rate of no-shows for audiometric test appointments is more than about 5 percent?
- Are employees incurring STS notified in writing within at least 21 days? (NIOSH recommends immediate notification if retest shows 15-dB significant threshold shift, same ear, same frequency.)

Referrals

Referrals to outside sources for consultation or treatment are sometimes in order, but they can be an expensive element of the hearing loss prevention program and should not be undertaken unnecessarily.

- Are referral procedures clearly specified?
- Have letters of agreement between the company and consulting physicians or audiologists been executed?

- Have mechanisms been established to ensure that employees needing evaluation or treatment actually receive the service (e.g., transportation, scheduling, reminders)?
- Are records properly transmitted to the physician or audiologist, and back to the company?
- If medical treatment is recommended, does the employee understand the condition requiring treatment, the recommendation, and methods of obtaining such treatment?
- Are employees being referred unnecessarily?

Hearing Protection Devices

When noise control measures are not feasible, or until such time as they are installed, hearing protection devices are the only way to prevent hazardous levels of noise from damaging the inner ear. Making sure that these devices are worn effectively requires continuous attention on the part of supervisors and program implementers as well as noise-exposed employees.

- Have hearing protectors been made available to all employees whose daily average noise exposures are 85 dBA or above? (NIOSH recommends requiring HPD use if noises equal or exceed 85 dBA regardless of exposure time.)
- Are employees given the opportunity to select from a variety of appropriate protectors?
- Are employees fitted carefully with special attention to comfort?
- Are employees thoroughly trained, not only initially but at least once a year?
- Are the protectors checked regularly for wear or defects and replaced immediately if necessary?
- If employees use disposable hearing protectors, are replacements readily available?
- Do employees understand the appropriate hygiene requirements?
- Have any employees developed ear infections or irritations associated with the use of hearing protectors? Are there any employees who are unable to wear these devices because of medical conditions? Have these conditions been treated promptly and successfully?
- Have alternative types of hearing protectors been considered when problems with current devices are experienced?
- Do employees who incur noise-induced hearing loss receive intensive counseling?
- Are those who fit and supervise the wearing of hearing protectors competent to deal with the many problems that can occur?
- Do workers complain that protectors interfere with their ability to do their jobs? Do they interfere with spoken instructions or warning signals? Are these complaints followed promptly with counseling, noise control, or other measures?
- Are employees encouraged to take home their hearing protectors if they engage in noisy nonoccupational activities?
- Are potentially more effective protectors considered as they become available?
- Is the effectiveness of the hearing protector program evaluated regularly?

- Have at-the-ear protection levels been evaluated to ensure that either over- or underprotection has been adequately balanced according to the anticipated ambient noise levels?
- Is each hearing protector user required to demonstrate that he or she understands how to use and care for the protector? Are the results documented?

Administration

Keeping organized and current on administrative matters will help the program run smoothly.

- Have there been any changes in federal or state regulations? Have hearing loss prevention programs' policies been modified to reflect these changes?
- Are copies of company policies and guidelines regarding the hearing loss prevention program available in the offices that support the various program elements? Are those who implement the program elements aware of these policies? Do they comply?
- Are necessary materials and supplies being ordered with a minimum of delay?
- Are procurement officers overriding the hearing loss prevention program implementer's requests for specific hearing protectors or other hearing loss prevention equipment? If so, have corrective steps been taken?
- Is the performance of key personnel evaluated periodically? If such performance is found to be less than acceptable, are steps taken to correct the situation?
- Has the failure to hear warning shouts or alarms been tied to any accidents or injuries? If so, have remedial steps been taken?

Key Terms and Concepts

- Active noise reduction (ANR)
- Administrative controls
- Air pressure
- ANSI S12.13
- Attenuation
- Audiogram
- Audiometric database analysis (ADBA)
- Audiometric testing
- Baseline audiogram
- Continuous noise
- Decibel (dB)
- Dosimeter
- Duration
- Electronic earmuffs
- Earplugs
- Enclosures
- Engineering controls
- Exchange rate
- Follow-up
- Formable earplugs
- Frequency
- Hand-arm vibration syndrome (HAV)
- Hazardous noise
- Hearing conservation
- Hearing protection devices (HPDs)
- Hearing threshold level (HTL)
- Hertz (Hz)
- Impulse noise
- Low-vibration tools
- Material hearing impairment
- Medical surveillance
- Narrow band noise
- Noise
- Noise controls
- Noise-induced hearing loss
- Noise surveys
- Optimization
- Passive earmuffs
- Path
- Percent better or worse sequential
- Percent worse sequential
- Personal protection
- Receiver
- Sound
- Sound level meter
- Sound pressure level
- Source
- Standard threshold shift (STS)
- Superaural caps
- Threshold of hearing
- Threshold of pain
- Time-weighted average (TWA)
- Uniform attenuation earmuffs
- Vibration
- Wide band noise

Review Questions

1. Define the term sound.
2. What is the difference between sound and noise?
3. Differentiate between sound and vibration.
4. Describe the relationship between the pitch of sound and the cycle of sound waves.
5. List and briefly explain the three broad types of industrial noise.
6. Describe the various physiological problems associated with excessive noise.
7. List four factors that affect the risk of hearing loss from exposure to excessive noise.
8. At what sound level is it necessary to begin using some type of personal protection?
9. Give a brief description of ANSI S12.13.
10. Give a brief description of OSHA 29 CFR 1910.95, Hearing Conservation Amendment.
11. What factors do medical professionals consider in determining causal relationships of hearing loss?
12. Define the following terms: follow-up, noise survey, and audiometric testing.
13. List three appropriate follow-up activities when an audiometric test reveals hearing loss in an employee.
14. Differentiate between engineering and administrative controls.
15. What is HAV? How can it be prevented?
16. Explain the four classifications of HPDs that are widely used.
17. Explain the main components of an evaluation of a hearing loss prevention program.

Endnotes

1. National Institute for Occupational Safety and Hearing (NIOSH), *A Practical Guide to Preventing Hearing Loss,* Publication 96–110 (Washington, DC: 2006), 2. Retrieved from http://www.cdc.gov/niosh/96–110a.html.
2. Ibid.
3. NIOSH, *A Practical Guide*, 6.
4. NIOSH, *A Practical Guide*, 4.
5. Ibid., 6.
6. Ibid.
7. J. L. Ruck, "Hidden Risk of Noise Standard," *Occupational Hazards*, April 2000, 27.
8. B. Sokol, "Ear Muffs: A Field Guide," *Occupational Health & Safety* 74, no. 6: 66–68.
9. NIOSH, *A Practical Guide*, 37.
10. NIOSH, *A Practical Guide*, 8.
11. Ibid., 12.
12. Ibid.
13. M. Eastman, "Vibration Shakes Workers," *Safety & Health* 143, no. 5; 32.
14. Ibid.
15. Ibid., 32–33.
16. Ibid., 34.
17. Ibid., 33.
18. Ibid., 35.
19. NIOSH, *A Practical Guide*, 1–2.
20. Ibid., Appendix B, 1–6.

Preparing for Emergencies and Terrorism

Major Topics

- Rationale for Emergency Preparation
- Emergency Planning and Community Right-to-Know Act
- Organization and Coordination
- OSHA Standards
- First Aid in Emergencies
- How to Plan for Emergencies
- Planning for Workers with Disabilities
- Evacuation Planning
- Customizing Plans to Meet Local Needs
- Emergency Response
- Computers and Emergency Response
- Dealing with the Psychological Trauma of Emergencies
- Recovering from Disasters
- Terrorism in the Workplace
- Resuming Business after a Disaster

Despite the best efforts of all involved, emergencies do sometimes occur. The potential for human-caused emergencies has increased significantly with the rise of worldwide terrorism. It is very important to respond to such emergencies in a way that minimizes harm to people and damage to property. To do so requires plans that can be implemented without delay. This chapter provides prospective and practicing safety and health professionals with the information they need to prepare for emergencies in the workplace. Everything in this chapter pertains to all kinds of emergencies including natural disasters and terrorism. A special section relating specifically to terrorism is included at the end of the chapter.

FIGURE 19-1 Elements of emergency preparation.

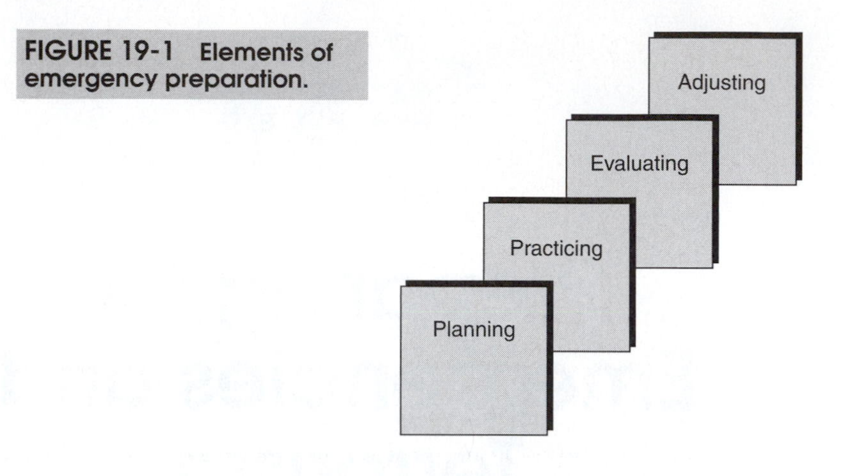

Rationale for Emergency Preparation

An *emergency* is a potentially life-threatening situation, usually occurring suddenly and unexpectedly. Emergencies may be the result of natural or human causes. Have you ever witnessed the timely, organized, and precise response of a professional emergency medical crew at an automobile accident? While passers-by and spectators may wring their hands and wonder what to do, the emergency response professionals quickly organize, stabilize, and administer. Their ability to respond in this manner is the result of preparation. As shown in Figure 19-1, preparation involves a combination of **planning**, **practicing**, **evaluating**, and **adjusting** to specific circumstances.

When an emergency occurs, immediate reaction is essential. Speed in responding can mean the difference between life and death or between minimal damage and major damage. Ideally, all those involved should be able to respond properly with a minimum of hesitation. This can happen only if all exigencies have been planned for and planned procedures have been practiced, evaluated, and improved.

A quick and proper response—which results because of proper preparation—can prevent panic, decrease the likelihood of injury and damage, and bring the situation under control in a timely manner. Because no workplace is immune to emergencies, preparing for them is critical. An important component of preparation is planning.

Emergency Planning and Community Right-to-Know Act

Title III of the Superfund Amendments and Reauthorization Act of 1986 (SARA) is also known as the **Emergency Planning and Community Right-to-Know Act (EPCRA).** This law is designed to make information about hazardous chemicals available to a community where they are being used so that residents can protect themselves in the case of an emergency. It applies to all companies that use, make, transport, or store chemicals.

Safety and health professionals involved in developing emergency response plans for their companies should be familiar with the act's requirements for emergency planning. As shown in Figure 19-2, the EPCRA includes the four major components discussed in the following paragraphs.

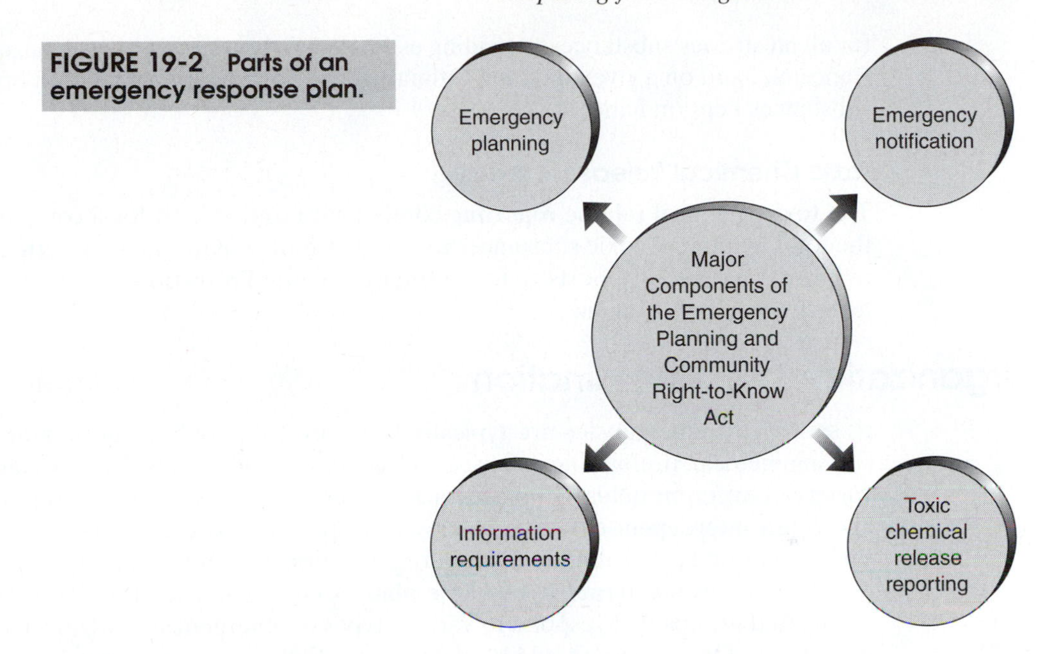

FIGURE 19-2 Parts of an emergency response plan.

Emergency Planning

The emergency planning component requires that communities form **local emergency planning committees (LEPCs)** and that states form **state emergency response commissions (SERCs)**. LEPCs are required to develop emergency response plans for the local communities, host public forums, select a planning coordinator for the community, and work with the coordinator in developing local plans. SERCs are required to oversee LEPCs and review their emergency response plans. Plans for individual companies in a given community should be part of that community's larger plan. Local emergency response professionals should use their community's plan as the basis for simulating emergencies and practicing their responses.

Emergency Notification

The **emergency notification** component requires that chemical spills or releases of toxic substances that exceed established allowable limits be reported to appropriate LEPCs and SERCs. Immediate notification may be verbal as long as a written notification is filed promptly. Such reports must contain at least the following information: (1) the names of the substances released, (2) where the release occurred, (3) when the release occurred, (4) the estimated amount of the release, (5) known hazards to people and property, (6) recommended precautions, and (7) the name of a contact person in the company.

Information Requirements

Information requirements mean that local companies must keep their LEPCs and SERCs and, through them, the public informed about the hazardous substances that the companies store, handle, transport, or use. This includes keeping comprehensive records of such substances on file, up-to-date, and readily available; providing copies of material safety data sheets for all hazardous substances; giving general storage locations for all hazardous substances; providing estimates of the amount of each hazardous substance on hand on a given day; and estimating the average annual amount of hazardous substances kept on hand.

Toxic Chemical Release Reporting

The **toxic chemical release reporting** component requires that local companies report the total amount of toxic substances released into the environment as either emissions or hazardous waste. Reports go to the Environmental Protection Agency and the state-level environmental agency.

Organization and Coordination

Responses to emergencies are typically from several people or groups of people including medical, firefighting, security, and safety personnel as well as specialists from a variety of different fields. People in each of these areas have different but interrelated and often interdependent roles to play in responding to the emergency. Because of their disparate backgrounds and roles, organization and coordination are critical.

A company's **emergency response plan** should clearly identify the different personnel and groups that respond to various types of emergencies and, in each case, who is in charge. One person should be clearly identified and accepted by all emergency responders as the **emergency coordinator**. This person should be knowledgeable, at least in a general sense, of the responsibilities of each individual emergency responder and how each relates to those of all other responders. This knowledge must include the **order of response** for each type of emergency set forth in the plan.

A company's safety and health professional is the obvious person to organize and coordinate emergency responses. However, regardless of who is designated, it is important that (1) one person is in charge, (2) everyone involved knows who is in charge, and (3) everyone who has a role in responding to an emergency is given ample opportunities to practice in simulated conditions that come as close as possible to real conditions.

Union Carbide's Texas City plant employs 1,500 workers and several hundred additional contract personnel. Safety is a high priority. Consequently, emergency planning is a fundamental part of the company's safety and health program. Emergency responders include emergency medical technicians, nurses, physicians, and production workers who are assigned specific emergency response duties. They are coordinated by a designated emergency director.[1]

Union Carbide's administrator of health services stresses the need for organization and coordination of emergency response teams. According to Marley, "Often staff physicians and nurses don't know the quickest, safest route to an emergency site. That's why emergency medical teams need to be organized and trained."[2]

Workers in the plant are obviously also knowledgeable about traffic flow, access routes, exit points, emergency approaches, and shortcuts. Because of this, Union Carbide's approach to emergency response "provides a method for dispatching EMTs and ambulance drivers to the scene while nursing and medical staff remain in radio communication and available for consultation or to receive patients if necessary."[3]

Another unique aspect of emergency planning and response at Union Carbide is how responses are coordinated: "The planning and coordination of all emergency response belongs to the **emergency response management team (ERMT)**. Composed of shift emergency directors and a full-time fire chief, this group unifies all emergency groups and equipment into a single, coordinated effort" (emphasis added).[4]

OSHA Standards

All OSHA standards are written for the purpose of promoting a safe, healthy, accident-free, hence emergency-free workplace. Therefore, OSHA standards play a role in emergency prevention and should be considered when developing emergency plans. For example, exits are important considerations when planning for emergencies. Getting medical personnel in and employees and injured workers out quickly is critical when responding to emergencies. The following sections of OSHA's standards deal with emergency preparedness:

Emergency Action Plan	29 CFR 1910.38
Exit arrangements	29 CFR 1910.37(e)
Exit capacity	29 CFR 1910.37(c),(d)
Exit components	29 CFR 1910.37(a)
Exit workings	29 CFR 1910.37(q)
Exit width	29 CFR 1910.37(c)
Exterior exit access	29 CFR 1910.37(g)
Occupational Health and Environmental Controls	29 CFR 1926.65, Appendix C
Hazardous Materials	29 CFR 1910.120, Appendix C

A first step for companies developing emergency plans is to review these OSHA standards. This can help safety and health personnel identify and correct conditions that may exacerbate emergency situations before they occur.

First Aid in Emergencies

Workplace emergencies often require a medical response. The immediate response is usually first aid. First aid consists of lifesaving measures taken to assist an injured person until medical help arrives.

Because there is no way to predict when first aid may be needed, providing first-aid training to employees should be part of preparing for emergencies. In fact, in certain cases, OSHA requires that companies have at least one employee on-site who has been trained in first aid (CFR 1910.151). Figure 19-3 contains a list of the topics that may be covered in a first-aid class for industrial workers.

First-Aid Training Program

First-aid programs are usually available in most communities. The continuing education departments of community colleges and universities typically offer first-aid training. Classes can often be provided on-site and customized to meet the specific needs of individual companies.

The American Red Cross provides training programs in first aid specifically geared toward the workplace. For more information about these programs, safety and health professionals may contact the national office of the American Red Cross at 202-639-3200 (http://www.redcross.org).

FIGURE 19-3 Sample course outline for first-aid class.

Basic First Aid

- Cardiopulmonary resuscitation
- Severe bleeding
- Broken bones and fractures
- Burns
- Choking on an obstruction
- Head injuries and concussion
- Cuts and abrasions
- Electric shock
- Heart attack
- Stroke recognition
- Moving an injured person
- Drug overdose
- Unconscious victim
- Eye injuries
- Chemical burns
- Rescue

The National Safety Council also provides first-aid training materials. The First Aid and Emergency Care Teaching Package contains a slide presentation, overhead transparencies, a test bank, and an instructor's guide. The council also produces a book titled *First Aid Essentials*. For more information about these materials, safety and health professionals may contact the NSC at 800-832-0034 (http://www.nsc.org).

Beyond Training

Training employees in first-aid techniques is an important part of preparing for emergencies. However, training is only part of the preparation. In addition, it is important to do the following:

1. Have well-stocked first-aid kits available. First-aid kits should be placed throughout the workplace in clearly visible, easily accessible locations. They should be properly and fully stocked and periodically checked to ensure that they stay fully stocked. Figure 19-4 lists the minimum recommended contents for a workplace first-aid kit.

SAFETY TIP

Failure to Warn the Community Can Be Costly

The hazards associated with accidents or incidents can extend beyond the walls of the facility in question to the surrounding community. This is why companies must have emergency plans in place for notifying the community and regulatory agencies when an accidental release poses a threat. Failure to notify its neighbors cost a food manufacturing company in Zanesville, Ohio, $43,829 in fines. When a refrigeration system's pressure relief valve malfunctioned, 820 pounds of anhydrous ammonia was released into the atmosphere. The company failed to notify local community officials and the National Response Center until three hours after the discharge.

SOURCE: From "Didn't Warn Neighbors Promptly about Hazard," *Facility Manager's Alert* 9, no. 187: 3.

FIGURE 19-4 Minimum recommended contents of workplace first-aid kits.

- Sterile gauze dressings (individually wrapped in sizes 4 × 4 and 8 × 10 inches)
- Triangular bandages
- Roll of gauze bandages (at least 2 inches wide)
- Assorted adhesive bandages
- Sealed moistened towelettes
- Adhesive tape
- Absorbent cotton
- Sterile saline solution
- Mild antiseptic for minor wounds
- Ipecac syrup to induce vomiting
- Powdered activated charcoal to absorb swallowed poisons
- Petroleum jelly
- Baking soda (bicarbonate of soda)
- Aromatic spirits of ammonia
- Elastic wraps
- Scissors
- Tweezers
- Needles
- Sharp knife or stiff-backed razor blades
- Medicine dropper (eye dropper)
- Measuring cup
- Oral thermometer
- Rectal thermometer
- Hot water bag
- Wooden safety matches
- Flashlight
- Rubber surgical gloves
- Face masks or mouthpieces
- Blanket
- Splint

2. Have appropriate personal protective devices available. With the concerns about AIDS and hepatitis, administering first aid has become more complicated than in the past. The main concerns are with bleeding and other body fluids. Consequently, a properly stocked first-aid kit should contain rubber surgical gloves and facemasks or mouthpieces for CPR.

3. Post emergency telephone numbers. The advent of 911 service has simplified the process of calling for medical care, police, or firefighting assistance. If 911 services are not available, emergency numbers for ambulance, hospital, police, fire department, LEPC, and appropriate internal personnel should be posted at clearly visible locations near all telephones in the workplace.

4. Keep all employees informed. Some companies require all employees to undergo first-aid training; others choose to train one or more employees in each department. Regardless of the approach used, it is important that all employees be informed and kept up-to-date concerning basic first-aid information. Figures 19-5 and 19-6 are first-aid fact sheets of the type used to keep employees informed.

How to Plan for Emergencies

Developing an **emergency action plan (EAP)** is a major step in preparing for emergencies. A preliminary step is to conduct a thorough analysis to determine the various types of emergencies that may occur. For example, depending on geography and the types of products and processes involved, a company may anticipate such emergencies

FIGURE 19-5 First-aid fact sheet.

First-Aid Fact Sheet No. 16
Moving an Injured Person

If a victim has a neck or back injury, do not move him unless it must be done to prevent additional injuries. If it is absolutely essential to move the victim, remember the following rules of thumb:

1. Call for professional medical help.
2. Always pull the body lengthwise, never sideways.
3. If there is time, slip a blanket under the victim and use it to pull him to safety.
4. If the victim must be lifted, support all parts of the body so that it does not bend or jackknife. (Use a spine board and cervical collar.)

FIGURE 19-6 First-aid fact sheet.

Global Technologies

500 Anchors Street • Industrial Park • Fort Walton Beach, Florida 32748

First-Aid Fact Sheet No. 12
ABCs of First Aid

If a fellow employee is injured and you are the first person to respond, remember the ABCs of first aid.

A = Airway
Is the airway blocked? If so, clear it quickly.

B = Breathing
Is the victim breathing? If not, begin administering artificial respiration.

C = Circulation
Is the victim bleeding severely? If so, stop the bleeding.
Is there a pulse? If not, begin administering CPR.

SAFETY FACT

Emergency Response Regulations

There are numerous government regulations requiring the development of emergency response plans. Following are the most important of these:

- Environmental Protection Agency's (EPA) Spill Prevention Control and Countermeasures (SPCC) and Facility Response Plans—40 CFR Part 112(d), 112.20–21.
- EPA's Resource Conservation and Recovery Act (RCRA) Contingency Plan—40 CFR Part 264, Subpart D; Part 265, Subpart D and 279.52.
- EPA's Risk Management Plan (RMP)—40 CFR Part 68.
- EPA's Emergency Planning and Community Right-to-Know (EPCRA, also known as SARA Title III).
- Comprehensive Emergency Response Compensation and Liabilities Act (CERCLA).
- Occupational Safety and Health Administration's OSHA Emergency Action Plan—29 CFR 1910.38.
- OSHA's Process Safety Standard (PSS)—29 CFR 1910.119.
- OSHA's Hazardous Waste Operations and Emergency Response (HAZWOPER) Regulation—29 CFR 1910.120.
- Department of Transportation's (DOT) Research and Special Programs Administration (RSPA) Pipeline Response Plans—49 CFR Part 194.
- U.S. Coast Guard's (USCG) Facility Response Plans—33 CFR Part 154, Subpart F.
- Minerals Management Services' (MMS) Facility Response Plans—30 CFR Part 254.

as the following: fires, chemical spills, explosions, toxic emissions, train derailments, hurricanes, tornadoes, lightning, floods, earthquakes, or volcanic eruptions.

A company's EAP should be a collection of small plans for each anticipated or potential emergency. These plans should have the following components:

1. Procedures. Specific, step-by-step emergency response procedures should be developed for each potential emergency.

2. Coordination. All cooperating agencies and organizations and emergency responders should be listed along with their telephone number and primary contact person.

3. Assignments and responsibilities. Every person who will be involved in responding to a given emergency should know his or her assignment. Each person's responsibilities should be clearly spelled out and understood. One person may be responsible for conducting an evacuation of the affected area, another for the immediate shutdown of all equipment, and another for telephoning for medical, fire, or other types of emergency assistance. When developing this part of the EAP, it is important to assign a backup person for each area of responsibility. Doing so ensures that the plan will not break down if a person assigned a certain responsibility is one of the victims.

4. Accident prevention strategies. The day-to-day strategies to be used for preventing a particular type of emergency should be summarized in this section of the EAP. In this way, the strategies can be reviewed, thereby promoting prevention.

DISCUSSION CASE

What Is Your Opinion?

"Forget it! We are not going to train our employees in first aid. Emergencies notwithstanding, I don't want a bunch of amateur doctors running around the company doing more harm than good." Mary Vo Dinh, safety director for Gulf Coast Manufacturing, was getting nowhere trying to convince her boss that the company should have employees trained in first aid in the event of an emergency. "But John, we have had three hurricanes in just two years. Tornadoes are not uncommon here on the coast." "I will repeat myself just one more time," said her boss. "No first-aid training." Who is right in this case? What is your opinion?

5. Schedules. This section should contain the dates and times of regularly scheduled practice drills. It is best to vary the times and dates so that practice drills don't become predictable and boring. Figure 19-7 is a checklist that can be used for developing an EAP.

FIGURE 19-7 Emergency planning checklist.

Type of Emergency

____Fire	____Explosion
____Chemical spill	____Toxic emission
____Train derailment	____Hurricane
____Tornado	____Lightning
____Flood	____Earthquake
____Volcanic eruption	

Procedures for Emergency Response

1. Controlling and isolating?
2. Communication?
3. Emergency assistance?
4. First aid?
5. Shutdown/evacuation/protection of workers?
6. Protection of equipment/property?
7. Egress, ingress, exits?
8. Emergency equipment (e.g., fire extinguishers)?
9. Alarms?
10. Restoration of normal operations?

Coordination

1. Medical care providers?
2. Fire service providers?
3. LEPC personnel?
4. Environmental protection personnel?
5. Civil defense personnel (in the case of public evacuations)?
6. Police protection providers?
7. Communication personnel?

FIGURE 19-7 *(continued)*

Assignments and Responsibilities

1. Who cares for the injured?
2. Who calls for emergency assistance?
3. Who shuts down power and operations?
4. Who coordinates communication?
5. Who conducts the evacuation?
6. Who meets emergency responders and guides them to the emergency site?
7. Who contacts coordinating agencies and organizations?
8. Who is responsible for ensuring the availability and upkeep of fire extinguishers?
9. Who is responsible for ensuring that alarms are in proper working order?
10. Who is responsible for organizing cleanup activities?

Accident Prevention Strategies

1. Periodic safety inspections?
2. Industrial hygiene strategies?
3. Personal protective equipment?
4. Ergonomic strategies?
5. Machine safeguards?
6. Hand and portable power tool safeguards?
7. Material handling and storage strategies?
8. Electrical safety strategies?
9. Fire safety strategies?
10. Chemical safety strategies?

Schedules

1. Dates of practice drills:________________________________
2. Times of practice drills:________________________________
3. Duration of practice drills: ________________________________

Planning for Workers with Disabilities

When developing emergency action plans it is important to consider the special needs of workers with disabilities. Do not develop a separate EAP specifically for workers with disabilities—they should be included as a normal part of the organization's plan. An excellent set of guidelines is available to assist organizations in considering the needs of workers with disabilities when developing EAPs.[5]

Titled *Preparing the Workplace for Everyone*, these guidelines were developed for the use of federal agencies but are available to private organizations and other public organizations as well. To obtain a copy of the guidelines, contact the Office of Disability Employment Policy at the following address:

United States Department of Labor
Office of Disability Employment Policy
200 Constitution Ave. NW, Room S-1303
Washington, DC 20210

202-693-7880
http://www.dol.gov/odep

The guidelines are presented under numerous subheadings with questions listed relating to each subheading. What follows in the remainder of this section are categories of questions similar to and based on those presented in the actual guidelines. Safety and health professionals are encouraged to obtain a complete copy of the guidelines to ensure that their EAPs account for workers with disabilities appropriately.

Involving Key Personnel

Under this heading, the guidelines pose questions similar to the following:

- Are key personnel familiar with the EAP?
- Are personnel with disabilities involved in all aspects of the development of the EAP?
- Do senior executives support the development and updating of the EAP?
- Has the EAP been reviewed by first responders and facility personnel?
- Does any part of the plan conflict with procedures established by other applicable agencies?

Implementing Shelter-in-Place (SIP) Plans

Shelter-in-place (SIP) means that rather than evacuate in a disaster or emergency, immediate shelter is sought. SIP comes into play when attempting to evacuate might increase the risk of harm or injury. There should be a SIP provision built into the larger EAP containing provisions that speak to the following concerns:

- Are there provisions for shutting down the building's ventilation system quickly?
- Are there provisions for turning off the elevators?
- Are there provisions for closing all exits and entrances and for securing the loading dock and garage areas (as applicable)?
- Are there provisions for notifying all occupants—including visitors—of emergency procedures?
- Are there provisions for asking all occupants—including visitors—to remain in the building until it is safe to leave?

Evaluating Personnel Needs

It is important to determine the needs of workers with disabilities when developing an EAP, but to do so in strict accordance with the Rehabilitation Act. Before evaluating the needs of personnel with disabilities, consult your human resources department to develop guidelines that are consistent with the Rehabilitation Act. Then ask the following types of questions during development of the EAP:

- Has all applicable information about the needs of personnel with disabilities been collected in accordance with the Rehabilitation Act?
- Has the information collected been appropriately protected by sharing it only with those who need to know it as part of emergency planning (e.g., first-aid personnel, safety and health professionals, those responsible for carrying out the EAP during an emergency, etc.)?

- In selecting equipment, have appropriate agencies as well as individuals with disabilities been consulted as to the ability of personnel with disabilities to use the equipment?
- Have the needs of personnel with service animals been considered in the development of the plan?
- Has the establishment of personal support networks been built into the plan?
- Has the appropriate training been built into the plan for people who need assistance during an emergency as well as for people who will provide the assistance?

Distributing and Communicating the Plan

In developing an EAP, a critical element that should not be overlooked concerns how the completed plan will be distributed to personnel and how those who developed the plan will communicate with personnel about it. The following questions will help ensure proper development of this section of the plan:

- How will the EAP be communicated with the same level of detail to all personnel?
- Is the plan available on the organization's Web site as well as in hard copy? Are there text explanations of all graphic material contained in the plan?
- Are training sessions contained in the plan offered in accessible locations?
- Are the learning aids necessary to accommodate personnel with disabilities readily available in all training locations (e.g., sign-language interpreters, listening devices, etc.)?
- Are hard copies of the EAP placed in various easily accessible locations throughout the facility?
- Are appropriate sections of the plan (evacuation and SIP information) given to frequent visitors to the facility?

Balancing Employer Responsibilities and Employee Right to Self-Determination

On one hand, employers have specific responsibilities for ensuring the safety of their personnel. On the other hand, employees with disabilities have certain rights of self-determination. To ensure that these two issues do not collide when developing the EAP, ask the following questions:

- Are there employees who impede the evacuation of others during practice drills? If so, has the issue been dealt with privately and have appropriate solutions been developed?
- In determining if an employee with a disability represents a direct threat in the event of an emergency, have the following factors been considered: nature and severity of the potential harm, likelihood that the potential harm will occur during an emergency, during of the risk posed by the individual in question, the imminence of the potential harm, and the availability of reasonable accommodations that would mitigate or eliminate the risk.
- Have personnel with disabilities been included in the emergency planning process?
- Has every effort been made to ensure that personnel with disabilities have not been segregated as part of the emergency planning process?

- Have employees with disabilities made requests for reasonable accommodations in the event of an emergency? Have those accommodations been built into the EAP? If not, can the organization show clearly that the requested accommodation would pose an undue hardship?

Working with First Responders

It is important to coordinate emergency planning and the resulting EPA with first-responder agencies and organizations. These personnel and those who carry out the EAP should coordinate closely on all matters pertaining to implementation of the plan.

- Have first responders been made aware of any special issues relating to personnel with disabilities?
- Does the organization have a policy regarding evacuation built into its plan? Are all stakeholders aware of the policy and its ramifications? Has anyone expressed opposition to the policy? Has the opposition been properly addressed?
- Have first responders been included in all steps of the emergency planning process? If not, how will they be made aware of all elements of the EAP?

Implementing an Elevator Policy

Should the organization use its elevators during an emergency or close them down? This can become a critical issue when the needs of personnel with disabilities are considered.

- Does the organization have a policy concerning elevator use during emergencies? Has the policy been built into the EAP?
- Were first responders consulted during the development of the elevator policy?
- Under what circumstances may elevators be used during an emergency? Who may operate the elevators under these conditions?
- Who gets priority in the use of elevators during an emergency?
- How are personnel, including those with disabilities, to be evacuated in the event that elevators are inoperable during an emergency?

Developing Emergency Notification Strategies

Developing the EAP is just one step in an ongoing process. Once the plan is developed, all stakeholders must be notified of its contents and what they mean. The following questions will help ensure appropriate notification:

- What steps have been built into the plan for notifying stakeholders in the event of an emergency and for ensuring that personnel with disabilities have access to the same information that any other stakeholder has?
- Have a variety of notification or communication methods been built into the EAP?
- Do notification or communication methods account for personnel who are away from their desks or the office?
- Do notification strategies take into account the presence of visitors?

Practicing and Maintaining the EAP

The emergency action plan should be viewed as a living document that is updated and maintained continually. In addition, all aspects of the plan should be practiced periodically to ensure that all parties know how to carry out their responsibilities. The following questions will help ensure that the plan is properly maintained and practiced:

- Does the organization have a policy for the regular practice and maintenance of the EAP?
- Does the organization comply with the policy and even exceed compliance requirements?
- Have first responders been involved in all practice drills and have they been consulted to ensure that all applicable equipment is appropriate and in proper working order?
- Have drills been varied in terms of time of day and type of drill?
- Have various unforeseen problems, roadblocks, and inhibitors been built into the drills so that emergency personnel have opportunities to practice improvising in a realistic way?
- Are personnel with disabilities included in all drills?
- Does the organization have a policy built into its EAP for dealing with employees or visitors who might need to leave the facility in the middle of an emergency drill?

Evacuation Planning

The OSHA standard for evacuation planning is 29 CFR 1910.38. This standard requires a written plan for evacuating the facility in the event of an emergency. Critical elements of the plan are marking of exit routes, communications, outside assembly, and training.

Marking of Exit Routes

Clearly identified and marked routes of egress are critical during a time of crisis (fire, natural disaster, terrorist attack, etc.). To ensure that routes of egress and all related evacuation response items are clearly marked, safety professionals should answer the following questions about the facilities for which they are responsible:

1. Are all exit, emergency exit, and nonexit doors clearly identified and marked?
2. Are there up-to-date evacuation route maps mounted at strategic locations throughout the facility?
3. Are all egress route aisles, hallways, and stairs marked clearly and can the markings be seen in the event of darkness?
4. Are there low-level markings posted strategically throughout the facility that can be viewed in the event that smoke fills the facility?
5. Are all items of firefighting equipment clearly marked with directional signs so they can be easily located in the event of an emergency?
6. Is all emergency first-aid equipment clearly marked with directional signs so that it can be easily located in the event of an emergency?
7. Are all electrical, chemical, and physical hazards identified and clearly marked?
8. Are all physical obstructions clearly outlined?
9. Are all critical shutdown procedures and equipment identified and clearly marked?
10. Are the handrails, treads, and risers on all stairs clearly marked?
11. In the case of multicultural workforce settings, are all signs and markings provided in a bilingual or pictogram format?[6]

Ensuring that all signs, pictograms, and other markings relating to facility evacuation are visible during power outages is critical. Battery backup systems are one approach that is widely used. However, there are environments in which even the smallest spark from a battery might set off an explosion. For this reason, some facilities find the use of photoluminescent signs a better alternative.

Photoluminescent signs and markings absorb normal light energy from their surroundings and then release this energy in the form of light during periods of darkness. Some of the better photoluminescent materials will give off light for as long as 24 hours and have a maintenance-free life expectancy of up to 25 years.

Communication and Alarm Procedures

People are so accustomed to false alarms in their lives that when the *real thing* occurs, it can be difficult to convince them that it's not just another drill. In addition, people tend to trust what they can see, smell, and hear. Consequently, if they cannot physically sense an emergency, they tend to ignore the warning. The communication component of a facility's evacuation plan should include procedures for early detection of a problem, procedures for reporting an emergency, procedures for initiating an evacuation, and procedures for providing the necessary information to employees who are being evacuated.[7]

Notifying employees of the emergency is the function of the facility's alarm system. However, just pulling the alarm switch is not sufficient communication. Once the alarm has been given, verbal instruction should be broadcast so that people know that the alarm is real (not just another drill) as well as specific actions they should take immediately. There must also be procedures for informing evacuated employees that the emergency is over and they can return to their work.

External communication procedures are also important. All employees should know how to notify outside authorities of the emergency. With the advent of 911 service, this problem has been simplified. However, do not assume that all employees will remember the number when under the stress of an emergency, or that they will remember to dial "9" or some other code to gain access to an outside line. This problem can be solved by placing clearly marked signs above or near telephones containing such messages as "In an emergency dial 911" or "In an emergency dial 9911" (if it is necessary to first access an outside line).

Outside Assembly

The company's evacuation plan should include an assembly area to which employees go once evacuated.[8] This area should be well known by all employees, and employees should understand that it is critical to assemble there so that a headcount can be taken. In addition, there should be a backup assembly area known to all employees so they know where to go if the primary assembly area has been rendered inaccessible or hazardous.

Part of the evacuation plan relating to assembly areas must be devoted to transient personnel—nonemployees such as vendors, visitors, contractors, and so on. How will they know where to assemble? Who will check to see that they have been notified of the emergency? These issues should be spoken to in the evacuation plan.

Training

Training for evacuations is a critical element of the evacuation plan.[9] Developing a plan and then letting it just sit on the shelf gathering dust is a formula for disaster.

Everything contained in the plan that requires action or knowledge on the part of employees should be part of the required evacuation training. Training should be provided when employees are first hired, and retraining should be provided periodically as various elements of the plan are updated.

Drills should be a major part of the training provided for employees. The old sports adage that says "What you do in practice you will do in the game" applies here. When an emergency actually occurs should not be the first time employees have gone through the action required of them in an emergency.

Customizing Plans to Meet Local Needs

Emergency plans must be **location-specific**. General plans developed centrally and used at all plant locations have limited effectiveness. The following rules of thumb can be used to ensure that EAPs are location-specific:

1. A map of the plant. A map of the specific plant helps localize an EAP. The map should include the locations of exits, access points, evacuation routes, alarms, emergency equipment, a central control or command center, first-aid kits, emergency shutdown buttons, and any other important elements of the EAP.

2. Chain of command. An organization chart illustrating the chain of command—who is responsible for what and who reports to whom—also helps localize an EAP. The chart should contain the names and telephone numbers (internal and external) of everyone involved in responding to an emergency. It is critical to keep the organization chart up-to-date as personnel changes occur. It is also important to have a designated backup person shown for every position on the chart.

3. Coordination information. All telephone numbers and contact names of people in agencies with which the company coordinates emergency activities should be listed. Periodic contact should be maintained with all these people so that the EAP can be updated as personnel changes occur.

4. Local training. All training should be geared toward the types of emergencies that may occur in the plant. In addition, practice drills should take place on-site and in the specific locations where emergencies are most likely to happen.

Emergency Response

An **emergency response team (ERT)** is a special team that responds to emergencies to ensure proper personnel evacuation and safety, shut down building services and utilities, work with responding civil authorities, protect and salvage property, and evaluate areas for safety prior to reentry" (emphasis added).[10] The ERT is typically composed of representatives from several different departments such as maintenance, security, safety and health, production and processing, and medical. The actual composition of the team depends on the size and type of company. The ERT should be contained in the assignments and responsibilities section of the EAP.

Not all ERTs are company based. Communities also have ERTs for responding to emergencies that occur outside of a company environment. Such teams should be included in a company's EAP in the coordinating organizations section. This is especially important for companies that use hazardous materials.

Most dangerous spills occur on the road and the greatest number of hazardous materials (hazmat) incidents occur on the manufacturer's loading dock. More than 80 percent of chemical releases are caused by errors in loading and unloading procedures.[11] Community-based teams typically include members of the police and fire departments who have had special training in handling hazmat emergencies.

Another approach to ERTs is the **emergency response network (ERN)**. An ERN is a network of ERTs that covers a designated geographical area and is typically responsible for a specific type of emergency. For example, the Chlorine Institute (CI), an association of more than 130 firms that produce chlorine and manufacture related products, implemented its Chlorine Emergency Plan (CHLOREP), a network for responding to chlorine emergencies. Sixty-two teams from 28 companies are assigned geographic areas where they are on call to help during chlorine emergencies.[12]

The following example shows how a CHLOREP team responds to a chlorine-related emergency. A chlorine leak was discovered in a cement pit near Arlington National Cemetery. The pit had once been used to chlorinate water for irrigating the cemetery. Initial reports were that as many as half of the ten 150-pound chlorine cylinders were leaking. Local emergency response teams asked CHLOREP for help.[13] The CHLOREP team covering Washington, DC, and vicinity responded and found that only one of the ten cylinders was leaking. They took the following action:

> The team used a recently designed recovery vessel, a coffin-like steel container for enclosing leaking cylinders. After the nine unaffected cylinders were moved aside, the 64-inch-long recovery vessel was lowered into the pit. The team slid the leaking cylinder into the coffin and secured it.[14]

Whether the ERT is a local company team or a network of teams covering a geographical region, it should be included in the EAP. In-house teams are included in the assignments and responsibilities section. Community-based teams and networks are included in the coordinating organizations section.

Computers and Emergency Response

Advances in chemical technology have made responding to certain types of emergencies particularly complicated.

Fortunately, the complications brought by technology can be simplified by technology. Expert computer systems especially programmed for use in emergency situations can help meet the challenge of responding to a mixed-chemical emergency or any other type of emergency involving multiple hazards.

An **expert system** is a computer programmed to solve problems. Such systems rely on a database of knowledge about a very particular area, an understanding of the problems addressed in that area, and expertise in solving problems in that area. Talking to an expert system is like sitting at a terminal and keying in questions to an expert who is sitting at another terminal responding to the questions. The expert in this case is a computer program that pulls information from a database and uses it to make decisions based on **heuristics** or suppositional rules stated in an if-then format.[15]

Human thought processes work in a similar manner. For example, if our senses provide the brain with input suggesting the stovetop is hot, the decision is made not to touch it. In an if-then format, this may read as follows:

IF hot, THEN do not touch.

SAFETY FACT

Responding to Chemical Spills

To respond quickly and effectively to a chemical spill, the emergency response team must have the right equipment. Standard equipment should include a portable spill cart that can be quickly rolled to the site of a chemical spill. This cart should contain at least the following items:

- Spill suppression and absorption materials such as pads, blankets, and pillows
- Mops and brooms
- Acid neutralizers
- Barricade devices or materials such as mesh or tape
- Appropriate personal protective gear (e.g., gloves, eye protection, aprons, coveralls, and so on)

This similarity to human thought processes is why the science on which expert systems are based is known as *artificial intelligence*.

A modern expert system used for responding to chemical emergencies provides information such as the following:

- Personal protective equipment needed for controlling and cleaning up
- Methods to be used in cleaning up the spill or toxic release
- Decontamination procedures
- Estimation of the likelihood that employees or the community will be exposed to hazard
- Reactions that may result from interaction of chemicals
- Combustibility of chemicals and other materials on hand
- Evacuation information
- Impact of different weather conditions on the situation
- Recommended first-aid procedures[16]

Expert systems can be user-friendly so that computer novices have no difficulty interacting with them. The advantages of expert systems are they do not panic or get confused, they consider every possible solution in milliseconds, they are not biased, they do not become fatigued, and they are detailed.

Dealing with the Psychological Trauma of Emergencies

In addition to the physical injuries and property damage that can occur in emergencies, modern safety and health professionals must also be prepared to deal with potential psychological damage. Psychological trauma among employees involved in workplace disasters is as common as it is among combat veterans. According to Johnson, "Traumatic incidents do not affect only immediate survivors and witnesses. Most incidents result in layers of victims that stretch far beyond those who were injured or killed."[17]

Trauma is psychological stress. It occurs as the result of an event, typically a disaster or some kind of emergency, so shocking that it impairs a person's sense of security or well-being. Johnson calls trauma response "the normal reactions of normal people to an abnormal event."[18] **Traumatic events** are typically unexpected and shocking, and they involve the reality or threat of death.

Dealing with Emergency-Related Trauma

The typical approach to an emergency can be described as follows: control it, take care of the injured, clean up the mess, and get back to work. Often, the psychological aspect is ignored. This leaves witnesses and other coworkers to deal on their own with the trauma they've experienced. "Left to their own inadequate resources, workers can become ill or unable to function. They may develop resentment toward the organization, which can lead to conflicts with bosses and co-workers, high employee turnover—even subconscious sabotage."[19]

It is important to respond to trauma quickly, within 24 hours if possible and within 72 hours in all cases. The purpose of the response is to help employees get back to normal by enabling them to handle what they have experienced. This is best accomplished by a team of people who have had special training. Such a team is typically called the **trauma response team (TRT)**.

Trauma Response Team

A company's trauma response team may consist of safety and health personnel who have undergone special training or fully credentialed counseling personnel, depending on the size of the company. In any case, the TRT should be included in the assignments and responsibilities section of the EAP.

The job of the TRT is to intervene as early as possible, help employees acknowledge what they have experienced, and give them opportunities to express how they feel about it to people who are qualified to help. The *qualified to help* aspect is very important. TRT members who are not counselors or mental health professionals should never attempt to provide care that they are not qualified to offer. Part of the trauma training that safety and health professionals receive involves recognizing the symptoms of employees who need professional care and referring them to qualified care providers.

In working with employees who need to deal with what they have experienced, but are not so traumatized as to require referral for outside professional care, a group approach is best. According to Johnson, the group approach offers several advantages:

- It facilitates public acknowledgment of what the employees have experienced.
- It keeps employees informed, thereby cutting down on the number of unfounded rumors and horror stories that will inevitably make the rounds.
- It encourages employees to express their feelings about the incident. This alone is often enough to get people back to normal and functioning properly.
- It allows employees to see that they are not alone in experiencing traumatic reactions (e.g., nightmares, flashbacks, shocking memories, and so on) and that these reactions are normal.[20]

Convincing Companies to Respond

Modern safety and health professionals may find themselves having to convince higher management of the need to have a TRT. Some corporate officials may not believe that trauma even exists. Others may acknowledge its presence but view trauma as a personal problem that employees should handle on their own.

In reality, psychological trauma that is left untreated can manifest itself as **post-traumatic stress disorder**, the same syndrome experienced by some veterans of Vietnam and other wars. This disorder is characterized by "intrusive thoughts and flashbacks of the stressful event, the tendency to avoid stimulation, paranoia, concentration difficulties, and physiological symptoms such as rapid heartbeat and irritability."[21]

The American Psychiatric Association included posttraumatic stress disorder in its *Diagnostic and Statistical Manual* as far back as 1980.[22] Jeffrey T. Mitchell, president of the American Critical Incident Stress Foundation, likens preventing posttraumatic stress disorder to working with cement. Wet cement can be molded, shaped, manipulated, and even washed away. However, once it hardens, there is not much one can do with it.[23] This is the rationale for early intervention.

In today's competitive marketplace, companies need all their employees operating at peak performance levels. Employees experiencing trauma-related disorders will not be at their best. Safety and health professionals should use this rationale when it is necessary to convince higher management of the need to provide a company-sponsored trauma response team.

Recovering from Disasters

Many organizations put a great deal of effort into planning for disaster response including emergency evacuation. But what about after the disaster? According to John Kauffman, 43 percent of U.S. companies that experience a disaster have no recovery plan.[24] Recovering quickly is the key to staying in business. Approximately 70 percent of businesses that close down for a month or more as a result of a disaster either will never reopen or will fail within three years.[25]

A comprehensive disaster recovery plan should have at least the following components: recovery coordinator, recovery team, recovery analysis and planning, damage assessment and salvage operations, recovery communications, and employee support and assistance. The overall goal of a disaster recovery plan is to get an organization fully operational again as quickly as possible.

Recovery Coordinator

There must be one person who has ultimate responsibility and authority for disaster recovery. This person must have both the ability and the authority to take command of the situation, assess the recovery needs, delegate specific responsibilities, approve the necessary resources, interact with outside agencies, and activate the organization's overall response.

Recovery Team

The recovery team consists of key personnel to whom the disaster coordinator can delegate specific responsibilities. These responsibilities include facility management, security, human resources, environmental protection (if applicable), communications, and the various personnel needed to restart operations.

Recovery Analysis and Planning

This phase involves assessing the impact of the disaster on the organization and establishing both short- and long-term recovery goals. The more recovery analysis and planning that can be done, the better. One of the ways to do this is to consider various predictable scenarios and plan for them. This is the business equivalent of the war-gaming activities that take place in the military.

Damage Assessment and Salvage

This component of the plan has two elements: preparedness and recovery. The preparedness element should include the following information: (1) a comprehensive inventory of all property at the facility in question; (2) a checklist of the items on the inventory that are essential for maintaining the facility; (3) a list of all personnel who will aid in the recovery (make sure to have fully trained and qualified backup personnel in case a primary player is not available or is injured during the emergency); (4) a list of all vendors, contractors, and so on whose assistance will be needed during the damage assessment and salvage phase of recovery; (5) a worksheet that can be used to document all actions taken during recovery operations; and (6) procedures for quickly establishing a remote operational site.

The recovery element should include procedures for securing workspace for the recovery team and coordinator; identifying areas of the facility that must be accessible; maintaining security at the facility against looting and vandalism; analyzing and inspecting damage to the facility and reporting it to the recovery coordinator; assessing the extent of damage to goods, supplies, and equipment; photographing and videotaping damage to the facility; taking appropriate action to prevent additional damage to the facility; repairing, restoring, and resetting fire detection and suppression equipment; and investigating accidents.

Recovery Communications

Communication is one of the most important considerations in disaster recovery. This component of the plan should deal with both who is to be notified and how that is to take place. The "how" aspects concern backup procedures for telephone service, e-mail, and so forth. Will cell phones be used? Will radio stations be part of the mix? Will "walkie-talkie" type radios be used for communicating on-site? The following list contains the types of entities who might have to be contacted as part of the disaster recovery operation:

- Customers
- Vendors and suppliers
- Insurance representatives
- Employees' families
- Appropriate authorities
- Media outlets (radio and television stations and newspapers)

Employee Assistance

After a disaster, employees are likely to need various types of assistance including financial, medical, and psychological. Kauffman recommends the following steps for developing this component of the disaster recovery plan:

1. Determine postdisaster work schedules and provide them to employees. Include overtime work if it will be necessary, and make sure employees know that flexibility in scheduling work hours will be important until the recovery is complete.
2. Plan for the whole range of employee-assistance services that might be needed including medical, transportation, financial, shelter, food, water, clothing, and psychological services (trauma, shock, and stress counseling).

3. Plan for the provision of grief counseling. The best way to handle this is to assign grief counseling to the company's employee assistance program provider.
4. Plan for the possible need to relocate the facility as part of disaster recovery.
5. Plan to give employees opportunities to participate in personal actions taken on behalf of fatally injured employees and their families. Work for employee consensus before deciding what to do for these families.
6. Plan to fully inform all employees about what happened, why, how the company is responding in the short term, and how it will respond in the long term. Be sure to build in ways to let employees know the company cares about them and will do everything possible to protect their safety.[26]

Terrorism in the Workplace

"The events of Sept. 11 and the ongoing threat of bioterrorism in American workplaces have many people living and working in fear. Unquestionably, the world has changed, and few of us are happy with the direction. It's important to put everything in perspective, however, and consider the proper role of the employer."[27] This section describes the roles that employers and their safety and health personnel can play in preparing for, preventing, and responding to terrorist attacks in the workplace.

Role of the Employer

There is no question that the threat of terrorist attacks has become an ever-present reality in today's workplace. Because this is the case, employers clearly have a role to play in preparing for terrorist attacks, taking all prudent precautions to prevent them, and in responding properly should an attack occur. Because terrorism threatens the safety and health of employees, it is more than just a security issue, it is also an occupational safety and health issue.

Chip Dawson summarizes the roles of employers and safety and health professionals relating to terrorism in the workplace:

Run a safe and caring operation. Employees watch the nightly news and read their morning newspapers. They know what is going on in their world. As a result, many are discomfited by the possibility that they or their workplace might become the target of a terrorist attack. Consequently, the first responsibility of the employer is to run a safe operation in which employees know their safety is a high priority. Many of the engineering, administrative, training, and enforcement actions taken to make the workplace safe from occupational hazards will also help mitigate the threat of terrorism.

Listen to employees. Employees are concerned about the threat of terrorism, and they have a right to be. Employers should take the concerns of employees seriously, and deal with them. Answer questions, communicate openly and frequently, and refer employees who need professional help to the employee assistance program.

Train employees. Security and safety procedures do little good unless employees know what they are and how to use them. In addition, personnel in certain positions need to have specialized knowledge relating to terrorism. For example, mailroom personnel need to be trained in how to screen incoming mail for biohazards and explosives.

Communicate. Talk with employees openly and frequently. Let them know what the company knows. It is better for employees to hear news from the company than to receive it in the form of third-party gossip and rumors. Before giving out information, however, it is a good idea to verify it. Check with local authorities or go online to one of the following Web sites: http://www.snopes.com or http://www.urbanlegends.com.

Know your personnel. Institute background checks as part of the hiring process. Make sure that supervisors get to know their direct reports well enough to sense when something is wrong with one of them. If inconsistencies in normal behavioral patterns occur, address them right away.

Empower personnel. Empower employees to back away from a situation that does not feel right or that makes them uncomfortable. Be flexible in allowing them to have time off for family activities. This can be especially important following particularly busy times for employees who have worked longer than normal hours.

Harden the site against external threats and restrict access. Insulate the workplace from negative outside influences, control who has access to the workplace, and take all necessary steps to reduce the exposure of employees to potential threats. Call in security experts if necessary to help develop and implement the necessary controls.

Remove any barriers to clear visibility around the facility. The better employees can see around them, the less likely it will be that a terrorist will be able to pull off a surprise attack. The trees, shrubs, and bushes that make the perimeter of the facility and parking lot so attractive can be used by unauthorized personnel to hide while attempting to gain access.

Have and enforce parking and delivery regulations. Arrange parking spaces so that no car is closer to a building than 100 feet. This will lessen the likelihood of damage from a car bomb. In addition, have strict delivery procedures so that terrorists posing as delivery personnel do not gain access through this means.

Make sure that visitors can be screened from a distance. Arrange the facility so that visitors are channeled into a specific area, an area in which they can be viewed by company personnel from a distance. This will lessen the likelihood of a terrorist gaining access by overpowering or killing access-control personnel.

Keep all unstaffed entrance doors locked and alarmed. Employees need to be able to get out of the building through any exit door, but access into the building should be channeled through doors staffed by access-control personnel.

Make air intakes and other utilities inaccessible to all but designated maintenance personnel. Releasing toxic material into the air intake is one way terrorists could harm the highest possible number of people. The likelihood of this happening can be decreased by locating air intakes in inaccessible locations.

Prevent access to roofs and upper stories. Terrorists who cannot gain access on the ground floor might simply gain entry through doors and other openings on the roof. Consequently, it is important to keep roof doors locked from the outside and alarmed. They should open from the inside, but doing so should trigger an alarm. It is also important to establish control procedures for emergency escape routes from the roof and upper stories so that these avenues are not used by terrorists trying to gain access into the building.

Secure trash containers. The grounds should be kept free of debris and clutter. Further, trash containers should be secured either by keeping them inside the building's wall or, if kept outside, at a distance from the building.

Ensure that employees, contractors, and visitors wear badges. Establish a system in which all employees, contractors, and visitors must wear badges in order to gain access to the facility. Require identification of all visitors and an internal sponsor before providing them with badges. Have visitors sign in and out on every visit.

Have an emergency response plan and practice it periodically. Plan for all predictable exigencies and practice the various components of the plan on a regular basis.

Be cautious of information placed on your company's Web site. Make sure that your company's Web site does not contain information that can be used by terrorists such as detailed maps, floor plans, or descriptions of hazardous materials stored on-site.

Keep up-to-date with the latest safety and security strategies. Crime Prevention Through Environmental Design (CPTED) should become part of the knowledge base of safety and health professionals. For the latest information concerning CPTED, visit the following Web site periodically: http://www.arch.vt.edu/crimeprev/pages/home.html.

Protect the integrity of your facility's key system. Terrorists can use any key from your facility to make their own master key if they know how, and many do. To make matters worse, they share information about how to make master keys via the Internet. The following procedures will help protect the integrity of your facility's key system: (a) restrict access to keys (including restroom keys); (b) consider not using master keys; and (c) switch from a key system to an electronic keycard or fingerprint recognition system.[28]

Securing Hazardous Materials

According to Dale Petroff

> You must understand that the hazmats under your control may be used by terrorists to create weapons of mass destruction. It is important to note that the first poison gas used in warfare was chlorine gas, and that fertilizer mixed with fuel oil became a devastating explosive device that destroyed the Murrah Federal Building in Oklahoma City. More recently . . . a truck used to transport natural gas was turned into a bomb in Tunisia that destroyed a synagogue and killed 17 people. All of these materials and many others that can be converted into weapons are found in large quantities throughout the United States.[29]

Clearly, one of the tactics of terrorists is to convert hazardous materials used in the workplace into weapons of mass destruction. Consequently, it is important for facilities that produce, use, or store any type of hazardous materials to develop, implement, and enforce a security program that denies terrorists access to these materials.[30] "The goal of the security plan should be to implement measures that deter, detect, delay, or defeat the

SAFETY TIP

Sources of Help on Terrorism for Safety and Health Professionals

- *Pentagon*
  http://www.dtic.mil/ref/biochem/Training/pent/pent_slidel.html
- *U.S. Department of Defense*
  http://www.cbiac.apgea.army.mil/
- *Centers for Disease Control and Prevention*
  http://www.bt.cdc.gov/Agent/Agentlist.asp
- *Central Intelligence Agency (CIA)*
  http://www.cia.gov/terrorism/index.html
- *U.S. State Department*
  http://www.ojp.usdoj.gov/odp/

threat. Deterrence can be improved by using highly visible measures and randomness. This is cost-efficient and complicates the threatening person or group's planning."[31]

A hazmat security plan should have two broad components: *personnel security* and *physical security*. The fundamental elements of the personnel component of the plan are to (1) determine who should be granted access to the materials; (2) conduct comprehensive background checks on all individuals who require access; (3) submit employees who require access to psychological screening to ensure stability; and (4) require identification badges with photographs or fingerprints of those authorized access.

The physical security component consists of measures taken to prevent or control access to the hazardous materials in question. Security practices in this component should be integrated and layered by using a combination of measures including fences, lights, electronic alarm systems, guards, reaction forces, and the *two-person rule* that requires two employees to be present in order to gain access to the hazardous materials. A good source of help concerning how to develop a plan for keeping hazardous materials out of the hands of terrorists is as follows:

American Chemistry Council
703-741-5000
http://www.americanchemistry.com

Resuming Business after a Disaster

After a disaster in the workplace—regardless of whether the source of the disaster was natural, accidental, or terrorist related—a comprehensive hazard assessment should be completed before business is resumed. Before resuming business, an organization should consider the following factors:[32]

- **Structural integrity.** Has the structural integrity of the building been checked by competent engineering professionals to ensure that it is safe to enter?
- **Utility checks.** Have all utilities—gas, electricity, water, sewer—been checked by competent professionals to ensure that there are no leaks, cracks, loose wires, and so on? Have the appropriate utility companies given their approval for reopening?

Remember that if reopening involves the use of electric generators, these devices should not be used inside the building because they can create a carbon monoxide hazard.

- **Cleanup protection.** Make sure that cleanup crews are properly protected from any hazardous materials or conditions that might have been created by the disaster. Make sure they properly use the appropriate personal protective gear, and comply with all applicable safety and health regulations and procedures as they clean the facility.
- **Health and sanitation.** Kitchens, bathrooms, and any area in which food or potentially hazardous or toxic substances are stored should be checked by competent professionals and thoroughly cleaned to prevent the exposure of employees and customers to hazardous conditions.
- **Air quality.** Make sure the air quality in the facility is tested by competent professionals before allowing personnel to enter the building. Certain types of disasters such as hurricanes, for example, might have caused the proliferation of mold and mildew. The air should be checked for any potentially hazardous biological or chemical agents that could be harmful to humans.
- **Ventilation.** Make sure that all types of ductwork and ventilation have been checked for the presence of potentially harmful biological and chemical agents as well as for dust and debris that might impede airflow. Once it appears that ventilations systems are clean, have the air conditioning and heating systems started up and all ventilation systems checked again before allowing personnel back into the building. When the air-conditioning and heating systems are restarted, blow cold air through them, even in winter. This will help prevent the growth of mold in the ventilation ducts.
- **Walls, ceilings, and floors.** Check walls and ceilings to ensure that no materials are in danger of falling off—inside for occupants and outside for pedestrians. Check floors for any hazards that might contribute to slipping and falling.
- **Safety equipment.** Check all fire extinguishers, all types of alarms, and any other safety equipment to determine if it has been damaged. Make sure that all safety equipment is in proper working order before allowing personnel to reenter the facility.
- **Lighting.** Ensure that all illumination devices are in proper working order and that all personnel have the required amount of illumination to do their jobs. Employees should not return to work unless the necessary illumination is available and working.
- **Hazardous waste removal.** Any type of potentially hazardous material that is left lying around after the disaster should be collected and properly disposed of. Broken glass, debris, litter, and sharp-edged material should be removed before employees are allowed to return to work.
- **Machines and equipment.** All machines and equipment should be checked carefully before they are reenergized. All electrical, gas, hydraulic, fill, drain, and plumbing lines should be checked for leaks and proper connections before the machines are energized for use.
- **Furniture.** Check all furniture to ensure its structural integrity. Make sure that fasteners, braces, and supports have not been damaged during the emergency or that furniture has not become unstable due to water damage.

SAFETY FACT

Federal Guidelines for Resuming Business

Businesses may obtain comprehensive federal guidelines for resuming business after a disaster from the National Institute for Occupational Safety and Health (NIOSH) or the Federal Emergency Management Agency (FEMA) at these Web sites:

http://www.cdc.gov.niosh
http://www.fema.gov

Summary

1. An emergency is a potentially life-threatening situation, usually occurring suddenly and unexpectedly. Emergencies may be the result of natural or human causes.
2. Preparing for emergencies involves planning, practicing, evaluating, and adjusting. An immediate response is critical in emergencies.
3. The Emergency Planning and Community Right-to-Know Act has the following four main components: emergency planning, emergency notification, information requirements, and toxic chemical release reporting.
4. For proper coordination of the internal emergency response, it is important that one person be in charge and that everyone involved knows who that person is.
5. Because there is no way to predict when first aid may be needed, part of preparing for emergencies should include training employees to administer first aid. In certain cases, OSHA requires that companies have at least one employee on-site who has been trained in first aid.
6. In addition to providing first-aid training, it is important to have well-stocked first-aid kits readily available, have personal protective devices available, post emergency telephone numbers, and keep all employees informed.
7. A company's emergency action plan should be a collection of small plans for each anticipated emergency. These plans should have the following components: procedures, coordination, assignments and responsibilities, accident prevention strategies, and schedules.
8. The OSHA standard for evacuation planning is 29 CFR 1910.38. This standard requires a written plan for evaluating the facility in the event of an emergency. Critical elements of the plan are as follows: marking of exit routes, communications, outside assembly, and training.
9. EAPs should be customized to be location-specific by including a map, an organization chart, local coordination information, and local training schedules. They should consider the needs of all personnel, including those with disabilities.
10. An emergency response team is a special team to handle general and localized emergencies, facilitate evacuation and shutdown, protect and salvage company property, and work with civil authorities.
11. An emergency response network is a network of emergency response teams that covers a designated geographical area.
12. Computers can help simplify some of the complications brought by advances in technology. Expert systems mimic human thought processes in making decisions on an if–then basis regarding emergency responses.
13. Trauma is psychological stress. It typically results from exposure to a disaster or emergency so shocking that it impairs a person's sense of security or well-being. Trauma left untreated can manifest itself as posttraumatic stress disorder. This disorder is characterized by intrusive thoughts, flashbacks, paranoia, concentration difficulties, rapid heartbeat, and irritability.
14. A disaster recovery plan should have at least the following components: recovery coordinator, recovery team, recovery analysis and planning, damage assessment and salvage operations, recovery communications, and employee support and assistance.

15. Employers can help decrease the likelihood of a terrorist attack on their facilities by taking the following actions: run a safe and caring operation; listen to employees; train employees; communicate; know your personnel; empower personnel; harden the site against external threats and restrict access; remove any barriers to clear visibility around the facility; have and enforce parking and delivery regulations; make sure that visitors can be screened from a distance; keep all unstaffed entrance doors locked from the outside and alarmed; make air intakes and other utilities inaccessible to all but designated personnel; prevent access to roofs and upper stories; secure trash containers; ensure that employees, contractors, and visitors wear badges; have an emergency response plan and practice it on a regular basis; be cautious of what information is placed on your company's Web site; keep up-to-date with the latest safety and security strategies; and protect the integrity of your facility's key system.
16. Secure hazardous materials so that terrorists cannot gain access to them for use in making bombs and other weapons of mass destruction. A hazmat security plan should have two components: personnel security and physical security.
17. All systems, conditions, and potential hazards should be checked and corrected as appropriate before resuming business after a disaster.

Key Terms and Concepts

- Adjusting
- Chain of command
- Elevator policy
- Emergency action plan (EAP)
- Emergency coordinator
- Emergency notification
- Emergency planning
- Emergency Planning and Community Right-to-Know Act (EPCRA)
- Emergency response management team (ERMT)
- Emergency response network (ERN)
- Emergency response plan
- Emergency response team (ERT)
- Evaluating
- Expert system
- First-aid training
- Heuristics
- Information requirements
- Local emergency planning committee (LEPC)
- Location-specific
- Order of response
- OSHA standards
- Planning
- Posttraumatic stress disorder
- Practicing
- Shelter-in-place (SIP) plans
- Spill cart
- State emergency response commission (SERC)
- Toxic chemical release reporting
- Trauma
- Trauma response team (TRT)
- Traumatic event

Review Questions

1. Define the term *emergency*.
2. Explain the rationale for emergency preparation.
3. List and explain the four main components of the Emergency Planning and Community Right-to-Know Act.
4. Describe how a company's emergency response effort should be coordinated.
5. How do OSHA standards relate to emergency preparation?
6. Explain how you would provide first-aid training if you were responsible for setting up a program at your company.
7. Besides training, what other first-aid preparation should a company take?
8. What are the critical elements of OSHA's standard for evacuation planning?
9. Describe the essential components of an EAP and explain how to build the needs of personnel with disabilities into the plan.
10. How can a company localize its EAP?
11. Define the following emergency response concepts: ERT, ERN, and TRT.
12. What is an expert system? How can one be used in responding to an emergency?
13. What is trauma?
14. Why should a company include trauma response in its EAP?
15. Describe how a company may respond to the trauma resulting from a workplace emergency.
16. What elements should a disaster recovery plan contain?
17. How can employers prepare for the threat of terrorism?
18. Explain the precautions that should be taken before resuming business after a disaster.

ENDNOTES

1. L. Marley, "Emergency Medical Teams Need Organization and Administration," *Safety & Health* 142, no. 5: 28.
2. Ibid.
3. Ibid.
4. Ibid., 30.
5. United States Department of Labor, Office of Disability Employment Policy, "Preparing the Workplace for Everyone: Accounting for the Needs of People with Disabilities." Retrieved June 2006 from http://www.dol.gov.odep/pubs/ep/preparing/workplace_final.pdf.
6. S. Larson, "Heading for the Exits," *Occupational Health & Safety* 72, no. 2: 60.
7. C. Schroll, "Evacuation Planning: A Matter of Life and Death," *Occupational Hazards*, June 2002, 50–54.
8. Ibid., 54.
9. Ibid., 54.
10. H. Christen, *The EMS Incident Management System* (Upper Saddle River, NJ: Prentice Hall, 1998), 8.
11. Ibid., 9.
12. Ibid., 10.
13. Ibid., 16.
14. Ibid., 17.
15. Ibid., 168.
16. Ibid., 169.
17. E. Johnson, "Where Disaster Strikes," *Safety & Health* 145, no. 2: 29.
18. Ibid., 28.
19. Ibid., 29.
20. Ibid., 30.
21. National Safety Council, "Trained for Trauma," *Safety & Health* 145, no. 2: 32.
22. Ibid.
23. Ibid.
24. J. Kauffman, "Recovering from Disaster," *Occupational Hazards*, January 2003, 69–71.
25. Ibid., 71.
26. Ibid.
27. C. Dawson, "The Role of the Employer in Domestic Security," *Occupational Hazards*, January 2002, 31.
28. Ibid., 31–32.
29. D. M. Petroff, "Security of Hazardous Materials," *Occupational Health & Safety* 72, no. 4: 46.
30. Ibid., 44–48.
31. Ibid., 48.
32. American Society of Safety Engineers, "ASSE Offers Business Resumption Safety Checklist," Occupational Health & Safety Online, September 9, 2005. Retrieved from http://www.ohsonline.com/stevens/ohspub.nsf/d3d5b4f938b22b6e8625670c006bc58/6600.

Bloodborne Pathogens in the Workplace

Major Topics

- Facts about AIDS
- Symptoms of AIDS
- AIDS in the Workplace
- Legal Concerns
- AIDS Education
- Counseling Infected Employees
- Easing Employees' Fears about AIDS
- Protecting Employees from AIDS
- Hepatitis B (HBV) and Hepatitis C (HCV) in the Workplace
- OSHA's Standard on Occupational Exposure to Bloodborne Pathogens
- Preventing and Responding to Needlestick Injuries

Acquired immunodeficiency syndrome (AIDS) has become one of the most difficult issues that safety and health professionals are likely to face today. It is critical that they know how to deal properly and appropriately with this controversial disease. The major concerns of safety and health professionals with regard to AIDS are knowing the facts about AIDS; knowing the legal concerns associated with AIDS; knowing their role in AIDS education and related employee counseling; and knowing how to ease unfounded fears concerning the disease while simultaneously taking the appropriate steps to protect employees from infection. In addition to AIDS, the modern safety and health manager must be concerned with other bloodborne pathogens including **human immunodeficiency virus (HIV)** and **hepatitis B (HBV)**.

Facts about AIDS

Consider the following facts about HIV and AIDS:

- The total number of new AIDS cases diagnosed annually among men in the United States exceeds 33,000.[1]

FIGURE 20-1 Trend for people living with AIDS in the United States.

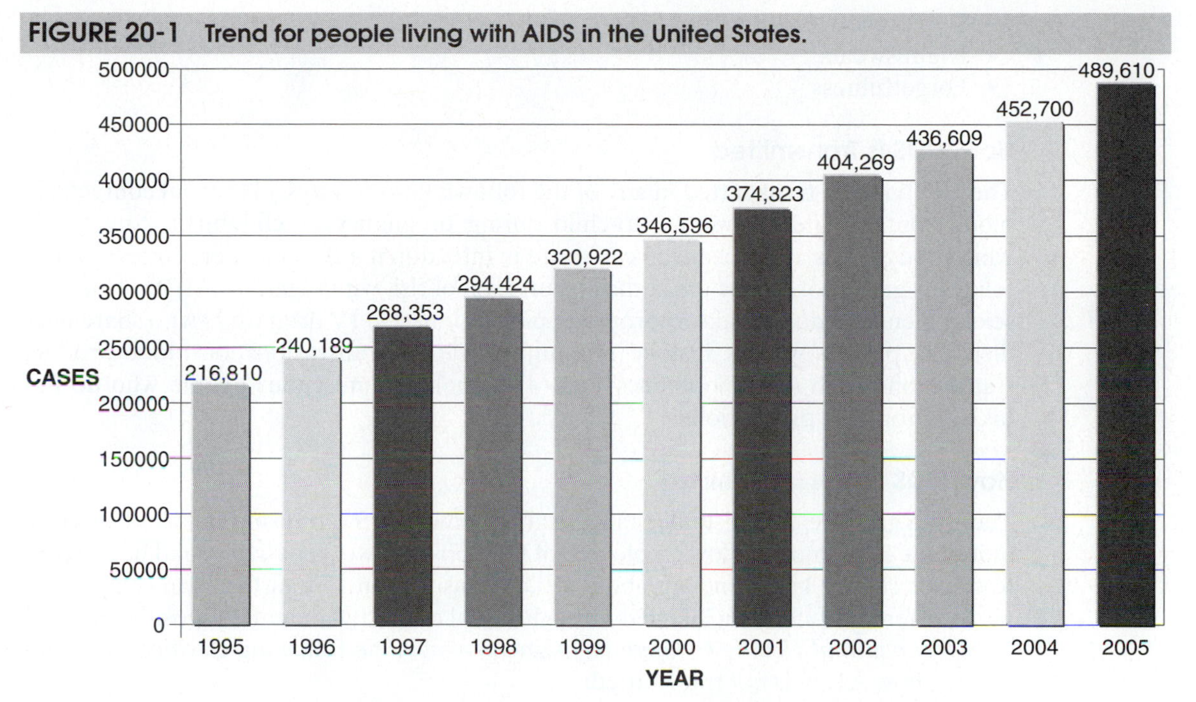

Source: Centers for Disease Control and Prevention, National Center for HIV, STD, and TB Prevention.

- The total number of new AIDS cases diagnosed annually among women in the United States exceeds 10,000.[2]
- The estimated cumulative number of people in the United States living with AIDS has risen steadily over the years (see Figure 20-1). The increase over the years is attributed in large measure to declines in the death rate due to advances in medical technology and treatments.[3]

AIDS is feared, misunderstood, and controversial. Modern safety and health professionals need to know the facts about AIDS and be prepared to use these facts to make the workplace safer.

Symptoms of AIDS

AIDS and various related conditions are caused when humans become infected with the human immunodeficiency virus (HIV). This virus attacks the human immune system, rendering the body incapable of repelling disease-causing microorganisms. Symptoms of the onset of AIDS are

- Enlarged lymph nodes that persist
- Persistent fevers
- Involuntary weight loss
- Fatigue
- Diarrhea that does not respond to standard medications
- Purplish spots or blotches on the skin or in the mouth

- White, cheesy coating on the tongue
- Night sweats
- Forgetfulness

How AIDS Is Transmitted

The HIV virus is transmitted in any of the following three ways: (1) sexual contact, (2) blood contact, and (3) mother-to-child during pregnancy or childbirth. Any act in which body fluids are exchanged can result in infection if either partner is infected. The following groups of people are at the highest level of risk with regard to AIDS: (1) homosexual men who do not take appropriate precautions; (2) IV drug users who share needles; (3) people with a history of multiple blood transfusions or blood-product transfusions, such as hemophiliacs; and (4) sexually promiscuous people who do not take appropriate precautions.

How AIDS Is *Not* Transmitted

There is a great deal of misunderstanding about how AIDS is transmitted. This can cause inordinate fear among fellow employees of HIV-positive workers. Safety and health professionals should know enough about AIDS transmission so that they can reduce employees' fears about being infected through casual contact with an HIV-positive person.

Occupational Health & Safety magazine provides the following clarifications concerning how AIDS is *not* transmitted:

> AIDS is a blood-borne, primarily sexually transmitted disease. It is not spread by casual social contact in schools, workplaces, public washrooms, or restaurants. It is not spread via handshakes, social kissing, coughs, sneezes, drinking fountains, swimming pools, toilet facilities, eating utensils, office equipment, or by being next to an infected person.
>
> No cases of AIDS have been reported from food being either handled or served by an infected person in an eating establishment.
>
> AIDS is not spread by giving blood. New needles and transfusion equipment are used for every donor.
>
> AIDS is not spread by mosquitoes or other insects.
>
> AIDS is not spread by sexual contact between uninfected individuals—whether homosexual or heterosexual—if an exclusive sexual relation has been maintained.[4]

AIDS in the Workplace

The first step in dealing with AIDS at the company level is to develop a comprehensive AIDS policy. Safety and health professionals should be part of the team that drafts the initial policy and updates an existing policy. If a company has no AIDS policy, the safety and health professional should encourage the company to develop one. In all likelihood, most companies won't take much convincing.

According to Peter Minetos,

> Industry is doing its part to eliminate any unnecessary fear: Nearly half of companies surveyed offer their employees literature or other materials to keep them informed on the disease; more than half have an Employee Assistance Program

FIGURE 20-2 Components of a corporate policy for bloodborne pathogens.

- Employee rights
- Testing
- Education

(EAP) to deal with emotional problems concerning AIDS; two-thirds of those who have not yet addressed AIDS with employees plan to do so in the future.[5]

AIDS is having a widely felt impact in the workplace, particularly on employers. According to Minetos, employers are feeling the impact of AIDS in increased insurance premiums and health care costs, time-on-the-job losses, decreased productivity, AIDS-related lawsuits, increased stress, and related problems that result from misconceptions about AIDS.[6]

The starting point for dealing with AIDS in the workplace is the development of a company policy that covers AIDS and other bloodborne pathogens. The policy should cover the following areas at a minimum: employee rights, testing, and education (Figure 20-2).

Employee Rights

An AIDS policy should begin by spelling out the rights of employees who have tested positive for the disease. The following recommendations can be used for developing the **employee rights** aspects of an AIDS policy:

- Treat HIV-positive employees compassionately, allowing them to work as long as they are able to perform their jobs.
- Develop your company's AIDS policy and accompanying program before learning that an employee is HIV positive. This allows the company to act instead of having to react.
- Make reasonable allowances to accommodate the HIV-positive employee. The U.S. Supreme Court has recognized AIDS as a disabling condition. Consequently, reasonable allowances must include modified work schedules and special adaptations to the work environment.
- Ensure that HIV-positive employees have access to private health insurance that covers the effects of AIDS. Also, work with state and federal government insurance providers to gain their support in helping cover the costs of health care for HIV-positive employees.
- Include provisions for evaluating the work skills of employees to determine if there has been any degradation of ability caused by the disease.[7]

Testing

According to the Centers for Disease Control and Prevention (CDC), there is no single test that can reliably diagnose AIDS.[8] However, there is a test that can detect antibodies produced in the blood to fight the virus that causes AIDS. The presence of these antibodies does not necessarily mean that a person has AIDS, as the CDC states:

> Presence of HTLV-III antibodies [now called HIV antibodies] means that a person has been infected with that virus. . . . The antibody test is used to screen

> donated blood and plasma and assist in preventing cases of AIDS resulting from blood transfusions or use of blood products, such as Factor VIII, needed by patients with hemophilia. For people who think they may be infected and want to know their health status, the test is available through private physicians, most state or local health departments and at other sites. Anyone who tests positive should be considered potentially capable of spreading the virus to others.[9]

Whether a company can, or even should, require AIDS tests of employees or potential employees is widely debated. The issue is contentious and controversial. However, there is a growing body of support for mandatory testing. The legal and ethical concerns surrounding this issue are covered in the next section. The testing component of a company's AIDS policy should take these concerns into account.

Education

The general public is becoming more sophisticated about AIDS. People are beginning to learn how AIDS is transmitted. However, research into the causes, diagnosis, treatment, and prevention of this disease is ongoing. The body of knowledge changes continually. Consequently, it is important to have an ongoing education program to keep employees up-to-date and knowledgeable. According to Minetos,

> AIDS education campaigns can be conducted in many forms. Literature, slide shows, and video presentations are all communication vehicles. Presentations by health professionals are one of the most popular and effective methods of communicating information on AIDS. The primary purpose of each is to convey basic knowledge and, subsequently, eliminate unnecessary fear among co-workers.[10]

Once a comprehensive AIDS policy has been developed and shared with all employees, a company has taken the appropriate and rational approach for dealing with this deadly and controversial disease. If an employer has not yet taken this critical step, safety and health professionals should encourage such action immediately. It is likely that most companies either employ now, or will employ in the future, HIV-positive employees. A poll conducted by *U.S. News and World Report* found that 48 percent of the companies responding indicated that AIDS was a concern.[11]

Minetos writes, "AIDS is having an effect on the workplace. Yet . . . only 5 percent of all employers have a written corporate policy on AIDS. For a disease that according to some estimates is reaching epidemic proportions, this is an extremely low percentage."[12] Clearly, one of the major challenges facing safety and health professionals is convincing their employers to develop an AIDS policy.

Legal Concerns

There are legal considerations relating to AIDS in the workplace with which safety and health professionals should be familiar. These grow out of several pieces of federal legislation, including the Rehabilitation Act, the Occupational Safety and Health Act, and the Employee Retirement Income System Act.

The **Rehabilitation Act** was enacted in 1973 to give protection to people with disabilities (then referred to as handicaps), including workers. Section 504 of the act makes discrimination on the basis of a disability unlawful. Any agency, organization, or company that receives federal funding falls within the purview of the act. Such entities may not discriminate against individuals who have disabilities but are otherwise qualified. Through various court actions, this concept has been well defined. A person with a disability is "otherwise qualified" when he or she can perform what the courts have described as the **essential functions** of the job.

When the disability that a worker has is a contagious disease such as AIDS, it must be shown that there is no significant risk of the disease being transmitted in the workplace. If there is a significant risk, the infected worker is not considered "otherwise qualified." Employers and the courts must make these determinations on a case-by-case basis.

Another concept associated with the Rehabilitation Act is the concept of **reasonable accommodation**. In determining if a worker with a disability can perform the essential functions of a job, employers are required to make reasonable accommodations to help the worker. This concept applies to workers with any type of disabling condition, including a communicable disease such as AIDS. What constitutes "reasonable accommodation," just as what constitutes "otherwise qualified," must be determined on a case-by-case basis.

The concepts growing out of the Rehabilitation Act give the supervisor added importance when dealing with AIDS-infected employees. The supervisor's knowledge of the various jobs in his or her unit is essential in helping company officials make an otherwise qualified decision. The supervisor's knowledge that AIDS is transmitted only by exchange of body fluids coupled with his or her knowledge of the job tasks in question is helpful in determining the likelihood that AIDS may be transmitted to other employees. Finally, the supervisor's knowledge of the job tasks in question is essential in determining what constitutes reasonable accommodation and what the actual accommodations should be. Therefore, it is critical that safety and health professionals work closely with supervisors and educate them in dealing with AIDS in the workplace.

In arriving at what constitutes reasonable accommodation, employers are not required to make fundamental changes that alter the nature of the job or result in undue costs or administrative burdens. Clearly, good judgment and a thorough knowledge of the job are required when attempting to make reasonable accommodations for an AIDS-infected employee. Safety and health professionals should involve supervisors in making such judgments.

The **Occupational Safety and Health Act (OSH Act)** requires that employers provide a safe workplace free of hazards. The act also prohibits employers from retaliating against employees who refuse to work in an environment they believe may be unhealthy (Section 654). This poses a special problem for employers of AIDS-infected employees. Other employees may attempt to use Section 654 of the OSH Act as the basis for refusing to work with such employees. For this reason, it is important that companies educate their employees about AIDS and how it is transmitted. If employees know how AIDS is transmitted, they will be less likely to exhibit an irrational fear of working with an infected colleague.

Even when a comprehensive AIDS education program is provided, employers should not automatically assume that an employee's fear of working with the infected individual is irrational. Employers have an obligation to treat each case individually. Does the complaining employee have a physical condition that puts him or her at greater risk of contracting AIDS than other employees? If so, that employee's fears may not be irrational. However, a fear of working with an AIDS-infected coworker is usually irrational, making it unlikely that Section 654 of the OSH Act could be used successfully as the basis of a refusal to work.

The **Employee Retirement Income Security Act (ERISA)** protects the benefits of employees by prohibiting actions taken against them based on their eligibility for benefits. This means that employers covered by ERISA cannot terminate an employee with AIDS or who is suspected of having AIDS as a way of avoiding expensive medical costs. With ERISA, it is irrelevant whether the employee's condition is considered a disability because the act applies to all employees regardless of condition.

Testing Issue

Perhaps the most contentious legal concern growing out of the AIDS controversy is the issue of testing. Writing in the *AAOHN Journal,* Beatrice Crofts Yorker says,

> Few topics have generated the amount of controversy that currently exists in the area of testing for Acquired Immune Deficiency Syndrome (AIDS). Proponents and opponents have strong arguments, often based on emotional reactions to this deadly epidemic. In the workplace, the issues of AIDS testing are very specific and have implications for health policies in occupational settings. Few clear laws or statutes specifically regulate AIDS testing in the workplace.[13]

Issues regarding testing for AIDS and other diseases with which safety and health professionals should be familiar are summarized as follows (see Figure 20-3):

1. State laws. Control of communicable diseases is typically considered to be the province of the individual state. In response to the AIDS epidemic, several states have passed legislation. Some states prohibit the use of preemployment AIDS tests to deny employment to infected individuals. Because of the differences among states with regard to AIDS-related legislation, safety and health professionals should familiarize themselves with the laws of the state in which their company is located.

2. Federal laws and regulations. The laws protecting an individual's right to privacy and due process apply to AIDS testing. These laws fall within the realm of constitutional law. They represent the primary federal contribution to the testing issue.

3. Civil suits. Case law serves the purpose of establishing precedents that can guide future decisions. One precedent-setting case was taken all the way to the Supreme Court (*School Board of Nassau County v. Arline,* 1987), where it was decided that an employer cannot discriminate against an employee who has a communicable disease.

FIGURE 20-3 Disease testing issues.

Health and safety professionals should be familiar with how the following factors might affect the issue of AIDS testing at their companies:

- Applicable state laws
- Applicable federal laws and regulations
- Case law from civil suits
- Company policy

4. Company policy. It was stated earlier that companies should have an AIDS policy that contains a testing component. This component should include at least the following: a strong rationale, procedures to be followed, employee groups to be tested, the use and dissemination of results, and the circumstances under which testing will be done. Safety and health professionals should be knowledgeable about their company's policy and act in strict accordance with it.[14]

On one side of the testing controversy are the issues of fairness, accuracy, and confidentiality or, in short, the rights of the individual. On the other side of the controversy are the issues of workplace safety and public health. Individual rights' proponents ask such questions as: What tests will be used? How do test results relate to the maintenance of workplace safety? How will test results be used, and who will see them? Workplace safety proponents ask such questions as: What is the danger of transmitting the disease to other employees? Can the safety of other workers be guaranteed?

Bayer, Levine, and Wolf recommend the following guidelines for establishing testing programs that satisfy the concerns of both sides of the issue:

- The purpose of screening must be ethically acceptable.
- The means to be used in the screening program and the intended use of the information must be appropriate for accomplishing the purpose.
- High-quality laboratory services must be used.
- Individuals must be notified that the screening will take place.
- Individuals who are screened must have a right to be informed of the results.
- Sensitive and supportive counseling programs must be available before and after screening to interpret the results, whether they be positive or negative.
- The confidentiality of screened individuals must be protected.[15]

Facts about Testing for AIDS and Other Diseases

In addition to understanding the legal concerns associated with disease testing, safety and health professionals should also be familiar with the latest facts about AIDS tests and testing. A concerned employee may ask for recommendations concerning AIDS testing.

Ensuring the accuracy of an **HIV antibody test** (there is no such thing as an AIDS test) requires two different tests, one for initial screening and one for confirmation. The screening test currently used is the **enzyme linked immunosorbent assay (ELISA)** test. The confirmation test is the **immuno-florescent (IFA)**, or the **Western Blot test**. The ELISA test is relatively accurate, but it is susceptible to both false positive and false negative results. A **false positive** test is one that shows the presence of HIV antibodies when no such antibodies exist. A **false negative** result is one that shows no HIV antibodies in people who actually are infected. A negative result indicates that no infection exists at the time of the test. A confirmed positive result indicates that HIV antibodies are present in the blood.[16]

The American College Health Association makes the following recommendations concerning the HIV antibody test:

- The test is not a test for AIDS, but a test for antibodies to HIV, the virus that can cause AIDS.
- Talk to a trained, experienced, sensitive counselor before deciding whether to be tested.

- If you decide to be tested, do so *only* at a center that provides both pre- and posttest counseling.
- If possible, use an *anonymous* testing center.
- Be sure that the testing center uses two ELISA tests and a Western Blot or IFA test to confirm a positive result.
- A positive test result is *not* a diagnosis of AIDS. It does mean you have HIV infection and that you should seek medical evaluation and early treatment.
- A positive test result *does* mean that you can infect others and that you should avoid risky or unsafe sexual contact and IV needle sharing.
- It can take six months (and—although rarely—sometimes even longer) after infection to develop antibodies, so the test result may not indicate whether you have been infected during that period.
- A negative test result *does not* mean that you are immune to HIV or AIDS, or that you cannot be infected in the future.[17]

Safety and health professionals need to share this type of information with employees who ask about AIDS tests and testing. Employees who need more detailed information should be referred to a health care professional.

AIDS Education

The public is becoming more knowledgeable about AIDS and how the disease is spread. However, this is a slow process, and AIDS is a complex and controversial disease. Unfortunately, many people still respond to the disease out of ignorance and inaccurate information. For this reason, it is imperative that a company's safety and health program include an **AIDS education program**.

A well-planned AIDS education program can serve several purposes: (1) It can give management the facts needed to develop policy and make informed decisions with regard to AIDS; (2) it can result in changes in behavior that will make employees less likely to contract or spread the disease; (3) it can prepare management and employees to respond appropriately when a worker falls victim to the disease; and (4) it can decrease the likelihood of legal problems resulting from an inappropriate response to an AIDS-related issue. Consequently, safety and health professionals should be prepared to participate in developing AIDS education programs.

Planning an AIDS Education Program

The first step in planning an AIDS education program is to decide its purpose. A statement of purpose for an AIDS education program should be a broad conceptual declaration that captures the company's reason for providing the education program. Following is an example:

> The purpose of this AIDS education program is to deal with the disease in a positive proactive manner that is in the best interests of the company and its employees.

The next step in the planning process involves developing goals that translate the statement of purpose into more specific terms. The goals should tell specifically what the AIDS education program will do. Sample goals are as follows:

- The program will change employee behaviors that may otherwise promote the spread of AIDS.

FIGURE 20-4 Outline for AIDS education course.

Statement of Purpose

The purpose of this course is to give employees the knowledge they need to deal with AIDS in a positive, proactive manner.

Major Topics

- What is AIDS?
- What causes AIDS?
- How is AIDS transmitted?
- Who is most likely to get AIDS?
- What are the symptoms of AIDS?
- How is AIDS diagnosed?
- Who should be tested for AIDS?
- Where can I get an AIDS test?
- How can I reduce my chances of contracting AIDS?
- How is AIDS treated?
- Can AIDS be prevented?
- What are common myths about AIDS?

- The program will help the company's management team develop a rational, appropriate AIDS policy.
- The program will help managers make responsible decisions concerning AIDS issues.
- The program will help employees protect themselves from the transmission of AIDS.
- The program will alleviate the fears of employees concerning working with an AIDS-infected coworker.
- The program will help managers respond appropriately and humanely to the needs of AIDS-infected workers.

Once goals have been set, a program is developed to meet the goals. The various components of the program must be determined. These components may include confidential one-on-one counseling, referral, posters, a newsletter, classroom instruction, self-paced multimedia instruction, group discussion sessions, printed materials, or a number of other approaches. Figure 20-4 shows a suggested outline for a course on AIDS.

Counseling Infected Employees

The employee who learns that he or she has AIDS will be angry, frightened, and confused. Safety and health professionals who are faced with such an employee should proceed as follows:

- Listen.
- Maintain a nonjudgmental attitude.

- Make the employee aware of the company's policy on AIDS.
- Respond in accordance with company policy.

Listen carefully as you would with an employee who comes to you with any problem. If you must ask a question for clarification, do so, but be objective, professional, and nonjudgmental. Make the employee aware of the company's policy on AIDS and respond in strict accordance with the policy.

The U.S. Public Health Service recommends the following steps for persons who have determined they are HIV positive.

- Seek regular medical evaluation and follow-up.
- Either avoid sexual activity or inform your prospective partner of your antibody test results and protect him or her from contact with your body fluids during sex. (Body fluids include blood, semen, urine, feces, saliva, and women's genital secretions.) Use a condom, and avoid practices that may injure body tissues (for example, anal intercourse). Avoid oral-genital contact and open-mouthed intimate kissing.
- Inform your present and previous sex partners, and any persons with whom needles may have been shared, of their potential exposure to HTLV-III [HIV] and encourage them to seek counseling and antibody testing from their physician or at appropriate health clinics.
- Don't share toothbrushes, razors, or other items that could become contaminated with blood.
- If you use drugs, enroll in a drug treatment program. Needles and other drug equipment must never be shared.
- Don't donate blood, plasma, body organs, other body tissue, or sperm.
- Clean blood or other body fluid spills on household or other surfaces with freshly diluted household bleach—1 part bleach to 10 parts water. (Don't use bleach on wounds.)
- Inform your doctor, dentist, and eye doctor of your positive HTLV-III status so that proper precautions can be taken to protect you and others.
- Women with a positive antibody test should avoid pregnancy until more is known about the risks of transmitting HTLV-III from mother to infant.[18]

Safety and health professionals can pass along these recommendations to employees who have contracted the virus. Any information that is requested beyond this should be provided by a qualified professional. Safety and health professionals should be prepared to make appropriate referrals. The Public Health Service's AIDS hotline number is 1-800-342-AIDS.

Employee Assistance Programs

Company-sponsored **employee assistance programs (EAPs)** should have an AIDS component so that employees can seek confidential advice and counseling about the disease. EAPs may provide on-site services or contract for them through a private organization or agency. In either case, confidential counseling, referral, the provision of AIDS-related information, seminars, and other forms of assistance should be provided. From an employee assistance perspective, AIDS should be treated like other health problems such as stress, depression, substance abuse, and so on.[19]

Easing Employees' Fears about AIDS

During his first inaugural address on March 4, 1933, in the depths of the Great Depression, President Franklin D. Roosevelt said, "So first let me assert my firm belief that the only thing we have to fear is fear itself."[20] This is a message that safety and health professionals should be prepared to spread among employees. Although the level of sophistication concerning AIDS is increasing, people tend to react to the disease from an emotional perspective. Typically, their fears are unfounded and are more likely to cause them problems than does AIDS.

Fear, panic, and even hysteria are all common reactions to AIDS. Fellow employees are likely to respond this way when they discover that a coworker has AIDS. Consequently, safety and health professionals need to know how to ease the fears and misconceptions associated with AIDS. According to Brown and Turner, the following strategies will help:

- Work with higher management to establish an AIDS education and awareness program that covers the following topics at a minimum: (1) how HIV is transmitted, (2) precautions that workers can take, and (3) concerns about AIDS testing.
- Conduct group round-table discussions that allow employees to express their concerns.
- Correct inaccuracies, rumors, and misinformation about AIDS as soon as they occur.[21]

Protecting Employees from AIDS

Safety and health professionals should be familiar with the precautions that will protect employees from HIV infection on and off the job. OSHA's guidelines for preventing exposure to HIV infection identify three categories of workrelated tasks: Categories I, II, and III. Jobs that fall into Category I involve routine exposure to blood, body fluids, or tissues that may be HIV infected. Category II jobs do not involve routine exposure to blood, body fluids, or tissues, but some aspects of the job may involve occasionally performing Category I tasks. Category III jobs do not normally involve exposure to blood, body fluids, or tissues.

Most industrial occupations fall into Category III, meaning there is very little risk of contracting AIDS on the job. However, regardless of the category of their job, employees should know how to protect themselves, and safety and health professionals should be prepared to tell them how.

The U.S. Public Health Service recommends the following precautions for reducing the chances of contracting AIDS:

- Abstain from sex or have a mutually monogamous marriage or relationship with an infection-free partner.
- Refrain from having sex with multiple partners or with a person who has multiple partners. The more partners you have, the greater the risk of infection.
- Avoid sex with a person who has AIDS or whom you think may be infected. However, if you choose not to take this recommendation, the next logical course is to take precautions against contact with the infected person's body fluids (blood, semen, urine, feces, saliva, and female genital secretions).
- Do not use intravenous drugs or, if you do, do not share needles.[22]

Safety and health professionals should make sure that all employees are aware of these common precautions. They should be included in the company-sponsored AIDS education program, available through the company's EAP, and posted conspicuously for employee reading.

CPR and AIDS

It is not uncommon for an employee to be injured in a way that requires resuscitation. Consequently, many companies provide employees with CPR training. But what about AIDS? Is CPR training safe? Writing for *Safety & Health,* Martin Eastman said,

> In the early 1960s when CPR training procedures were being developed, some thought was given to the cleaning and disinfection of manikins. This was before acquired immune deficiency syndrome (AIDS) and hepatitis-B became headline diseases. In the fearful climate that has followed the spread of these diseases, practices that once appeared merely unsanitary now seem truly life-threatening.[23]

The HIV virus has been found in human saliva. Because CPR involves using your fingers to clear the airway and placing your mouth over the victim's, there is concern about contracting AIDS while trying to resuscitate someone. Although there is no hard evidence that HIV can be transmitted through saliva, there is some legitimacy to the concern.

Because of the concern, disposable face masks and various other types of personal protective devices are now being manufactured. Typical of these devices is the rescue key ring produced by Ambu, Inc., of Hanover, Maryland.[24] The ring contains a face mask made of a transparent film material that has a oneway valve. The valve prevents the passage of potentially contaminated body fluids from the victim to the rescuer.

Safety and health professionals should ensure that such devices are used in both training and live situations involving CPR. These devices should be readily available in many easily accessible locations throughout the company.

Safety Needles

Employees who work with needles for taking blood, giving injections, or inserting intravenous systems should be considered at high risk for becoming infected with HIV. Whether inserting an IV, taking blood, or giving a vaccination, medical personnel live with the scary reality of bloodborne pathogens exposure every minute of the day. Studies consistently have shown that 50 to 85 percent of health care workers who have contracted a bloodborne disease cannot identify when or how they were exposed to bloodborne pathogens.

In addition, whereas more than one million accidental needlesticks are reported each year, it is estimated that 66 percent of accidental needlesticks go unreported. With the use of injury prevention devices, the Centers for Disease Control and Prevention estimates that the number of accidental needlesticks could be reduced by 76 percent. Not all devices that are called "safe" are safer. Injury prevention devices on syringes, IVs, catheters, blood-drawing equipment, vaccination instruments, lancets, and scalpels must be one-handed to be considered safer. Some sheathing devices require two-handed operation and often result in an increase in needlestick injuries.

FIGURE 20-5 Examples of syringes with safety features.

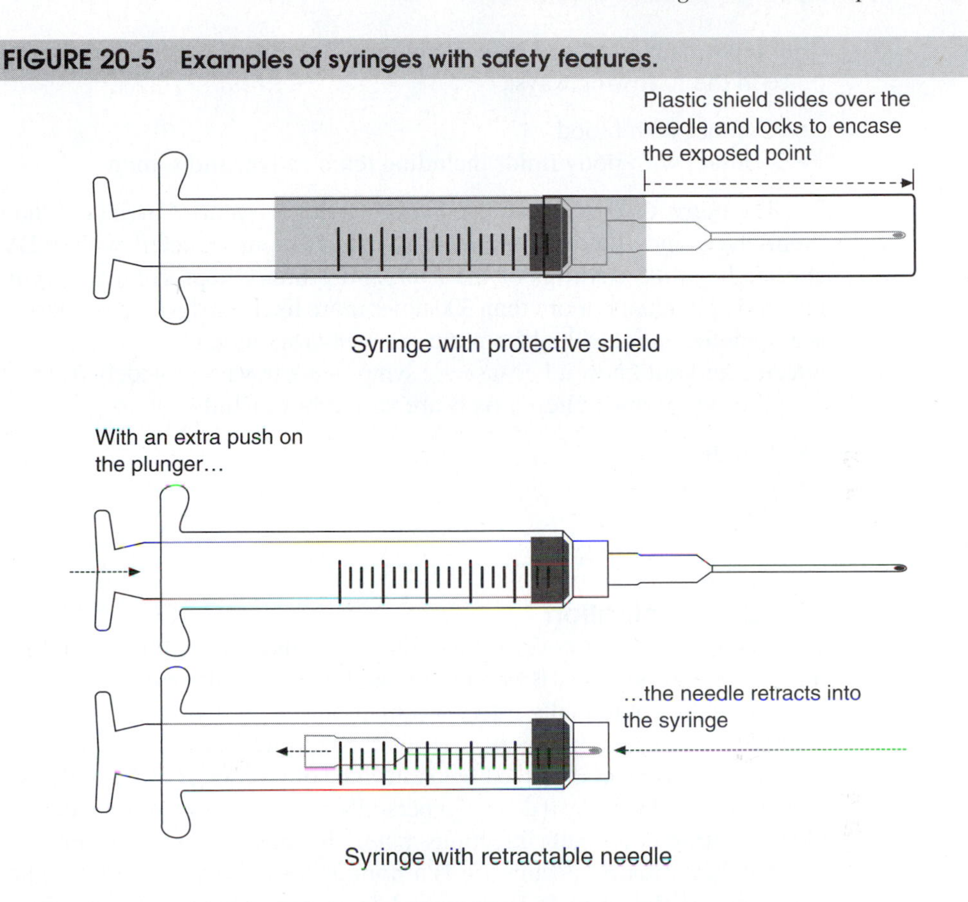

Ellis writes in *Occupational Health & Safety,* "OSHA currently addresses safer needles in the Bloodborne Pathogen Standard through the Engineering and Work Practice Controls clause [1910.1030, Section d(2)(i)]. Although this clause does not specifically require the use of safer needles, inspection guidelines state that the compliance officer shall determine whether engineering controls are used. Most preferable is the use of devices which offer an alternative to needles being used to perform the procedure. Examples of such devices include . . . needle-protected systems or needleless systems."[25] Figure 20-5 contains examples of syringes with safety features.

Hepatitis B (HBV) and Hepatitis C (HCV) in the Workplace

Hepatitis B (HBV)

Although the spread of HIV receives more attention, a greater risk is from the spread of **hepatitis B (HBV)**. This bloodborne virus averages approximately 300,000 new cases per year compared with AIDS, which averages approximately 43,000 new cases. The hepatitis B virus is extremely strong compared with HIV. For example, it can live on surfaces for up to a week if it is exposed to air. Hepatitis B is also much more concentrated than HIV.

Hepatitis B is caused by a double-shelled virus. It can be transmitted in the workplace in the following ways:

- Contact with blood
- Contact with body fluids including tears, saliva, and semen

The hepatitis B virus can live in body fluids for years. Carriers of the virus are at risk themselves, and they place others at risk. Persons infected with HBV may contract chronic hepatitis, cirrhosis of the liver, and primary heptocellular carcinoma. An HBV-infected individual is more than 300 times more likely to develop primary liver cancer than is a noninfected individual from the same environment. Unfortunately, it is possible to be infected and not know it because the symptoms can vary so widely from person to person.

The symptoms of hepatitis B are varied but include

- Jaundice
- Joint pain
- Rash
- Internal bleeding

HBV Vaccination

The OSHA standard covering bloodborne pathogens requires employers to offer the three-injection hepatitis B vaccination series free to all employees who are exposed to blood or other potentially infectious materials as part of their job duties. This includes health care workers, emergency responders, morticians, first-aid personnel, law enforcement officers, correctional facilities staff, and launderers, as well as others. The vaccination must be offered within 10 days of a person's initial assignment to a job where exposure to blood or other potentially infectious materials can be "reasonably anticipated."

The hepatitis B vaccination is a noninfectious, yeast-based vaccine given in three injections in the arm. It is prepared from recombinant yeast cultures, rather than human blood or plasma. Thus, there is no risk of contamination from other bloodborne pathogens, nor is there any chance of developing HBV from the vaccine.

The second injection should be given one month after the first, and the third injection six months after the initial dose. More than 90 percent of those vaccinated develop immunity to the hepatitis B virus. To ensure immunity, it is important for individuals to receive all three injections. At this point it is unclear how long the immunity lasts, so booster shots may be required at some point in the future.

The vaccine causes no harm to those who are already immune or to those who may be HBV carriers. Although employees may opt to have their blood tested for antibodies to determine the need for the vaccine, employers may not make such screening a condition of receiving the vaccination, nor are employers required to provide prescreening.

Each employee should receive counseling from a health care professional when the vaccination is offered. This discussion will help an employee determine whether inoculation is necessary.

Hepatitis C (HCV)

Hepatitis C virus (HCV) infection is the most common chronic bloodborne infection in the United States, with approximately 36,000 new cases diagnosed every year.[26] Most of these persons are chronically infected and may be unaware of their infection because they are not clinically ill. Infected persons serve as a source of transmission to others and are at risk for chronic liver disease or other HCV-related chronic diseases during the first two or more decades following the initial infection.

Chronic liver disease is the 10th leading cause of death among adults in the United States, and accounts for approximately 25,000 deaths annually, or approximately 1 percent of all deaths. Population-based studies indicate that 40 percent of chronic liver disease is HCV related, resulting in an estimated 8,000 to 10,000 deaths each year. Because most HCV-infected persons are ages 30 to 49, the number of deaths attributable to HCV-related chronic liver disease may increase substantially during the next 10 to 20 years as this group of infected persons reaches the age at which complications from chronic liver disease typically occur.

HCV is transmitted primarily through large or repeated direct percutaneous exposures to blood. In the United States, the two most common exposures associated with transmission of HCV are blood transfusion and injecting-drug use. Figure 20-6 summarizes tests that are used to detect HCV infection.

FIGURE 20-6 Various tests available for detecting the presence of HCV.

Tests for Hepatitis C Virus (HCV) Infection

Test	Application
Hepatitis C virus antibody (anti-HCV) • EIA (enzyme immunoassay) • Supplemental assay	• Indicates past or present infection, but does not differentiate between acute, chronic, or resolved infection • All positive EIA results should be verified with a supplemental assay
HCA RNA (hepatitis C virus ribonucleic acid) **Qualitative tests** • Reverse transcriptase polymerase chain reaction (RT-PCR) amplification of HCV RNA by in-house or commercial assays	• Detect presence of circulating HCV RNA • Monitor patients on antiviral therapy
Quantitative tests • RT-PCR amplification of HCV RNA by in-house or commercial assays • Branched chain DNA (bDNA) assays	• Determine concentration of HCV RNA • Might be useful for assessing the likelihood of response to antiviral therapy
Genotype • Several methodologies available (e.g., hybridization, sequencing)	• Group isolates of HCV based on generic differences, into 6 genotypes and 90 subtypes • With new therapies, length of treatment might vary based on genotype
Serotype • EIA based on immunoreactivity to synthetic peptides	• No clinical utility

Source: Centers for Disease Control and Prevention.

Preventing and controlling HCV requires a comprehensive strategy comprised of at least the following activities:

- Primary prevention activities
  - Screening and testing blood, plasma, organ, tissue, and semen donors
  - Virus inactivation of plasma-derived products
  - Risk-reduction counseling and services
  - Implementation and maintenance of infection control practices
- Secondary prevention activities
  - Identification, counseling, and testing of persons at risk
  - Medical management of infected persons
- Education and training
- Monitoring the effectiveness of prevention activities to develop improved prevention methods

The next section explains OSHA's standard on occupational exposure to bloodborne pathogens. This standard applies to all bloodborne pathogens including HBV and HIV.

OSHA'S Standard on Occupational Exposure to Bloodborne Pathogens

OSHA's standard on occupational exposure to bloodborne pathogens is contained in 29 CFR Part 1910.1030. The purpose of the standard is to limit the exposure of personnel to blood and to serve as a precaution against bloodborne pathogens that may cause diseases.

Scope of Application

This standard applies to all employees whose job duties may bring them in contact with blood or other potentially infectious material. There is no attempt on OSHA's part to list all occupations to which 1910.1030 applies. The deciding factor is the *reasonably anticipated* theory. If it can be reasonably anticipated that employees may come in contact with blood in the normal course of performing their job duties, the standard applies. It does not apply in instances of *good Samaritan* acts in which one employee attempts to assist another employee who is bleeding.

The standard covers blood and other infectious materials. These other materials include the following:

- Semen
- Vaginal secretions
- Cerebrospinal fluid
- Synovial fluid
- Pleural fluid
- Peritoneal fluid
- Amniotic fluid
- Saliva
- Miscellaneous body fluids mixed with blood

In addition to these fluids, other potentially infectious materials include the following:

- Unfixed human tissue, or organs other than intact skin
- Cell or tissue cultures
- Organ cultures
- Any medium contaminated by human immunodeficiency virus (HIV) or hepatitis B (HBV)

Exposure Control Plan

OSHA's 1910.1030 requires employers to have an **exposure control plan** to protect employees from exposure to bloodborne pathogens. It is recommended that such plans have at least the following major components:

Part 1 Administration

Part 2 Methodology

Part 3 Vaccinations

Part 4 Postexposure investigation and follow-up

Part 5 Labels and signs

Part 6 Information and training

Administration

This component of the plan should clearly define the responsibilities of employees, supervisors, and managers regarding exposure control. It should also designate an exposure control officer (usually the organization's safety and health manager or a person who reports to this manager). The organization's exposure control plan must be readily available to all employees, and the administration component of it must contain a list of locations where copies of the plan can be examined by employees. It should also describe the responsibilities of applicable constituent groups and individuals. These responsibilities are summarized as follows:

- **Employees.** All employees are responsible for the following: knowing which of their individual and specific job tasks may expose them to bloodborne pathogens; participating in training provided concerning bloodborne pathogens; carrying out all job duties in accordance with the organization's various control procedures; and practicing good personal hygiene.
- **Supervisors and managers.** Supervisors and managers are responsible for coordinating with the exposure control officer to implement and monitor exposure control procedures in their areas of responsibility.
- **Exposure control officer.** This individual is assigned overall responsibility for carrying out the organization's exposure control plan. In addition to the administrative duties associated with it, this position is also responsible for employee training. In small organizations, exposure control officers may double as the training coordinator. In larger organizations, a separate training coordinator may be assigned exposure control as one more training responsibility.

In either case, the exposure control officer is responsible for the following duties: overall development and implementation of the exposure control plan; working with management to develop other exposure-related policies; monitoring and updating the plan; keeping up-to-date with the latest legal requirements relating to exposure; acting as liaison with OSHA inspectors; maintaining up-to-date training files and documentation showing training; developing the needed training program; working with other managers to establish appropriate control procedures; establishing a hepatitis B vaccination program as appropriate; establishing a postexposure evaluation and follow-up system; displaying labels and signs as appropriate; and maintaining up-to-date, confidential medical records of exposed employees.

Methodology

This section of the plan describes the procedures established to protect employees from exposure. These procedures fall into one of the following five categories:

- General precautions
- Engineering controls
- Work practice controls
- Personal protection equipment
- Housekeeping controls

General precautions include such procedures as assuming that all body fluids are contaminated, and acting accordingly. Engineering controls are design and technological precautions that protect employees from exposure. Examples of engineering controls include self-sheathing needles, readily accessible handwashing stations equipped with antiseptic hand cleaners, leakproof specimen containers, and puncture-proof containers for sharp tools that are reusable. Work practice controls are precautions that individual employees take, such as washing their hands immediately after removing potentially contaminated gloves or refraining from eating or drinking in areas where bloodborne pathogens may be present. Personal protection equipment includes any device designed to protect an employee from exposure. Widely used devices include the following:

- Gloves (disposable and reusable)
- Goggles and face shields
- Respirators
- Aprons, coats, and jackets

Examples of housekeeping controls include the use, disposal, and changing of protective coverings; decontamination of equipment; and regular cleaning of potentially contaminated areas.

Vaccinations

The OSHA standard requires that employers make hepatitis B vaccinations available to all employees for whom the reasonable anticipation rule applies. The vaccination procedure must meet the following criteria: available at no cost to employees; administered at a reasonable time and place within 10 days of assignment to a job with exposure potential; and administered under the supervision of an appropriately licensed health care professional, according to the latest recommendations of the U.S. Public Health Service.

Employees may decline the vaccination, but those who do must sign a *declination form* stating that they understand the risk to which they are subjecting themselves. Employees who decline are allowed to change their minds. Employees who do must be allowed to receive the vaccination.

Part of the exposure control officer's job is to keep accurate, up-to-date records showing vaccinated employees, vaccination dates, employees who declined, and signed declination forms.

Postexposure Investigation and Follow-Up

When an employee is exposed to bloodborne pathogens, it is important to evaluate the circumstances and follow up appropriately. How did it happen? Why did it happen? What should be done to prevent future occurrences? The postexposure investigation should determine at least the following:

- When did the incident occur?
- Where did the incident occur?
- What type of contaminated substances or materials were involved?
- What is the source of the contaminated materials?
- What type of work tasks were being performed when the incident happened?
- What was the cause of the incident?
- Were the prescribed precautions being observed when the incident occurred?
- What immediate action was taken in response to the incident?

Using the information collected during the investigation, an incident report is written. This report is just like any other accident report. Keeping the exposed employee fully informed is important. The employee should be informed of the likely avenue of exposure and the source of the contaminated material, even if the source is another employee. If the source is another employee, that individual's blood should be tested for HBV or HIV, and the results should be shared with the exposed employee. Once the exposed employee is fully informed, he or she should be referred to an appropriately certified medical professional to discuss the issue. The medical professional should provide a written report to the employer containing all pertinent information and recommendations. The exposed employee should also receive a copy.

Labels and Signs

This section of the plan describes the procedures established for labeling potential biohazards. Organizations may also use warning signs as appropriate and color-coded containers. It is important to label or designate with signs the following:

- Biohazard areas
- Contaminated equipment
- Containers of potentially contaminated waste
- Containers of potentially contaminated material (i.e., a refrigerator containing blood)
- Containers used to transport potentially contaminated material

Information and Training

This section of the plan describes the procedures for keeping employees knowledgeable, fully informed, and up-to-date regarding the hazards of bloodborne pathogens. The key to satisfying this requirement is training. Training provided should cover at least the following:

- OSHA Standard 1910.1030
- The exposure control plan

- Fundamentals of bloodborne pathogens (e.g., epidemiology, symptoms, and transmission)
- Hazard identification
- Hazard prevention methods
- Proper selection and use of personal protection equipment
- Recognition of warning signs and labels
- Emergency response techniques
- Incident investigation and reporting
- Follow-up techniques
- Medical consultation

It is important to document training and keep accurate up-to-date training records on all employees. These records should be available to employees and their designated representatives (e.g., family members, attorneys, and physicians), and to OSHA personnel. They must be kept for at least three years and include the following:

- Dates of all training
- Content of all training
- Trainers' names and qualifications
- Names and job titles of all participants

Record Keeping

OSHA Standard 1910.1030 requires that medical records be kept by employers on all employees who are exposed to bloodborne pathogens. These records must be confidential and should contain at least the following information:

- Employee's name and social security number
- Hepatitis B vaccination status
- Results of medical examinations and tests
- Results of incident follow-up procedures
- A copy of the written opinions of health care professionals
- A copy of all information provided to health care professionals following an exposure incident

Preventing and Responding to Needlestick Injuries

Needlestick injuries are not a major concern in most workplaces outside the health care industry, but they are enough of a concern that safety and health professionals should know how to prevent them and how to respond when they occur. An excellent source of help for safety and health professionals concerning needlestick injuries is the National Institute for Occupational Safety and Health (NIOSH). NIOSH maintains a Web site dedicated specifically to the prevention of needlestick injuries. The Web address is as follows:

http://www.cdc.gov/niosh/topics/bbp/safer

This Web site recommends a five-step model for developing, establishing, and maintaining a needlestick-prevention program:

1. Form a sharps injury prevention team.
2. Identify priorities.

3. Identify and screen safer medical devices.
4. Evaluate safer medical devices.
5. Institute and monitor the use of the safer medical devices selected.[27]

The NIOSH site also provides detailed information that safety and health professionals can use in implementing each step in the development of a needlestick prevention program.

Responding to Needlestick Incidents

When in spite of your best efforts at prevention a needlestick injury does occur, the following steps are recommended:

1. Encourage bleeding where the skin is penetrated.
2. Wash the penetration area thoroughly with copious amounts of warm, soapy water (do not use a scrub brush).
3. If the eyes are somehow involved, wash them immediately with water.
4. If the mouth is somehow involved, rinse it immediately with water, but do not swallow.
5. Get the injured employee to the hospital as soon as possible.
6. Contact a clinical virologist.
7. Make sure that management personnel for the company in question are fully informed of the situation.[28]

Key Terms and Concepts

- Acquired immunodeficiency syndrome (AIDS)
- AIDS education program
- AIDS policy
- Blood transfusions
- Body fluids
- Case law
- Confidentiality
- Confirmation test
- Counseling
- Disposable face mask
- Employee assistance programs (EAPs)
- Employee Retirement Income Security Act (ERISA)
- Employee rights
- Enzyme linked immunosorbent assay (ELISA) test
- Essential functions
- False negative
- False positive
- Hepatitis B (HBV)
- Hepatitis C (HCV)
- HIV antibody test
- HIV positive
- Human immunodeficiency virus (HIV)
- Immuno-florescent (IFA) test
- Intravenous drugs
- IV drug users
- Mutually monogamous relationship
- Occupational Safety and Health Act (OSH Act)
- Otherwise qualified
- Reasonable accommodation
- Rehabilitation Act
- Screening test
- Sexual contact
- Western Blot test

Review Questions

1. What does the acronym "AIDS" mean?
2. List five symptoms of AIDS.
3. What are the three known ways that AIDS is transmitted?
4. List the groups of people who are considered high risk with regard to AIDS.
5. List five ways that AIDS is *not* transmitted.
6. Describe the ways in which AIDS is having an impact in the workplace.
7. Briefly explain the minimum components of a corporate AIDS policy.
8. Explain how the following legal concepts relate to AIDS: *otherwise qualified, essential functions,* and *reasonable accommodation.*

9. Briefly explain both sides of the AIDS testing controversy.
10. What are the most widely used HIV antibody tests for screening and confirming AIDS?
11. Name four purposes of a well-planned AIDS education program.
12. How should a safety and health professional proceed when faced with an employee who thinks he or she has AIDS?
13. Briefly explain the steps that a company can take to alleviate the fears of employees about AIDS.
14. Explain how OSHA categorizes work tasks relative to exposure to HIV infection.
15. List four ways to guard against contracting AIDS.
16. How can CPR be administered safely?
17. Explain why HBV poses more of a problem for safety personnel than HIV does.
18. Explain the primary prevention strategies for HCV.
19. What are the recommended steps in a needle-stick prevention program?

Endnotes

1. Centers for Disease Control and Prevention (CDC), *HIV/AIDS Surveillance Report* 11, no. 2: 35–37.
2. Ibid.
3. Ibid.
4. "AIDS—The Basic Facts," *Occupational Health & Safety* 10, no. 3: 6.
5. P. Minetos, "Corporate America vs. AIDS," *Safety & Health* 138, no. 6: 34.
6. Ibid.
7. CDC, *HIV/AIDS Surveillance Report,* 37.
8. Centers for Disease Control and Prevention (CDC), U.S. Department of Health and Human Services, *Facts About AIDS,* Spring 2002, 6.
9. Ibid.
10. Minetos, "Corporate America vs. AIDS," 36.
11. Ibid.
12. Ibid.
13. B. C. Yorker, "AIDS Testing," *AAOHN Journal* 36, no. 5: 231.
14. Ibid.
15. R. Bayer, C. Levine, and S. M. Wolf, "HIV Antibody Screening: An Ethical Framework for Evaluating Proposed Programs," in B. C. Yorker, "AIDS Testing," *AAOHN Journal* 36, no. 5: 232.
16. American College Health Association, *The HIV Antibody Test* (Rockville, MD: American College Health Association, 1989).
17. Ibid.
18. CDC, *Facts About AIDS,* 10.
19. National Safety Council, "Is the AIDS Fear a Threat at Work?" *Safety & Health* 139, no. 6: 52.
20. G. Seldes, *The Great Quotations* (Secaucus, NJ: Castle Books, 1966), 590.
21. K. C. Brown and I. C. Turner, *AIDS: Policies and Programs for the Workplace* (New York: Van Nostrand Reinhold), 106–107, 116–117.
22. CDC, *Facts About AIDS,* 8.
23. M. Eastman, "CPR Training: Is It Still Safe?" *Safety & Health* 142, no. 5: 36.
24. Ibid., 39.
25. T. Ellis, "Toward Safer Needles," *Occupational Health & Safety* 68, no. 3: 74.
26. Centers for Disease Control and Prevention, "Recommendations for Prevention and Control of Hepatitis C Virus (HCV) Infection and HCV-Related Chronic Disease," *Morbidity and Mortality Weekly Report* 47, no. RR-19: 1–16.
27. "Web-Based Information Could Help Prevent Needlestick Injuries," *Occupational Health & Safety NEWS* 17, no. 11: 3.
28. Bhavini Lad, "Pricked by a Needle," *Occupational Health & Safety* 71, no. 4: 60.

Glossary

abatement period The amount of time given an employer to correct a hazardous condition that has been cited.

ability to pay Applies when there are a number of defendants in a case but not all have the ability or means to pay financial damages.

absorption Passage through the skin and into the bloodstream.

acceleration Increase in the speed of a falling object before impact.

accident/incident theory Theory of accident causation in which overload, ergonomic traps, or a decision to err lead to human error.

accident prevention The act of preventing a happening that may cause loss or injury to a person.

accident rate A fixed ratio between the number of employees in the workforce and the amount who are injured or killed every year.

accident report Records the findings of an accident investigation, the cause or causes of an accident, and recommendations for corrective action.

accident scene The area where an accident occurred.

accidents Unexpected happenings that may cause loss or injuries to people who are not at fault for causing the injuries.

acclimatization Process by which the body becomes gradually accustomed to heat or cold in a work setting.

accommodation The ability of the eye to become adjusted after viewing the VDT to be able to focus on other objects, particularly objects at a distance.

adjustable guard A device that provides a barrier against a variety of different hazards associated with different production operations.

administrative controls Procedures that are adopted to limit employee exposure to hazardous conditions.

aerosols Liquid or solid particles so small that they can remain suspended in air long enough to be transported over a distance.

aerospace engineering Program of study incorporating a solid foundation of physical and mathematical fundamentals that provides the basis for the development of the engineering principles essential to the understanding of both atmospheric and extra-atmospheric flight.

agreement settlement The injured employee and the employer or its insurance company work out an agreement on how much compensation will be paid and for how long.

air dose Dose measured by an instrument in the air at or near the surface of the body in the area that has received the highest dosage of a toxic substance.

altitude sickness A form of hypoxia associated with high altitudes.

ampere The unit of measurement for current.

analysis and evaluation All potential solutions to a problem are subjected to scientific analysis and careful evaluation.

ancestry A person's line of descent.

anesthetics In carefully controlled dosages, anesthetics can inhibit the normal operation of the central nervous system without causing serious or irreversible effects.

appeals process The process of challenging an OSHA standard as it goes through the adoption process and after formal adoption.

artificial environment One that is fully created to prevent a definite hazardous condition from affecting people or material.

aseptic necrosis A delayed effect of decompression sickness.

asphyxiants Substances that can disrupt breathing so severely as to cause suffocation.

associate degree A two-year college degree.

assumption of risk Based on the theory that people who accept a job assume the risks that go with it. It says employees who work voluntarily should accept the consequences of their actions on the job rather than blaming the employer.

audiogram The results of an audiometric test to determine the noise threshold at which a subject responds to different test frequencies.

audiometric testing Measures the hearing threshold of employees.

auto-ignition temperature The lowest temperature at which a vapor-producing substance or a flammable gas will ignite even without the presence of a spark or a flame.

automatic ejection A system that ejects work pneumatically and mechanically.

automatic feed A system that feeds stock to the machine from rolls.

bachelor's degree A four-year college degree.

barometer Scientific device for measuring atmospheric pressure.

bends Common name for decompression sickness.

best-ratio approach People are basically good and under the right circumstances behave ethically.

biological hazards Hazards from molds, fungi, bacteria, and insects.

black-and-white approach Right is right, wrong is wrong, and circumstances are irrelevant.

blackball To ostracize an employee.

body surface area (BSA) The amount of surface area that is covered with burns.

bonding Used to connect two pieces of equipment by a conductor. Also, involves eliminating the difference in static charge potential between materials.

Boyle's law States that the product of a given pressure and volume is constant with a constant temperature.

cancer A malignant tumor.
carcinogen Any substance that can cause a malignant tumor or a neoplastic growth.
carpal tunnel syndrome (CTS) An injury to the median nerve inside the wrist.
case law Serves the purpose of establishing precedents that can guide future decisions.
causal relationship A situation in which an action leads to a certain result.
ceiling The level of exposure that should not be exceeded at any time for any reason.
central factor The main issue or factor in a problem or act.
chemical burn injuries Burn damage to the skin caused by chemicals such as acids and alkalies.
chemical engineer An engineer concerned with all the physical and chemical changes of matter to produce economically a product or result that is useful to humans.
chemical hazards Include mists, vapors, gases, dusts, and fumes.
chokes Coughing and choking, resulting from bubbles in the respiratory system.
circadian rhythm Biological clock.
circuit tester An inexpensive piece of test equipment with two wire leads capped by probes and connected to a small bulb.
claim notice Notice filed to indicate an expectation of workers' compensation benefits owed.
closed environment One that is completely or almost completely shut off from the natural environment.
closing conference Involves open discussion between the compliance officer and company and employee representatives.
code A set of standards, rules, or regulations relating to a specific area.
Code of Hammurabi Developed around 2000 B.C. during the time of the Babylonians, the ruler Hammurabi developed this code which encompassed all the laws of the land at that time. The significant aspect is that it contained clauses dealing with injuries, allowable fees for physicians, and monetary damages assessed against those who injured others.
coefficient of friction A numerical correlation of the resistance of one surface against another surface.
cold stress Physical or mental stress that results from working in cold conditions.
combination theory The actual cause of an accident may be explained by combining many models.
combustible substance Any substance with a flash point of 37.8°C (100°F) or higher.
combustion A chemical reaction between oxygen and a combustible fuel.
combustion point The temperature at which a given fuel can burst into flame.
commonsense test Requires person to listen to what instincts and common sense are telling him or her.
competitiveness The ability to succeed and prosper consistently in the marketplace whether it is local, regional, national, or global.
conduction The transfer of heat between two bodies that are touching or from one location to another within a body.
conductors Substances that have many free electrons at room temperature and can pass electricity.
conference method A problem-solving teaching method in which the trainer serves as a facilitator rather than a teacher.
confined space An area with limited means of egress that is large enough for a person to fit into, but is not designed for occupancy.
consistency The rules are enforced in the same manner every time with no regard to any outside factors.
continuity tester May be used to determine whether a conductor is properly grounded or has a break in the circuit.
contributory negligence An injured worker's own negligence contributed to the accident. If the actions of employees contributed to their own injuries, the employer is absolved of any liability.
controlled environment A natural or induced environment that has been changed in some way to reduce or eliminate potential environmental hazards.
convection The transfer of heat from one location to another by way of a moving medium.
convergence The coordinated turning of the eyes inward to focus on a nearby point or object.
Cooperative Safety Congress (CSC) The CSC was a result of the Association of Iron and Steel Electrical Engineers' (AISEE) desire to have a national conference on safety. The first meeting took place in Milwaukee in 1912. The meeting planted the seeds for the eventual establishment of the National Safety Council (NSC).
corium The inner layer of human skin.
cost The amount of money needed to produce a product, not to be confused with "price," which is the amount of money needed to purchase a product after the cost has been marked up.
cost allocation Spread the cost of workers' compensation appropriately and proportionately among industries ranging from the most to the least hazardous.
creeps Caused by bubble formation in the skin, which causes an itchy, crawling, rashy feeling in the skin.
critical burns Second-degree burns covering more than 30 percent of the body and third-degree burns covering over 10 percent are considered critical.
crushing Occurs when a part of the body is caught between two hard surfaces that progressively move together, thereby crushing anything between them.
cutis The inner layer of human skin.
cutting Occurs when a body part comes in contact with a sharp edge.
Dalton's law of partial pressures States that in a mixture of theoretically ideal gases, the pressure exerted by the mixture is the sum of the pressures exerted by each component gas of the mixture.
damages Financial awards assigned to injured parties in a lawsuit.
DBBS See Division of Biology and Biomedical Science.
death rates A fixed ratio between the number of employees in the workforce and the number that are killed each year.
decibel The unit applied when measuring sound. One-tenth of a bel. One decibel is the smallest difference in the level of sound that can be perceived by the human ear.

decompression sickness Can result from the decompression that accompanies a rapid rise from sea level to at least 18,000 feet or a rapid ascent from around 132 to 66 feet underwater.

demonstration method The instructor shows students how to perform certain skills or tasks.

dermis The inner layer of human skin.

design flaw A defect in a product.

design process A plan of action for reaching a goal.

direct settlement The employer or its insurance company begins making what it thinks are the prescribed payments.

discovery period Period in which evidence is collected, depositions are taken, and products are examined.

Division of Biology and Biomedical Science (DBBS) Conducts research in the areas of toxicology, behavioral science, ergonomics, and the health consequences of various physical agents.

Division of Training and Manpower Development (DTMD) Implements Section 21 of the OSH Act, which sets forth training and education requirements.

document and communicate Engineering drawings, detailed calculations, and written specifications document the design of a product and communicate its various components to interested parties.

domino theory Injuries are caused by the action of preceding factors. Removal of the central factor negates the action of the preceding factors and, in so doing, prevents accidents and injuries.

dose The amount of ionizing radiation absorbed per unit of mass by part of the body or the whole body.

dose threshold The minimum dose required to produce a measurable effect.

dosimeter Provides a time-weighted average over a period such as one complete work shift.

double insulation A means of increasing electrical equipment safety.

drowning The act of suffocating due to submersion in water.

DTMD See Division of Training and Manpower Development.

dusts Various types of solid particles that are produced when a given type of organic or inorganic material is scraped, sawed, ground, drilled, heated, crushed, or otherwise deformed.

dysbarism The formation of gas bubbles due to rapid ambient pressure reduction.

ego strength An employee's ability to undertake self-directed tasks and to cope with tense situations.

electrical engineering A science-oriented branch of engineering primarily concerned with all phases and development of the transmission and utilization of electric power and intelligence.

electrical hazards Potentially dangerous situations related to electricity (e.g., a bare wire).

electrical system grounding Achieved when one conductor of the circuit is connected to the earth.

electricity The flow of negatively charged particles called electrons through an electrically conductive material.

electrolytes Minerals that are needed for the body to maintain the proper metabolism and for cells to produce energy.

electromechanical devices Contact bars that allow only a specified amount of movement between the worker and the hazard.

electrons Negatively charged particles.

emergency action plan (EAP) A collection of small plans for every anticipated emergency (e.g., fire, hurricane, chemical spill, and so on).

emergency coordinator A person who is clearly identified in a company emergency response plan as the responsible party for a specific type of emergency situation.

emergency notification Requires that chemical spills or releases of toxic substances that exceed established allowable limits be reported to appropriate LEPCs and SERCs.

emergency planning Requires that communities form local emergency planning committees and that states form state emergency response commissions.

emergency response management team Composed of shift emergency directors and a full-time fire chief. Unifies all emergency groups and equipment into a single, coordinated effort.

emergency response network A network of emergency response teams that covers a designated geographical area and is typically responsible for a specific type of emergency.

emergency response plan A written document that identifies the different personnel or groups that respond to various types of emergencies and, in each case, who is in charge.

emergency response team A special team that responds to general and localized emergencies to facilitate personnel evacuation and safety, shut down building services and utilities as needed, work with responding civil authorities, protect and salvage company property, and evaluate areas for safe reentry.

employee A person who is on the company's payroll, receives benefits, and has a supervisor.

employee responsibilities Specific obligations of employees relating to safety and health as set forth in OSHA 2056 (revised).

Employee Retirement Income Security Act (ERISA) Protects the benefits of employees by prohibiting actions taken against them based on their eligibility for benefits.

employee rights Protections that an employee has against punishment for complaining to an employer union, OSHA, or any governmental agency about hazards on the job.

employer-biased laws A collection of laws that favored employers over employees in establishing a responsibility for workplace safety.

employer liability In 1877, the Employer's Liability Law was passed and established the potential for employers to be liable for accidents that occurred in the workplace.

employer responsibilities Specific responsibilities of employers as specified in OSHA 2056 (revised).

employer rights The rights of an employer as specified in OSHA 2056 (revised).

environment The aggregate of social and cultural conditions that influence the life of an individual.

environmental engineer An individual whose job is to protect and preserve human health and the well-being of the environment.

environmental factors Characteristics of the environment in which an employee works that can affect his or her state of mind or physical conditions such as noise or distractions.
environmental heat Heat that is produced by external sources.
environmental management system (EMS) That component of an organization with primary responsibility for leading, planning, organizing, and controlling as these functions relate specifically to an organization's processes, products, or services and the impact that they have on the environment.
enzyme linked immunosorbent assay (ELISA) Screening test currently used to ensure the accuracy of an HIV antibody test.
epidemiological theory Holds that the models used for studying and determining epidemiological relationships can also be used to study causal relationships between environmental factors and accidents or diseases.
epidermis The outer layer of human skin.
ergonomic hazards Workplace hazards related to the design and condition of the workplace. For example, a workstation that requires constant overhead work is an ergonomic hazard.
ergonomic trap An unsafe condition unintentionally designed into a workstation.
ergonomics The science of conforming the workplace and all its elements to the worker.
ERISA See Employee Retirement Income Security Act.
ethical behavior That which falls within the limits prescribed by morality.
ethics The application of morality within a context established by cultural and professional values, social norms, and accepted standards of behavior.
expert system A computer programmed to solve problems.
expiration Occurs when air leaves the lungs and the lung volume is less than the relaxed volume, increasing pressure within the lungs.
explosion A very rapid, contained fire.
explosive range Defines the concentrations of a vapor or gas in air that can ignite from a source.
exposure ceiling Refers to the concentration level of a given substance that should not be exceeded at any point during an exposure period.
exposure threshold A specified limit on the concentration of selected chemicals. Exposure to these chemicals that exceeds the threshold is considered hazardous.
extraterritorial employees Those who work in one state but live in another.
failure mode and effects analysis A formal, step-by-step analytical method that is a spin-off of reliability analysis, a method used to analyze complex engineering systems.
falls Category of accidents in which an employee unintentionally drops under the force of gravity from a surface that is elevated.
false negative A result that shows no HIV antibodies in people who, in reality, are infected.
false positive A result that shows the presence of HIV antibodies when, in reality, no such antibodies exist.
fault tree analysis An analytical methodology that uses a graphic model to display visually the analysis process.
feedback Employee's opinions of a project or an employer's opinion about an employee's job performance.
fellow servant rule Employers are not liable for workplace injuries that result from negligence of other employees.
fire A chemical reaction between oxygen and a combustible fuel.
fire hazards Conditions that favor the ignition and spread of fire.
fire point The minimum temperature at which the vapors or gas in air can ignite from a source of ignition.
fire-related losses Loss of life caused by fires and instances related to fires such as burns, asphyxiation, falls, and so on.
first-degree burn Results in a mild inflammation of the skin known as erythema.
fixed guards Provide a permanent barrier between workers and the point of operation.
flame-resistant clothing Special clothing made of materials or coated with materials that are able to resist heat and flames.
flammable substance Any substance with a flash point below 37.8°C (100°F) and a vapor pressure of less than 40 pounds per square inch at this temperature.
flash point The lowest temperature for a given fuel at which vapors are produced in sufficient concentrations to flash in the presence of a source of ignition.
flashing back When a flame consumes a flammable material such as a gas traveling quickly back to its source.
flexible hours A work/time scheduling system in which employees are allowed to adopt a work schedule that more closely matches their personal needs.
fluid loss Depletion of necessary body fluids, primarily through perspiration.
foreign object Any object that is out of place or in a position to trip someone or to cause a slip.
foreseeability Concept that a person can be held liable for actions that result in damages or injury only when risks could have been reasonably foreseen.
four-step teaching method Preparation, presentation, application, and evaluation.
four-to-one ratio Base one foot away from the wall for every four feet between the base and the support point of a ladder.
free environment One that does not interfere with the free movement of air.
freeze The inability to release one's grip voluntarily from a conductor of electricity.
frequency The number of cycles per second.
friable asbestos Asbestos that is in a state of crumbling deterioration. When asbestos is in this state, it is most dangerous.
front-page test Encourages persons to make a decision that would not embarrass them if it were printed as a story on the front page of their hometown newspaper.
full-potential approach People are responsible for realizing their full potential within the confines of morality.
fuses Consist of a metal strip or wire that will melt if a current above a specific value is conducted through the metal.
gases Formless fluids that are airborne and can be toxic.
gates Provide a barrier between the danger zone and workers.

global marketplace The worldwide economic market in which many companies must compete for business.
good housekeeping Proper cleaning and maintenance of a work area.
ground fault When the current flow in the hot wire is greater than the current in the neutral wire.
ground fault circuit interrupter Can detect the flow of current to the ground and open the circuit, thereby interrupting the flow of current.
grounded conductor Neutral wire.
grounding conductor The groundwire.
hand-arm vibration (HAV) syndrome A form of Raynaud's syndrome that afflicts workers who use vibrating power tools frequently over time.
harmful equipment A work environment in which physical or psychological factors exist that are potentially hazardous.
hazard A condition with the potential of causing injury to personnel, damage to equipment or structures, loss of material, or lessening of the ability to perform a prescribed function.
hazard analysis A systematic process for identifying hazards and recommending corrective action.
hazard and operability review An analysis method that was developed for use with new processes in the chemical industry.
hazardous condition A condition that exposes a person to risks.
hazardous waste reduction Reducing the amount of hazardous waste generated and, in turn, the amount introduced into the waste stream through the process of source reduction and recycling.
health physicist Concerned primarily with radiation in the workplace.
hearing conservation Systematic procedures designed to reduce the potential for hearing loss in the workspace. Employers are required by OSHA to implement hearing conservation procedures in settings where the noise level exceeds a time-weighted average of 85 dBA.
heart disease An abnormal condition of the heart that impairs its ability to function.
heat burn injuries Burn damage to the skin caused by high temperatures from some of the following activities: welding, cutting with a torch, and handling tar or asphalt.
heat cramps A type of heat stress that occurs as a result of salt and potassium depletion.
heat exhaution A type of heat stress that occurs as a result of water or salt depletion.
heat rash A type of heat stress that manifests itself as small raised bumps or blisters that cover a portion of the body and give off a prickly sensation that can cause discomfort.
heat stroke A type of heat stress that occurs as a result of a rapid rise in the body's core temperature.
heat transfer The spread of heat from a source to surrounding materials or objects by conduction, radiation, or convection.
high-radiation area Any accessible area in which radiation hazards exist that could deliver a dose in excess of 100 millirem within one hour.
horizontal work area One that is designed and positioned so that it does not require the worker to bend forward or to twist the body from side to side.
human error A mistake that is made by a human, not a machine.
human error analysis Used to predict human error and not as an after-the-fact process.
human factors theory Attributes accidents to a chain of events ultimately caused by human error.
humidification Adding moisture to the air to reduce electrical static.
hyperoxia Too much oxygen or oxygen breathed under too high a pressure.
hypothermia The condition that results when the body's core temperature drops to dangerously low levels.
ignition temperature The temperature at which a given fuel can burst into flame.
impact accidents Involve a worker being struck by or against an object.
impulse noise Consists of transient pulses that can occur repetitively or nonrepetitively.
inanimate power Power that is lacking life or spirit. During the Industrial Revolution, humans and animals were replaced with inanimate power (e.g., steam power).
inappropriate activities Activities undertaken with disregard for established safety procedures.
inappropriate response A response in which a person disregards an established safety procedure.
income replacement Replacement of current and future income (minus taxes) at a ratio of two-thirds (in most states).
independent contractor A person who accepts a service contract to perform a specific task or set of tasks and is not directly supervised by the company.
indirect costs Costs that are not directly identifiable with workplace accidents.
induced environments Environments that have been affected in some way by human action.
industrial hygiene An area of specialization in the field of industrial safety and health that is concerned with predicting, recognizing, assessing, controlling, and preventing environmental stressors in the workplace that can cause sickness or serious discomfort. Concerns environmental factors that can lead to sickness, disease, or other forms of impaired health.
industrial hygiene chemist Often hired by companies that use toxic substances to test the work environment and the people who work in it. Industrial hygiene chemists have the ability to detect hazardous levels of exposure or unsafe conditions so that corrective and preventive measures can be taken at an early stage.
industrial hygienist A person having a college or university degree in engineering, chemistry, physics, medicine, or related physical or biological sciences who, by virtue of special studies and training, has acquired competence in industrial hygiene.
industrial medicine A specialized field that is concerned with work-related safety and health issues.
industrial place accidents Accidents that occur at one's place of work.
industrial safety engineer An individual who is a safety and health professional with specialized education and training. Responsible for developing and carrying out those aspects of a company's overall safety and health program relating to his or her area of expertise.

industrial safety manager An individual who is a safety and health generalist with specialized education and training. Responsible for developing and carrying out a company's overall safety and health program including accident prevention, accident investigation, and education and training.

industrial stress Involves the emotional state resulting from a perceived difference between the level of occupational demand and a person's ability to cope with this demand.

infection The body's response to contamination by a disease-producing microorganism.

ingestion Entry through the mouth.

inhalation Taking gases, vapors, dust, smoke, fumes, aerosols, or mists into the body by breathing in.

inspection tour A facility tour by an OSHA compliance officer to observe, interview, and examine as appropriate.

inspiration When atmospheric pressure is greater than pressure within the lungs, air flows down this pressure gradient from the outside into the lungs.

instructional approach A brief action plan for carrying out the instruction.

interlocked guards Shut down the machine when the guard is not securely in place or is disengaged.

interlocks Automatically break the circuit when an unsafe situation is detected.

internal factors Factors that can add a burden on a person and interfere with his or her work, such as personal problems.

ionizer Ionizes the air surrounding a charged surface to provide a conductive path for the flow of charges.

ionizing radiation Radiation that becomes electrically charged or changed into ions.

irritants Substances that cause irritation to the skin, eyes, and the inner lining of the nose, mouth, throat, and upper respiratory tract.

islands of automation Individual automated systems lacking electronic communication with other related systems.

job autonomy Control over one's job.

job descriptions Written specifications that describe the tasks, duties, reporting requirements, and qualifications for a given job.

job safety analysis A process through which all the various steps in a job are identified and listed in order.

job security A sense—real or imagined—of having the potential for longevity in a job and having a measure of control concerning that longevity.

kinetic energy The energy resulting from a moving object.

lacrimination The process of excreting tears.

learning objectives Specific statements of what the learner should know or be able to do as a result of completing the lesson.

let-go current The highest current level at which a person in contact with a conductor of electricity can release the grasp of the conductor.

liability A duty to compensate as a result of being held responsible for an act or omission.

lifting hazard Any factor that if not properly dealt with may lead to an injury from lifting.

lightning Static charges from clouds following the path of least resistance to the earth, involving very high voltage and current.

line authority The safety and health manager has authority over and supervises certain employees.

litigation The carrying on of legal matters by judicial process.

load A device that uses currents.

lockout/tagout system A system for incapacitating a machine until it can be made safe to operate. "Lockout" means physically locking up the machine so that it cannot be used without removing the lock. "Tagout" means applying a tag that warns employees not to operate the machine in question.

locus of control The perspective of workers concerning who or what controls their behavior.

lost time The amount of time that an employee was unable to work due to an injury.

lost wages The amount that an employee could have earned had he or she not been injured.

Machiavellianism The extent to which an employee will attempt to deceive and confuse others.

malpractice Negligent or improper practice.

material safety data sheets (MSDSs) These sheets contain all the relevant information needed by safety personnel concerning specific hazardous materials.

meaninglessness The feeling that workers get when their jobs become so specialized and so technology-dependent that they cannot see the meaning in their work as it relates to the finished product or service.

means of egress A route for exiting a building or other structure.

mechanical engineering The professional field that is concerned with motion and processes whereby other energy forms are converted into motion.

mechanical hazards Those associated with power-driven machines, whether automated or manually operated.

mechanical injuries Injuries that have occurred due to misuse of a power-driven machine.

medical expenses Money paid to cover the costs of emergency medical response and follow-up treatment when an employee is injured.

Merit Program Seen as a stepping-stone to recognize companies that have made a good start toward Star Program recognition.

metabolic heat Produced within a body as a result of activity that burns energy.

mindlessness The result of the process of dumbing down the workplace.

minor burns All first-degree burns are considered minor as well as second-degree burns covering less than 15 percent of the body.

mirror test Encourages people to make choices based on how they will feel about their decision when they look in the mirror.

mists Tiny liquid droplets suspended in air.

moderate burns Second-degree burns covering less than 30 percent of the body and third-degree burns covering less than 10 percent are considered moderate.

monetary benefits Actual money owed to injured employees or their relatives under workers' compensation laws.

morning-after test Encourages people to make choices based on how they will feel about their decision the next day.

MSDSs See Material safety data sheets.
NACOSH See National Advisory Committee on Occupational Safety and Health.
narrow band noise Noise that is confined to a narrow range of frequencies.
National Advisory Committee on Occupational Safety and Health (NACOSH) Makes recommendations for standards to the secretary of health and human services and to the secretary of labor.
National Council of Industrial Safety (NCIS) Established in 1913, one year after the first meeting of the CSC. In 1915, the NCIS changed its name to the National Safety Council. It is now the premier safety organization in the United States.
National Electrical Code (NEC) Specifies industrial and domestic electrical safety precautions.
National Institute for Occupational Safety and Health (NIOSH) This organization is part of the Centers for Disease Control and Prevention (CDC) of the Department of Health and Human Services (DHHS). It is required to publish annually a comprehensive list of all known toxic substances. It will also provide on-site tests of potentially toxic substances so that companies know what they are handling and what precautions to take.
natural disasters Incidents prompted by nature such as earthquakes, hurricanes, floods, and tornadoes.
natural environment Not human-made. It is the environment that we typically think of as Earth and all its natural components, including the ground, the water, flora and fauna, and the air.
NCIS See National Council of Industrial Safety.
NEC See National Electrical Code.
negative pressures Caused by pressures below atmospheric level.
negligence Failure to take reasonable care or failure to perform duties in ways that prevent harm to humans or damage to property.
negligent manufacture The maker of a product can be held liable for its performance from a safety and health perspective.
neoplastic growth Cancerous tissue or tissue that might become cancerous.
neutrons Particles that neutralize the negative charge of electrons. They act as temporary energy repositories between positively charged particles called protons and electrons.
NIOSH See National Institute for Occupational Safety and Health.
noise Unwanted sound.
nonionizing radiation Radiation on the electromagnetic spectrum that has a frequency of 1015 or less and a wavelength in meters of 3 3 1027 or less.
nonskid footwear Shoes that have special nonskid soles.
normlessness The phenomenon in which people working in a highly automated environment can become estranged from society.
notice of contest A written note stating that an employer does not wish to comply with a citation, an abatement period, or a penalty.
notice of proposed rule making Explains the terms of the new rule, delineates the proposed changes to existing rules, or lists rules that are to be revoked.
nuclear engineering Concerned with the release, control, and safe utilization of nuclear energy.
objectivity Rules are enforced equally regardless of who commits an infraction from the newest employee to the chief executive officer.
occupational diseases Pathological conditions brought about by workplace conditions or factors.
occupational health nurse One whose job it is to conserve the health of workers in all occupations.
Occupational Safety and Health Act (OSH Act) Passed by the United States Congress in 1970 and updated periodically since that time.
Occupational Safety and Health Administration (OSHA) The government's administrative arm for the Occupational Safety and Health Act. It sets and revokes health and safety standards, conducts inspections, investigates problems, issues citations, assesses penalties, petitions the courts to take appropriate action against unsafe employers, provides safety training, provides injury prevention consultation, and maintains a database of safety and health statistics.
Occupational Safety and Health Review Commission (OSHRC) An independent board whose members are appointed by the president and given quasi-judicial authority to handle contested OSHA citations.
ohms Measure resistance.
opening conference An initial meeting in which an OSHA compliance officer informs company officials and employee representatives of the reason that an on-site inspection is going to occur and what to expect.
organized labor A group of employees who joined together to fight for the rights of all employees (i.e., unions).
OSHA See Occupational Safety and Health Administration.
OSH Act See Occupational Safety and Health Act.
OSHA Form 101 Supplementary Record of Occupational Injuries and Illnesses.
OSHA Form 300 Log of Work-Related Injuries and Illnesses.
OSHA Form 300A Summary of Workplace Injuring and Illnesses.
OSHA Form 301 Injury and Illness Incident Report.
OSHA Poster 2203 Explains employee rights and responsibilities as prescribed in the OSH Act.
OSHRC See Occupational Safety and Health Review Commission.
overexertion The result of employees working beyond their physical limits.
overload Amounts to an imbalance between a person's capacity at any given time and the load that person is carrying in a given state.
oxygen limit The amount of oxygen that must be present in a given substance in order for an explosion to occur.
participatory ergonomics An approach to intervention that combines outside experience and inside experience to design ergonomic interventions that are tailored to a specific workplace.
patent defect One that occurs in all items in a manufactured batch.
permanent partial disability The condition that exists when an injured employee is not expected to recover.
permanent total disability The condition that exists when an injured employee's disability is such that he or she cannot compete in the job market.

permanent variance An exemption from an OSHA standard awarded to an organization that can show it already exceeds the standards.

permissible exposure limit (PEL) OSHA-established exposure threshold.

personal monitoring devices Devices worn or carried by an individual to measure radiation doses received.

personal protective equipment (PPE) Any type of clothing or device that puts a barrier between the worker and the hazard (e.g., safety goggles, gloves, hard hats, and so on).

Petition for Modification of Abatement (PMA) Available to employers who intend to correct the situation for which a citation was issued, but who need more time.

photoelectric devices Optional devices that shut down the machine anytime the light field is broken.

photoelectric fire sensors Detect changes in infrared energy that is radiated by smoke, often by the smoke particles obscuring the photoelectric beam.

physical hazards Include noise, vibration, extremes of temperature, and excessive radiation.

PMA See Petition for Modification of Abatement.

point-of-operation guards Machine guards that provide protection right at the point where the user operates the machine.

poisoning To be injured or killed by a harmful substance.

powerlessness The feeling that workers have when they are not able to control the work environment.

preceding factors Factors that led up to an accident.

predispositional characteristics Human personality characteristics that can have a catalytic effect in causing an accident.

preliminary hazard analysis Conducted to identify potential hazards and prioritize them according to (1) the likelihood of an accident or injury being caused by the hazard; and (2) the severity of injury, illness, or property damage that could result if the hazard caused an accident.

pressure The force exerted against an opposing fluid or thrust distributed over a surface.

pressure hazard A hazard caused by a dangerous condition involving pressure.

private insurance Workers' compensation coverage purchased from a private insurance company.

problem identification Engineers must draft a description of a problem before anything else can be done.

produce and deliver Shop or detail drawings are developed, and the design is produced, usually as a prototype. The prototype is analyzed and tested. Design changes are made. The product is then produced and delivered.

product literature Tells users about hazards that cannot be removed by design or controlled by guards and safety devices.

product safety auditor Responsible for evaluating the overall organization and individual departments within it.

product safety committee Consists of representatives within a company having a wide range of expertise. Offers support to the product safety coordinator on safety issues.

product safety coordinator Responsible for coordinating and facilitating a product safety management program in all departments of a company.

product safety program Purpose is to limit as much as possible a company's exposure to product liability litigation and related problems.

productivity The concept of comparing output of goods or services to the input of resources needed to produce or deliver them.

professional societies Typically formed for the purpose of promoting professionalism, adding to the body of knowledge, and forming networks among colleagues in a given field.

proof pressure tests Tests in which containers are "proofed" by subjecting them to specified pressures for specified periods.

property damage Facilities, equipment, or other nonpersonnel items damaged as a result of an accident.

proposed penalty The initial penalty proposed by the OSHA compliance officer after an inspection tour.

protons Positively charged particles.

proximate cause The cause of an injury or damage to property.

psychophysiological techniques Require simultaneous measurement of heart rate and brain waves, which are then interpreted as indexes of mental workload and industrial stress.

psychosocial questionnaires Evaluate workers' emotions about their jobs.

public hearing If an injured worker feels that he or she has been inadequately compensated or unfairly treated, a public hearing can be requested.

pullback devices Pull the operator's hands out of the danger zone when the machine starts to cycle.

puncturing Results when an object penetrates straight into the body and pulls straight out, creating a wound in the shape of the penetrating object.

quality A measure of the extent to which a product or service meets or exceeds customer expectations.

quality management (QM) A way of managing a company that revolves around a total and willing commitment of all personnel at all levels to quality.

quality product One that meets or exceeds customer standards and expectations.

rad A measure of the dose of ionizing radiation absorbed by body tissue stated in terms of the amount of energy absorbed per unit of mass of tissue.

radiant heat The result of electromagnetic nonionizing energy that is transmitted through space without the movement of matter within that space.

radiation Consists of energetic nuclear particles and includes alpha rays, beta rays, gamma rays, X-rays, neutrons, high-speed electrons, and high-speed protons.

radiation area Any accessible area in which radiation hazards exist that could deliver doses (1) within one hour, a major portion of the body could receive more than 5 millirem; or (2) within five consecutive days, a major portion of the body could receive more than 100 millirem.

radiation control specialist Monitors the radiation levels to which workers may be exposed, tests workers for levels of exposure, responds to radiation accidents, develops companywide plans for handling radiation accidents, and implements decontamination procedures when necessary.

radioactive material Material that emits corpuscular or electromagnetic emanations as the result of spontaneous nuclear disintegration.
radio-frequency devices Capacitance devices that brake the machine if the capacitance field is interrupted by a worker's body or another object.
reasonable risk Exists when consumers (1) understand risk, (2) evaluate the level of risk, (3) know how to deal with the risk, and (4) accept the risk based on reasonable risk–benefit considerations.
receptacle wiring tester A device with two standard plug probes for insertion into an ordinary 110-volt outlet and a probe for the ground.
reclamation A process whereby potentially hazardous materials are extracted from the by-products of a process.
rehabilitation Designed to provide the needed medical care at no cost to the injured employee until he or she is pronounced fit to return to work.
Rehabilitation Act Enacted to give protection to people with disabilities.
rem A measure of the dose of ionizing radiation to body tissue stated in terms of its estimated biological effect relative to a dose of one roentgen of X-rays.
repeat violation A violation of any standard, regulation, rule, or order where, upon reinspection, a substantially similar violation is found.
repetitive motion Short-cycle motion that is repeated continually.
repetitive strain injury (RSI) A broad and generic term that encompasses a variety of injuries resulting from cumulative trauma to the soft tissues of the body.
resistance A tendency to block flow of electric current.
response time The amount of time between when an order is placed and the product is delivered.
restraint devices Hold back the operator from the danger zone.
restricted area Any area to which access is restricted in an attempt to protect employees from exposure to radiation or radioactive materials.
retrofit Renovating rather than replacing.
risk analysis An analytical methodology normally associated with insurance and investments.
role ambiguity The condition that occurs when an employee is not clear concerning the parameters, reporting requirements, authority, or responsibilities of his or her job.
role-reversal test Requires a person to trade places with the people affected by the decision that he or she made and to view the decision through their eyes.
safeguarding Machine safeguarding was designed to minimize the risk of accidents of machine–operator contact.
safety and health movement Began during World War II when all the various practitioners of occupational safety and health began to see the need for cooperative efforts. This movement is very strong today.
safety and health professional An individual whose profession (job) is to be concerned with safety and health measures in the workforce.
safety-friendly corporate culture Exists when the tacit assumptions, beliefs, values, attitudes, expectations, and behaviors that are widely shared and accepted in an organization support the establishment and maintenance of a safe and healthy work environment for all personnel at all levels.
safety policy A written description of an organization's commitment to maintaining a safe and healthy workplace.
safety trip devices Include trip wires, trip rods, and body bars that stop the machine when tripped.
SARA See Superfund Amendments and Reauthorization Act.
second-degree burn Results in blisters forming on the skin.
self-insurance Workers' compensation coverage in which a company insures itself by building its own fund.
semiautomatic ejection A system that ejects the work using mechanisms that are activated by the operator.
semiautomatic feed A system that uses a variety of approaches for feeding stock to the machine.
semiconductors Substances that are neither conductors nor insulators.
shift work Employees work at different times of the day instead of during the same hours.
shock A depression of the nervous system.
short circuit A circuit in which the load has been removed or bypassed.
short-term exposure limit The maximum concentration of a given substance to which employees may be safely exposed for up to 15 minutes without suffering irritation, chronic or irreversible tissue change, or narcosis to a degree sufficient to increase the potential for accidental injury, impair the likelihood of self-rescue, or reduce work efficiency.
sick-building syndrome An internal environment that contains unhealthy levels of biological organisms in the air. A common cause is the introduction of unhealthy outdoor air that is brought in and circulated through the cooling system.
simulation Involves structuring a training activity that simulates a line situation.
situational characteristics Factors that can change from setting to setting and can have a catalytic effect in causing an accident.
situational factors Environmental factors that can affect an employee's safety and that can differ from situation to situation.
smoke The result of the incomplete combustion of carbonaceous materials.
social environment The general value system of the society in which an individual lives, works, grows up, and so on.
sound Any change in pressure that can be detected by the ear.
sound level meter Produces an immediate reading that represents the noise level at a specific instant in time.
spraining The result of torn ligaments.
staff authority The safety and health manager is the staff person responsible for a certain function, but he or she has no line authority over others involved with that function.
standards and testing organizations Conduct research, run tests, and establish standards that identify the acceptable levels for materials, substances, conditions, and mechanisms to which people may be exposed in the modern workplace.

standpipe and hose systems Provide the hose and pressurized water for firefighting.

Star Program Recognizes companies that have incorporated safety and health into their regular management system so successfully that their injury rates are below the national average for their industry.

state funds Workers' compensation coverage provided by the state.

static electricity A surplus or deficiency of electrons on the surface of a material.

step and fall An accident that occurs when a person's foot encounters an unexpected step down.

straining The result of overstretched or torn muscles.

stress A pathological, and therefore generally undesirable, human reaction to psychological, social, occupational, or environmental stimuli.

stress claims Workers' compensation claims that are based on stress-induced disabilities.

stressors Stimuli that cause stress.

stump and fall An accident that occurs when a worker's foot suddenly meets a sticky surface or a defect in the walking surface.

subjective ratings Ratings that are less than objective and can be affected by emotions, human biases, presumptions, and perceptions.

Superfund Amendments and Reauthorization Act (SARA) Act designed to allow individuals to obtain information about hazardous chemicals in their community so that they are able to protect themselves in case of an emergency.

synthesis Second step in the design process wherein engineers combine systematic, scientific procedures with creative techniques to develop initial solutions.

systems theory Views a situation in which an accident may occur as a system comprised of the following components person (host), machine (agency), and environment.

technic of operation review An analysis method that allows supervisors and employees to work together to analyze workplace accidents, failures, and incidents.

technological alienation The frame of mind that results when employees come to resent technology and the impact that it has on their lives.

technology access Access to time- and work-saving devices, processes, or equipment that are up to date.

temporary emergency standards OSHA standards that can be adopted on a temporary basis without undergoing the normal adoption procedures.

temporary partial disability The injured worker is incapable of certain work for a period of time but is expected to recover fully.

temporary total disability The injured worker is incapable of any work for a period of time but is expected to recover fully.

temporary variance Employers may ask for this when they are unable to comply with a new standard but may be able to if given time.

thermal expansion detectors Use a heat-sensitive metal link that melts at a predetermined temperature to make contact and ultimately sound an alarm.

third-degree burn Penetrates through both the epidermis and the dermis. It may be fatal.

three E's of safety Engineering, education, and enforcement.

threshold limit values (TLVs) The levels of exposure to which all employees may be repeatedly exposed to specified concentrations of airborne substances without fear of adverse effects. Exposure beyond TLVs is considered hazardous.

threshold of hearing The weakest sound that can be heard by a healthy human ear in a quiet setting.

threshold of pain The maximum level of sound that can be perceived without experiencing pain.

time-weighted average The level of exposure to a toxic substance to which a worker can be repeatedly exposed on a daily basis without suffering harmful effects.

tort An action involving a failure to exercise a reasonable care that may, as a result, lead to civil litigation.

total safety management (TSM) The principles of total quality management (TQM) applied to safety management.

toxic substance One that has a negative effect on the health of a person or animal.

trade associations Promote the trade that they represent.

trip and fall An accident that occurs when a worker encounters an unseen foreign object in his or her path.

two-hand controls Require the operator to use both hands concurrently to activate the machine.

ultraviolet detectors Sound an alarm when the radiation from fire flames are detected.

Underwriters Laboratories (UL) Determines whether equipment and materials for electrical systems are safe in the various NEC location categories.

unreasonable risk Exists when (1) consumers are not aware that a risk exists; (2) consumers are not able to judge adequately the degree of risk even when they are aware of it; (3) consumers are not able to deal with the risk; and (4) risk could be eliminated at a cost that would not price the product out of the market.

unrestricted area Any area to which access is not controlled because no radioactivity hazard is present.

unsafe act An act that is not safe for an employee.

unsafe behavior The manner in which people conduct themselves that is unsafe to them or to another.

useful consciousness A state of consciousness in which a person is clear-headed and alert enough to make responsible decisions.

vacuum mentality Workers think that they work in a vacuum and don't realize that their work affects that of other employees and vice versa.

vacuums Caused by pressures below atmospheric level.

value added The difference between what it costs to produce a product and the value the marketplace puts on it.

vapor A mist state into which certain liquids and solids can be converted (e.g., gasoline fumes are vaporized petroleum).

vertical work area One that is designed and positioned so that workers are not required to lift their hands above their shoulders or bend down in order to perform any task.

vocational rehabilitation Involves providing the education and training needed to prepare the worker for a new occupation.

volatility The evaporation capability of a given substance.

voltage Measures the potential difference between two points in a circuit.

wage-loss theory Requires a determination of how much the employee could have earned had an injury not occurred.

water hammer A series of loud noises caused by liquid flow suddenly stopping.

wellness program Any program designed to help and encourage employees to adopt a healthier lifestyle.

whistle-blowing Act of informing an outside authority or media organ of alleged illegal or unethical acts on the part of an organization or individual.

whole-person theory What the worker can do after recuperating from the injury is determined and subtracted from what he or she could do before the accident.

wide band noise Noise that is distributed over a wide range of frequencies.

willful or reckless conduct Involves intentionally neglecting one's responsibility to exercise reasonable care.

work envelope The total area within which the moving parts of a robot actually move.

work injuries Injuries that occur while an employee is at work.

work stress A complex concept involving physiological, psychological, and social factors.

worker negligence Condition that exists when an employee fails to take necessary and prudent precautions.

workers' compensation Developed to allow injured employees to be compensated appropriately without having to take their employer to court.

workplace accidents Accidents that occur at an employee's place of work.

workplace inspection An on-site inspection conducted by OSHA personnel.

workplace stress Human reaction to threatening situations at work or related to the workplace.

Index